D1918498

Handbook of

THERAPEUTIC BIOMARKERS IN CANCER

edited by
Sherry X. Yang
Janet E. Dancey

Handbook of THERAPEUTIC BIOMARKERS IN CANCER

PAN STANFORD PUBLISHING

Published by

Pan Stanford Publishing Pte. Ltd.
Penthouse Level, Suntec Tower 3
8 Temasek Boulevard
Singapore 038988

Email: editorial@panstanford.com
Web: www.panstanford.com

British Library Cataloguing-in-Publication Data
A catalogue record for this book is available from the British Library.

Handbook of Therapeutic Biomarkers in Cancer

Copyright © 2013 by Pan Stanford Publishing Pte. Ltd.

All rights reserved. This book, or parts thereof, may not be reproduced in any form or by any means, electronic or mechanical, including photocopying, recording or any information storage and retrieval system now known or to be invented, without written permission from the publisher.

For photocopying of material in this volume, please pay a copying fee through the Copyright Clearance Center, Inc., 222 Rosewood Drive, Danvers, MA 01923, USA. In this case permission to photocopy is not required from the publisher.

ISBN 978-981-4364-65-2 (Hardcover)
ISBN 978-981-4364-66-9 (eBook)

Printed in the USA

Contents

Preface

The advent of the era of the molecularly targeted therapy in oncology in addition to conventional multimodality management signifies more hope for cancer patients. The discovery, validation, and clinical applications of biomarkers of prognosis and prediction are advancing the promise of personalized medicine. The clinically validated therapeutic (predictive) biomarkers for targeted and chemotherapy agents approved for use or having potential to be approved by the regulatory agencies such as the United States Food and Drug Administration facilitate the evolution of empiric therapy to individually tailored treatment. In essence, therapeutic biomarkers and appropriately validated clinical assays facilitate treatment decision-making. We have clearly entered the epoch that patients can receive the right drugs with the right doses at the right time with greater assurance of maximal benefits and reduced risks.

In editing and organizing the *Handbook of Therapeutic Biomarkers in Cancer*, we have made every attempt to cover the growing numbers of promising predictive biomarkers and associated assays in the fields of oncology and cancer research. We hope that many readers—oncologists, health professionals, patients, scientists involved in basic, translational, and clinical research, educators, and students both medical and undergraduate—will find each chapter of this book a valuable source of information and guidance.

It has been a great privilege to be involved in editing this book. We express our sincere thanks to all authors who have contributed their expertise, experience, and hard work to this book for publication. In addition, we welcome comments for planning future editions.

Sherry X. Yang

Janet E. Dancey

Chapter 1

Overview: Therapeutic Biomarkers in Cancer

Sherry X. Yang[a] **and Janet E. Dancey**[b]
[a]*Division of Cancer Treatment and Diagnosis,*
National Cancer Institute National Institutes of Health, Bethesda, MD 20892, USA
[b]*High Impact Clinical Trials, Ontario Institute for Cancer Research, 790 Bay Street, Suite 401, Canada*
Sherry.Yang@nih.gov, janet.dancey@oicr.on.ca

Cancer affects people of all ages and remains one of the leading causes of death worldwide. The American Cancer Society estimates that about 1,660,290 Americans new cancer diagnosis, and 580,350 cancer death are expected to occur in 2013.[1] Encouragingly, a significant reduction of cancer death has been achieved over the last 20 years.[2] According to the Annual Report to the Nation on the Status of Cancer, the overall rate of both new cancer incidence and cancer deaths continued to decrease between 2003 and 2007 in the United States.[3] The decrease in cancer death rates of 1.6% per year continues a trend that began in the early 1990s. The achievements are largely attributed to our growing understanding of cancer genetics and

Handbook of Therapeutic Biomarkers in Cancer
Edited by Sherry X. Yang and Janet E. Dancey
Copyright © 2013 Pan Stanford Publishing Pte. Ltd.
ISBN 978-981-4364-65-2 (Hardcover), 978-981-4364-66-9 (eBook)
www.panstanford.com

biology, cancer prevention, and improvements in cancer diagnosis and treatment. Of particular note is the expanding body of molecular solutions in the treatment of cancer, which opens up avenues for tailored cancer management. Concurrently, these have created significant needs for predictive (therapeutic) biomarkers to identify subsets of patients who are most likely or unlikely to benefit from a particular treatment. This book covers current advances on the therapeutic biomarkers that predict response or benefit or adverse events from a given antineoplastic treatment and, to a lesser degree, on markers of prognosis that influence the choice of treatment option. This review volume, in general, does not cover markers for cancer diagnosis, staging of any cancer types, and classification of extent of disease or disease monitoring. Each chapter provides a comprehensive overview of the molecular targets, a class of anti-cancer drugs and in depth insights of the relevant biomarkers.

1.1 Introduction

A biomarker is a characteristic that is objectively measured and evaluated as an indicator of normal biological processes, pathologic processes or pharmacologic responses to a therapeutic intervention.[4,5] The biomarkers can be amino and nucleic acids, lipids, carbohydrates, proteins, cells, enzymes or hormones, which are measured through some types of tests such as genomic, biochemical or immunologic. A prognostic marker provides information about overall outcome irrespective of therapy, and a predictive factor predicts the differential effect of a therapeutic intervention.[6] The latter is frequently utilized in making treatment recommendations or decisions to a specific treatment, and thus is also referred to as a therapeutic biomarker. It serves as an indicator of likelihood of benefit, at risk of developing severe adverse events, or chance of response or resistance. For example, breast cancer patients with estrogen receptor (ER) and/or progesterone receptor (PR) positive tumors most likely respond to and benefit from endocrine therapy, whereas patients with ER negative and/or PR negative tumors derive little or no benefit from this type of treatment. It is important to note that a biomarker can be both prognostic and predictive as is the case for ER in breast cancer.

As such, it not only predicts benefit of endocrine therapy but also provides prognostic information for a more favorable outcome in the absence of treatment.[7,8] In essence, the molecular therapeutic biomarkers can aid in selecting which patient receives or spares from a therapeutic intervention, or receives a normal or lower dose of a drug before the initiation of treatment.

1.2 Classification of Therapeutic Biomarkers

The molecular biomarkers that are in clinical practice or have the potential for clinical use can be classified on the basis of their applications. The markers can be utilized, for example, to identify responders and non-responders to a particular treatment, and adjust the dosage of drugs to optimize their efficacy and safety. As such, biomarkers can be classified as (i) clinical benefit or response; (ii) resistance; (iii) risk identification; and (iv) dose selection guidance or toxicity (Tables 1.1 and 1.2). According to therapy types, they can be categorized into three major ones: (i) molecularly targeted therapy; (ii) conventional chemotherapy; and (iii) targeted therapeutics in combination with chemotherapy (Table 1.3). The biomarkers by these categories will be discussed in greater detail in the following sections.

1.3 Chemotherapy Agents and Therapeutic Biomarkers

In oncology, chemotherapy is the treatment of cancer by cytotoxic agents and combination of a number of those into cytotoxic regimens that are usually administered systemically. The types of chemotherapy agents or regimens given to patients are primarily determined by cancer histology types derived from distinct organ sites.[9] Chemotherapy agents may not specifically kill cancer cells but, rather, affect all proliferative and non-proliferative cells. Therefore, almost all chemotherapy agents have steep dose-response curves for both therapeutic and toxic effects.[9] In addition, not all patients benefit from systemic chemotherapy.

Table 1.1 Therapeutic biomarker categories based on their applications

Biomarker / drug	Benefit	Resistance	Toxicity	Risk identification
c-Kit expression or mutations in GIST / Imatinib	X			
BCR-ABL fusion gene (Ph+) in leukemia / Anti-BCR-ABL agents	X			
BCR-ABL/T315I mutation / anti-BCR-ABL agents		X		
$BRAF^{V600E}$ mutation in melanoma / BRAF kinase inhibitors	X			
BRCA dysfunction in solid tumors / PARP inhibitors	X			
DPD deficiency / 5-fluorouracil			X	
EGFR gene mutations in NSCLC / EGFR tyrosine kinase inhibitors	X			
EGFR expression in colorectal cancer / anti-EGFR antibodies	X			
EML4-ALK fusion gene in lung cancer / ALK inhibitors	X			
ER and/or PR in breast cancer / endocrine therapy	X			

Biomarker / drug	Benefit	Resistance	Toxicity	Risk identification
HER2 overexpression or gene amplification in breast cancer and gastric cancer / anti-HER2 agents, anthracycline	X			
KRAS mutations in codon 12 or 13 in colorectal cancer / anti-EGFR antibodies		X		
MammaPrint Test in breast cancer / endocrine or chemotherapy	X			X
Onco*type* DX in breast cancer / endocrine or chemotherapy	X			X
PML/RAR(α) gene expression / tretinoin, arsenic oxide	X			
TPMT deficiency or low activity / azathiopurine, thioguanine			X	
UGT1A1 polymorphisms and mutation / irinotecan, nilotinib			X	

Ph+, Philadelphia-positive chromosome; DPD, dihydropyrimine dehydrogenase; EGFR, epidermal growth factor receptor; ER, estrogen receptor; PR, progesterone receptor; echinoderm microtubule-associated protein-like 4 anaplastic lymphoma kinase; HER2, human epidermal growth receptor 2; GIST, gastrointestinal stromal tumors; ALL, acute lymphocytic leukemia; NSCLC, Non-Small Cell Lung Cancer; Topo II, topoisomerase II alpha; TPMT, thiopurine S-methyltransferase.

It is increasingly recognized that genomic or molecular basis of cancer are important to whether a patient with cancer will derive a greater benefit from a particular treatment. There are more and more numbers of examples that demonstrate how our improved understanding of cancer biology may lead to new treatments and how the selective application of treatments based on the results of the markers that correlate with greater benefit (Table 1.2). One such example is the overexpression or amplification of human epidermal growth receptor 2 (HER2; 3+ by protein or ≥ 2.0 FISH ratio by gene amplification), which identifies patients who have a poorer prognosis with standard chemotherapy and also a greater benefit from anthracycline-based adjuvant chemotherapy (Chapter 4). American Society of Clinical Oncology (ASCO) recommends that an anthracycline should be strongly considered for patients with HER2-positive breast cancer.[10] The recommendation is based on the Level II evidence, that is, prospective therapeutic trials in which biomarker utility is a secondary study objective in clinical trials (see Chapter 2).[11,12] Level I evidence is referred to as prospective controlled study that is specifically designed to test marker or evidence from meta-analysis. Please refer to the publications of Hayes *et al.* and Simon *et al.* for the tumor marker utility grading system in detail.[11,12] Recently, level II evidence has found that Akt phosphorylation at Ser473 (pAkt-S473) predicts a disease-free survival benefit from the addition of paclitaxel to adjuvant doxorubicin plus cyclophosphamide chemotherapy in node-positive breast cancer.[13]

Moreover, growing evidence has suggested that not all patients derive same magnitude of benefit from chemotherapy in patients with early stage I and II ER-positive and node-negative breast cancer. Onco*type* DX assay developed by Genomic Health, Inc., and the National Surgical Adjuvant Breast and Bowel Project (NSABP) can predict a large chemotherapy benefit (specifically the combination of cyclophosphamide, methotrexate, 5-fluorouracil or CMF regimen) in patients with high-recurrence score tumors.[14] Patients with low-recurrence score tumors have an excellent long-term survival with adjuvant tamoxifen treatment and may not need to receive adjuvant chemotherapy.

Many genetic factors correlate with the likelihood of severe and life-threatening drug toxicities. For instance, dihydropyrimine dehydrogenase (DPD) deficiency predispose patients to 5-fluorouracil (5-FU) toxicity as DPD is the initial and rate-limiting enzyme in the catabolism of 5-FU.[15] It may improve efficacy and reduce toxicity by tailoring dosage of 5-FU through screening the DPD activity before the initiation of 5-FU treatment (Chapter 11).[16] The uridine diphosphoglucuronosyl transferase 1A1 (UGT1A1) gene genotypes have been identified as a major pharmacogenetic predictor of severe hematological toxicity (neutropenia) after irinotecan administration (Chapter 12).[17] UGT1A1 genetic testing can detect the UGT1A1 6/6 (wildtype), UGT1A1 6/7 (heterozygous) and UGT1A1 7/7 (homozygous) genotypes in patients. A reduced initial irinotecan dose has been suggested for patients known to be homozygous for patients with 7/7 genotype (UGT1A1*28 allele).[18] The enzyme thiopurine S-methyltransferase (TPMT) is implicated in the metabolism of thiopurine drugs such as mercaptopurine or azathioprine.[19] TPMT deficiency and heterozygosity have been identified in patients who are intolerant to thiopurine drug therapy.[20] Prior to initiation of therapy, TPMT testing is recommended and dose adjustments should be considered for patients with TPMT mutant genotype and phenotype of low TPMT activity. Patients can be treated with these drugs without acute dose-limiting toxicity if the dosages were appropriately adjusted.

1.4 Targeted Cancer Therapeutics and Biomarkers

1.4.1 Targeted Cancer Therapeutics

The advances in the understanding of cancer genetics and biology have unveiled many key molecular alterations that play critical roles in cancer pathogenesis, driving tumor growth, and cancer cell survival. These alterations are the molecular basis of targeted cancer therapy. Drugs developed to target specific molecules with defined biological functions in cancer are referred to as the targeted cancer therapeutics.[21] This class of drugs block the growth and metastasis of cancer by interfering with the molecular targets that are involved in the pathogenesis of cancer, tumor growth and progression.[21] Over the past 20 years, the numbers of targeted drugs approved for use in the treatment of cancer has increased compared to cytotoxic chemotherapy drugs, with many more agents in pre-clinical, early and late stages of clinical development. These agents include small-molecule drugs or biologics, which include monoclonal antibodies, vaccines, and recombinant therapeutic proteins.[22]

1.4.2 Biomarker Validation

With the advances in targeted cancer treatment, much effort is being focused on the analytical and clinical validation of molecular biomarkers. The latter includes the application of analytically validated assays in patient samples and the demonstration of clinical utility (clinical qualification). The validation of analytical methods includes defining standard operating procedures and quality control measures to ensure optimal assay performance including assay precision and accuracy, limit of detection, clinical cut-off value, and sample or specimen preparation.[23] In addition, standardized procedures for pre-analytical variables such as the time from tissue acquisition to fixation and the duration of fixation as well as the type of fixative selected for use have been progressively implemented in clinical practice.[24,25] The Reporting Recommendations for tumor MARKer prognostic studies (REMARK) published in 2005 provide an excellent guidance for planning prognostic and predictive marker studies (Chapter 2).[26] Clinical validation includes the application of analytically validated assays to human samples derived from early phases of clinical trials (Chapter 3).[27–29] As such, a marker relative to drug activity, response or toxicity can be assessed in Phase 0, I or II studies of cancer therapy. The demonstration of clinical utility of a biomarker requires, in general, assessing a biomarker in relation to clinical outcomes in the context of randomized studies either prospectively or retrospectively.[11,12,30] Thus, the potential predictive value of a biomarker to clinical outcome from a specific treatment can be determined (Chapter 2).

1.4.3 Therapeutic Biomarkers of Targeted Therapy

1.4.3.1 Direct drug targets as therapeutic biomarkers

The first well-validated cancer therapeutic biomarkers are ER and/or PR, which are assessed routinely in breast cancer and used to select patients to receive hormonal treatment (Chapter 5).[10] ER is expressed in approximately 70% of breast cancer cases, through which proliferation of breast cancer cells is often driven by binding to its ligand estrogen, consequently leading to the activation of ER-regulated transcription in target genes. Therefore, agents that block the action of estrogen may interfere with, or even prevent, the growth of breast cancer cells. The selective estrogen receptor modulators (SERMs) such as tamoxifen are developed to block the action of estrogen in the breast tissue.[31] It has been well documented over several decades that ER and likely PR contents are associated with a favorable clinical outcome and predict benefit from endocrine treatment in both early breast cancer and metastatic breast disease (Chapter 5).[10]

Human epidermal growth factor receptor 2 (HER2), encoded by the *ERBB2* gene, activates multiple cellular signaling pathways that are involved in the cellular proliferation and survival. It is a driving factor in tumor progression and confers a more aggressive tumor behavior and an unfavorable prognosis.[32] Overexpression and/or amplification of the *HER2* gene predict response as well as benefit of treatment with trastuzumab, a recombinant humanized monoclonal antibody to the extracellular domain of HER2, in early and metastatic breast cancer (Chapter 4).[33] Lapatinib, a duel kinase inhibitor of HER2 and EGFR, is approved for use in the treatment of advanced or metastatic HER2-positive breast cancer. Recently, trastuzumab has been approved by the U.S. Food and Drug Administration (FDA) for the treatment of metastatic HER2-overexpressing gastric cancer.[34] Therefore, ER/PR and HER2, the molecular determinants of tumor growth and progression, are the direct drug targets and represent the therapeutic biomarkers of treatments.

Over the last decade, small-molecule kinase inhibitors have emerged as an effective treatment in cancer patients whose tumors harbor genomic alterations such as gene mutation, amplification or chromosomal translocation in the target kinases. The fusion gene of *the c-ABL* proto-oncogene from chromosome 9 with *the breakpoint cluster region* (*BCR*) on chromosome 22 (BCR-ABL fusion gene on chromosome 22) is named as the Philadelphia chromosome (Ph+). It encodes a constitutively active tyrosine kinase in chronic myelogenous leukemia (CML), some acute lymphoblastic leukemia, and acute myelogenous leukemia (Chapter 8). Patients with Ph+ disease dramatically respond to the tyrosine kinase inhibitors targeting the BCR-ABL enzyme such as imatinib, dasatinib or nilotinib. c-Kit or CD117, encoded by the *Kit* gene, is a cytokine receptor expressed on the cell membrane of hematopoietic stem cells and other cell types.

Mutations in the *Kit* gene are associated with gastrointestinal stromal tumors (GIST) and other types of tumors (Chapter 9). Most patients with gain-of-function mutations respond to imatinib, which is indicated for patients with c-Kit-positive unresectable and/or metastatic malignant GIST.[35]

Crizotinib, a small-molecule inhibitor of the receptor tyrosine kinases c-Met and an anaplastic lymphoma kinase (ALK), has been recently developed for the treatment of several cancer types (Chapter 16).[36,37] It has significant anti-tumor activity in non-small cell lung cancer (NSCLC) with the *echinoderm microtubule-associated protein-like 4 anaplastic lymphoma kinase (EML4-ALK)* fusion gene demonstrated by early clinical trials.[38] The *EML4-ALK* fusion gene is detected for up to 6% of NSCLC, which likely represents a valid therapeutic biomarker for a subset of NSCLC patients with *EML4-ALK* fusion gene. Another signaling kinase inhibitor vemurafenib (PLX4032), which selectively targets the activating $BRAF^{V600E}$ mutation, has been shown to have dramatic anti-tumor activity in metastatic melanoma.[39,40] A recent Phase III study (known as the BRIM3 trial) demonstrated that patients with metastatic melanoma treated with vemurafenib had an improved overall survival (OS) and progression-free survival (PFS) compared with those treated with dacarbazine.[41] The oncogenic $BRAF^{V600E}$ mutation is present in about 50% of melanoma, and accordingly represents a promising therapeutic biomarker for patients with melanoma (Chapter 17).[42]

As anticipated, in August 2011, the FDA approved vemurafenib tablets (ZELBORAF™, Hoffmann-La Roche Inc.) for the treatment of patients with unresectable or metastatic melanoma with the BRAF V600E mutation.[43] The mutation is detected by the approved companion diagnostic—the COBAS® 4800 BRAF V600E Mutation Test (Roche Molecular Systems, Inc.). Also in August 2011, the FDA granted accelerated approval to crizotinib (XALKORI® Capsules, Pfizer, Inc.) for the treatment of patients with locally advanced or metastatic NSCLC that is ALK-positive. Concurrently with the drug approval, the FDA approved its companion diagnostic test—the Vysis ALK Break-Apart FISH Probe Kit (Abbott Molecular, Inc.).

The epidermal growth factor receptor (EGFR) tyrosine kinase inhibitor erlotinib is prescribed for patients with advanced-stage NSCLC whose tumors have not progressed after initial treatment with certain types of chemotherapy or as second-line and third-line treatment after failure of at least one prior chemotherapy regimen.[44]

It is also indicated for patients with locally advanced, unresectable or metastatic pancreatic cancer in combination with gemcitabine. Several lines of evidence have shown that somatic mutations in the tyrosine kinase domains of the *EGFR* gene predict response to single agent erlotinib as well as to another EGFR tyrosine kinase inhibitor gefitinib (Chapter 6).[45–47] Recently, ASCO issued a provisional clinical opinion recommending EGFR mutation testing for patients with advanced NSCLC who are considered for first-line therapy with erlotinib and gefitinib.[48]

Cetuximab is a recombinant human/mouse chimeric monoclonal antibody against EGFR that is approved for administration in combination with irinotecan in the treatment of EGFR-expressing metastatic colorectal cancer in patients who are refractory to irinotecan-based chemotherapy.[49] Cetuximab is also indicated as a single agent in patients who are intolerant to irinotecan-based chemotherapy or after failure of both irinotecan and oxaliplatin-based regimens.[50,51] Panitumumab, a fully humanized monoclonal antibody to EGFR, is prescribed for metastatic colorectal cancer patients with progressive refractory disease on or following certain types of chemotherapy.[52] Assessment for EGFR expression is required for the identification of patients to receive cetuximab treatment although its anti-tumor activity has been observed in some patients with EGFR-negative tumors (Chapter 7).[53,54] The assessment for EGFR expression is, however, not required for squamous cell carcinoma of the head and neck cancer (SCCHN) patients with locoregionally advanced and metastatic SCCHN to receive cetuximab treatment as EGFR is overexpressed in the majority of SCCHN cases.[55]

Other targeted therapies include those that modify functions of proteins and other cellular processes. For example, PML/RARα is the aberrant protein product generated by the acute promyelocytic leukemia (APL)-specific t(15; 17) translocation, and functions as a potent repressor of the retinoid acid signalling pathway. Treatment with all *trans* retinoic acid (ATRA) dissociates the repressive complex from RAR and allows DNA transcription and differentiation of the immature leukemic promyelocytes into mature granulocytes (Chapter 10). ATRA induces the terminal differentiation of the leukemic promyelocytes, and is able to induce complete remission in almost all patients with APL harboring the PML/RARα fusion gene.[56,57] Patients who lack the genetic marker should be considered for alternative treatment.

In summary, many direct drug targets that are critical in the pathogenesis of cancer, tumor growth and progression are valid therapeutic biomarkers. It is expected that with the advances in the development of more selective and effective targeted agents, this class of biomarkers will steadily expand.

1.4.3.2 Indirect drug targets as therapeutic biomarkers

Sometimes, biomarkers are not the direct drug targets themselves but are important regulatory components upstream or downstream of the targets in the molecular signaling pathways. The EGFR signaling pathway includes the downstream hydrolase enzymes that can bind and hydrolyze guanosine triphosphate (GTPases) encoded by the *RAS* genes such as *KRAS, NRAS*, and *HRAS*. Adenocarcinomas in several cancer types contain activating mutations in the *RAS* family member *KRAS*. These mutations are most frequently found in codons 12 and 13 in exon 2.[58,59] Many lines of evidence demonstrated that *KRAS* mutations in metastatic colorectal cancer predict lack of efficacy to a class of monoclonal antibody drugs targeting EGFR such as cetuximab and panitumumab.[60–63] However, *KRAS* mutations occur in approximately 30% of colorectal cancer, which account for about 30% to 40% of non-responsiveness for cetuximab and panitumumab-based therapy.[64] Among colorectal tumors with wild-type KRAS, mutations of *BRAF* have been shown to associate with resistance to cetuximab and panitumumab treatment although this requires further validation before its application in the clinic.[65,66] *KRAS* and *BRAF* gene mutation status on the sensitivity and resistance to anti-EGFR agents are discussed in greater details in Chapter 6 and Chapter 7.

Recently, another interesting development demonstrated that *KRAS* mutations abrogate the response to everolimus, an inhibitor of the mammalian target of rapamycin (mTOR) in the PI3K/AKT/mTOR signaling pathway, in solid tumors with *PIK3CA* mutations. That is, *PIK3CA* mutations sensitize cancer cells to mTOR inhibitor everolimus while KRAS mutations negate the effect of PIK3CA mutations on everolimus in patients with advanced solid tumors.[67,68] *KRAS* mutation may thus be a potential predictive factor resistant to the inhibitors of PI3K/AKT/mTOR signaling pathway although further investigations are required to validate the results.

Moreover, drugs that target the base excision repair (BER) pathway by inhibitors of poly-ADP ribose polymerase (PARP), an enzyme critical to BER, have shown promising activity in patients with *BRCA* mutations (Chapter 15).[69,70] Both BRCA1 and BRCA2, which are involved in the process of homologous recombination (HR) that mediates the repair of double-stranded DNA breaks, are not the direct targets of PARP inhibitors. The incurred killing by PARP inhibitors in cells with BRCA dysfunctions is largely ascribed to a mechanism of synthetic lethality, a type of genetic interaction that one (or more) mutation or defect is not lethal, but two (or more) are lethal to a cell.[71] Anti-tumor activity of PARP inhibitors has been observed in advanced breast cancer and ovarian cancer patients with BRCA deficiencies.[69,72] In this scenario, BRCA deficiency is considered to be a surrogate therapeutic biomarker for anti-PARP agents. It, however, remains to be determined if pretreatment levels of poly-ADP ribose (PAR) and expression levels of the PARP enzymes—direct targets of PARP inhibitors—have significant impacts on efficacy or treatment outcome.[28] Moreover, PARP inhibitors' anti-tumor activity has also been observed in BRCA-wild-type tumors.[72] The variability in responses suggests that the mechanisms of activity of PARP inhibitors are not yet fully understood. Further preclinical and clinical research is needed to elucidate the mechanisms of anti-tumor activity of PARP inhibitors.

1.4.3.3 Anti-angiogenesis therapy and biomarkers

During the past decade, agents targeting the processes that lead to new blood vessel formation—angiogenesis, and drugs that disrupt the existing tumor vasculature have evolved as an effective cancer treatment when combined with conventional chemotherapy. Bevacizumab, a humanized antibody to vascular endothelial growth factor-A (VEGF-A), and other anti-angiogenic agents have been integrated into standard management of several tumor types over the past decade. However, a significant number of patients do not benefit from anti-angiogenesis therapy and/or in combination with chemotherapy, the search for biomarkers for appropriate selection of patients has been challenging and is an intense area of translational and clinical research.[6,73]

Table 1.2 Therapeutic or promising biomarkers and their applications in cancer

Biomarker	Application	Drug*	Chapter
c-Kit expression or mutations	Kit (CD117) positive unresectable or metastatic malignant GIST	Imatinib	**9**
BCR-ABL fusion gene (Ph+) in leukemia	Newly diagnosed Ph+ CML; Ph+ CML after failure interferon-α therapy; Ph+ ALL; Ph+ CML and ALL with resistance or intolerance to prior therapy including imatinib	Imatinib, dasatinib, nilotinib	**8**
BCR-ABL/T315I mutation	Resistance to first or second generation anti-BCR-ABL agents	Imatinib, dasatinib, nilotinib	**8**
$BRAF^{V600E}$ mutation	Selective $BRAF^{V600E}$ kinase inhibitor for metastatic melanoma treatment	Vemurafenib	**17**
BRCA dysfunction	Metastatic ovarian cancer and breast cancer with BRCA mutations; metastatic triple- negative breast cancer, & high grade ovarian cancer in clinical trials	Olaparib, valiparib, iniparib	**15**
DPD deficiency	Rarely, unexpected, severe toxicity associated with 5-fluorouracil has been attributed to DPD deficiency	5-Fluorouracil	**11**
EGFR gene mutations	Treatment with EGFR TKI for locally advanced, unresectable or metastatic NSCLC	Erlotinib, gefitinib	**6**

Biomarker	Application	Drug*	Chapter
EGFR expression	Treatment with anti-EGFR antibodies for metastatic colorectal cancer	Cetuximab, panitumumab	**7**
EML4-ALK fusion gene	Treatment with ALK kinase inhibitor for metastatic NSCLC	Crizotinib	**16**
ER and/or PR	Endocrine therapy for early and metastatic breast cancer	Tomoxifen, letrozole	**5**
HER2 overexpression or gene amplification	Anti-HER2 or anthracycline-based chemotherapy for early and metastatic breast cancer; treatment of metastatic gastric or gastroesophageal junction adenocarcinoma	Trastuzumab, doxorubicin	**4**
KRAS mutations in codon 12 or 13	Cetuximab or panitumumab is not recommended for anti-EGFR treatment in metastatic colorectal cancer with these mutations	Cetuximab, panitumumab	**7**
MammaPrint test	Adjuvant chemotherapy plus endocrine therapy for high risk while adjuvant endocrine therapy for low risk early node-negative and three or few-node-positive invasive breast cancer	Chemotherapy or tamoxifen	**14**

(*Continued*)

Table 1.2 (*Continued*)

Biomarker	**Application**	**Drug***	**Chapter**
Onco*type* DX assay	Adjuvant chemotherapy plus endocrine therapy for high risk while adjuvant endocrine therapy for low risk early ER-positive and node-negative breast cancer	CMF or tamoxifen	**13**
PML/RAR(α) gene expression	Treatment with ATRA may be based on its expression for acute promyelocytic leukemia	ATRA; arsenic oxide	**10**
TPMT deficiency or low activity	At increased risk of myelotoxicity. TPMT testing is recommended and consideration can be given to genotype or phenotype patients for TPMT	Mercaptopurine, azathioprine	-
UGT1A1 polymorphisms and mutation	A reduced initial dose for homozygous UGT1A1*28 allele	Irinotecan	**12**

*Prototype drug.

ATRA, all trans retinoic acid; Ph+, philadelphia-positive chromosome; DPD, dihydropyrimine dehydrogenase; EGFR, epidermal growth factor receptor; ER, estrogen receptor; PR, progesterone receptor; echinoderm microtubule-associated protein-like 4 anaplastic lymphoma kinase; HER2, human epidermal growth receptor 2; GIST, gastrointestinal stromal tumors; ALL, acute lymphocytic leukemia; NSCLC, Non-Small Cell Lung Cancer; Topo II, topoisomerase II alpha; TPMT, thiopurine S-methyltransferase.

1.5 Targeted Therapeutics in Combination with Chemotherapy and Therapeutic Biomarkers

As discussed above, HER2 overexpression or amplification in breast cancer identifies patients who have greater benefit from anthracycline-based adjuvant chemotherapy and derive benefit from trastuzumab treatment (Table 1.3). BRCA deficiency, on the other aspect, identifies patients who are more likely to respond to anti-PARP agents and perhaps to platinum chemotherapy.[69] Thus, HER2 overexpression or amplification, and BRCA deficiency act as a category of biomarkers with dual functions that predict response to both targeted therapy and chemotherapy.

Targeted treatment has offered an opportunity of inhibiting or killing cancer cells relative to normal healthy cells by potentially increasing efficacy and reducing systemic toxicity compared to traditional cytotoxic chemotherapy. Up to date, most of the targeted agents are not curative as single agents with some exceptions such as anti-BCR-ABL agents in CML. The targeted therapeutics are, therefore, frequently combined with the established chemotherapy agents or regimens to increase efficacy. These types of combination regimens, as are other new drugs developed in oncology, are first tested in patients with advanced metastatic diseases after failure of the established first-line treatment, and followed by evaluation as the first-line treatment.[74] The drugs or regimens that demonstrated improvement in efficacy and/or benefit in advanced diseases will likely be evaluated in patients with early-stage cancers in adjuvant (after surgery) and neoadjuvant (before surgery) settings.

In the circumstances of developing predictive biomarkers for multi-drug combination therapy, two or more markers may be needed to select patients—one for targeted treatment and the other for chemotherapy. For instance, trastuzumab in combination with taxane-based chemotherapy has been initially shown to have significant activity in HER2-overexpressing metastatic breast cancer compared with chemotherapy alone;[75] and subsequently demonstrated to be effective in adjuvant setting for early breast cancer.[76-78] HER2 is the molecular determinant of therapeutic efficacy of trastuzumab, and taxanes are considered to be active in both HER2-positive and HER2-negative breast tumors.[79] Although this approach is highly effective in HER2-positive breast cancer, given the fact that not all patients benefit from a taxane, using a biomarker for taxanes may further increase efficacy or add additional benefit. Recently, pAkt-S473 has been identified as a predictive factor for paclitaxel chemotherapy benefit in node-positive breast cancer. It would be legitimate to use both pAkt and HER2 for selecting patients to receive trastuzumab in combination with taxane chemotherapy although this approach remains to be evaluated.[13]

Rational use of multiple biomarkers in the context of targeted and chemotherapy regimens hold promise towards a more personalized medicine. The deployment of this approach potentially increases the odds for patients to benefit from therapy and avoid unnecessary toxicity from exposure to an ineffective or unnecessary treatment.

Table 1.3 Therapeutic biomarker categories based on therapy types

Biomarker	Chemotherapy	Targeted therapy	Chemotherapy plus targeted therapy
c-Kit expression or mutations		+++	
BCR-ABL fusion gene (Ph+)		+++	
BCR-ABL/T315I mutation		+++	
BRAF V600E mutation		+++	
BRCA dysfunction	++	++	++
DPD deficiency	++		
EGFR expression or gene mutations		+++	
EML4-ALK fusion gene		+++	
ER and/or PR	+	+++	
HER2 overexpression or gene amplification	++	+++	++

Biomarker	Chemotherapy	Targeted therapy	Chemotherapy plus targeted therapy
KRAS mutations in codon 12 or 13		++	
MammaPrint Test	++	++	
Onco*type* DX assay	++	++	
PML/RAR(a) gene expression	++	+++	
TPMT deficiency or low activity	++		
UGT1A1 polymorphisms and mutation	++		

+++, High level of relevance in relation to treatment outcome, which may reflect a degree of capturing or a linkage to the treatment effect or toxicity; ++, intermediate level of relevance; + low level of relevance.

1.6 Multi-Gene Expression or Signatures for Cancer Prognosis and Treatment

Patient and tumor characteristics at diagnosis or at baseline provide important information about a likely patient outcome and influence the treatment decisions across many tumor types. However, these subjects are not extensively discussed in this book. Some patients with early breast cancer may be cured with surgery alone or with surgery plus standard endocrine therapy. The existing guidelines recommend adjuvant chemotherapy in breast cancer patients whose tumors are at least 1.0 cm, which has led to the concerns of overtreatment. A critical question is which patients with early breast cancer require chemotherapy and which do not. Multi-gene expression and gene expression signatures such as Oncotype DX and MammaPrint have been evolved as valid prognostic and predictive tools of outcome for patients with early stage breast cancer. Onco*type* DX breast cancer test, a 21-gene recurrence score (RS), is an RT-PCR-based assay that utilizes paraffin-embedded tissue and measures the expression of 21 genes that include 16 cancer-related genes and five reference genes (Chapter 13). It estimates the risk of recurrence with distant metastasis, recently the likelihood of locoregional recurrence, in patients with hormone receptor-positive and node-negative breast cancer.[80,81]

The 21-gene RS assay also predicted chemotherapy benefit among patients with ER-positive and node-negative early stage breast cancer in the NSABP B-20 trial. The ongoing randomized phase III TAILORx trial, which recently completed accrual, is designed to prospectively evaluate the ability of the 21-gene assay to predict chemotherapy benefit in hormone receptor-positive and node-negative early-stage breast cancer.

In contrast to Oncotype DX, which has been evaluated in hormone receptor positive, node negative breast cancer, MammaPrint, a 70-gene signature derived from the gene expression profiling approach, has been validated as a prognostic tool for patients with hormone receptor and lymph node positive or negative early-stage breast cancer patients who received adjuvant endocrine therapy with or without chemotherapy (Chapter 14). The ongoing MINDACT (Microarray in Node-Negative Disease May Avoid Chemotherapy) study is designed to prospectively compare the 70-gene prognostic signature with standard clinical-pathologic variables in patients with node-negative or node-positive breast cancer (1–3 positive nodes) to evaluate its utility as a predictive assay. Recently, the consensus from the 2011 St. Gallen Conference on Primary Therapy of Early Breast Cancer agreed that Onco*type* DX rather than MammaPrint may be used to determine whether patients with endocrine-responsive breast cancer should receive chemotherapy.[82]

Rotterdam signature, a 76-gene signature, is specifically studied in node-negative breast cancer patients, regardless of age, tumor size and grade, or ER/PR status, which may be distinguished from Onco*type* DX for hormone receptor-positive cases and MammaPrint for young female cases. This signature is significantly prognostic in patients with early-stage breast cancer but to date it has not been investigated as a predictor of chemotherapy benefit using phase III clinical trial.

Similar to the management of early stage breast cancer, it is a challenge to make treatment decisions (using or not using chemotherapy) in stage II colon cancer. The Onco*type* DX colon cancer assay is also an RT-PCR-based molecular diagnostic test that quantitatively measures 12-gene expression and assess the recurrence risk in stage II colon cancer.[83] It includes five reference genes and seven cancer-related genes that are associated with the activated stroma, cell cycle and early response or genotoxic stress pathway. The assay generates a RS for individual patients from the reference gene normalized tumor expression of the 7-cancer related genes. It predicts the recurrence risk in patients with stage II colorectal cancer after surgery. However, it did not significantly predict benefit of 5-fluorouracil/leucovorin chemotherapy.

1.7 Diagnostic Techniques for Therapeutic Biomarkers

Details of diagnostic techniques used for detecting an individual therapeutic biomarker are discussed in relevant chapters. Table 1.4 presents a summary of the clinical diagnostic tests that include technical methods, brief detection mechanisms, and their therapeutic biomarker applications.

1.8 Conclusions and Perspectives

The steady advance in the development of targeted cancer therapy continues to assure the increased success and reduced toxicity in the management of cancer patients. Besides the therapeutic biomarkers already in the clinic, most of the promising biomarkers discussed in this book are very likely to evolve as valid therapeutic biomarkers in the next few years. The US National Institutes of Health and the FDA provide the scientific and regulatory framework for the co-development of drug-biomarkers.[84] In particular, the AACR (American Association of Cancer Research)-FDA-NCI (National Cancer Institute) Cancer Biomarker collaborative is bringing all national efforts together by integrating pharmacodynamic and diagnostic biomarkers into the cancer drug development enterprise.[5] We anticipate that more targeted cancer therapeutics and their companion therapeutic biomarkers will be developed and validated by the academic, government and private sectors in the years to

Table 1.4 Technology platforms for diagnostic tests of therapeutic biomarkers

Method	Principle	Application	Chapters
Immunohistochemistry	Antibodies binding specifically to tumor-associated antigens in human tissues	ER/PR, c-Kit, EGFR, EML4-ALK, HER2	**5, 9, 6, 7, 16, 4**
RT-PCR	mRNA is reverse transcribed into its cDNA, which is followed by amplification using PCR	BCR-ABL, EML4-ALK, Onco*type* DX, PML/RARa	**8, 16, 13, 10**
FISH or CISH	Detect and localize specific DNA or RNA sequences on a given chromosome in cells, tissues, and tumors	BCR-ABL, EML4-ALK, EGFR, HER2, PML/RARa, TOPII	**8, 16, 6, 7, 4, 10**
PCR/DNA sequencing	Detect mutation and SNP involves PCR amplification and DNA sequencing	BCR-ABL kinase domain mutation, KIT, *DYPD*, KRAS, BRAF, EGFR, UGT1A1	**8, 9, 11, 7, 17, 7, 12**
cDNA Microarray	Measure the expression levels of a large numbers of genes at the RNA level	MammaPrint test, Rotterdam signature	**14**
Parasequencing	Determine the order of nucleotides in DNA based on the sequencing by synthesis	BRAF codon 600 mutation	**17**

CISH, chromogenic *in situ* hybridization; FISH, fluorescence *in situ* hybridization; DPD is encoded by the *DYPD* gene; RT-PCR, reverse transcriptase-polymerase chain reaction; PCR, polymerase chain reaction; SNP, single nucleotide polymorphisms.

come as we move towards fulfilling the promise of a personalized medicine.

References

1. Siegel, R., Naishadham, D., and Jemal, A. Cancer statistics, 2013. *CA Cancer J Clin* **63**, 11–30, (2013).
2. Jemal, A., Siegel, R., Xu, J., and Ward, E. Cancer statistics, 2010. *CA Cancer J Clin* **60**, 277–300 (2010).
3. Kohler, B.A., *et al.* Annual report to the Nation on the status of cancer, 1975–2007, featuring tumors of the brain and other nervous system. *J Natl Cancer Inst* **103**, 714–736 (2011).
4. Biomarkers Definitions Working Group. Biomarkers and surrogate endpoints: preferred definitions and conceptual framework. *Clin Pharmacol Ther* **69**, 89–95 (2001).
5. Khleif, S.N., Doroshow, J.H., and Hait, W.N. AACR-FDA-NCI cancer biomarkers collaborative consensus report: advancing the use of biomarkers in cancer drug development. *Clin Cancer Res* **16**, 3299–3318 (2010).
6. Jain, R.K., *et al.* Biomarkers of response and resistance to antiangiogenic therapy. *Nat Rev Clin Oncol* **6**, 327–338 (2009).
7. Clark, G.M., and McGuire, W.L. Steroid receptors and other prognostic factors in primary breast cancer. *Semin Oncol* **15**, 20–25 (1988).
8. Early Breast Cancer Trialists' Collaborative Group. Effects of chemotherapy and hormonal therapy for early breast cancer on recurrence and 15-year survival: an overview of the randomised trials. *Lancet* **365**, 1687–1717 (2005).
9. Cavalli, F., Kaye, S.B., Hansen, H.H., Armitage, J.O., and Piccart-Gebhart, M.J. (eds.). *Textbook of Medical Oncology*, 462 (Informa Healthcare, London, 2009).
10. Harris, L., *et al.* American society of clinical oncology 2007 update of recommendations for the use of tumor markers in breast cancer. *J Clin Oncol* **25**, 5287–5312 (2007).
11. Hayes, D.F., *et al.* Tumor marker utility grading system: a framework to evaluate clinical utility of tumor markers. *J Natl Cancer Inst* **88**, 1456–1466 (1996).
12. Simon, R.M., Paik, S., and Hayes, D.F. Use of archived specimens in evaluation of prognostic and predictive biomarkers. *J Natl Cancer Inst* **101**, 1446–1452 (2009).

13. Yang, S.X., *et al.* Akt phosphorylation at Ser473 predicts benefit of paclitaxel chemotherapy in node-positive breast cancer. *J Clin Oncol* **28**, 2974–2981 (2010).
14. Paik, S., *et al.* Gene expression and benefit of chemotherapy in women with node-negative, estrogen receptor-positive breast cancer. *J Clin Oncol* **24**, 3726–3734 (2006).
15. van Kuilenburg, A.B. Dihydropyrimidine dehydrogenase and the efficacy and toxicity of 5-fluorouracil. *Eur J Cancer* **40**, 939–950 (2004).
16. Ciccolini, J., *et al.* A rapid and inexpensive method for anticipating severe toxicity to fluorouracil and fluorouracil-based chemotherapy. *Ther Drug Monit* **28**, 678–685 (2006).
17. Innocenti, F., *et al.* Genetic variants in the UDP-glucuronosyltransferase 1A1 gene predict the risk of severe neutropenia of irinotecan. *J Clin Oncol* **22**, 1382–1388 (2004).
18. Camptosar (irinotecan HCL) injection: http://www.fda.gov/Safety/MedWatch/SafetyInformation/ucm215480.htm.
19. Ford, L.T., and Berg, J.D. Thiopurine S-methyltransferase (TPMT) assessment prior to starting thiopurine drug treatment; a pharmacogenomic test whose time has come. *J Clin Pathol* **63**, 288–295 (2010).
20. Evans, W.E., *et al.* Preponderance of thiopurine S-methyltransferase deficiency and heterozygosity among patients intolerant to mercaptopurine or azathioprine. *J Clin Oncol* **19**, 2293–2301 (2001).
21. Hait, W.N., and Hambley, T.W. Targeted cancer therapeutics. *Cancer Res* **69**, 1263–1267; discussion 1267 (2009).
22. Higgins, M.J., and Baselga, J. Breast cancer in 2010: Novel targets and therapies for a personalized approach. *Nat Rev Clin Oncol* **8**, 65–66 (2011).
23. Chau, C.H., Rixe, O., McLeod, H., and Figg, W.D. Validation of analytic methods for biomarkers used in drug development. *Clin Cancer Res* **14**, 5967–5976 (2008).
24. Wolff, A.C., *et al.* American society of clinical oncology/college of american pathologists guideline recommendations for human epidermal growth factor receptor 2 testing in breast cancer. *Arch Pathol Lab Med* **131**, 18–43 (2007).
25. Hammond, M.E., Hayes, D.F., Wolff, A.C., Mangu, P.B., and Temin, S. American society of clinical oncology/college of american pathologists guideline recommendations for immunohistochemical testing of

estrogen and progesterone receptors in breast cancer. *J Oncol Pract* **6**, 195–197 (2010).

26. McShane, L.M., *et al.* Reporting recommendations for tumor marker prognostic studies. *J Clin Oncol* **23**, 9067–9072 (2005).
27. Kummar, S., *et al.* Compressing drug development timelines in oncology using phase '0' trials. *Nat Rev Cancer* **7**, 131–139 (2007).
28. Yang, S.X., *et al.* Immunohistochemical detection of poly(ADP-ribose) polymerase inhibition by ABT-888 in patients with refractory solid tumors and lymphomas. *Cancer Biol Ther* **8**, 2004–2009 (2009).
29. Wedam, S.B., *et al.* Antiangiogenic and antitumor effects of bevacizumab in patients with inflammatory and locally advanced breast cancer. *J Clin Oncol* **24**, 769–777 (2006).
30. Rubinstein, L., Crowley, J., Ivy, P., Leblanc, M., and Sargent, D. Randomized phase II designs. *Clin Cancer Res* **15**, 1883–1890 (2009).
31. Wolff, A.C., and Davidson, N.E. Use of SERMs for the adjuvant therapy of early-stage breast cancer. *Ann N Y Acad Sci* **949**, 80–88 (2001).
32. Slamon, D.J., *et al.* Studies of the HER-2/neu proto-oncogene in human breast and ovarian cancer. *Science* **244**, 707–712 (1989).
33. Dybdal, N., *et al.* Determination of HER2 gene amplification by fluorescence *in situ* hybridization and concordance with the clinical trials immunohistochemical assay in women with metastatic breast cancer evaluated for treatment with trastuzumab. *Breast Cancer Res Treat* **93**, 3–11 (2005).
34. HERCEPTIN (Trastuzumab) Lablel: http://www.accessdata.fda.gov/drugsatfda_docs/label/2010/103792s5256lbl.pdf.
35. GLEEVAC (imatinib mesylate) tablets Label: http://www.accessdata.fda.gov/drugsatfda_docs/label/2008/021588s024lbl.pdf.
36. Gerber, D.E., and Minna, J.D. ALK inhibition for non-small cell lung cancer: from discovery to therapy in record time. *Cancer Cell* **18**, 548–551 (2010).
37. Choi, Y.L., *et al.* EML4-ALK mutations in lung cancer that confer resistance to ALK inhibitors. *N Engl J Med* **363**, 1734–1739 (2010).
38. Kwak, E.L., *et al.* Anaplastic lymphoma kinase inhibition in non-small-cell lung cancer. *N Engl J Med* **363**, 1693–1703 (2010).
39. Bollag, G., *et al.* Clinical efficacy of a RAF inhibitor needs broad target blockade in BRAF-mutant melanoma. *Nature* **467**, 596–599 (2010).
40. Flaherty, K.T., *et al.* Inhibition of mutated, activated BRAF in metastatic melanoma. *N Engl J Med* **363**, 809–819 (2010).

41. Chapman, P.B. *et al.* Improved survival with vemurafenib in melanoma with BRAF V600E mutation. *N Engl J Med* 364, 2507–2516 (2011).

42. Nazarian, R., *et al.* Melanomas acquire resistance to B-RAF(V600E) inhibition by RTK or N-RAS upregulation. *Nature* **468**, 973–977 (2010).

43. ZELBORAF™ (vemurafenib) label: http://www.accessdata.fda.gov/drugsatfda_docs/label/2011/202429s000lbl.pdf.

44. Tarceva (erlotinib) Label: http://www.accessdata.fda.gov/drugsatfda_docs/label/2010/021743s14s16lbl.pdf.

45. Lynch, T.J., *et al.* Activating mutations in the epidermal growth factor receptor underlying responsiveness of non-small-cell lung cancer to gefitinib. *N Engl J Med* **350**, 2129–2139 (2004).

46. Rosell, R., Viteri, S., Molina, M.A., Benlloch, S., and Taron, M. Epidermal growth factor receptor tyrosine kinase inhibitors as first-line treatment in advanced nonsmall-cell lung cancer. *Curr Opin Oncol* **22**, 112–120 (2010).

47. Dahabreh, I.J., *et al.* Somatic EGFR mutation and gene copy gain as predictive biomarkers for response to tyrosine kinase inhibitors in non-small cell lung cancer. *Clin Cancer Res* **16**, 291–303 (2010).

48. Keedy, V.L., *et al.* American society of clinical oncology provisional clinical opinion: epidermal growth factor receptor (EGFR) mutation testing for patients With advanced non-small-cell lung cancer considering first-line EGFR tyrosine kinase inhibitor therapy. *J Clin Oncol* **22**, 2121–2127 (2011).

49. Erbitux (cetuximab) Lablel: http://www.accessdata.fda.gov/drugsatfda_docs/label/2009/125084s168lbl.pdf.

50. Wong, S.F. Cetuximab: an epidermal growth factor receptor monoclonal antibody for the treatment of colorectal cancer. *Clin Ther* **27**, 684–694 (2005).

51. Goldberg, R.M. Cetuximab. *Nat Rev Drug Discov* **Suppl**, S10-1 (2005).

52. Vectibix (panitumumab) Label: http://www.accessdata.fda.gov/drugsatfda_docs/label/2009/125147s080lbl.pdf.

53. Moroni, M., *et al.* Gene copy number for epidermal growth factor receptor (EGFR) and clinical response to antiEGFR treatment in colorectal cancer: a cohort study. *Lancet Oncol* **6**, 279–286 (2005).

54. Chung, K.Y., *et al.* Cetuximab shows activity in colorectal cancer patients with tumors that do not express the epidermal growth factor receptor by immunohistochemistry. *J Clin Oncol* **23**, 1803–1810 (2005).

55. Ang, K.K., *et al.* Impact of epidermal growth factor receptor expression on survival and pattern of relapse in patients with advanced head and neck carcinoma. *Cancer Res* **62**, 7350–7356 (2002).

56. Grignani, F., *et al.* Formation of PML/RAR alpha high molecular weight nuclear complexes through the PML coiled-coil region is essential for the PML/RAR alpha-mediated retinoic acid response. *Oncogene* **18**, 6313–6321 (1999).

57. Meani, N., *et al.* Molecular signature of retinoic acid treatment in acute promyelocytic leukemia. *Oncogene* **24**, 3358–3368 (2005).

58. Rodenhuis, S., *et al.* Incidence and possible clinical significance of K-ras oncogene activation in adenocarcinoma of the human lung. *Cancer Res* **48**, 5738–5741 (1988).

59. Pao, W., *et al.* KRAS mutations and primary resistance of lung adenocarcinomas to gefitinib or erlotinib. *PLoS Med* **2**, e17 (2005).

60. Allegra, C.J., *et al.* American society of clinical oncology provisional clinical opinion: testing for KRAS gene mutations in patients with metastatic colorectal carcinoma to predict response to anti-epidermal growth factor receptor monoclonal antibody therapy. *J Clin Oncol* **27**, 2091–2096 (2009).

61. Karapetis, C.S., *et al.* K-ras mutations and benefit from cetuximab in advanced colorectal cancer. *N Engl J Med* **359**, 1757–1765 (2008).

62. Freeman, D.J., *et al.* Association of K-ras mutational status and clinical outcomes in patients with metastatic colorectal cancer receiving panitumumab alone. *Clin Colorectal Cancer* **7**, 184–190 (2008).

63. Normanno, N., *et al.* Implications for KRAS status and EGFR-targeted therapies in metastatic CRC. *Nat Rev Clin Oncol* **6**, 519–527 (2009).

64. Di Fiore, F., *et al.* Clinical relevance of KRAS mutation detection in metastatic colorectal cancer treated by Cetuximab plus chemotherapy. *Br J Cancer* **96**, 1166–1169 (2007).

65. Di Nicolantonio, F., *et al.* Wild-type BRAF is required for response to panitumumab or cetuximab in metastatic colorectal cancer. *J Clin Oncol* **26**, 5705–5712 (2008).

66. Siena, S., Sartore-Bianchi, A., Di Nicolantonio, F., Balfour, J., and Bardelli, A. Biomarkers predicting clinical outcome of epidermal growth factor receptor-targeted therapy in metastatic colorectal cancer. *J Natl Cancer Inst* **101**, 1308–1324 (2009).

67. Di Nicolantonio, F., *et al.* Deregulation of the PI3K and KRAS signaling pathways in human cancer cells determines their response to everolimus. *J Clin Invest* **120**, 2858–2866 (2010).

68. Tabernero, J., *et al.* Dose- and schedule-dependent inhibition of the mammalian target of rapamycin pathway with everolimus: a phase I tumor pharmacodynamic study in patients with advanced solid tumors. *J Clin Oncol* **26**, 1603–1610 (2008).
69. Fong, P.C., *et al.* Inhibition of poly(ADP-ribose) polymerase in tumors from BRCA mutation carriers. *N Engl J Med* **361**, 123–134 (2009).
70. Fong, P.C., *et al.* Poly(ADP)-ribose polymerase inhibition: frequent durable responses in BRCA carrier ovarian cancer correlating with platinum-free interval. *J Clin Oncol* **28**, 2512–2519 (2010).
71. Ashworth, A.A synthetic lethal therapeutic approach: poly(ADP) ribose polymerase inhibitors for the treatment of cancers deficient in DNA double-strand break repair. *J Clin Oncol* **26**, 3785–3790 (2008).
72. Gelmon, K.A., *et al.* Olaparib in patients with recurrent high-grade serous or poorly differentiated ovarian carcinoma or triple-negative breast cancer: a phase 2, multicentre, open-label, non-randomised study. *Lancet Oncol* **12**, 852–861 (2011).
73. Yang, S.X. Bevacizumab and breast cancer: current therapeutic progress and future perspectives. *Expert Rev Anticancer Ther* **9**, 1715–1725 (2009).
74. Soria, J.C., *et al.* Added value of molecular targeted agents in oncology. *Ann Oncol* **22**, 1703–1716 (2011).
75. Marty, M., *et al.* Randomized phase II trial of the efficacy and safety of trastuzumab combined with docetaxel in patients with human epidermal growth factor receptor 2-positive metastatic breast cancer administered as first-line treatment: the M77001 study group. *J Clin Oncol* **23**, 4265–4274 (2005).
76. Romond, E.H., *et al.* Trastuzumab plus adjuvant chemotherapy for operable HER2-positive breast cancer. *N Engl J Med* **353**, 1673–1684 (2005).
77. Piccart-Gebhart, M.J., *et al.* Trastuzumab after adjuvant chemotherapy in HER2-positive breast cancer. *N Engl J Med* **353**, 1659–1672 (2005).
78. Smith, I., *et al.* 2-year follow-up of trastuzumab after adjuvant chemotherapy in HER2-positive breast cancer: a randomised controlled trial. *Lancet* **369**, 29–36 (2007).
79. De Laurentiis, M., *et al.* Taxane-based combinations as adjuvant chemotherapy of early breast cancer: a meta-analysis of randomized trials. *J Clin Oncol* **26**, 44–53 (2008).

80. Paik, S., *et al.* A multigene assay to predict recurrence of tamoxifen-treated, node-negative breast cancer. *N Engl J Med* **351**, 2817–2826 (2004).

81. Mamounas, E.P., *et al.* Association between the 21-gene recurrence score assay and risk of locoregional recurrence in node-negative, estrogen receptor-positive breast cancer: results from NSABP B-14 and NSABP B-20. *J Clin Oncol* **28**, 1677–1683 (2010).

82. Gnant, M., Harbeck, N., and Thomssen, C. St. Gallen 2011: Summary of the consensus discussion. *Breast Care (Basel)* **6**, 136–141 (2011).

83. Clark-Langone, K.M., Sangli, C., Krishnakumar, J., and Watson, D. Translating tumor biology into personalized treatment planning: analytical performance characteristics of the Oncotype DX Colon Cancer Assay. *BMC Cancer* **10**, 691 (2010).

84. Hamburg, M.A., and Collins, F.S. The path to personalized medicine. *N Engl J Med* **363**, 301–304 (2010).

Chapter 2

Statistical Considerations in the Development and Evaluation of Therapeutic Biomarkers in Cancer

Lisa M. McShane, Edward L. Korn, and Boris Freidlin
Biometric Research Branch,
Division of Cancer Treatment and Diagnosis,
National Cancer Institute, Bethesda, MD 20892
mcshanel@ctep.nci.nih.gov, korne@ctep.nci.nih.gov, freidlinb@ctep.nci.nih.gov

2.1 Introduction

Many new anti-cancer therapies target tumors with specific biological characteristics. It is important in these situations to not only establish efficacy of the new therapy but also to accurately identify the subgroup of patients who will benefit. The premise underlying molecularly targeted anti-cancer therapy development is that an understanding of the biological characteristics of these tumors (as well as germline genetic make-up) and of how an anti-cancer agent interacts with those characteristics should help to optimize therapy for patients. Often a new therapy will benefit only a subset of

Handbook of Therapeutic Biomarkers in Cancer
Edited by Sherry X. Yang and Janet E. Dancey
Copyright © 2013 Pan Stanford Publishing Pte. Ltd.
ISBN 978-981-4364-65-2 (Hardcover), 978-981-4364-66-9 (eBook)
www.panstanford.com

patients or be effective against only a subset of tumor types, namely those exhibiting the biological characteristics that can be targeted by a therapy having a specific mechanism of action. This paradigm of biologically driven therapy selection presents challenges for the statistical design of clinical trials to establish clinical benefit of new therapies because the evaluation of a therapy cannot be separated from the evaluation of a biomarker test that will identify the patients having the characteristics predictive of benefit. In this chapter we will assume that the biological characterization is accomplished by use of biomarker-based tests, although the principles discussed here apply also to other methods of biological characterization such as imaging.

The outline of this chapter is as follows. The chapter begins with a brief discussion of some considerations in the evaluation of analytical performance of a biomarker-based test; a biomarker assay should meet some minimal analytical performance criteria before being used in a clinical trial where it might influence the therapy a patient receives. Next, the distinction between prognostic and predictive biomarkers is explained in order to stress the importance of considering potential prognostic effects of biomarkers in clinical trial designs for evaluating use of those biomarkers for therapy selection. The focus shifts then to a discussion of designs for phase I, II, and III clinical trials incorporating biomarkers. Several design options for phase II and III studies are presented with particular attention to the increasingly important role of randomization when only a biomarker-defined subgroup of patients is eligible to receive the new therapy.

2.2 Analytical Performance of a Biomarker-Based Test

Ideally the development of the biomarker assay forming the basis for a biomarker-based test would be synchronized with the development of the targeted therapy, but this does not always happen. However, it is important that by the time the therapeutic development process reaches the stage of large phase II trials or phase III trials, the biomarker assay has undergone sufficient analytical validation. In particular the assay should be reproducible and accurately reflect the biologic characteristic of interest. Table 2.1 presents considerations for determination of the readiness of a biomarker assay for use in a clinical trial. For the specific situation of a dichotomous biomarker (e.g., mutation present/absent) or a test based on a continuous biomarker measurement to which a cutpoint is applied to report a dichotomous result, a more in-depth discussion on analytical validation considerations is given elsewhere [1]. To achieve an acceptable level of analytical performance prior to use of a biomarker assay in a trial may require multiple iterations of assay refinement. The reader is referred to documents published by CLSI (Clinical and Laboratory Standards Institute) and the U.S. Food and Drug Administration (FDA) for further useful information on the topic of assay analytical validation [2–5].

Table 2.1 Considerations for biomarker assays that play an integral role in a clinical trial

Assays that must be performed for a trial to proceed are denoted here as "integral" assays. When an assay is used in some way that could potentially affect patient care, it must be performed in a laboratory with at least a CLIA Certificate of Compliance, and an investigational device exemption from the FDA will usually be required. The types of evaluations that should be performed and the information that should be reported prior to use of a biomarker-based test in a large Phase II or Phase III trial include the categories of data that would typically be required for submission for FDA clearance (510k – substantial equivalence) or approval (premarket application). Additional regulatory approvals may be required
The requirements and the information to support the use of the assay are as follows:
A. The role the assay will play in the trial
Roles would include use for eligibility determination, assignment to therapy, or risk stratification for randomization to improve the balance of biomarker values between treatment arms. The intended use of the assay/test in clinical practice may differ from its role in the trial. The intended use for which the assay is being evaluated in the trial should be described. Examples of intended clinical uses would include prognostic indicator, predictive variable for benefit from a particular treatment or class of therapeutic agents, or indicator of favorable response or toxic reaction to a specific drug.

(Continued)

Table 2.1 (*Continued*)

B.	Measurements of precision and reproducibility (within lab and between labs if more than one lab will be performing the assay) Precision and reproducibility address closeness of agreement between independent test results obtained under stipulated conditions. The precision of an assay procedure refers to repeatability of measurements under essentially unchanged assay conditions, often referred to as "within-series precision" or "within-run precision." For imaging tests, "within-patient test-retest" reproducibility data would be relevant. Intermediate precision refers to measurements taken when there is variation in one or more factors, such as time, calibration, operator, and equipment - usually within a laboratory. Reproducibility generally refers to inter-laboratory precision and relates to changes in conditions such as different operators and measuring systems (including different calibrations and reagent batches). Independent test results refer to results obtained in a manner that is not influenced by previous results obtained on the same or similar test samples. Information to be provided should include the assay technical protocol followed, the conditions of the study, what factors were varied, and summary metrics including calculations of standard deviation (SD), coefficient of variation (CV) and descriptions of relationships between variation measures and means. Precision studies will optimally be performed in ranges of assay values corresponding to important clinical decision points (e.g., near a cutpoint that separates different clinical states).
C.	Data to support proposed cutpoint(s) if assay results are not reported as a continuous variable Cutpoints are thresholds that are applied to continuous or semi-quantitative assay measurements for purposes of reducing the assay or imaging test result to a positive/negative determination or perhaps to a few categories (e.g., low, medium, high). Any cutpoint(s) must be clearly pre-specified because the statistical strength of the association between the categorized marker and a clinical endpoint, and the clinical interpretation of the assay result, may vary depending on the particular cutpoint(s) used.

The cutpoints to be applied to assay measurements, the rationale and the background data for the selection as it relates to the intended clinical use must be provided. In the case of a continuous marker that will be used to predict a binary outcome (e.g., treatment response or toxicity), cutpoint rationale might be based on ROC analysis aimed at achieving a desired level of sensitivity or specificity. For time-to-event endpoints cutpoints might be selected to achieve a specified separation of survival curves. In general, choosing cutpoints based on observed data can lead to biased results, and operating characteristics of the cutpoint (e.g., sensitivity, specificity, predictive values) should be demonstrated on data sets independent of the ones used to derive them.

D. Analytic specificity (cross-reacting substances, interfering substances, etc.) and analytic sensitivity (limits of detection)

Analytic sensitivity is the ability of a test to detect an analyte or entity when it is present. When the output of a test is binary, sensitivity traditionally refers to the proportion of positive test results obtained on cases that are truly positive for the entity or analyte of interest. For tests with quantitative output, the sensitivity refers to the change in the test output relative to the change in the actual amount of analyte, and this relation may depend on the absolute amount of analyte present. The limit of detection is defined as the smallest amount of analyte that an analytical method can detect with a specified probability. A related term is limit of quantitation, the smallest amount of an analyte in a sample that can be quantitatively determined with acceptable precision, and trueness as measured by bias. Analytic specificity is the ability of a test or procedure to correctly indicate absence of an analyte or entity when it is truly absent or to accurately quantify an entity or analyte in the presence of interfering or cross-reacting substances. Almost all assays demonstrate potential for false positive results due to interfering substances. (Sensitivity may also be affected by interfering substances.) When the output of a test is binary, specificity traditionally refers to the proportion of negative test results obtained on cases that truly do not possess the entity or analyte of interest.

(Continued)

Table 2.1 (*Continued*)

	Information to be provided about the design of the sensitivity and specificity studies that were performed should include characteristics of the samples and positive and negative controls, the rationale for interfering substances studied, analyte or entity (e.g., tumor cells harboring a particular mutation) spike-in amounts and matrices used in any dilution experiments. Summary results such as sensitivity and specificity rates over the range of test samples considered, and calibration or dilution curves should be presented, as appropriate.
E.	Accuracy measurements
	Accuracy is defined as the closeness of agreement between the test results obtained using the new biomarker test and results obtained using a reference standard method widely accepted as producing "truth" for the analyte. For example, a reference method considered standard for detection of DNA mutations is sequencing. The observed level of agreement will depend on both the bias and precision of the new test. Bias is the amount by which an average of many repeated measurements made using the new test systematically over- or underestimates the reference standard method result. Precision is discussed separately above. For many new biomarker tests, there will not be a universally accepted reference standard method.
	The reference standard method, if any exists, should be clearly stated. Accuracy measures such as overall percent agreement, and sensitivity and specificity relative to the reference standard results should be reported for tests that yield binary results. For continuous marker values, accuracy measures such as average bias, mean absolute deviation or mean squared deviation should be reported over the relevant range of true (reference standard method) values.
	In addition to the information indicated above, the general background and description of the biomarker-based test should include:
F.	Statement indicating whether the test is quantitative (includes degrees of positivity), semi-quantitative, or qualitative (positive vs. negative); what platform will be used; what is to be measured; controls; scoring procedures, including the values that will be used (e.g., pos vs. neg; 1+, 2+ 3+); interpretation, etc. If the biomarker-based test combines values for multiple biomarkers into a composite score, then the exact algorithm for combining the values must be specified, including any cutpoints applied to the composite score.

G. Specimen type(s) and preparation/handling/shipping SOPs, including definition for acceptability of a sample

H. Information about the statistical design used to establish the correlation of the assay result with the clinical phenotype or outcome of interest

I. Procedures to be used when results are not interpretable or are discrepant; this is especially important for assays used for eligibility or assignment to treatment arms

Source: Adapted from http://www.cancerdiagnosis.nci.nih.gov/scientificPrograms/pacct/assay_standards.htm.

2.3 Prognostic versus Predictive Biomarkers

A *prognostic factor* is a measurement that is associated with clinical outcome in the absence of therapy or with the application of a standard therapy that all patients are likely to receive. It can be viewed as an indicator of the natural history of the disease that is independent of the effect of a particular therapy [6, 7]. In contrast, a *predictive factor* is a measurement that is associated with response or lack of response to a particular therapy, where response can be defined using any of the clinical endpoints commonly used in clinical trials. A predictive factor implies a benefit from the therapy that depends on the status of the predictive biomarker. (In statistical terms, this constitutes an interaction between treatment benefit and biomarker status.)

To evaluate whether a biomarker is prognostic, one can examine the outcomes for a group of patients treated with a standard therapy to see if the patients who have a positive biomarker perform better (or worse) than the patients who have a negative biomarker.
To evaluate whether a biomarker is predictive, one typically needs data from a randomized trial of two treatments in which the relative efficacy of the two treatments can be examined separately in the biomarker positive and negative patients.

An example of a predictive biomarker is EGFR-mutation status for treatment with gefitinib (an EGFR inhibitor) for patients with advanced non-small-cell lung cancer. A randomized trial [8] of gefitinib versus chemotherapy (carboplatin + docetaxel) in this population showed some benefit from the gefitinib but with crossing progression-free survival curves (Fig. 2.1A). However, when the results were stratified by EGFR-mutation status (mutated versus wild type), the interpretation became clearer: Gefitinib worked better than the chemotherapy in the EGFR-mutation-positive patients (Fig. 2.1B) and worse than the chemotherapy in the EGFR-mutation-negative patients (Fig. 2.1C).

Sometimes initial clues for what biomarkers might identify patients who benefit from a new therapy are obtained from examination of the relationship between biomarkers and clinical outcomes in a group of patients uniformly treated with the new therapy. In particular, if the patients with a positive biomarker value do better than patients with a negative biomarker value, this might lead one to suggest that the biomarker is predictive. This approach is reasonable only if one is using a clinical endpoint for which it is known that standard therapy has no activity. For example, suppose patients given standard therapy never have complete responses, but biomarker-positive patients have many complete responses when given the new therapy and biomarker-negative patient have no complete responses when given the new therapy. This would suggest the biomarker is predictive. Care is required here with other clinical endpoints. For example, suppose when given a new therapy that the survival is better for biomarker-positive patients than for biomarker-negative patients. It is possible that the survival would be better for biomarker-positive patients than for biomarker-negative patients given the standard therapy too and the biomarker is not associated with the sensitivity to the new therapy, implying that the biomarker is prognostic but not predictive.

Obviously, a predictive biomarker can be useful in directing treatments for individual patients. A prognostic biomarker can also sometimes be useful, e.g., when suggesting a patient will do so poorly with a standard treatment that other options should be explored, or suggesting a patient will do so well with a standard base therapy (e.g., surgery or surgery plus radiation) that additional (adjuvant) therapy need not be considered.

2.4 Biomarker Evaluations in Phase I Trials

Phase I trials are dose escalation trials that are designed to determine the dose (and schedule) of a new therapy for further

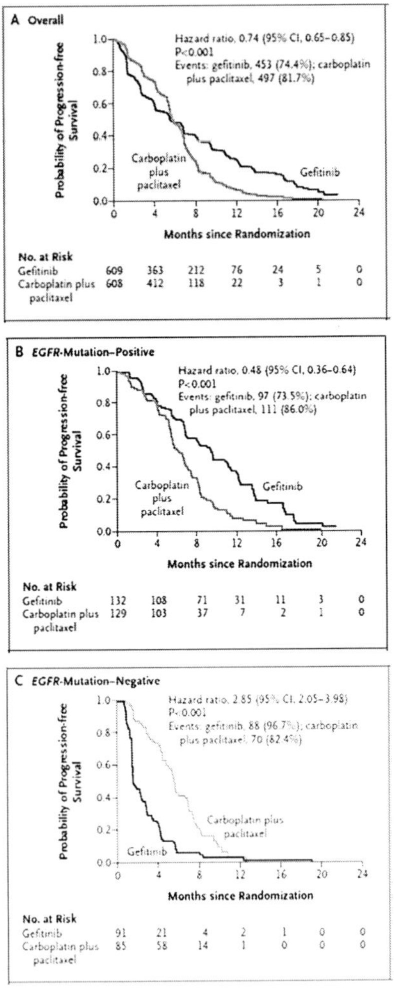

Figure 2.1 Progression-free survival for patients with advanced adenocarcinoma of the lung receiving first-line therapy with gefitinib or carboplatin-paclitaxel in the IPASS trial. (A) Overall population. (B) EGFR-mutation-positive patients (C) EGFR-mutation-negative patients. From *The New England Journal of Medicine*, Mok, T. S., Wu, Y-L., Thongprasert, S., Yang, C-H., Chu, D-T, Saijo, N., Sunpaweravong, P., Han, B., Margono, B., Ichinose, Y., Nishiwaki, Y., Ohe, Y., Yang, J-J., Chewaskulyong, B., Jiang, H., Duffield, E. L., Watkins, C. L., Armour, A. A., Fukuoka, M., Gefitinib or Carboplatin–Paclitaxel in Pulmonary Adenocarcinoma, 361, 947–957. Copyright © (2009) Massachusetts Medical Society. Reprinted with permission from Massachusetts Medical Society.

evaluation. Historically, when evaluating new chemotoxic agents, this recommended dose was considered the highest dose with acceptable toxicity (maximum tolerated dose). This was based on the assumptions that (a) the clinical benefit of an agent increases with increasing dose, (b) the toxicity of an agent increases with increasing dose, and (c) there is a dose with acceptable toxicity that provides clinical benefit [9]. Assumptions (a) and (b) can be questioned when the therapy involves a targeted agent. This has lead to proposed phase I trial designs that are based on finding a dose that has adequate (or "optimal") target effect (with acceptable toxicity) [10–14]. However, designs that use target effects rather than toxicity (or pharmacokinetics of the agent) to choose the recommended dose have rarely been used [15, 16]. There may be a number of good reasons for this, including the fact that target effects seen in a biomarker may not be the same as the target effect seen in the tumor, tumor target effects may not correlate with clinical benefit, and the agent may work by mechanisms other than the effect on the putative target [17]. However, after a number of doses have been explored up to a maximum tolerated dose, it is appropriate to evaluate how well the target is being affected at the recommended dose for future evaluation. Also, a phase I trial is good platform for developing and refining the relevant biomarker assay [18].

Another dose escalation design using a biomarker endpoint is the phase 0 design [19]. In this design the dose levels tested are assumed to be in a range that is pharmacodynamically active, but so low as not to have toxicity or clinical benefit. The aim of the design is to demonstrate dose levels where the agent has sufficient pharmacodynamic effects on the target, as a method to eliminate inactive agents or potentially choose among active ones.

2.5 Biomarker Evaluations in Phase II Trials

Phase II trials assess clinical activity of a new agent for determining whether further development in definitive phase III trials is warranted. A variety of trial designs are used for phase II investigations. Historically, single-arm trials using objective response rate as the endpoint have been widely used for cytotoxic agents. However, there are situations in which a single-arm phase II design is not appropriate. For example, one cannot use a response endpoint for agents that may not be expected to shrink tumors but only delay their growth (i.e., are cytostatic), and targeting benchmarks for other endpoints (e.g., progression-free survival or overall survival) may not be reliable [20] without an extensive meta-analysis of trial-to-trial variability [21]. Another situation in which a single-arm trial would be hard to interpret is when the new agent is given together with an agent that has activity, making it difficult-to-impossible to define a response rate target. In cases where single-arm trials would be inappropriate, randomized phase II trial designs have been proposed [22]. These trial designs are similar to definitive phase III trials except that (a) the endpoint will typically not be a definitive clinical benefit endpoint like overall survival but instead be an early measure of clinical activity like response rates or progression-free survival, and (b) the type 1 error rate (alpha level, significance level) will typically be larger than the 0.05 used in a phase III trial.

There are different ways biomarkers can be used in phase II trials. One possibility is to use a biomarker as the outcome rather than tumor shrinkage or survival. However, one would need to be convinced that sufficient "biomarker responses" in the patients was indicative that the new therapy works (i.e., would demonstrate clinical utility in a follow-up phase III trial) and insufficient biomarker responses would be indicative that the new therapy does not work. Biomarkers can seldom meet this standard. The use of levels and changes in prostate-specific antigen (PSA) as a measure or response in phase II prostate cancer trials illustrates some of the issues. It has been discussed at length as to whether PSA should be used this way [23, 24], especially because it is known that some agents can affect PSA levels without affecting prostate tumors [25]. Even if one has evidence that a biomarker is a reasonable surrogate for clinical response for a set of treatments (perhaps from a meta-analysis [26]), there will typically be an extrapolation to assume that it will be a reasonable surrogate for a new agent, especially if the new agent has a different mechanism of action than the previous set of treatments.

The most promising application of biomarkers in phase II trials is to help define what population would be most appropriate to be treated by a new agent. Two main goals for this application are (1) to assess the new therapy in a biomarker-defined patient subset where it is expected to work (an "enriched" population), and (2) to determine if the new therapy should be developed for all patients or an enriched population only. (Sometimes the definition of enrichment is relaxed to allow enrichment on other patient characteristics thought to be positively associated with a biomarker of interest.) We consider these goals in the settings of a single biomarker of interest (Section 2.5.1) and when there are multiple biomarkers of interest (Section 2.5.2).

2.5.1 Designs Involving Single Biomarkers

To assess whether a new therapy works in a biomarker-enriched population (goal 1) using a well-defined biomarker, one can perform a standard phase II trial restricted to this population. Some care is required here: Whereas a single-arm phase II trial might normally be appropriate in a certain setting, when the population is enriched the historical targets may be inappropriate because the biomarker may be prognostic, falsely leading one to conclude that the new therapy is beneficial when any therapy might perform well on the enriched population [27]. In this situation, or whenever a single-arm trial would not be appropriate, a randomized phase II trial design can be used with restriction to the enriched subpopulation. With either the enriched single-arm or randomized phase II designs, a large number of patients may need to be screened to obtain the required number of biomarker-positive patients to be treated if the prevalence of the enrichment subpopulation in the general patient population is low.

To determine if the new therapy should be developed for all patients or an enriched population only (goal 2), both biomarker-negative and biomarker-positive patients will need to be treated with the new agent. This can sometimes be done retrospectively by evaluating the biomarker on stored specimens of patients treated with the new agent. For example, the first demonstrations that EGFR mutation status was predictive of response to gefitinib were done retrospectively and revealed that the responses were confined to EGFR-mutation positive tumors [28, 29].

With a response rate endpoint and a single biomarker, a prospective trial of a new agent could again compare response rates in the biomarker-positive and biomarker-negative subgroups. One might want to consider enrolling a sufficient number of biomarker-positive patients for evaluation in that subgroup. This might be done in the context of a two-stage design that enrolls differing numbers of patients in the subgroups in the second stage of accrual based on the number of responders in the two subgroups in the first stage [30,

31]. When a phase II randomized comparison with a control therapy is required, one can compare the treatment effect in the biomarker positive and negative subgroups to get a suggestion as to whether the biomarker is predictive, but a definitive analysis will require a phase III biomarker-stratified design (Section 2.6.1).

2.5.2 Designs Involving Multiple Biomarkers

With each biomarker having its own associated targeted therapy, one can assign patients to the targeted therapy that is appropriate for them. If the setting is such that a control comparison is needed then one would randomize patients with a particular biomarker to their associated targeted therapy versus a control therapy. If a control comparison is not required, e.g., seeking positive response rates with single agents, then no randomization to control therapies would be needed. An example of the latter situation is a National Cancer Institute trial (NCT01306045) in advanced lung and thymic cancer, where depending upon the patients' biomarkers, they are assigned one of five targeted therapies [32]. In this setting, no standard active therapy is available and thus observing any tumor responses would be interesting so a control therapy is not required. Another example was a trial [33] for patients with refractory metastatic cancer where a panel of 64 biomarker targets is used to choose an individual treatment targeted for each patient. The outcome was the number of patients whose time to progression was ≥1.3 times longer than their time to progression before entering the trial, a potentially problematic outcome [34, 35].

When the multiple biomarkers do not each have an associated targeted therapy, there are different approaches that can be used. A retrospective approach examines a number of biomarkers to see which, if any, are associated with good responses in a group of patients given the same (or a set of) treatments. Care is required if thousands of potential biomarkers are considered (e.g., expression of many individual genes) in that an important biomarker could easily be missed because of the background statistical noise of all the biomarkers being considered [31]. A prospective approach randomizes patients to different treatments and evaluates a (small) set of biomarkers on the patients to look for associations. This type of exploratory approach would be appropriate when one has little idea about the relationships between the biomarkers and the therapies.

An example of this is the BATTLE trial, where chemorefractory non-small cell lung cancer patients were randomly assigned to four therapies and were simultaneously categorized into five biomarker subgroups; the outcome was eight-week disease control [36]. Another example is the I-SPY 2 trial, where locally advanced breast cancer patients are randomly assigned neoadjuvant therapies and were simultaneously categorized with multiple biomarkers; the outcome was (modeled) pathological complete response rate [37]. Both the BATTLE and I-SPY 2 trials use outcome-adaptive randomization, in which treatment arms that are doing better receive a higher proportion of patients randomized to them. However, outcome-adaptive randomization leads to trials with larger required sample sizes and a larger absolute number of patients having bad outcomes, making its use controversial [38–41].

2.6 Biomarker Evaluations in Phase III Trials

For trials with a pre-specified binary biomarker, the three main types of trial designs are biomarker-stratified designs, enrichment designs, and biomarker strategy designs [42], although numerous hybrid versions and extensions are possible. For trials in which biomarker has not been completely specified ("locked down") at the time of the initiation of the trial, we briefly discuss some possible phase III trial design approaches.

2.6.1 Biomarker-Stratified Designs

Biomarker-stratified designs allow for full information about the relative treatment efficacy in each biomarker-defined subgroup and in the overall patient population. When each patient is pre-registered to the trial, the patient's biomarker status is assessed. If a biomarker result is successfully obtained, the patient is then offered full registration, and consenting patients are randomized to the treatments within the appropriate biomarker stratum (possibly stratified to ensure balance of treatment arms within each biomarker subgroup). This design schema is presented in Fig. 2.2. If one of the treatment arms is standard of care, the prognostic value of the marker can be assessed in this design by comparing the standard-of-care arms between the biomarker-positive and biomarker-negative subgroups. Predictive value of the biomarker can be initially assessed by using a statistical test to determine if the treatment-by-marker interaction is statistically significantly different from zero. If the interaction test is significant, it can be assessed whether the treatment effect in the two biomarker subgroups point in the same direction but differ in magnitude (quantitative interaction), or if the treatment effects point in opposite directions (qualitative interaction).

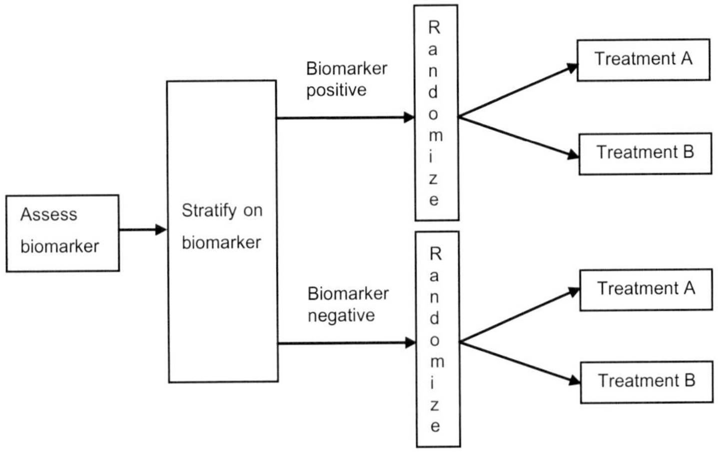

Figure 2.2 Biomarker-stratified design. Treatment A is directed toward the biomarker and Treatment B is standard therapy. The biomarker is used to stratify treatment randomization but does not influence what treatments patients receive.

Qualitative interactions are far more useful than quantitative interactions because they can provide a clearer indication of optimal choice of therapy for the biomarker-positive and biomarker-negative subgroups. Because the treatment differences point the same direction in both biomarker subgroups in the case of a quantitative interaction, one might conclude that all patients should receive the same (superior) treatment, regardless of biomarker status. However, it is important to consider whether a more modest treatment benefit in the biomarker-negative subgroup is still worthwhile if there are added toxicities or other costs of the new treatment. This must be balanced against the realization that if it is decided to forego the modest benefit of the new therapy in the biomarker-negative subgroup, a cost will be incurred to perform the biomarker testing to identify the biomarker-negative patients. Further complicating this decision is the fact that biomarker-stratified designs are often not sufficiently powered statistically to distinguish between modest treatment effects in opposite directions within any biomarker subgroup. This makes a decision based on an observed direction of modest treatment effects even riskier.

An example of a biomarker-stratified design is the S1201 trial [43], which is randomly assigning gastric cancer patients to FOLFOX or docetaxel+irintecan, stratified by ERCC1 (excision repair cross-complementing 1) status; low ERCC1 has been associated with platinum sensitivity.

In some situations, it will be more practical to obtain the biomarker assessment after randomization rather than before (Fig. 2.3); this is sometime known as the "all comers" design. As not all patients will have a successful biomarker assessment (e.g., because of insufficient tumor or assay failure), the analyses are limited to those patients with assessments; these analyses provide an unbiased evaluation of the predictive value of the biomarker provided that the biomarker assessments are done blinded with respect to the randomized treatment assignment (and, of course, treatment outcome). The advantage of obtaining the biomarker assessment before randomization is that only patients who will be used in the analyses will be enrolled on the trial and the eligibility requirement of having a biomarker assessment may encourage appropriate specimen handling to allow the assessment to be made. With either of the biomarker-stratified designs (Fig. 2.2 or 2.3), a retrospective evaluation of additional biomarkers or the refinement of the initial biomarker assay is possible.

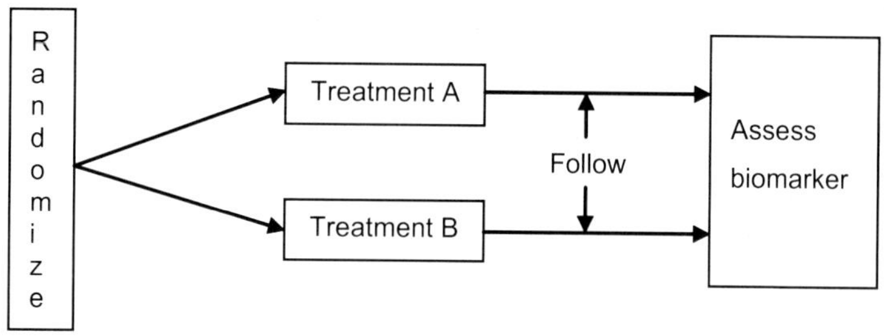

Figure 2.3 Biomarker-stratified alternative design ("all comers" design). Treatment A is directed toward the biomarker and Treatment B is standard therapy. The biomarker is not measured until after patients are randomized to treatment and are followed for some period of time to observe outcome.

In situations where there is not very much preliminary evidence that the biomarker will be predictive for the experimental treatment, an alternative to analyzing treatment effects in the biomarker positive and negative subgroups is to analyze the treatment effect in the biomarker-positive subgroup and in the whole randomized population. This strategy is focused on broadening the population for which the benefit of the new therapy can be formally demonstrated.

An example of a biomarker-stratified design using this alternative analysis approach is the S0819 trial [44], which is testing whether the addition of cetuximab to chemotherapy improves outcomes for patients with non-small cell lung cancer. The biomarker is EGFR status, and the co-primary analyses involve the whole randomized population and the EGFR-positive subgroup.

2.6.2 Enrichment Designs

If there is strong scientific evidence that the benefit of the new treatment is likely to be limited to only patients whose tumors harbor a certain biomarker, it is appropriate to restrict eligibility to that biomarker-defined subset of patients. This design, displayed in (Fig. 2.4), is called an enrichment design. It is motivated by ethical considerations to avoid treating patients who are thought unlikely to benefit from the new therapy. It can be a far more efficient design than one with broad eligibility that analyzes all patients together regardless of marker status. If a large number of patients who will not benefit from the new therapy are randomized, this will mask the treatment benefit in the biomarker-positive subset when an analysis is performed in the entire patient cohort. A larger number of patients would have to be randomized to detect this diluted treatment effect in the overall population.

In practice, it can be difficult to decide when the background information is sufficiently compelling to exclude biomarker-negative patients from a trial. Prematurely excluding some patients could unnecessarily limit the indication for the new therapy and deny some patients access to a therapy that could benefit them. For example, there could be off-target effects of a new targeted agent that would be missed using an enrichment design. In addition, a positive trial cannot directly prove the utility of the biomarker, and prognostic effects of the marker cannot be assessed using this design. Nonetheless, the enrichment design does allow one to definitively establish whether or not there is a benefit of the new treatment in the biomarker-positive subgroup.

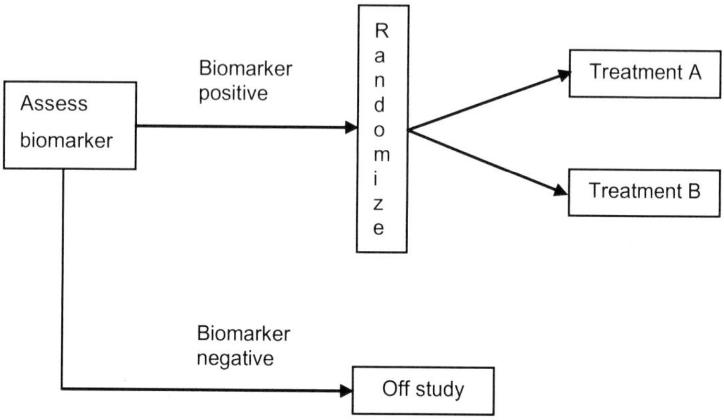

Figure 2.4 Enrichment design. Treatment A is directed toward the biomarker and Treatment B is standard therapy. Only biomarker positive patients are randomized to treatment. Biomarker negative patients are off study.

A tension can develop between the goal of a drug developer and the goal of a biomarker test developer when an enrichment design is used A drug developer may be less concerned about precisely defining the benefiting patient subgroup as long as the group is sufficiently enriched for benefiting patients that a treatment effect will be detected in the subgroup. In contrast, the biomarker developer might have strong interest in accurately defining the subgroup of patients who benefit. An enrichment design does not allow for estimation of negative predictive value of the biomarker test, i.e., the probability that patients with a negative biomarker value will not have a (good) response. This can sometimes represent an obstacle for regulatory approval of a biomarker test, particularly when there is legitimate difference of opinion about the strength of the preliminary data suggesting benefit of the new therapy will be restricted to the biomarker-positive subgroup.

An example of an enrichment design is the CALGB-10603 trial [45], which restricted entry to acute myeloid leukemia patients who have a documented FLT3 mutation and then randomly assigned them to standard chemotherapy or standard chemotherapy plus the FLT3 kinase inhibitor midostaurin [46].

2.6.3 Biomarker-Strategy Designs

In the biomarker-strategy design, patients are randomly assigned to a control-arm standard treatment *B* or a biomarker-directed treatment arm. In the simplest case of a binary biomarker, patients in the biomarker-directed arm receive treatment *A* if the biomarker is positive and treatment *B* if the biomarker is negative (Fig. 2.5). At first glance, this type of design has appeal because it provides a direct estimation of benefit from using the biomarker in clinical practice (the biomarker-directed treatment arm) relative to the standard treatment (the control arm). However, the design is not efficient because many patients on both arms will be treated with the same therapy, treatment *B*. Furthermore, one can use the arm-specific outcomes of a biomarker-stratified or enrichment design to estimate the same between-arm difference that would be seen in a biomarker-strategy design, but with a smaller required sample size. For example, suppose the biomarker positivity rate was one-half and one was considering performing a trial with 800 patients using a biomarker-strategy design. A calculation shows that one could perform a biomarker-stratified design with 400 patients (or screen 400 patients and randomize 200 patients in an enrichment design) to be able to detect a given treatment effect with the same power as the 800 patient biomarker-strategy trial.

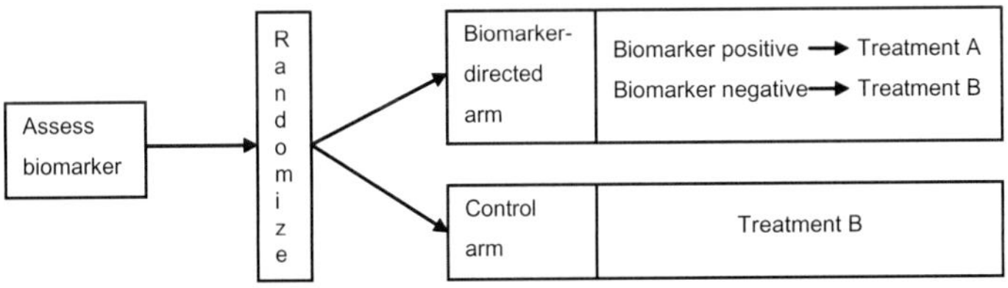

Figure 2.5 Biomarker-strategy design. Treatment A is directed toward the biomarker and Treatment B is standard therapy. Patients randomized to the biomarker-directed arm receive Treatment A if they are positive for the biomarker and Treatment B otherwise. All patients randomized to the control arm receive Treatment B.

Although we see no advantage to using the biomarker-strategy design in the simplest case (Fig. 2.5), does it have a role when the biomarker is choosing among many possible treatments rather than just two? With, say, a dozen treatments being chosen by the biomarker, it may be impractical to perform a stratified design with a dozen strata or enrichment design that has sufficient numbers of patient in each biomarker category. However, in this setting, the biomarker-strategy design will be hard to interpret: (1) The overall results of the trial can be negative even if the biomarker identifies an agent (or a few agents) for a small subset of patients where it is highly effective if the agents in most of the other identified subsets are no better than the control treatment (alternatively, the overall results of the trial may be positive even though the biomarker-identified therapy is worse than the control treatment in a small subset of patients). (2) The overall results of the trial can be positive even if the biomarker has no predictive ability if the agents being used in the biomarker-directed arm are better agents than those used in the control arm. (3) There can be a large overlap in the agents that patients received on the biomarker-directed arm and what they would have received if they had been on the control arm.

An example of a trial using a biomarker-strategy design was a trial of docetaxel/cisplatin (standard arm) versus a biomarker-directed treatment arm for patients with metastatic non-small cell lung cancer [47]. In the biomarker-treatment directed arm, patients with low ERCC1 received docetaxel/cisplatin and patients with high ERCC1 received docetaxel/gemcitabine. Fifty-seven percent of the patients randomly assigned to the biomarker-directed arm received the same therapy (docetaxel/cisplatin) they would have received had they been assigned to the standard arm.

2.6.4 Designs in Which the Biomarker Has Not Been Completely Specified

In some situations, there is insufficient preliminary information about a biomarker to design a clinical trial using a biomarker-stratified, enrichment or biomarker-strategy approach. For example, there may be uncertainty about the most appropriate cutpoint for defining positive or negative biomarker assay results, or there may be uncertainty about how to combine information from more than one biomarker into a single binary biomarker-based test. Ideally, a locked-down version of the biomarker-based test would be defined based on phase I and II trials. However, sometimes the experimental agent appears so promising that there is interest in beginning a phase III trial before the biomarker-based test has been completely specified.

Consider first the case of a single continuous biomarker where it is unknown what the cutpoint should be for defining biomarker positivity. One approach would be to use an all-comers randomized design and then, after all the data have been collected, try many different possible cutpoints and choose the cutpoint that showed the best predictive ability of the biomarker. However, assuming that this "optimal" cutpoint represented the true predictive ability of the biomarker with this cutpoint would be a mistake; determining a cutpoint and then assessing it on the same dataset leads to an overestimate of its predictive ability because of the implicit multiple comparisons with different possible cutpoints [48]. One way around this problem would be to perform a second randomized trial with a biomarker-stratified design (using the biomarker cutpoint determined from the first trial) to assess unbiasedly its predictive ability. This is time consuming and may be difficult to do if the first trial and analysis strongly suggested that biomarker negative group would not be helped by the experimental therapy. A second approach would be to randomly split the (original) trial data in half before considering cutpoints. The best observed cutpoint would be determined using one-half of the data, and when decided upon, the other one-half of the data could be used to estimate unbiasedly the predictive ability of the biomarker with this cutpoint. This approach is inefficient in that only one-half of the data is being used to determine the cutpoint and only one-half of the data are being used to evaluate the predictive ability of the biomarker with this cutpoint. A third approach, which is preferred, is to choose the cutpoint based on all the data but use statistical methods to account for the implicit multiple comparisons when estimating the predictive ability of the biomarker with that cutpoint [49].

In some situations, it is not known what the appropriate biomarker cutpoint should be, but it is believed that the experimental treatment would not benefit patients with a very low value of the biomarker, say lower than X. In this situation, a possible design randomizes patients with biomarker values $\geq X$ but restricts the primary analysis to patients with higher biomarkers, say with values $\geq Y$ (where $Y > X$). A secondary analysis can examine the effect of the experimental treatment for those patients with biomarker value between X and Y.

An example of a trial using such a design is the CALGB-30801 trial [50], which restricts entry to non-small cell lung cancer patients who have a COX-2 index ≥ 2 and then randomly assigns them to standard chemotherapy or standard chemotherapy plus the COX-2 inhibitor celecoxib; the primary analysis will be on patients with COX-2 index ≥ 4.

When there are multiple biomarkers that need to be combined in a single binary biomarker-based test (the component biomarkers might have been measured by a high-dimensional genomic assay), the same approaches as used with a single biomarker and unknown cutpoint can be used. One can use all of the trial data to try to optimize the combination of biomarkers to form a single binary biomarker-based test that has the most predictive ability and then evaluate the derived biomarker-based test on a new trial. Alternatively, one can use one-half the trial data to construct the biomarker and evaluate it on the other one-half of the data, or one can use statistical methods that allow one to construct and assess a biomarker on the original trial data [51,52].

It should be noted that all these trial-design approaches using a biomarker are less efficient than when the biomarker and its cutpoints has been completely defined before the trial.

2.7 Summary

Biostatistical considerations are important during all phases of biomarker evaluation. It is essential that the assay methodology and relevant cutpoints be validated so that subjects/patients are correctly classified by the biomarker. Ideally, this is done during the phase I and phase II trial assessments of the relevant targeted agents. The appropriate phase II trial design depends on whether one is considering a single biomarker and targeted agent or multiple targets and agents. With multiple targeted agents, one can direct therapy based on the biomarker evaluations or can be in an exploratory mode where patients randomly receive therapies regardless of their biomarkers (with promising biomarker-agent combinations sorted out retrospectively). The other major design consideration in phase II is whether a control arm can be omitted, e.g., when examining response rates for single agents. In phase III, the biomarker-stratified design offers the richest information.

In situations where it is believed to be inappropriate to use the targeted therapy on biomarker negative patients, the enrichment design can still be used to demonstrate that the targeted therapy benefits patients who are biomarker positive. Approaches have been suggested to perform phase III trials when the biomarker has not been completely defined, but it is preferable to have the biomarker locked-down before entering phase III. In the future, as potentially small biomarker subsets become defined, it will be important for efficient trial design that specimens for concurrent and future biomarker evaluation be collected so that multiple questions can be addressed in each trial.

References

1. Clark, G. M. McShane, L. M. (2011). Biostatistical considerations in development of biomarker-based tests to guide treatment decisions, *Stat. Biopharm. Res.*, **3**, 549–560.
2. CLSI. Protocols for Determination of Limits of Detection and Limits of Quantitation; Approved Guideline. NCCLS document EP17-A (ISBN 1-56238-551–8). Clinical and Laboratory Standards Institute (formerly NCCLS), 940 West Valley Road, Suite 1400, Wayne, Pennsylvania 19087–1898 USA, 2004.
3. CLSI. Interference Testing in Clinical Chemistry; Approved Guideline. CLSI document EP7-A2 (ISBN 1-56238-584–4). Clinical and Laboratory Standards Institute, 940 West Valley Road, Suite 1400, Wayne, Pennsylvania 19087–1898 USA, 2005.
4. FDA International Conference on Harmonization; Guideline on Validation of Analytical Procedures: Definitions and Terminology; Availability. Docket No. 94D-0016, March 1995.
5. FDA International Conference on Harmonization; Guideline on the Validation of Analytical Procedures: Methodology; Availability. Docket No. 96D–0030, May 1997.
6. Clark, G. M., Zborowski, D. M., Culbertson, J. L., Whitehead, M., Savoie, M., Seymour, L., Shepherd, F. A. (2006). Clinical utility of epidermal growth factor receptor expression for selecting patients with advanced non-small cell lung cancer for treatment with erlotinib, *J. Thorac. Oncol.*, **1**, 837–846.
7. Clark, G. M. (2008). Prognostic factors versus predictive factors: examples from a clinical trial of erlotinib, *Mol. Oncol.*, **1**, 406–412.

8. Mok, T. S., Wu, Y-L., Thongprasert, S., Yang, C-H., Chu, D-T, Saijo, N., Sunpaweravong, P., Han, B., Margono, B., Ichinose, Y., Nishiwaki, Y., Ohe, Y., Yang, J-J., Chewaskulyong, B., Jiang, H., Duffield, E. L., Watkins, C. L., Armour, A. A., Fukuoka, M. (2009). Gefitinib or carboplatin-paclitaxel in pulmonary adenocarcinoma, *N. Engl. J. Med.*, **361**, 947–957.
9. Korn, E. L. (2004). Nontoxicity endpoints in Phase I trial designs for targeted, non-cytotoxic agents, *J. Natl. Cancer Inst.*, **96**, 977–978.
10. Bekele, B. N., Shen, Y. (2005). A Bayesian approach to jointly modeling toxicity and biomarker expression in a phase I/II dose-finding trial, *Biometrics*, **61**, 344–354.
11. Friedman, H. S., Kokkinakis, D. M., Pluda, J., Friedman, A. H., Cokgor, I., Haglund, M. M., Ashley, D. M., Rich, J., Dolan, M. E., Pegg, A. E., Moschel, R. C., McLendon, R. E., Kerby, T. Herndon, J. E., Bigner, D. D., Schold, S. C. (1998). Phase I trial of O-6-benzylguanine for patients undergoing surgery for malignant glioma, *J. Clin. Oncol.*, **16**, 3570–3575.
12. Hunsberger, S., Rubinstein, L. V., Dancey, J., Korn, E. L. (2005). Dose escalation trial designs based on a molecularly targeted endpoint, *Statist. Med.*, **24**, 2171–2181.
13. Mandrekar, S. J., Cui, Y., Sargent, D. J. (2007). An adaptive phase I design for identifying a biologically optimal dose for dual agent drug combinations, *Statist. Med.*, **26**, 2317–2330.
14. Meany H., Balis, F. M., Aikin, A., Whitcomb, P., Murphy, R. F., Steinberg, S. M., Widemann, B. C., Fox, E. (2010). Pediatric phase I trial design using maximum target inhibition as the primary endpoint, *J. Natl. Cancer Inst.*, **102**, 909–912.
15. Goulart, B. H. L., Clark, J. W., Pien, H. H., Roberts, T. G., Finkelstein, S. N., Chabner, B. A. (2007). Trends in the use and role of biomarkers in Phase I oncology trials, *Clin. Cancer Res.*, **13**, 6719–6726.
16. Parulekar, W. R. Eisenhauer, E. A. (2004). Phase I trial design for solid tumor studies of targeted, non-cytotoxic agents: theory and practice, *J. Natl. Cancer Inst.*, **96**, 990–997.
17. Ratain, M. J. Glassman, R. H. (2007). Biomarkers in phase I oncology trials: signal, noise or expensive distraction? *Clin. Cancer Res.*, **13**, 6545–6548.
18. Booth, C. M., Calvert, A. H., Giaccone, G, Lobbezoo, M. W., Seymour, L. K., Eisenhauer, E. A. (2008). Endpoints and other considerations in phase I studies of targeted anticancer therapy: Recommendations from the task force on methodology for the development of innovative cancer therapies (MDICT), *Eur. J. Cancer*, **44**, 19–24.

19. Rubinstein, L. V., Steinberg, S. M., Kummar, S., Kinders, R., Parchment, R. E., Murgo, A. J., Tomaszewski, J. E., Doroshow, J. H. (2010). The statistics of phase 0 trials, *Statist. Med.*, **29,** 1072–1076.

20. Korn, E. L., Arbuck, S. G., Pluda, J. M., Simon, R., Kaplan, R. S., Christian, M. C. (2001). Clinical trial designs for cytostatic agents: Are new approaches needed? *J. Clin. Oncol.*, **19**, 265–272.

21. Korn, E. L., Liu, P-Y., Lee, S. J., Chapman, J-A. W., Niedzwiecki, D., Suman, V. J., Moon, J., Sondak, V. K., Atkins, M. B., Eisenhauer, E. A., Parulekar, W., Markovic, S. N., Saxman, S., Kirkwood, J. M. (2008). Meta-analysis of phase II Cooperative Group trials in metastatic stage IV melanoma to determine progression-free and overall survival benchmarks for future phase II trials, *J. Clin. Oncol.*, **26**, 527–534.

22. Rubinstein, L. V., Korn, E. L., Freidlin, B., Freidlin, B., Hunsberger, S., Ivy, S. P., Smith, M. A. (2005). Design issues of randomized phase II trials and a proposal for phase II screening trials, *J. Clin. Oncol.*, **23**, 7199–7206.

23. Bubley, G. J., Carducci, M., Dahut, W., Dawson, N., Daliani, D., Eisenberger, M., Figg, W. D., Freidlin, B., Halabi, S., Hudes, G., Hussain, M., Kaplan, R., Myers, C., Oh, W., Petrylak, D. P., Reed, E., Roth, B., Sartor, O., Scher, H., Simons, J., Sinibaldi, V., Small, E. J., Smith, M. R., Trump, D. L., Vollmer, R., Wilding, G. (1999). Eligibility and response guidelines for phase II clinical trials in androgen-independent prostate cancer: recommendations from the prostate-specific antigen working group, *J. Clin. Oncol.*, **17**, 3461–3467.

24. Scher, H. I., Halabi, S., Tannock, I., Morris, M., Sternberg, C. N., Carducci, M. A., Eisenberger, M. A., Higano, C., Bubley, G. J., Dreicer, R., Petrylak, D., Kantoff, P., Basch, E., Kelly, W. K., Figg, W. D., Small, E. J., Beer, T. M., Wilding, G., Martin, A., Hussain, M. (2008). Design and endpoints of clinical trials for patients with progressive prostate cancer and castrate levels of testosterone: recommendations of the prostate cancer clinical trials working group, *J. Clin. Oncol.*, **26**, 1148–1159.

25. Thompson, I. M., Tangen, C. M., Kristal, A. R. (2008). Prostate-specific antigen: a misused and maligned prostate cancer biomarker, *J. Natl. Cancer Inst.*, **100**, 1487–1488.

26. Korn, E. L., Albert, P. S., McShane, L. M. (2005). Assessing surrogates as trial endpoints using mixed models, *Statist. Med.*, **24**, 163–182.

27. McShane, L. M., Hunsberger, S., Adjei, A. A. (2009). Effective incorporation of biomarkers into Phase II trials, *Clin. Cancer Res.*, **15**, 1898–1905.

28. Lynch, T. J., Bell, D. W., Sordella, R., Gurubhagavatula, S., Okimoto, R. A., Brannigan, B. W., Harris, P. L., Haserlat, S. M., Supko, J. G., Haluska, F. G., Louis, D. N., Christiani, D. C., Settleman, J., Haber, D. A. (2004). Activating mutations in the epidermal growth factor receptor underlying responsiveness of non-small-cell lung cancer to gefitinib, *N. Eng. J. Med.*, **350**, 2129–2139.

29. Paez, J. G., Janne, P. A., Lee, J. C., Tracy, S., Greulich, H., Gabriel, S., Herman, P., Kaye, F. J., Lindeman, N., Boggon, T. J., Naoki, K., Sasaki, H., Fujii, Y., Eck, M. J., Sellers, W. R., Johnson, B. E., Meyerson, M. (2004). EGFR mutations in lung cancer: correlation with clinical response to gefitinib therapy, *Science*, **304**, 1497–1500.

30. Jones, C. L., Holmgren, E. (2007). An adaptive Simon two-stage design for phase 2 studies of targeted therapies, *Contemp. Clin. Trials*, **28**, 654–661.

31. Pusztai, L., Anderson, K., Hess, K. R. (2007). Pharmacogenomic predictor discovery in phase II clinical trials for breast cancer. *Clin. Cancer Res.*, **13**, 6080–6086.

32. National Cancer Institute Trial: *Molecular Profiling and Targeted Therapy for Advanced Non-Small Cell Lung Cancer, Small Cell Lung Cancer, and Thymic Malignancies*, http://www.clinicaltrials.gov/ct2/show/NCT01306045 (last accessed 3/19/12).

33. Von Hoff, D. D., Stephenson, J. J., Jr., Rosen, P., Loesch, D. M., Borad, M. J., Anthony, S., Jameson, G., Brown, S., Cantafio, N., Richards, D. A., Fitch, T. R., Wasserman, E., Fernandez, C., Green, S., Sutherland, W., Bittner, M., Alarcon, A., Mallery, D., Penny, R. (2010). Pilot study using molecular profiling of patients' tumors to find potential targets and select treatments for their refractory cancers, *J. Clin. Oncol.*, **28**, 4877–4883.

34. Buyse M, *et al.* (2011). Progression-free survival ratio as end point for phase II trials in advanced solid tumors, *J. Clin. Oncol.*, **29**, e451–e452.

35. Doroshow, J. H. (2010). Selecting systemic cancer therapy one patient at a time: is there a role for molecular profiling of individual patients with advanced solid tumors? *J. Clin. Oncol.*, **28**, 4869–4871.

36. Kim, E. S., Herbst, R. S., Wistuba, I. I., Lee, J. J., Blumenschein, G. R., Jr., Tsao, A., Stewart, D. J., Hicks, M. E., Erasmus, J., Jr., Gupta, S., Alden, C. M., Liu, S., Tang, X., Khuri, F. R., Tran, H. T., Johnson, B. E., Heymach, J. V., Mao, L., Fossella, F., Kies, M. S., Papadimitrakopoulou, V., Davis, S. E., Lippman, S. M., Hong, W. K. (2011). The BATTLE trial: Personalizing therapy for lung cancer. *Cancer Discov.*, **1**, 44–53.

37. Barker, A. D., Sigman, C. C., Kelloff, G. J., Hylton, N. M., Berry, D. A., Esserman. L. J. (2009). I-SPY 2: An adaptive breast cancer trial in the

setting of neoadjuvant chemotherapy, *Clin. Pharmacol. Ther.*, **86**, 97–100.

38. Berry, D. A. (2011). Adaptive clinical trials: The promise and the caution, *J. Clin. Oncol.*, **29**, 606–609.
39. Korn, E. L. Freidlin, B. (2011). Outcome-adaptive randomization: is it useful? *J. Clin. Oncol.*, **29**, 771–776.
40. Yuan, Y. Yin, G. (2011). On the usefulness of outcome-adaptive randomization. *J. Clin. Oncol.*, **29**, e390–e392.
41. Korn E. L. Freidlin, B. (2011). Reply to Y. Yuan *et al.*: on the usefulness of outcome-adaptive randomization, *J. Clin. Oncol.*, **29**, e393.
42. Freidlin, B., McShane, L. M., Korn, E. L. (2010). Randomized clinical trials with biomarkers: design issues, *J. Natl. Cancer Inst.*, **102**, 152–160.
43. S1201 Trial: *Combination Chemotherapy in Treating Patients With Advanced or Metastatic Esophageal Cancer, Gastric Cancer, or Gastroesophageal Junction Cancer*, http://clinicaltrials.gov/ct2/show/NCT01498289 (last accessed 3/19/12).
44. S0819 Trial: Carboplatin and Paclitaxel With or Without Bevacizumab and/or Cetuximab in Treating Patients With Stage IV or Recurrent Non-Small Cell Lung Cancer, http://www.clinicaltrials.gov/ct2/show/NCT00946712 (last accessed 3/19/12).
45. CALGB-10603Trial: *Daunorubicin, Cytarabine, and Midostaurin in Treating Patients With Newly Diagnosed Acute Myeloid Leukemia*, http://www.clinicaltrials.gov/ct2/show/NCT00651261 (last accessed 3/19/12).
46. Stone, R. M., Dohner, H., Ehninger, G., Villeneuve, M., Teasdale, T., Virkus, J. D., Bressler, L. R., Seiler, M. M., Marcucci, G., Larson, R. A. on behalf of all study investigators. (2011). CALGB 10603 (RATIFY): A randomized phase III study of induction (daunorubicin/cytarabine) and consolidation (high-dose cytarabine) chemotherapy combined with midostaurin or placebo in treatment-naive patients with FLT3 mutated AML. [Abstract], *J. Clin. Oncol.*, **29** (Suppl 15): A-TPS199.
47. Cobo, M., Isla, D., Massuti, B., Montes, A., Sanchez, J. M., Provencio, M., Vinolas, N., Paz-Ares, L., Lopez-Vivanco, G., Munoz, M. A., Felip, E., Alberola, V., Camps, C., Domine, M., Sanchez, J. J., Sanchez-Ronco, M., Danenberg, K., Taron, M., Gandara, D., Rosell, R. (2007). Customizing cisplatin based on quantitative excision repair cross-complementing 1 mRNA expression: a phase III trial in non-small-cell lung cancer, *J. Clin. Oncol.*, **25**, 2747–2754.

48. Altman, D. G., Lausen, G., Sauerbrei, W., Schumacher, M. (1994). Dangers of using "optimal" cutpoints in the evaluation of prognostic factors, *J. Natl. Cancer Inst.*, **86**, 829–835.

49. Jiang, W., Freidlin, B., Simon, R. (2007). Biomarker-adaptive threshold design: a procedure for evaluating treatment with possible biomarker-defined subset effect, *J. Natl. Cancer Inst.*, **99**, 1036–1043.

50. CALGB-30801Trial: *Gemcitabine Hydrochloride or Pemetrexed Disodium and Carboplatin With or Without Celecoxib in Treating Patients With Advanced Non-Small Cell Lung Cancer*, http://www.clinicaltrials.gov/ct2/show/NCT01041781 (last accessed 3/19/12).

51. Freidlin, B. Simon, R. (2005). Adaptive signature design: an adaptive clinical trial design for generating and prospectively testing a gene expression signature for sensitive patients, *Clin. Cancer Res.*, **11**, 7872–7878.

52. Freidlin, B., Jiang, W., Simon, R. (2010). The cross-validated adaptive signature design, *Clin. Cancer Res.*, **16**, 691–698.

Chapter 3

Role of Biomarkers in Clinical Development of Cancer Therapies

Helen X Chen
Investigational Drug Branch,
Cancer Therapy Evaluation Program, National Cancer Institute,
6130 Executive Blvd, EPN 7131, Bethesda, MD 20892, USA
Helen.Chen@nih.gov

3.1 Introduction

The rapid expansion of our knowledge about cancer genome and biology has provided hypothesis for an increasing number of potential therapeutic targets. Technical advancement in medicinal chemistry and biotechnology has also significantly enhanced the drug design for optimal potency and specificity against the intended targets. Successful development of molecularly targeted agents (MTA), however, must overcome the challenges posed by the very nature of cancer: the pathway complexity within tumors, the heterogeneity between individual patients and the exploitation by cancer cells of the essential survival mechanism of normal tissue.

Handbook of Therapeutic Biomarkers in Cancer
Edited by Sherry X. Yang and Janet E. Dancey
Copyright © 2013 Pan Stanford Publishing Pte. Ltd.
ISBN 978-981-4364-65-2 (Hardcover), 978-981-4364-66-9 (eBook)
www.panstanford.com

Given the high attrition rate of late-stage clinical trials with MTAs, there is a great need to enhance the robustness of early clinical trials to address the objectives key to future directions: credentialing the agent, optimizing the dose and schedule, obtaining preliminary proof of concept for the target, and identifying patient selection markers for further studies. For most MTAs, incorporation of biomarkers to characterize the tumors and the drug effect on molecular levels is not only desirable but also indispensible for successful transition from target discovery to target validation and clinical benefit.

In response to the need for improving the efficiency of drug development, FDA's critical pathway initiative was launched in 2004, with the primary objective to foster collaboration between academic and industry partners as well as regulatory authorities, in order to integrate preclinical scientific process with all stages of clinical trials. Key areas identified as high priority include better development and utilization of biomarker of safety and efficacy.

This chapter will provide definitions of key biomarkers used in drug development, the potential utility of the biomarkers in different stages of clinical trials and the current challenges and future directions.

3.2 A Few Definitions and General Concepts

Proper use of biomarkers should start with a clear understanding of the nature and the intended use of various markers in clinical trials and treatment decisions. Biomarkers are commonly defined as "...a characteristic that is objectively measured and evaluated as an indicator of normal biologic processes, pathogenic processes, or pharmacologic responses to a therapeutic intervention." [1] Definitions of various biomarkers and the terminology for steps in clinical and analytical validations have been reviewed in several guideline reports.[2,3] The main concepts for markers commonly used in clinical trials for cancer drugs are defined below.

Based on the purposes in clinical studies, biomarkers can be categorized in a few types:

- *Pharmacodynamics markers (PD Markers)*: Changes in molecular measurements due to the drug effect (e.g., downregulation of PAPR activity after administration of PARP inhibitors)

- *Predictive markers:* characteristics associated with response to a therapy (e.g., HER2 amplification as measured by FISH is predictive of activity with trastuzumab)
- *Prognostic markers:* characteristics associated with the inherent nature of the cancers and outcome, independent of therapy (e.g., the multi-gene OncoType Dx scores predicts survival outcome in patients with ER/PR positive breast cancer)
- *Surrogate markers:* measurements that can be used as an substitute of a clinically relevant endpoint

Biomarkers can be measured on tissue and blood specimens or imaging modalities. A few examples are included in Table 3.1. In general, these markers can be derived from the tumor or normal tissue/cells, assayed on single- or multi-analyte platforms, and based on DNA, RNA, protein, or microRNA measurements.

For biomarkers to be used for their intended purposes in clinical trials or clinical care, it is important to recognize the levels of analytical and clinical qualifications, in order to avoid misinterpretation or misuse of the data. In the REMARK report,[4] the author noted that there had been numerous publications on many "predictive" markers, but very few were confirmed to be clinically useful. Issues identified included lack of assay validation, inappropriate statistical designs and inadequate or improper specimen collections. For both the developer and user of biomarkers, a common language with respect to the status of marker development and validation is critical. Validation of a biomarker requires two major steps:

- *Assay validation:* process that ensure that the assay characteristics (e.g., precision, linearity and reproducibility) is sufficient for the marker to be measured in the intended context (e.g., on frozen tissue obtained from biopsies) and for the proposed purpose (e.g., to detect more than 50% target protein reduction)
- *Biomarker qualification:* establishment of the correlation between a biomarker and the clinical or biological significance (e.g., KRAS mutation in colon cancer is associated with resistance to anti-EGFR monoclonal antibody therapy).

Table 3.1 Biomarker classifications by assay categories

Specimens		Markers	Assay platform	Examples
Tumor tissue	Genetic alteration	• Mutation	PCR based single or multi-gene platform; Whole genome sequencing; Whole exome sequencing	• EGFR mutation; BRAFV600E; KRAS
		• Gene copy loss or gain	FISH, CGH	• HER2, PTEN
		• Translocation	FISH, PCR	• EML4-ALK
	RNA	• Gene expression profile	Microarray; In situ hybridization	• OncoType Dx
	Protein	• In situ analysis	IHC	• C-MET
		• Lysate	Proteomics ELISA	
	Imaging	• Proliferation		FLT-PET
		• Metabolism (glycolysis)		FDG-PET
		• Drug distribution	^{14}C-, ^{128}In- or other tracer-labeled drug entity	^{128}In-trastuzumab
Blood	CTC	• Numbers and markers		
	Circulating DNA			
	Protein	• ELISA • Proteomics		
	PBMC DNA	• Germline DNA polymorphism • Germline mutation	Genome wide or specific gene based	• UGR1A1

Depending on the goals of biomarker evaluations in clinic or drug development, the required levels of assay validation and marker qualification may differ but they should all "fit for purpose". Degree of validation should also be considered in interpreting the outcomes of biomarker analysis.

3.3 Role of Biomarkers in the Different Stages of Drug Development

Stages in the development of MTAs include basic science studies for target discovery and validation, translational studies for drug screening and validation in tumor models, and clinical evaluations of the safety and efficacy in patients (see figure below). Critical to successful transition from target identification to therapy are selection of the right agent, determination of the right dose and schedule, and identification of the right patients. Use of PD and predictive markers has the potential to improve the efficiency of these tasks and accuracy of the decisions. Specifically, PD markers can be used to confirm target engagement, define the dose PK-PD relationship and obtain early proof of principle for the biological effect downstream of the target inhibition. Predictive markers can be used to enrich or stratify patients for evaluation of the antitumor activities. In order to properly use the biomarker data to inform the drug development, it is important to recognize not only the potential values of various biomarkers but also their limitations.

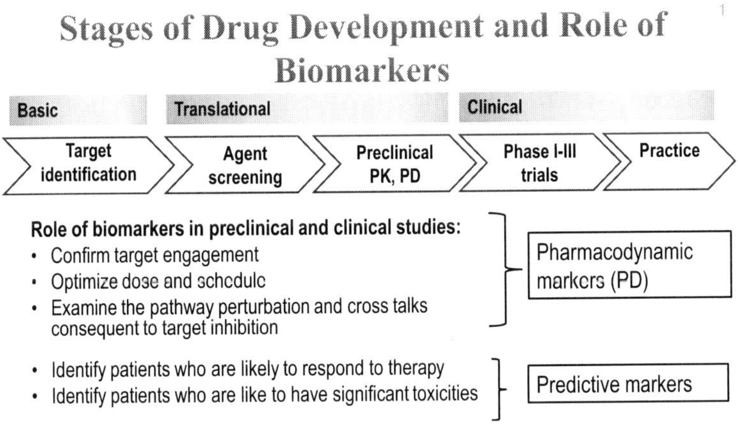

3.3.1 Use of PD Markers in Phase I and Early-Stage Proof of Principle Studies

The goals of phase I trials are to evaluate the pharmacokinetics and safety of the agent and to make recommendations for the dose that will be used in phase II efficacy evaluation. For a drug to reliable test the hypothesis of the target, it is essential to ensure that the drug hits the intended target, and that the selected dose is optimal for efficacy and tolerability. The traditional models based on PK and toxicities have worked well for cytotoxic chemotherapies, whose target toxicities on proliferative tissues (bone marrow suppression or diarrhea) often follow a similar dose-effect relationship as in tumor cells. Some MTAs are also associated specific and quantifiable target-related toxicities (e.g., skin rash after EGFR inhibition), and these adverse events are useful indicators of the target effects. For other MTAs, the target effects on host tissues can be low or nonspecific, in which case, measurements of molecular changes in tumor tissues before and after drug exposure would be critical to verifying the target engagement and estimating the required dose for efficacy. Although PK and safety endpoints should remain the primary objectives of phase I trials, PD markers can provide important ancillary information when properly used and interpreted. Various PD markers have been used in modern day early clinical trials (Table 3.2).

3.3.1.1 Role of PD markers in verifying target engagement

For agents without target toxicities or nonspecific target toxicities, PD markers would be the only means to confirm the target engagement and the presumed mechanisms of action. An excellent example is the PARP inhibitors, which only affect cells in the context of DNA repair deficiency (e.g., tumors with BRAC1/2 deficiency) and therefore had no expected target toxicities. PD markers played an essential role in the early development of veliparib and olaparib, providing confirmation of target inhibition. In contrast, another "PARP" inhibitor, BSI-201, progressed to phase II studies without clear demonstration of target inhibition. Although the agent in combination with chemotherapy demonstrated a significant improvement in response rate and progression free survival in patients with triple negative breast cancers,[5] subsequent preclinical

Table 3.2 Types of post-treatment markers

	Purpose	Examples
Direct target effect (proximal PD markers)	• Measure target engagement by the drug	• Skin rash after EGFR inhibitor • pEGFR after EGFR inhibition
Markers of target pathway modulation	• Measure impact on downstream signaling	• pERK and pAKT following inhibition of EGFR
Markers of biological effects	• Measure impact on cellular function, viability	• Tumor response, stabilization • Ki67, TUNEL, • FDG-PET, FLT-PET
Markers of alternate, off-target pathways	• Cross talks, resistance or escape mechanisms • Unknown mechanisms of action	• Activation of HER3 induced by EGFR inhibitor • Activation of AKT after MEK inhibition

studies with validated PAR assays revealed that the agent did not have any significant effect on the intended target, PARP (Ji *et al.*, AACR 2011). While BSI-201 may have antitumor activity and warrant further development, the clinical trial strategy would need to be modified based on different mechanisms of action. Many MTAs are associated with very non-specific side effects (e.g., mTOR inhibitors, multi-target TKIs or proteasome inhibitors). Although MTD can often be reached, inhibition of the intended targets may or may to be achieved and requires PD measurements to confirm the target effects.

3.3.1.2 Role of PD markers in decisions on the recommended phase II dose (RP2D): value and limitations

Because MTAs exerts their anti-tumor effects through modulation of the intended targets, it is conceivable that the optimal dose can be estimated based on the degree of post-treatment changes in the PD markers. PD marker played a critical role in the dose escalation and dose selection for bortezomib, a proteasome inhibitor. Because PK assay is not feasible, downregulation of proteasome 20 was used as the PD marker for guidance on dose. Preclinical studies defined that proteasome decrease by 80% is required for antitumor activity and that inhibition of more than 80% is lethal. Successful use of this marker in phase I trials also required development of a reliable assay and the validation that peripheral blood mononuclear cells (PBMC) is a surrogate of tumor cells for this marker. Phase I trials were conducted based toxicities and PD endpoint of achieving 80% reduction of the target in PBMC.[6,7]

However, solely replying on PD markers for dose selection can be misleading. In the phase I trial a specific VEGFR tyrosin kinase inhibition, PTK787,[8] vascular permeability (defined as Ki) derived from dynamic contrast enhanced MRI (DCE-MRI) was used as a PD marker. Target effect was confirmed by a significant decrease in Ki following drug administration. It was also observed that the degree of Ki decreases at the end of cycle 1 appeared to correlate with stable disease. While stable disease status in a single arm trial could be the results of the natural history of the tumor with or without the drug effect, a "desired degree" of Ki decrease was defined based on "stable disease" and used for selection of the PTK787 dose (1250 mg QD) for subsequent phase III trial in combination with FOLFOX. While the same trials also indicated a short half-life of the drug and decrease in drug exposure in cycle 2, the observation did not impact the dose selection. Both phase III trials failed to meet the primary endpoint, and further exploration of the dose and schedules had to be pursued. Among the many potential reasons for the trial failure, over-interpretation of the PD marker is probably the one that could have been avoided. This PD marker has not been validated in either preclinical or clinical settings for correlation with efficacy and should be used as an ancillary, hypothesis-generating, rather a decision-making endpoint.

A pre-requisite for PD markers in phase I trials is the availability of reliable assays. It is also important to obtain sufficient information from preclinical models for guidance for the timing of the biopsies and the degree/duration of target inhibition that should be used to define as the desirable PD response. However, it is important to recognize that due to the heterogeneity between tumors as well as the intrinsic difference between models and patients, the PK-PD modeling derived from one or a few preclinical models may not always accurately predict the optimal dose in patients.

3.3.1.3 Use of distal PD markers to measure the biological and molecular consequences of target inhibition

The hypothesis of molecularly targeted therapies assumes that if a valid target is inhibited, the downstream molecular and biological processes would be perturbed, leading to growth inhibition or death of the tumor cells. Although objective responses are the most direct and rapid readout of the antitumor activity, some potentially useful MTAs are cytostatic and requires randomized trials to demonstrate the therapeutic effect. It is hoped that PD markers distal to the target and reflective of the biological consequences may provide earlier and more sensitive indicators for the proof of principle. Examples of markers for biological activities include those for cellular proliferation (e.g., Ki67), apoptosis (TUNEL), among others (Table 3.2). Molecular and functional imaging methodologies have also been developed to monitor the biological effects in tumors, including 18F-FDG-PET for metabolic status (glycolysis), 18F-FLT-PET for proliferation and ^{124}I-annexin V scans for apoptosis. Because imaging modalities are noninvasive and feasible for serial measurements at multiple time points, they have been increasingly used in clinical trials with MTAs. Utility of functional imaging in drug development however has not been validated. For some targeted agents, such as inhibitors of the AKT-mTOR pathway, which is directly involved in glucose homeostasis, changes in FDG-PET may simply reflect modulation of the target, rather than impact on cellular viability and proliferation.[9]

The post-treatment tumor tissues also offer the opportunity to explore additional molecular changes outside the putative downstream pathways and biological effects. Such studies have the potential to uncover resistance and escape mechanisms or unanticipated mechanisms of action.

Continued effort in developing and qualifying markers for intermediate biological effect are warranted. At the current time, use of these markers in the development of investigational agents should be considered explorative. Because biological consequences of target inhibition depend on the biological relevance of the target, "distal" PD marker status can vary with the molecular context of the tumor being treated. Similar to clinical evaluation of efficacy, studies for markers of biological will be more efficient in patients with tumor of similar molecular background.

In summary, PD markers should be considered in early clinical evaluation of MTAs, and the value and limitations should be recognized in their application. To date, PD markers have been successfully used to validate the MOA and confirm target engagement, and lack of or weak PD effects should be a clear warning sign that the agent or the dose is inadequate. However, what constitutes an optimal or sufficient dose is difficult to predict due to lack of validated preclinical models. We recommend that for most agents, it would be prudent to escalate the dose to MTD and that decision on the phase 2 dose should take into account all factors including PK, toxicity as well as PD effects. When in doubt, testing two potential dose levels in expansion cohorts could be valuable for additional assessment of the PD markers and biological activities, before finalizing the dose selection for definitive trials.

3.3.2 Incorporation and Exploration of Patient Selection Markers in Early Clinical Trials

Cancer is an extraordinarily heterogeneous disease. Even within the same histology, individual patient may respond differently to a given therapy. For MTAs designed to exploit the unique molecular vulnerability in tumor cells, the therapeutic window for a targeted agent or combination may only exist in a small subset of patients. Predictive markers of response or resistance are essential for optimized drug development and patient care, as exemplified in a few FDA approved agents (Table 3.3). Without predictive markers, agents such as Crizotinib and trastuzumab would be likely to fail in the clinical development. Retrospective identification of predictive markers such as EGFR activation mutation and KRAS mutation has also provided significant value for optimizing the treatment choice for patients with NSCLC and colon cancer.

However, identification of predictive markers remains the most challenge task in oncology. To date, some of the most commonly used agents in oncology practice (e.g., chemotherapy, VEGF targeting agents) still do not have validated patient selection markers (Table 3.3). With few exceptions, many investigational MTAs also do not have known predictive markers or strong candidates of predictive markers when entering into clinical evaluation.

Table 3.3 Patient selection markers for FDA approved cancer drugs

Agent	Indication	Patient selection markers
Tamoxifen, AI	Breast cancer	ER
Imatinib, dasatinib, nilotinib	CML	BCR-ABL
Imatinib	GIST	c-KIT
Trastuzumab	Breast	HER2 amplification
Erlotinib	NSCLC	EGFR activation mutation
Crizotinib	NSCLC	EML4-ALK
Vemurafenib	Melanoma	BRAFV600E
Cetuximab	Colon cancer	KRAS mutation (exclusion marker)
	Head and Neck cancer	None
Chemotherapy	Many	None
IL-2	RCC	None
IFN-alpha	Melanoma	None
Bevacizumab	RCC, colon cancer, NSCLC, GBM	None
Sorafenib, Sunitinib, Pazopanib	RCC	None
Temsirolimus	RCC, Mantle cell lymphoma	None
Everolimus	PNET, RCC	None
Vironastat, romidepsin	CTCL	None
Ipilimumab	Melanoma	None

The high failure rate of late stage and phase III trials pointed to the importance of biomarker and drug co-development. Efforts in predictive marker discovery, assay development and validation have been increasingly included as objectives of preclinical testing. From the clinical trial perspective, there is also an increasingly recognized need to incorporate predictive marker studies in early clinical trials.

3.3.2.1 Trial design for patient selection markers

Depend on the level of qualification for the candidate predictive markers, a variety of trial designs have been proposed and will be discussed in detail in another chapter of this book. In summary, the ***enrichment design*** (i.e., only the marker positive patients are enrolled) can be used if the marker-efficacy associated has been clearly defined in previous clinical experience or preclinical models. If there are several candidates of predictive markers or the predictive value of the marker is uncertain, ***a stratification design*** (enrolling both marker positive and marker negative patients but in separate marker cohorts) should be used to validate the marker(s) or prioritize among marker candidates. The enrichment design is more efficient for obtaining the proof of principle for the drug but is less robust than the stratification design for validating the marker. Both enrichment and stratification designs require validated assays and prescreening prior to study entry.

For many investigational agents (e.g., agents that target the IGF-1R, VEGF, BCL-2 or immunotherapy), despite extensive preclinical work the potential predictive markers remain elusive at the time of clinical trials, or even long after the FDA approval. ***Retrospective studies*** of the tumor or blood specimens would be the only option for the exploration of predictive markers. Sufficient specimen collection has proved to be critically important for the discovery of predictive markers in oncology and should be emphasized in all clinical trials of MTAs. A well-known example of retrospective predictive marker discovery is KRAS mutation as the negative predictor of activity with the anti-EGFR monoclonal antibody therapies.[10–12] Retrospective studies in patients with remarkable response to Erlotinib also led to the discovery of EGFR activation mutation,[12,13] which was subsequently confirmed to be the predictive marker for Erlotinib and gefitinib. In a recently presented phase II trial in NSCLC of Erlotinib with or without MetMab, a monoclonal antibody against C-MET, while no improvement was observed in the overall population, a prospectively planned, retrospectively performed subset analysis revealed that the combination was significantly better than Erlotinib alone in patients with high C-MET IHC in tumors.[14]

In order for retrospective biomarker analysis to timely inform the drug development, it is important to ensure availability of validated assays and timely execution of the correlative studies such that biomarker hypothesis can be generated in time for the confirmatory trial. Ideally, quick turnaround of the biomarker data can streamline the hypothesis generating and hypothesis testing process, and as such, retrospective analysis on "all comers" and prospective enrichment design for the new identified marker candidates can be built in the same trial.

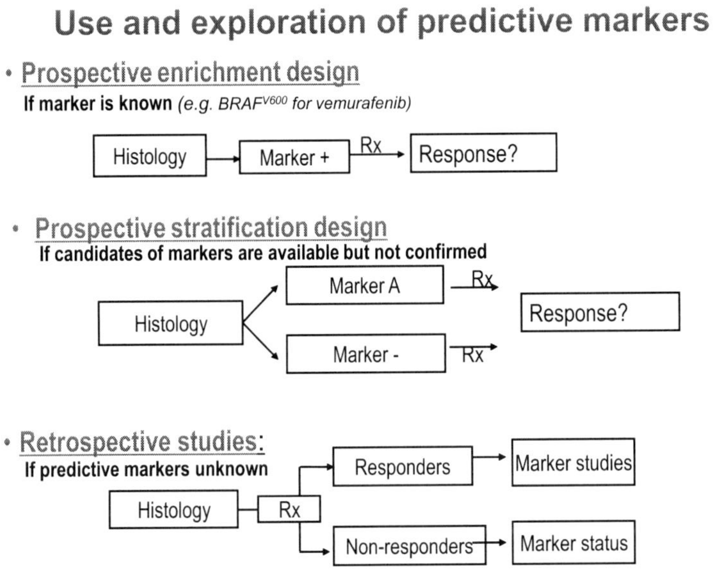

3.3.2.2 Scientific and technical challenges of predictive markers

With the exceptions of ER, validated predictive markers for all signaling pathway targeting drugs are uniformly based on single gene alterations. Emerging data from cancer genome projects and biological studies reinforce the notion that most solid tumors are driven by more than one genetic changes as well as epigenetic perturbations, and that predictive markers for therapies of most solid tumors may have to rely on multi-analyte, high-dimensional molecular signatures.

An extensively studied and published platform prognostic or predictive marker studies is the gene-expression-based signatures. Well-developed and validated genomic signatures have the promise to assist in personalized treatment decisions. An excellent example is the development of OncoType Dx for prognostic prediction of patients with breast cancer. In the context of MTA against key molecular targets and pathways, it is conceivable that gene expression signatures of pathway activations (e.g., MAPK or PI3K) may be potential candidates of predictive markers. However, development and validation of non-DNA based multi-analyte markers are technically and statistically challenging. In an analysis by Subramanian and Simon of 16 studies for gene expression-based prognostic signatures for non-small cell lung cancer,[15] there was little evidence that any of the reported gene expression signatures were ready for clinical application. More importantly, there were serious problems in the design and analysis of many of the studies. Guidelines have now been proposed to aid the proper trial design, validated assays and unbiased analysis and validation.[16]

Another challenge is the longitudinal evolutions of the molecular and clonal profiles through the tumor progression, particularly under the selective pressure of various drug therapies. Therefore tumor tissues obtained from the time of diagnosis, may not reflect the molecular features at the time of new drug clinical trials. Exploration of predictive markers based on these distant tumor specimens can be misleading. However, fresh biopsies for current tumor tissues are costly and not always feasible in patients in advanced stage of cancer.

With the increasing realization of the important of predictive markers for drug development and personalized medicine, there is an enhanced awareness and investment in the biomarker studies, including the search of predictive marker hypotheses in preclinical development of new drugs. The rapidly evolving sequencing technologies have also made it possible to perform genetic characterizations with increasing efficiency and decreasing cost. As exemplified by the BATTLE trials, it is now feasible to screen a patient for potential clinical trials and therapeutic options through multiplexed platforms of genetic characterizations of multiple genes. Currently such "snapshot" approach to patient characterizations can be established for hot-spot mutations in many "actionable" genes. It can be envisioned that in the future, such approach can be extended to characterization of whole-exome sequencing or RAN- and protein-based assays in a real time efficient manner.

3.4 Conclusions and Future Directions

It is now widely accepted that there must be a paradigm shift in the development of cancer therapies in the 21st century, and that incorporations of biomarkers, in addition to the traditional clinical trial endpoints, is critical for all stages of drug development. Early clinical trials should not only focus on PK and safety, but also seek to define the drug effects at the molecular levels, including target modulation, feedback loops and pathway switching. Early drug development should also be viewed as a crucial stage to test and begin to qualify potential predictive markers in preparation for larger confirmatory studies.

The strategy of incorporating PD and predictive/enrichment markers early in development has proven to be critical to the development of a few new MTAs including PARP inhibitor, Crizotinib and Vemurafenib. However, it is also recognized that many of the potentially useful targeted agents do not have a known predictive markers or strong candidates of predictive markers. Furthermore, the majority of tumors in patients are driven by multiple gene alterations and epigenetic changes. In order to allow efficient and reliable testing of the drug candidates and to optimize therapy for individual patients, a few essential tasks have been identified to fill the knowledge and technical gaps:

- Establishment of comprehensive and molecularly characterized preclinical models, to recapitulate the molecular signatures and heterogeneity of tumors in patients
- In-depth studies on the biology of individual targets and pharmacology of the agents, to better define the molecular contexts predictive of sensitivity or resistance, to examine the molecular interactions of the target with other pathways and to develop and test hypothesis of combination strategies.
- Systematic efforts in biomarkers discovery, assay development, and clinical qualification. Given the time, cost and expertise that would be required in the process, the effort should start as early as possible in the drug development.
- Collaborative efforts between public, industry and academic partners to establish publically accessible assay protocols and platforms for key targets and pathways.

References

1. Atkinson AJ, Colburn WA, DeGruttola VG, *et al.* Biomarkers and surrogate endpoints: Preferred definitions and conceptual framework. *Clin Pharmacol Ther* 2001; 69: 89–95.
2. Dancey JE, Dobbin KK, Groshen S, *et al.* Guidelines for the development and incorporation of biomarker studies in early clinical trials of novel agents. *Clin Cancer Res* 2010; 16: 1745–1755.
3. Taube SE, Clark GM, Dancey JE, McShane LM, Sigman CC, Gutman SI. A perspective on challenges and issues in biomarker development and drug and biomarker codevelopment. *J Natl Cancer Inst* 2009; 101: 1453–1463.
4. McShane LM, Altman DG, Sauerbrei W, Taube SE, Gion M, Clark GM. Reporting recommendations for tumor marker prognostic studies (REMARK). *J Natl Cancer Inst* 2005; 97: 1180–1184.
5. O'Shaughnessy J, Osborne C, Pippen JE, *et al.* Iniparib plus chemotherapy in metastatic triple-negative breast cancer. *N Engl J Med* 2011; 364: 205–214.
6. Lightcap ES, McCormack TA, Pien CS, Chau V, Adams J, Elliott PJ. Proteasome inhibition measurements: clinical application. *Clin Chem* 2000; 46: 673–683.
7. Adams J. Development of the proteasome inhibitor PS-341. *Oncologist* 2002; 7: 9–16.
8. Morgan B, Thomas AL, Drevs J, *et al.* Dynamic contrast-enhanced magnetic resonance imaging as a biomarker for the pharmacological response of PTK787/ZK 222584, an inhibitor of the vascular endothelial growth factor receptor tyrosine kinases, in patients with advanced colorectal cancer and liver metastases: results from two phase I studies. *J Clin Oncol* 2003; 21: 3955–3964.
9. Ma WW, Jacene H, Song D, *et al.* [18F]fluorodeoxyglucose positron emission tomography correlates with Akt pathway activity but is not predictive of clinical outcome during mTOR inhibitor therapy. *J Clin Oncol* 2009; 27: 2697–2704.
10. Amado RG, Wolf M, Peeters M, *et al.* Wild-type KRAS is required for panitumumab efficacy in patients with metastatic colorectal cancer. *J Clin Oncol* 2008; 26: 1626–1634.
11. Van Cutsem E, Kohne CH, Hitre E, *et al.* Cetuximab and chemotherapy as initial treatment for metastatic colorectal cancer. *N Engl J Med* 2009; 360: 1408–1417.

12. Lynch TJ, Bell DW, Sordella R, *et al.* Activating mutations in the epidermal growth factor receptor underlying responsiveness of non–small-cell lung cancer to gefitinib. *N Engl J Med* 2004; 350: 2129–2139.

13. Paez JG, Jänne PA, Lee JC, *et al.* EGFR mutations in lung cancer: correlation with clinical response to gefitinib therapy. *Science* 2004; 304: 1497–1500.

14. Spigel D, Ervin J, Ramlau R, *et al.* Final efficacy results from OAM4558g, a randomized phase II study evaluating MetMAb or placebo in combination with erlotinib in advanced NSCLC. In: American Society of Clinical Oncology; 2011: *J Clin Oncol* (suppl; abstr 7505); 2011.

15. Subramanian J, Simon R. Gene expression-based prognostic signatures in lung cancer: ready for clinical use? *J Natl Cancer Inst* 2010; 102: 464–474.

16. Subramanian J, Simon R. What should physicians look for in evaluating prognostic gene-expression signatures? *Nat Rev Clin Oncol* 2010; 7: 327–334.

Chapter 4

HER-2 as a Prognostic and Predictive Biomarker in Cancer

Suparna B. Wedam[a] **and Stanley Lipkowitz**[b]
[a]*US Food and Drug Administration, Center for Drug Evaluation and Research, Silver Spring, MD 20993, USA*
[b]*Laboratory of Cellular and Molecular Biology, Center for Cancer Research, National Cancer Institute, National Institutes of Health, Bethesda, MD 20892, USA*
Suparna.Wedam@fda.hhs.gov,lipkowis@mail.nih.gov

4.1 Introduction

Approximately 15–30% of breast cancers have amplified HER-2. Anti-HER-2 directed therapies are among the first and most successful targeted therapies. These therapies have made a dramatic impact on the outcomes in breast cancer patients whose tumors have amplified HER-2 and may be useful in other tumor types as well. The amplification of HER-2 serves as both a robust prognostic and a predictive biomarker in breast cancer and will be reviewed in this chapter. More recent data suggest similar prognostic and predictive value to HER-2 amplification in gastric cancer. Also, there is evidence that HER-2 serves as a prognostic biomarker and may be a predictive biomarker in other cancers and this will be discussed more briefly.

Handbook of Therapeutic Biomarkers in Cancer
Edited by Sherry X. Yang and Janet E. Dancey
Copyright © 2013 Pan Stanford Publishing Pte. Ltd.
ISBN 978-981-4364-65-2 (Hardcover), 978-981-4364-66-9 (eBook)
www.panstanford.com

4.2 Biology of HER-2

HER-2 is a member of the epidermal growth factor receptor (EGFR) family of receptor tyrosine kinases (RTKs). EGFR, the founding member of this family, was first identified as the receptor for the epidermal growth factor (EGF) and shown to have EGF stimulated tyrosine kinase activity [1–5]. Subsequently, the EGFR was found to be the cellular homologue of the *v-ErbB* (*a.k.a.*, *v-ErbB-1*) oncogene, one of the two transforming genes of the avian erythoblastosis virus [5–7]. Hence the EGFR is known also as *c-ErbB-1* [6]. HER-2 was discovered by several laboratories using two different approaches. In the first, the Weinberg group identified DNA from chemically induced neuroblastomas in rats that could transform NIH 3T3 cells [8]. Subsequently, they cloned the transforming gene, which they called *neu* and found that it had significant homology to the EGFR [9, 10]. In the second approach, several other groups used probes for either EGFR or *v-ErbB* to identify and clone homologues of the EGFR, which they called Human EGF Receptor 2 (HER-2) or *c-ErbB-2* [11–13]. Sequence analysis demonstrated that *neu*, HER-2, and *c-ErbB-2* were identical [9, 11–13]. The human gene will be referred to as HER-2 for the remainder of this chapter. Two related RTKs (HER-3 or *c-ErbB-3* and HER-4 or *c-ErbB-4*) were identified by their homology to the EGFR [14–16].

The four members of the EGFR RTK family (EGFR, HER-2, HER-3, HER-4) are single pass type I transmembrane proteins that contain extracellular cysteine-rich domains and an intracellular tyrosine kinase (TK) domain (Fig. 4.1A) [17]. HER-2 comprises 1255 amino acids with a molecular weight of 185 kDa [9, 11–13]. Ligands that bind directly to these RTKs have been identified for the EGFR, HER-3 and HER-4 [17]. However, no ligand has been identified that binds directly to HER-2 [17]. Instead, HER-2 heterodimerizes with one of the other ligand-bound EGFR family receptors resulting in stimulation of the HER-2 TK activity (Fig. 4.1B) [18–21]. Interestingly, the HER-2-HER-3 heterodimer appears to be one of the most potent mitogenic pairs [18–21]. The TK domain of HER-3 is not catalytically active and binding of ligand stimulates heterodimerization of HER-3 with HER-2, which stimulates HER-2 TK activity. This leads to the stimulation of signaling pathways that mediate cell growth and survival (e.g., the RAS/ERK and PI3K/AKT pathways) [17].

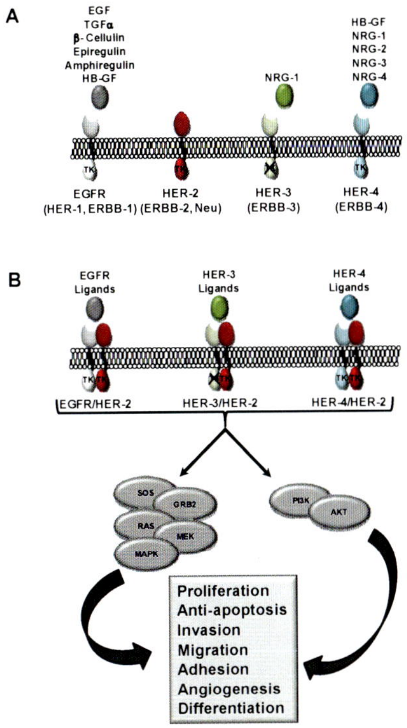

Figure 4.1 HER-2 and the EGFR family. A. Schematic diagram of the EGFR, HER-2, HER-3 and HER-4. Alternative names are listed in parentheses below each receptor. The high affinity ligands for each receptor are listed above the receptors. HER-2 has an active tyrosine kinase (TK) domain but no ligand has been identified that binds to HER-2. The TK domain of HER-3 is catalytically inactive. B. HER-2 is activated by dimerization with one of the other EGFR family members upon ligand binding, leading to activation of downstream signaling pathways. Activated receptors mediate many cell processes. Abbreviations: EGFR, epidermal growth factor receptor; EGF, epidermal growth factor; HB-GF, heparin binding growth factor; HER-1, human epidermal growth factor 1; HER-2, human epidermal growth factor 2; HER-3, human epidermal growth factor 4; HER-4, human epidermal growth factor 4; ERBB-1, erythroblastic leukemia viral oncogene homolog 1; ERBB-2, erythroblastic leukemia viral oncogene homolog 2; ERBB-3, erythroblastic leukemia viral oncogene homolog 3; ERBB-4, erythroblastic leukemia viral oncogene homolog 4; Neu, neuro/glioblastoma derived oncogene; NRG-1, neuregulin-1; NRG-2, neuregulin-2; NRG-3, neuregulin-3; NRG-4, neuregulin-4; TGFα, transforming growth factor alpha.

Homozygous deletion of HER-2 in mice results in embryonic lethality due to defective development of cardiac trebeculae [22]. Also, HER-2 null mice have abnormal development of the nervous system and the neuromuscular junction [22–25]. Homozygous deletion of neuregulin, which binds to HER-3 and HER-4 and stimulates heterodimerization of these receptors with HER-2 and activation of the HER-2 kinase activity, causes similar developmental defects in the heart and nervous system [26]. In the developing heart, HER-2 is expressed in embryonic myocytes while the ligand neuregulin is expressed in the adjacent endocardium, consistent with a paracrine developmental signal (reviewed in [27]). Thus HER-2 is essential for cardiac and neural development. Also, HER-2 serves important functions in the adult heart. Mice with conditional deletion of HER-2 targeted to cardiomyocytes have normal embryonic heart development but develop dilated cardiomyopathy at 3 months of age [28]. Doxorubicin treatment of mice with a heterozygous deletion of neuregulin-1 results in increased cardiac damage and mortality and induces less phosphorylation of HER-2, HER-4, ERK 1 and 2, AKT, and 70S6K in the heterozygous neuregulin-deficient mice compared with wild-type mice [29]. Together these data demonstrate a role for HER-2 in normal homeostasis and response to damage in the adult heart. This is likely to underly the cardiac toxicity observed in breast cancer patients treated with the anti-HER-2 antibody trastuzumab (reviewed in [30]).

The homology to the v-*ErbB* oncogene and the identification of HER-2 as the transforming gene from chemically induced neuroblastomas in rats immediately indicated that HER-2 could function as an oncogene [6, 8–12]. Sequence analysis of HER-2 from normal rat tissues and the transformed cells identified activating mutations in the coding sequence that resulted in transforming activity of the mutant protein but not of the normal protein [31, 32]. Transgenic mice expressing activated HER-2 driven by the mouse mammary tumor virus (MMTV) promoter express high levels of the activated HER-2 transgene in breast tissue and develop multiple hyperplastic and dysplastic nodules throughout their mammary tissue [33]. All of the mice develop multiple mammary carcinomas with a short latency [33]. Transgenic mice expressing wild-type HER-2 driven by the MMTV promoter develop breast tumors with a longer latency than seen with activated HER-2 [34]. Interestingly, the tumors that arise in the transgenic mice expressing wild-type

HER-2 frequently harbor activating mutations in the transgene in the tumors that develop [35]. Expression of HER-2 in these animal models is driven by a strong viral promoter. Transgenic mice expressing activated HER-2 driven by the endogenous promoter develop tumors with a longer latency than those in which activated HER-2 expression is under the control of the MMTV promoter. However, the tumors that develop contain the amplification of the activated HER-2, suggesting that both activating mutation and amplification are necessary for the development of breast tumors in mice [36]. While the amplification of HER-2 is found in human breast tumors (see Section 4.4), activating mutations are not described [37, 38][1]. Overexpression of wild-type human HER-2 is able to transform mouse and human cells [39–42]. The reason for this requirement for activating mutations in mice but not in humans in HER-2 driven breast tumors has not been elucidated.

4.3 HER-2 Amplification and Overexpression: Methods of HER-2 Measurement

The amplification of the HER-2 gene (defined as an increased copy number of the gene within genomic DNA) was identified by Southern analysis in breast, salivary, and gastric adenocarcinoma DNA in the papers describing the cloning of the HER-2 gene [7, 12, 13]. Subsequent work indentified gene amplification of HER-2 from a variety of adenocarcinomas such as bladder, breast, colon, gastric, non-small cell lung, ovarian, and salivary cancer [43–51].

Initial studies of HER-2 amplification were based on Southern DNA blotting techniques, which measured copy number of the HER-2 gene in the DNA from patient samples [7, 12, 13, 50–53]. Subsequently, Slamon *et al.* demonstrated that Southern analysis of DNA amplification correlated with the high expression of HER-2 mRNA by Northern analysis, high expression of the HER-2 protein by immunoblot, and high expression by immunohistochemistry (IHC) [51]. Coincident with the studies looking at HER-2 as prognostic biomarker and as a predictive biomarker for HER-2 targeted therapies for the treatment of cancer, methods were developed to define those patients with HER-2 amplification or overexpression. Most of the earliest studies and trials used one of several different antibodies to determine HER-2 overexpression by IHC with varying criteria of overexpression (reviewed in [54]). For example, in some studies testing was performed using IHC with an antibody to HER-2 developed by the investigators known as the clinical trials assay (CTA). The CTA assay was composed of two monoclonal antibodies, CB11 and 4D5. The latter antibody is the mouse monoclonal antibody that was humanized to create trastuzumab [55]. Scoring was IHC 0–1+ (negative), IHC 2+ (weakly positive membrane), or IHC 3+ (strongly positive membrane staining) (e.g., Fig. 4.2A). Other studies using the A085 polyclonal antibody from DAKO scored tumors as IHC 0–1+ (negative), IHC 2+ (weakly positive membrane staining on more than 10% of the cells), or IHC 3+ (strongly positive membrane staining on more than 10% of the cells). The A085 polyclonal antibody (DAKO) and the CB11 monoclonal antibody (Ventana) are approved by the FDA for IHC measurement of HER-2 overexpression [54]. Several studies have shown that the different antibodies vary in terms of specificity, sensitivity and accuracy [56, 57]. In addition, all of the IHC based definitions of HER-2 overexpression are subjective.

Fluorescent *in situ* hybridization (FISH) was developed to measure gene amplification in formalin fixed, paraffin embedded tissues by quantifying the number of copies of the HER-2 gene in interphase nuclei of cells (Fig. 4.2B) [58]. This method uses fluorescently labeled HER-2-specific DNA probes or biotin-labeled HER-2-specific DNA probes, which are then detected by fluorescently labeled avidin [54, 58]. Amplification can be determined by two scoring criteria. In one, more than four copies of HER-2/cell was considered gene amplification [58]. The second method normalized the number of copies of HER-2/cell to the number of copies of chromosome 17 measured by a probe to the centromere for chromosome 17 (CEP17). In this method, a HER-2/CEP17 ratio ≥ 2.0 was considered gene amplification [58]. A number of studies have found that overexpression based on IHC is rarely seen in the absence of gene amplification [58–60]. Currently two commercial kits, the PathVysion HER-2 DNA probe kit (Abbott), which measures the HER-2/CEP17 ratio, and the Inform HER2/*neu* probe (Ventana), which measures the number of HER-2 copies/cell, are FDA approved for the measurement of HER-2 amplification [54]. When tested on 117 tumor samples for which HER-2 amplification status had been previously determined by DNA blotting, these FISH assays were found to have high sensitivity, specificity, and accuracy [57]. Importantly, in comparisons of outside laboratory assessments of HER-2 amplification/overexpression by IHC and FISH with central

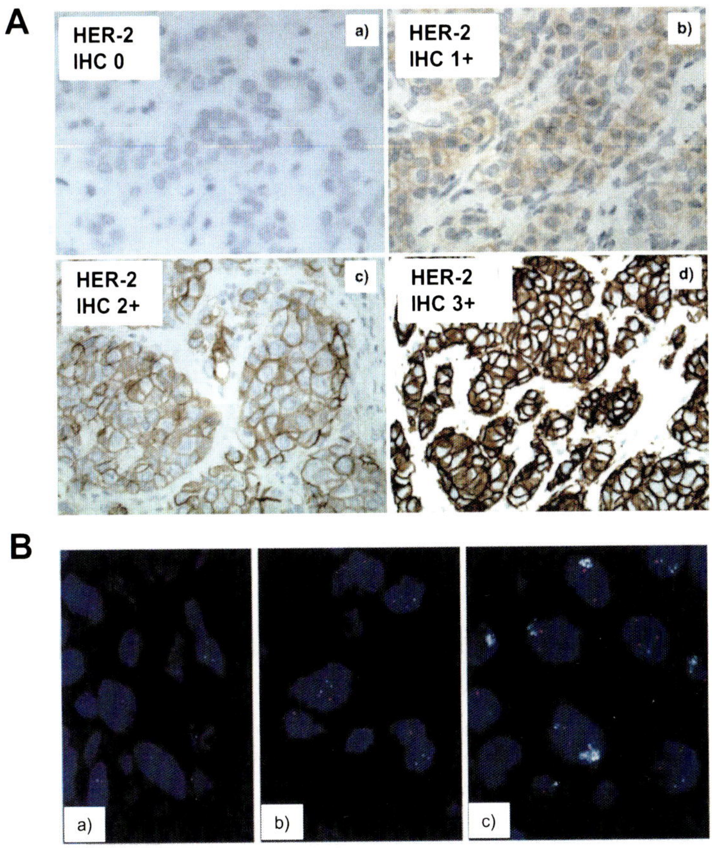

Figure 4.2 Detection of HER-2 amplification in breast cancer by IHC and FISH. A. IHC staining of HER-2 using the CB11 monoclonalantibody (Ventana) demonstrating (a) IHC 0, no staining; (b) IHC 1+, weak staining of the incomplete membrane in more than 10% of the cells; (c) IHC 2+, weak to moderate staining of the entire membrane in more than 10% of the cells; and (d) IHC 3+, strong staining of the entire membrane in more than 30% of the cells. IHC3+ is considered positive for amplification. Reproduced with permission from Mayr *et al.* [196]. B. FISH analysis of HER-2 amplification using probes for HER-2 (green) and chromosome 17 (red) demonstrating (a) normal diploid content of HER-2/nucleus; (b) low-level amplification; and (c) high level amplification. Amplification is defined as HER-2/chromosome 17 > 2.0. Reproduced with permission from Press *et al.* [197].

laboratory determinations of HER-2 amplification by FISH, FISH assays have been found to be more accurate and reproducible [61]. This is likely due, at least in part, to the more quantitative nature of these measurements.

Two additional methods to determine HER-2 gene amplification have been developed. Chromogenic *in situ* hybridization (CISH) and silver enhanced *in situ* hybridization (SISH) are similar to FISH except that the HER-2 DNA probes are detected by using a peroxidase reaction [62, 63]. This allows detection using a light microscope rather than a fluorescent microscope, and the stained slides can be stored indefinitely (unlike FISH slides, which rapidly fade). Comparisons of the CISH and SISH methods with FISH demonstrated a 94% and 96% concordance, respectively [62, 63]. Thus the methods are reasonable alternatives to FISH, but they are not available in the United States for routine use [54].

To standardize HER-2 testing for both clinical practice and future clinical trials, the American Society for Clinical Oncology and the College of American Pathologists developed guidelines for HER-2 testing for use in breast cancer patients [54]. A positive HER-2 test using a validated IHC assay (e.g., the Herceptest or Pathway antibodies) was defined as a tumor expressing uniform intense membrane staining on >30% of the invasive tumor cells (IHC 3+). An equivocal HER-2 test by IHC was defined as a tumor expressing non-uniform or weak membrane staining on at least 10% of the cells (IHC 2+). Samples with less staining were defined as negative (IHC 0 or 1+). These guidelines recommend evaluation of samples with equivocal IHC staining (IHC 2+) by a validated assay for gene amplification. A positive test for gene amplification by FISH was defined as either a HER-2/CEP17 ratio of > 2.2 or > 6 copies of HER-2/nucleus if no chromosome 17 control is done. An equivocal FISH test was defined as a HER-2/CEP17 ratio from 1.8–2.2. The HER-2/CEP17 ratio is derived from the average signal number for each probe from 60 cells. The indeterminate range was derived from data suggesting that low level amplification could result in variable results when read by different pathologists (i.e., when a different set of 60 cells are scored) [64]. The guidelines noted that most clinical trials defined amplification by FISH as either HER-2/CEP17 >2.0 or ≥ 2.0; thus patients with HER-2/CEP17 from 2.0–2.2 were included on the trials [54]. Few patients (~3%) will have values between 2.0–2.2 and the guidelines note that the data do not support excluding such patients from anti-HER-2 targeted therapy. Also, these guidelines establish criteria for quality controls such as conditions for fixation [54].

4.4 HER-2 Amplification as a Prognostic Biomarker in Breast Cancer

Amplification of HER-2, identified by Southern analysis of genomic DNA, was first described to predict poor prognosis in breast cancer patients in a landmark paper by Slamon *et al.* [50]. In this study the authors evaluated 189 tumors and found that 53 (28%) had between 2- and >20-fold amplification of HER-2 (defined by Southern analysis of genomic DNA). In a cohort of 86 patients for whom the outcome data were available, there was a statistically significant inverse correlation between increasing copy number of HER-2 and both disease free and overall survival (Fig. 4.2) [50]. Also, a correlation existed between the number of lymph node metastases and the amplification of HER-2. Several other studies, using Southern analysis, confirmed that HER-2 amplification is found in 15–30% of tumors and that this correlates with worse prognostic features (e.g., the number of lymph node metastases) and poor outcome [52, 53, 65]. An important study by Slamon *et al.* demonstrated that amplification of HER-2 demonstrated by Southern analysis correlated with high expression of the mRNA for HER-2 by Northern analysis, high expression of the HER-2 protein by immunoblot, and high expression by immunohistochemistry [51]. Again, in this study HER-2 amplification was associated with a worse disease free and overall survival. A recent review of 107 studies sampling 39,700 patients found HER-2 amplification in 22% (range 10–74%) of the patients [66]. Ninety five of these studies found a correlation between HER-2 amplification and poor prognosis, and 68 of the 93 studies that performed multivariate analysis determined HER-2 amplification to be an independent prognostic variable (Fig. 4.3) [66]. These studies measured DNA amplification, protein expression, or in some cases both. Thirty seven of the 39 reports (95%) using techniques that measured DNA amplification (Southern analysis, slot blotting, FISH, and CISH) found a correlation between amplification and worse outcome. Seventy four of the 86 studies (86%) using methods that measured protein expression (e.g., IHC, Elisa, or Western Blotting) found a correlation between high expression and worse outcome. Thus, the direct measurement of gene amplification appears to be a more reliable approach. Overall, in studies where it was reported, the mean relative risk for HER-2 amplification was 2.74 for overall survival and 2.04 for disease free survival [66]. A recent molecular analysis of gene expression by cDNA microarray has identified HER-2 amplified breast cancer based on mRNA expression and found a similar poor prognosis for this patient subgroup [67]. Thus, the association of HER-2 amplification and poor prognosis in breast cancer has been a consistent finding over many studies using DNA, RNA and protein based approaches.

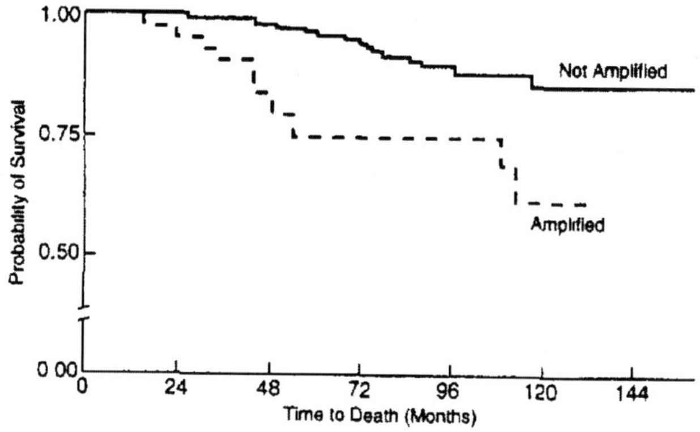

Figure 4.3 HER-2 as a prognostic marker. Survival of node negative early-stage breast cancer patients stratified for HER-2 amplification. Reproduced with permission from Press *et al.* [197].

4.5 HER-2 Amplification as a Predictive Biomarker for Response to HER-2 Targeted Agents in Breast Cancer

4.5.1 Trastuzumab

A predictive biomarker is one that predicts the likelihood of response to a specific agent [68]. The preclinical data described above demonstrating the transforming potential of HER-2 and the prognostic significance of the amplification or overexpression of HER-2 suggest that it would be a predictive biomarker for the use of HER-2 targeted agents. Preclinical studies of the anti-HER-2 mouse monoclonal antibody 4D5 and the humanized version of this antibody (trastuzumab) demonstrated anti-tumor effects of this antibody as a single agent and synergistic activity when the antibody was combined with chemotherapeutic agents in cancer cell lines in culture and in xenografts that had amplified HER-2 but showed no effect in cell lines lacking amplified HER-2 [69–72]. These preclinical studies led directly to the development of clinical trials of trastuzumab in breast cancer patients with amplified HER-2.

Phase I and phase II studies in patients with metastatic breast cancer overexpressing HER-2 demonstrated single agent activity of trastuzumab [73–76]. When tested in patients with disease that had progressed on prior chemotherapy, responses were seen in ~11–12% of patients (including 2 complete responses (CR) and 5 partial responses (PRs) out of 61 evaluable patients treated in the two studies) [74, 75]. When trastuzumab was used as first line therapy in patients with metastatic breast cancer higher response rates were seen (23–26% with 9 CRs and 40 PRs out of 197 evaluable patients treated in the two studies) [73, 76]. Phase I and II studies of trastuzumab in combination with cisplatin in metastatic breast cancer overexpressing HER-2 (as measured by IHC staining of 2+ or 3+) demonstrated activity in ~25% of patients (including 1 CR and 12 PRs out of 52 evaluable patients treated in these two studies) [77]. These studies led to a phase III study of the addition of trastuzumab to chemotherapy as first line treatment for patients with metastatic breast cancer overexpressing HER-2 (as measured by IHC staining of 2+ or 3+) [78]. The addition of trastuzumab to chemotherapy resulted in a statistically significant increase in objective response rate (ORR), progression free survival (PFS), and overall survival (OS) (Fig. 4.4A) [78]. Subsequent studies in the adjuvant setting have demonstrated that the addition of trastuzumab to chemotherapy increases recurrence free survival (RFS) and OS in patients with HER-2 overexpressing breast cancer (Fig. 4.4B) [79–83]. Three studies conducted by the National Surgical Adjuvant Breast and Bowel Project (NSABP B31), the North Central Cancer Treatment Group (NCCTG N9831), and the Breast International Group (BIG 1-01, HERA) defined HER-2 overexpression or amplification as FISH + (HER-2/CEP17 ≥ 2.0), IHC 3+, or IHC 2+ and FISH + [81–83]. A study conducted in Finland, defined HER-2 overexpression or amplification as IHC 2+ or 3+ and CISH positive (defined as ≥6 copies

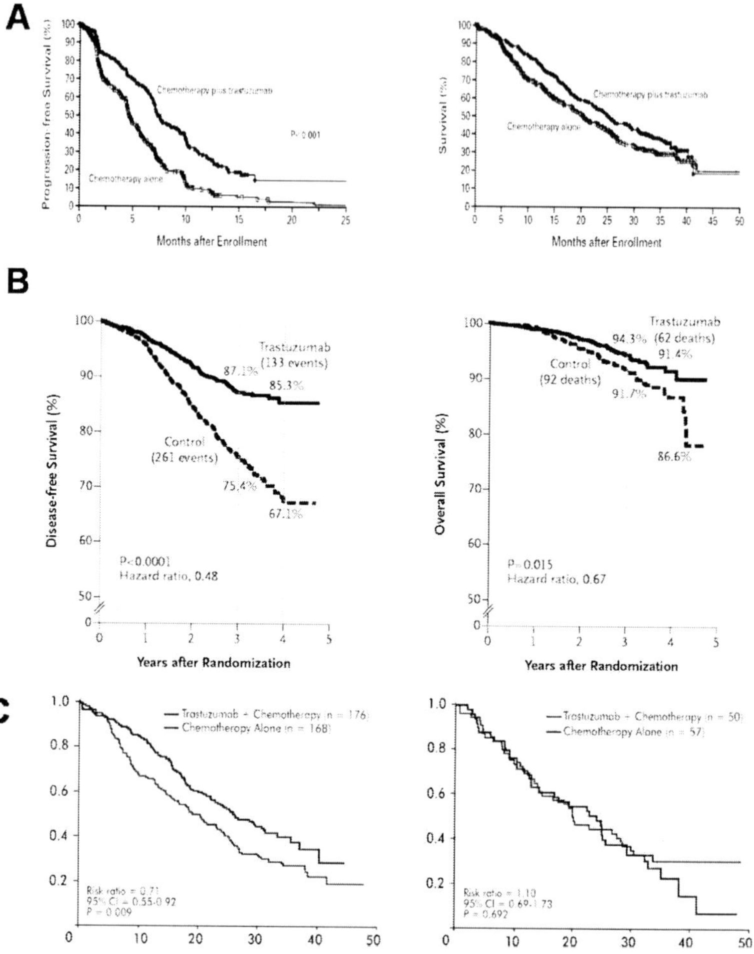

Figure 4.4 HER-2 as a predictive biomarker for trastuzumab. A. Progression-free survival (left) and overall survival (right) for patients with HER-2-amplified metastatic breast cancer treated with chemotherapy +/- trastuzumab. Reproduced with permission from Slamon *et al.* [78]. B. Disease-free survival (left) and overall survival (right) for patients with early-stage HER-2 positive breast cancer treated with chemotherapy +/- trastuzumab. Reproduced with permission from Romond *et al.* [82]. C. HER-2 amplification predicts the response to trastuzumab. Patients with metastatic cancer were treated with chemotherapy +/- trastuzumab. Overall survival for patients with HER-2 amplified tumors are shown on the right and with HER-2 non-amplified tumors are shown on the left. Reproduced with permission from Mass *et al.* [87].

of the HER-2 gene/cell) [79]. The BCIRG006 trial enrolled patients with amplification of HER2 measured by FISH (FISH+ defined as HER-2/CEP17 ≥ 2.0) [80]. A randomized trial of neoadjuvant chemotherapy with or without trastuzumab in patients with HER2+ (IHC 3+ or FISH+ defined by HER-2/CEP17 ≥ 2.0) operable breast cancer demonstrated a significantly higher pathological response rate when trastuzumab was added to chemotherapy (65% vs. 26%) and an improved RFS [84, 85].

All of these phase III studies demonstrated that the addition of trastuzumab to chemotherapy in either the metastatic or adjuvant setting was clearly beneficial in patients with tumors that overexpress HER-2. This randomized trial design is a targeted or enrichment design in which only patients expressing the biomarker are enrolled in the study [86]. The ideal study to validate a biomarker is a randomized controlled trial in which the biomarker is not an inclusion criteria (so that the study includes both patients with and without the biomarker) and that then assesses outcome of the experimental therapy in both biomarker positive and negative populations [86]. Such a study has not been done in breast cancer for trastuzumab. However, Mass *et al.* evaluated clinical outcomes based on amplification of HER-2 determined by FISH in a retrospective analysis of 765 patients with metastatic HER-2 amplified breast cancer originally determined by IHC who were enrolled in three studies sponsored by Genentech [87]. Overall, 78% of the patients initially characterized as HER-2 amplified by IHC were FISH positive (defined as a HER-2/CEP17 ratio ≥ 2.0) and 22% were FISH negative. Two studies included in this analysis (H0649g and H0650g) treated patients with metastatic breast cancer with trastuzumab alone [87]. The ORR rate in the FISH positive patients was 24% vs. 3% in the FISH negative patients for the two studies combined. In both studies, PFS was longer in the FISH positive patients than in the FISH negative patients. In the H0649g study, there was an increased OS in the FISH positive patients. In a third study analyzed comparing trastuzumab and chemotherapy with chemotherapy alone (H0648g), there was a significant increase in the ORR (54% vs. 30%), the median PFS (7.3 vs. 4.6 months), and the median OS (26.2 vs. 20.3 months) when trastuzumab was added to chemotherapy in the FISH positive patients (Fig. 4.4C) [87]. By contrast, there was no increase in the ORR (38% vs. 39%) or the OS when trastuzumab was added to chemotherapy in the FISH negative patients. However, there was a trend to increased median PFS (7.7 vs. 4.9 months) in the FISH negative group when trastuzumab was added to chemotherapy [87]. Overall, this retrospective analysis is consistent with HER-2 amplification as a predictive biomarker for the response to trastuzumab. However, the increased PFS in the FISH negative patients raises the possibility that there may be some clinical benefit to trastuzumab in patients whose tumors do not have HER-2 amplification.

Several analyses raise questions about the use of HER-2 amplification/overexpression as a predictive biomarker for trastuzumab response [88, 89]. In the adjuvant NSABP-B31 and NCCTG-N9831 studies, patients had to have HER-2 amplification measured by either 3+ staining by IHC or by FISH (defined as HER-2/CEP17 signal ratio ≥ 2.0) [82]. Paik *et al.* found that ~10% of the patients enrolled in the trial were negative for amplification as measured by FISH in the central laboratory [88]. The HER-2 negative patients appeared to have a similar recurrence free survival benefit compared with the HER-2 positive patients. In a similar analysis of the N9831 adjuvant trial, the authors found that approximately 6% of the patients did not have amplified HER-2 when measured in the central laboratory [89]. This analysis found a non-significant trend towards benefit from trastuzumab in the patients without amplified HER-2 [89]. These analyses raise the possibility that trastuzumab combined with chemotherapy may be beneficial in tumors that express HER-2 without gene amplification. However, these findings are limited due to the small number of patients without HER-2 amplification. In addition, the findings could be due to tumor heterogeneity in the samples assayed by the outside and central laboratories or could represent false negative results in the central laboratory. Thus, while it is clear that breast cancer patients with HER-2 amplification benefit from trastuzumab, it remains to be determined if there is benefit to trastuzumab in patients whose tumors express HER-2 without amplification. An ongoing trial (NSABP B47) is evaluating the benefit of adding trastuzumab to chemotherapy in patients with node positive or high risk node negative breast cancer that expresses HER-2 without HER-2 amplification [90]. To be eligible for this study, patients are required to have expression of HER-2 by IHC (IHC 1+ or 2+) but no amplification by FISH (HER-2/CEP17< 2.0 or < 4 copies of HER-2/nuclei if no ratio is performed) or by CISH (< 4 copies of HER-2/nuclei). Patients with IHC 3+ or amplification by FISH or CISH (HER-2/CEP17 ≥ 2.0 or ≥ 4 copies of HER-2/nuclei) are excluded from this trial.

4.5.2 Lapatinib

Studies of lapatinib, a dual TK inhibitor (TKI) of EGFR and HER-2 have demonstrated efficacy in the metastatic setting for patients with tumors that have amplified HER-2 and ongoing studies are investigating lapatinib in the adjuvant setting [91–101]. Importantly, two of these phase II studies included patients with HER-2 amplified and non-amplified tumors allowing assessment of amplification/overexpression as a biomarker for the efficacy of lapatinib. Johnson *et al.* treated two cohorts of patients with locally advanced or metastatic inflammatory breast cancer that had progressed on prior therapy [100]. In one cohort of 30 patients with HER-2 amplified tumors (defined as IHC 2+ or 3+ or FISH+ with HER-2/CEP19 ≥ 2.0), the ORR was 50% (2 CRs and 13 PRs), and responses were seen both in patients previously treated with trastuzumab and those not treated with trastuzumab. In the second cohort of 15 patients, who had tumors that did not have amplified/overexpressed HER-2 but did express EGFR, only one patient had a PR (6.7%) and this arm was closed to accrual [100]. An expanded cohort of 126 patients with HER-2 amplified locally advanced or metastatic inflammatory tumors (IHC 3+ or FISH + defined by HER-2/CEP17 ≥ 2.0) that had progressed on prior therapy was treated with lapatinib and demonstrated a 39% ORR [92]. In a study investigating single agent lapatinib in progressive, previously treated metastatic disease, Toi *et al.* treated 100 patients with HER-2 amplified/overexpressing tumors (IHC 3+ or IHC 2+ and FISH + defined by HER-2/CEP17 ≥ 2.0) and 22 patients with HER-2 negative tumors [102]. All of the patients with HER-2 amplified tumors had been treated with trastuzumab. Nineteen percent of the patients with HER-2 amplified tumors had a clinical response and 9% had prolonged stable disease for a clinical benefit of 28%. In contrast, only one (4.5%) of the 22 patients with HER-2 negative tumors had a response, and none had stable disease (for a clinical benefit or 4.5%). No relationship was found between EGFR expression and the response to lapatinib [100, 102]. These studies are consistent with HER-2 amplification/overexpression as a predictive biomarker for response to single agent lapatinib.

Two phase III studies have assessed the efficacy of adding lapatinib to chemotherapy in patients with metastatic disease [95–97]. One study enrolled patients with HER-2 amplified (IHC 3+ or IHC 2+ and FISH + defined by HER-2/CEP17 ≥ 2.0) metastatic breast cancer that had progressed on trastuzumab treatment and randomized them between lapatinib plus capecitabine and capecitabine alone [95, 97]. The patients treated with lapatinib plus capecitabine had an improved progression free survival and response rate compared with those treated with capecitabine alone [95, 97]. The second study enrolled patients with metastatic disease to first line treatment with lapatinib plus paclitaxel or paclitaxel alone [96]. This study included a preplanned retrospective analysis of HER-2 amplification/overexpression by a central laboratory. In patients whose tumors had amplified HER-2 (FISH+ defined by HER-2/CEP17 ≥ 2.0 or IHC 3+ if FISH was not available) lapatinib resulted in a statistically significant increase in the time to progression (36 vs. 25 weeks), event-free survival (35 vs. 22 weeks), and overall response rate (63 vs. 38%) compared with paclitaxel alone [96]. By contrast, in the patients whose tumors did not have amplified HER-2 lapatinib did not result in a statistically significant increase in the time to progression (25 vs. 24 weeks), event-free survival (23 vs. 23 weeks), and overall response rate (30 vs. 23%) compared with paclitaxel alone [96]. Similar to the retrospective analyses of the trastuzumab studies described above, Press *et al.* found that approximately 16% of the patients enrolled in the first phase III study of lapatinib and capecitabine did not have amplified/overexpressed HER-2 when the samples were centrally analyzed [103]. HER-2 amplification was determined by FISH (HER-2/CEP17 ≥ 2.0). Samples where FISH could not be performed were considered positive with IHC 3+ staining. When the data from both of the lapatinib phase III studies were analyzed, the overall response rate and progression free survival were significantly improved when lapatinib was added to chemotherapy in patients whose tumors had amplification or overexpression of HER-2 [103]. There was no benefit to lapatinib in the patients with tumors that did not have amplification or overexpression [103]. Also, this analysis found no relationship between the benefit of lapatinib and EGFR expression [103].

A phase III study randomized patients with hormone receptor positive metastatic breast cancer to first line treatment with either lapatinib plus letrozole or placebo plus letrozole [99]. This study included 219 patients with HER-2 amplification (FISH+ defined as HER-2/CEP17 ≥ 2.0, IHC 3+, or IHC 2+ and gene amplification by FISH) and 952 patients without HER-2 amplification. In the patients with HER-2 amplification, the addition of lapatinib increased PFS (8.2 vs. 3 months) and objective response rate (28 vs. 15%). In contrast, the addition of lapatinib did not result in a significant change in either PFS or ORR in the patients without HER-2 amplification/overexpression [99]. All together, these data are consistent with HER-2 amplification/overexpression as a predictive biomarker for response to lapatinib as a single agent, in combination with chemotherapy, and in combination with aromatase inhibitors.

4.5.3 Pertuzumab

Pertuzumab is an antibody that inhibits the dimerization of HER-2 with other members of the EGFR family, thus inhibiting activity of HER-2 [104–106]. Few studies have been reported in the literature to date[2]. Two phase II trials measured the efficacy of pertuzumab in combination with trastuzumab in patients that had HER-2 amplification in their tumors and whose tumors had progressed on trastuzumab based therapies [107, 108]. The two studies found objective responses in 16/66 (24%) and 2/11 (18%) and prolonged stable disease in 17/66 (26%) and 3/11 (27%) for an overall clinical benefit in approximately 50% of the patients. In contrast, a phase II study investigated the efficacy of pertuzumab in patients with HER-2 negative tumors (FISH negative and/or IHC 0, 1+ or 2+) and found very low rates of clinical benefit [109]. Partial responses were seen in 2 of 78 (2.6%) patients and stable disease lasting greater than 24 weeks in 4 of 78 (5%), resulting in a clinical benefit in approximately 7.7% of patients. The data are consistent with HER-2 amplification/overexpression as a predictive biomarker although more data will be needed to assess this fully.

Overall, the data support HER-2 amplification/overexpression as a predictive biomarker for response to HER-2 targeted therapies including trastuzumab, lapatinib, and pertuzumab in breast cancer.

4.6 HER-2 Amplification as a Predictive Biomarker for Response to Chemotherapy in Breast Cancer

HER-2 amplification or overexpression has been reported as a predictive biomarker for the response to anthracyclines and taxanes.

A number of studies have reported that HER-2 amplification/overexpression is associated with response to anthracycline based chemotherapy [110–113]. In early studies the overexpression was measured by IHC and not standardized as described above [110, 111, 113]. The Canada Clinical Trials Group (CCTG) MA.5 trial demonstrated that cyclophosphamide, epirubicin, and fluorouracil (CEF) was superior to cyclophosphamide, methotrexate, and fluorouracil (CMF) in node positive premenopausal women with early breast cancer [114]. In a retrospective analysis of that study, Pritchard *et al.* found that patients with HER-2 amplified tumors (FISH+ defined as HER-2/CEP17 ≥ 2.0) benefited from the CEF regimen while there was no difference between CEF and CMF in the patients whose tumors were HER-2 negative by FISH [112]. A pooled analysis of eight trials that randomized early breast cancer patients to anthracycline vs. non-anthracycline containing regimens (6564 randomized patients of whom 5354 had HER-2 data available and were included in the analysis) found that patients whose tumors had HER-2 amplification or overexpression had improved DFS and OS when treated with an anthracycline containing regimen compared with the non-anthracycline containing regimen [115]. In contrast, patients whose tumors did not have HER-2 amplification or overexpression had similar outcomes on either regimen. In five of the eight studies, IHC was used for the determination of HER-2 status with varying definitions of overexpression [115]. Two studies used FISH (HER-2/CEP17 ≥ 2.0) and one study used IHC of 3+, IHC 2+ and FISH+ (HER-2/CEP17 ≥ 2.0), or FISH+ [115].

Together these data indicate that HER-2 amplification is predictive of response to anthracyclines in the absence of HER-2 targeted therapies. The biological basis of this relationship was unclear since overexpression of transfected HER-2 in cancer cell lines which did not have amplified HER-2 did not alter the sensitivity to anthracyclines [116]. The amplicon on chromosome 17q that contains HER-2 often results in coamplification of other genes, including topoisomerase IIα (TOP2A) [25, 117–123]. A recent retrospective analysis of HER-2 and TOP2A amplification in a 339 breast cancers using FISH (amplification defined as HER-2/CEP17 ≥ 2.0 and TOP2A/CEP17 ≥ 2.0) found that TOP2A was amplified in 29% and deleted in 14% of 339 samples with HER-2 amplification [124]. In contrast, TOP2A was not amplified in any of the 60 patients without HER-2 amplification and 3% of the HER-2 non-amplified samples had TOP2A deletion. The authors extended this study to a larger set of samples from the

BCIRG 005 study and found no amplification of TOP2A and only 2.6% with TOP2A deletion in 1,614 HER-2 non-amplified breast cancer [124]. From this study it would appear that TOP2A amplification only occurs in the context of HER-2 amplification. A second study evaluated patients from the Danish Breast Cancer Cooperative Group (DBCCG) 89D study, which tested the efficacy of CEF compared with CMF in pre- and post-menopausal early-stage breast cancer patients [125]. HER-2 and TOP2A amplification was assessed by FISH as above, but using different reagents from the paper by Press *et al.* [125]. This group found TOP2A amplification in 32% and TOP2A deletion in 25% of 246 patients with HER-2 amplification but only found TOP2A amplification in 2.7% and TOP deletion in 4.9% of 527 samples without HER-2 amplification. A third group, using FISH to evaluate samples from the CCTG MA.5 study found TOP2A amplification in 28% and TOP2A deletion in 15% of 116 samples with HER-2 amplification but only found TOP2A amplification in 6% and TOP deletion in 3% of 314 samples without HER-2 amplification [126]. From these studies, TOP2A amplification in the absence of HER-2 amplification is rare. TOP2A is the target of anthracyclines, and cancer cell lines expressing higher levels of TOP2A are more sensitive to anthracyclines [127–129]. When the last two studies that both compared CAF with CEF were analyzed for response in relationship to TOP2A amplification, both found improvement in DFS and OS for patients with TOP2A amplification treated with CAF compared with CEF but no such improvement was seen in patients without TOP2A amplification [125, 126, 130, 131]. Thus, a biologically plausible hypothesis is that TOP2A amplification, and not HER-2 amplification, is responsible for the improved outcome when anthracyclines are used to treat patients with HER-2 amplification. All of these analyses are based on retrospective analyses and a prospective trial directly testing this has not been done. The importance of these observations given the cardiac and leukemogenic toxicities of anthracyclines is to identify patients in whom anthracyclines can safely be avoided. In the BCIRG 006 study, patients with early-stage HER-2 positive breast cancer were randomized to three treatment regimens; doxorubicin, cyclophosphamide and paclitaxel (ACT), ACT with trastuzumab (ACTH), or docetaxel, carboplatin, and trastuzumab (TCH) [80]. When Press *et al.* analyzed these treatment arms with regard TOP2A amplification for RFS and OS, they found that the patients with tumors containing TOP2A amplification had improved survival on ACT compared with those without TOP2A amplification

[124]. However, when trastuzumab was added to the therapy (ACTH or TCH), the patients did equally well whether or not the tumors had had TOP2A amplification and whether or not they received an anthracycline [124]. While the final analysis of the study is yet to be published, these results suggest that the addition of trastuzumab may obviate the need for an anthracycline when TOP2A is co-amplified with HER-2[3]. Since TOP2A amplification is rare in the absence of HER-2 amplification, the value of HER-2 and TOP2A amplification as a predictive biomarker for the benefit of anthracyclines may be limited.

Two studies suggest that HER-2 amplification may be a biomarker for the response to paclitaxel [132, 133]. The phase III CALGB 9344 trial demonstrated improved DFS and OS by the addition of paclitaxel to adriamycin and cytoxan in the adjuvant setting for node positive breast cancer patients [134]. A retrospective analysis of 1322 patients treated on this study found that patients with HER-2 amplification (defined by FISH with HER-2/CEP17 ratio ≥ 2.0) benefited from the addition of paclitaxel but the patients without HER-2 amplification did not benefit [132]. A retrospective analysis of 297 patients with metastatic breast cancer treated with either epirubicin and paclitaxel (ET) or epirubicin and cytoxan (EC) found that patients with HER-2 amplified tumors (measured by FISH with amplification defined by HER-2/CEP17 > 2.0) had a better DFS and OS when treated with ET [133]. In contrast, patients with non-amplified tumors had similar DFS and OS with either treatment. However, other retrospective studies have not found a relationship between HER-2 amplification and the response to taxanes in early breast cancer including a meta-analysis by Delaurentiis *et al.* [135–137]. Overall, in the absence of a prospective trial designed to assess the role of HER-2 amplification or overexpression as a predictive biomarker for taxane response, the data do not currently support this relationship.

4.7 HER-2 Amplification as a Predictive Biomarker for Response to Hormonal Therapy in Breast Cancer

There is evidence that HER-2 amplification and/or overexpression may lead to primary resistance to endocrine therapy and specifically to tamoxifen [138–142]. The exact mechanism for resistance is not clear but is likely due to crosstalk between the estrogen receptor (ER) and HER-2 pathways [143–145]. Tamoxifen is a selective estrogen receptor modulator (SERM) with mixed antagonist-agonist activity depending on the target tissue [146]. Preclinical studies have shown that in ER+/HER-2+ cell lines with elevated levels of the ER coactivator AIB1(SRC3), the agonist properties of tamoxifen are enhanced leading to resistance to tamoxifen therapy [145]. AIB1 appears to be activated by phosphorylation through HER-2 signaling pathways [147].

Resistance to endocrine therapy in women with ER+ and HER-2 overexpressing tumors has been noted in studies, although none of these have been prospective, randomized clinical trials [138–142]. SWOG 9314 is an ancillary study in which 349 specimens from the an earlier study where women received tamoxifen as initial treatment for ER+ advanced breast cancer were evaluated for HER-2 expression and outcome [139]. Overall response rate (47% vs. 56%) and time to treatment failure (5 months vs. 7 months) were lower for patients with amplified HER-2 tumors compared with tumors without HER-2 amplification. Also, there was a trend toward improved overall survival in patients with HER-2 non-amplified tumors (31 months vs. 25 months) [139].

However, other studies have failed to confirm the relationship between HER-2 overexpression and hormonal therapy resistance [148, 149]. Knoop *et al.* investigated whether HER-2, EGFR, and/or p53 expression are biologically important predictors of tamoxifen resistance [149]. They reviewed the expression of these markers in 1515 samples obtained from high risk postmenopausal breast cancer patients enrolled into the Danish Breast Cancer Cooperative Group's (DBCG) 77c protocol. In this prospective study women were randomly assigned to one year of adjuvant tamoxifen therapy or observation. Neither HER-2, EGFR, nor p53 status predicted tamoxifen benefit or resistance. DFS was similar in all patients treated with tamoxifen regardless of HER-2, EGFR or p53 status.

Given the conflicting data noted in the literature, at this time HER-2 status is not used to help guide endocrine therapy in breast cancer patients.

4.8 Serum HER-2 Extracellular Domain (ECD) as a Biomarker in Breast Cancer

The ECD of HER-2 was first detected in the supernatant from cultured breast cancer cells overexpressing HER-2 [150–152]. The ECD is generated by post-translational proteolytic cleavage of HER-2 at the cell surface resulting in a 110 kDa soluble glycosylated ECD and a 95 kDa transmembrane protein lacking most of the ECD [151, 153]. *In vitro* data suggest that the truncated transmembrane 95 kDa form of the HER-2 has higher transforming activity than the than the full length form of the protein [154, 155]. Langton *et al.* demonstrated that the ECD could be detected by ELISA assay in the serum of nude mice containing HER-2 expressing xenografts but not in the serum of mice without tumors or those that had xenografts that did not express HER-2 [156]. Subsequently, the ECD of HER-2 was detected in the serum of patients with advanced breast cancer using ELSIA assays [77, 157–159]. The measurement of the ECD of HER-2 in patient serum has sparked interest for several reasons. First, ELISA assays on serum samples are a potentially simple quantitative way to measure HER-2 and might be a surrogate means to measure HER-2 amplification. Second, the assay can be repeated easily and so used to follow disease progression or the response to treatment without sequential biopsies of the tumor. The ECD contains the trastuzumab binding site so that high levels of serum ECD could indicate that a significant number of the tumor cells might lack the binding site for trastuzumab and thus be resistant to trastuzumab. Finally, preclinical data suggest that the ECD can competitively inhibit the binding of trastuzumab to tumors that overexpress HER-2, thereby preventing the anti-tumor effects of trastuzumab [160].

The utility of serum ECD measurement as either a prognostic or predictive biomarker has been summarized in the ASCO overview on the use of biomarkers in breast cancer and more recently in a review by Leyland-Jones and Smith [161, 162]. ELISA assays have been used to measure serum ECD with a number of different available assays, there is no standardization of the cutoffs to define the upper limits of normal (and therefore what constitutes elevated serum ECD), and none of the tests has been approved for use by the FDA [162]. Thus comparison across studies is not straightforward. Studies investigating the relationship of baseline measurements of serum ECD have generally found elevated ECD is associated with poor prognostic features (e.g., high tumor burden, high grade, lack of hormone receptors, HER-2 amplification, and lymph node positivity in early-stage cancers) and as a result with poor outcomes [162]. The use of serum ECD measurements as a surrogate for determination of amplification has shown variable results. There are numerous studies finding correlation and others finding no correlation between serum ECD and HER-2 amplification [162]. A recent study by Finn *et al.* compared the elevation of serum ECD (defined as ≥ 16 ng/ml) with amplification directly determined on the tumor cells by approved FISH (positive defined as HER-2/CEP17 ≥ 2.0) or by IHC (positive defined as 3+ staining) measurements [163]. This study found that serum ECD measurements had a sensitivity of 62% and specificity of 76% for the detection of HER-2 amplification. Overall, the specificity and sensitivity of serum ECD is not adequate for serum ECD to serve as a surrogate measure of HER-2 amplification [162]. Also, attempts to use serum ECD measurements to predict response to trastuzumab or lapatinib have not generally shown an association between elevated ECD and clinical outcomes [161–163]. Similarly, neither baseline or serial serum ECD measurements have been consistently associated with clinical outcomes [161, 162].

Overall, the measurement of serum ECD of HER-2 is not recommended for use in breast cancer patients [161, 162].

4.9 HER-2 Amplification as a Prognostic Biomarker and a Predictive Biomarker for Response to HER-2 Targeted Agents in Other Cancers

4.9.1 Gastric Cancer

HER-2 amplification has been reported in between 9–38% of patients based on IHC and/or FISH analysis in gastric cancers (reviewed in [45]). The frequency of amplification or overexpression appears to be related to histologic subtype (e.g., 16% in intestinal vs. 7% in diffuse type) and location (e.g., 25% in gastro-esophageal junction cancers vs. 9% in gastric cancers). A number of studies using IHC, FISH, and CISH have found that HER-2 overexpression is a poor prognostic biomarker [45]. Two phase II studies have tested trastuzumab in combination with chemotherapy in patients with advanced gastric cancer with amplified HER-2 (defined by FISH+ (HER-2/CEP17 ≥ 2.0) or IHC 3+ (defined as strong membrane staining in ≥10% of the tumor cells)) and have shown object responses [45, 164]. A recent phase III study investigated the addition of trastuzumab to chemotherapy in patients with advanced gastric or gastro-esophageal junction cancer [165]. Patients were eligible if they had HER-2 amplification or overexpression as measured by FISH (HER-2/CEP17 ≥ 2.0) or IHC 3+ (defined as strong membrane staining in ≥10% of the tumor cells). Twenty-two percent of the tumors (810/3665) had amplification/overexpression of HER-2 and so were eligible for treatment on the study. The addition of trastuzumab to chemotherapy resulted in a statistically significant improvement in ORR (47 vs. 35%), PFS (6.7 vs. 5.5 months) and OS (16 vs. 11.8 months) [165]. Thus, there is a clear benefit to addition of trastuzumab to chemotherapy in patients with gastric cancers that have HER-2 amplification as defined in this study. Exploratory analyses found that patients with high expression of HER-2 protein (IHC 3+ or IHC 2+ and FISH +) benefited from trastuzumab[165]. By contrast, patients with low expression of HER-2 by IHC (IHC 0 or 1+) but amplified HER-2 by FISH did not benefit from the addition of trastuzumab. The phase II and III studies included only patients with tumors that had HER-2 amplification by either IHC or FISH, and so there are no data that allow assessment of the role of trastuzumab in patients without HER-2 amplification.

4.9.2 Ovarian Cancer

The amplification of the HER-2 gene has been reported in 11–26% of patients with ovarian cancer using Southern analysis, FISH, or CISH [51, 166–178]. A greater range (2–35%) has been reported for overexpression of the protein based primarily on IHC. In most studies, amplification or overexpression of HER-2 is associated with a worse outcome [51, 166–178]. There is limited clinical data with anti-HER-2 targeted agents primarily from phase I and single arm phase II studies of trastuzumab, pertuzumab, and lapatinib [179–184]. One randomized phase II study enrolled patients with platinum resistant ovarian, fallopian tube, or primary peritoneal cancer to treatment with either gemcitabine plus pertuzumab compared with gemcitabine alone [182]. Pertuzumab increased the

ORR compared with gemcitabine alone (16.7% vs. 2.4%), but there was no significant change in PFS or OS. In an exploratory analysis of biomarkers, the expression of EGFR family members was assessed by quantitative RT-PCR [182]. There was no correlation between HER-2 mRNA expression and response to pertuzumab. There was a correlation with response to pertuzumab and low HER-3 mRNA levels. Overall, it is not possible to assess HER-2 amplification or overexpression as a predictive biomarker for these therapies in ovarian cancer.

4.9.3 Non-Small Cell Lung Cancer (NSCLC)

Overexpression or amplification of HER-2 has been described in NSCLC [185, 186]. In a screening study for a phase II study of trastuzumab, 563 NSCLC tumors were screened by IHC (using the Herceptest), FISH (HER-2/CEP17 > 2.0), and ELISA (measuring serum extracellular domain (ECD) of HER-2) (CCR 03) [46]. Of the 410 tumors assessed by IHC, seven (1.7%) showed 3+ staining and 76 (18.5%) had 2+ staining [46]. Of the 378 samples assessed by FISH, seven (1.8%) had amplification [46]. ELISA was considered positive for HER-2 ECD in 16 (3.1%) of the tumors [46]. Overall, these data suggest that the frequency of HER-2 amplification or overexpression is very low in NSCLC. A number of phase II trials of trastuzumab, pertuzumab and lapatinib either alone or in combination with chemotherapy have shown little activity of in these patients [187–194]. However, some of these studies included patients with tumors that were IHC 1+ , 2+, or 3+ , and others included patients with tumors that were IHC 2+ and 3+. Thus, the patients with true HER-2 amplification or overexpression represented only a small fraction of the patients treated on these studies. Therefore, it is not possible to assess the efficacy of HER-2 targeted agents in patients with HER-2 amplification or the role of HER-2 as a biomarker.

4.9.4 Transitional Cell Carcinoma (TCC) of the Urothelium

HER-2 amplification has been described in TCC of the urothelium. In a study of 1005 patients using IHC and FISH, 51 (5.1%) were found to have IHC 3+ staining and gene amplification by FISH (HER-2/CEP17 > 2.2) [195]. 42 samples (4.2%) had IHC 2+ staining, but all were negative by FISH. There are studies to suggest that HER-2 overexpression is a poor prognostic biomarker (e.g., [49] and references therein). However, definitive studies that allow the assessment of HER-2 as a prognostic or predictive biomarker have not been done.

4.9.5 Colorectal Cancer

HER-2 amplification has been described in colorectal cancer, but the frequency of HER-2 amplification appears to be low. One study evaluated 77 tumors using IHC and one (1.3%) had 2+ staining and two (2.6%) had 3+ staining [48]. Another study using FISH (HER-2/CEP > 2.0) to assess HER-2 amplification in colorectal cancer found four of 404 (1.3%) had HER-2 amplification (defined as HER-2/CEP > 2.0) [43]. There are no data to assess HER-2 as a predictive or prognostic biomarker.

4.9.6 Other Tumors

Amplification of HER-2 has been observed in other tumors such as endometrial cancers and salivary gland tumors [44, 47]. However, convincing data to support the role of HER-2 as either a prognostic or predictive marker are not available for other tumors.

4.10 Conclusions

Robust data exist that HER-2 amplification and overexpression have both prognostic and predictive value in breast cancer. For most of the other tumor types, the frequency of HER-2 amplification or overexpression is low. There are data to support that other tumor types with HER-2 amplification have a poor prognosis and may benefit from HER-2 directed therapies. Oncology is undergoing a dramatic transition from a histological based treatment to a molecular-directed approach, where the treatment is selected based on the presence of a target (e.g., amplified HER-2, mutant constitutively active EGFR, activated V600E B-Raf, and so on). The challenge for the future will be to design clinical trials that incorporate the measurement of HER-2 amplification or overexpression, perhaps present in a small percentage of patients with a given tumor type, in a manner that allows rapid analysis of the utility of these measurements as a prognostic and/or predictive biomarker.

Acknowledgments

We thank Marion Nau for her critical reading and editing of this chapter.

Notes Added in Proofs

1. Recent sequencing data has identified activating mutations in HER-2 from patients without HER-2 amplification (Bose *et al.*, *Cancer Discovery* **3**: 224–237, 2013).
2. Since this chapter was submitted, three clinical studies of pertuzumab in HER-2 amplified breast cancer have been published: one demonstrating that pertuzumab in combination with trastuzumab and docetaxel in the first line treatment of metastatic disease improved PFS compared to trastuzumab and docetaxel (Cleopatra Study; Baselga *et al.*, *New Engl J Med* **366**: 109–119, 2012); a second demonstrating that the combination of pertuzumab and trastuzumab has more activity than pertuzumab alone in patients with metastatic disease whose disease had progressed on trastuzumab (Cortes *et al.*, *J Clin Oncol* **30**: 1594–1600, 2012); and a third neoadjuvant study demonstrating that adding pertuzumab to trastuzumab plus docetaxel improved the rate of pathological complete response compared to trastuzumab plus docetaxel (NeoSphere study; Gianni *et al.*, *Lancet Oncol* **13**: 25–32, 2012).
3. Since this chapter was submitted, the BCIRG006 study was published (Slamon *et al.*, *New Engl J Med* **365**: 1273–1283, 2011).

References

1. Carpenter, G., L. King, Jr., and S. Cohen, Epidermal growth factor stimulates phosphorylation in membrane preparations *in vitro*. *Nature*, 1978. **276**(5686): 409–410.

2. Cohen, S., G. Carpenter, and L. King, Jr., Epidermal growth factor-receptor-protein kinase interactions. Co-purification of receptor and epidermal growth factor-enhanced phosphorylation activity. *J Biol Chem*, 1980. **255**(10): 4834–4842.
3. Downward, J., P. Parker, and M.D. Waterfield, Autophosphorylation sites on the epidermal growth factor receptor. *Nature*, 1984. **311**(5985): 483–485.
4. Ushiro, H., and S. Cohen, Identification of phosphotyrosine as a product of epidermal growth factor-activated protein kinase in A-431 cell membranes. *J Biol Chem*, 1980. **255**(18): 8363–8365.
5. Xu, Y.H., *et al.*, Human epidermal growth factor receptor cDNA is homologous to a variety of RNAs overproduced in A431 carcinoma cells. *Nature*, 1984. **309**(5971): 806–810.
6. Downward, J., *et al.*, Close similarity of epidermal growth factor receptor and v-erb-B oncogene protein sequences. *Nature*, 1984. **307**(5951): 521–527.
7. Yamamoto, T., *et al.*, The erbB gene of avian erythroblastosis virus is a member of the src gene family. *Cell*, 1983. **35**(1): 71–78.
8. Shih, C., *et al.*, Transforming genes of carcinomas and neuroblastomas introduced into mouse fibroblasts. *Nature*, 1981. **290**(5803): 261–264.
9. Bargmann, C.I., M.C. Hung, and R.A. Weinberg, The neu oncogene encodes an epidermal growth factor receptor-related protein. *Nature*, 1986. **319**(6050): 226–230.
10. Schechter, A.L., *et al.*, The neu oncogene: an erb-B-related gene encoding a 185,000-Mr tumour antigen. *Nature*, 1984. **312**(5994): 513–516.
11. Coussens, L., *et al.*, Tyrosine kinase receptor with extensive homology to EGF receptor shares chromosomal location with neu oncogene. *Science*, 1985. **230**(4730): 1132–1139.
12. King, C.R., M.H. Kraus, and S.A. Aaronson, Amplification of a novel v-erbB-related gene in a human mammary carcinoma. *Science*, 1985. **229**(4717): 974–976.
13. Semba, K., *et al.*, A v-erbB-related protooncogene, c-erbB-2, is distinct from the c-erbB-1/epidermal growth factor-receptor gene and is amplified in a human salivary gland adenocarcinoma. *Proc Natl Acad Sci U S A*, 1985. **82**(19): 6497–6501.
14. Kraus, M.H., *et al.*, Isolation and characterization of ERBB3, a third member of the ERBB/epidermal growth factor receptor family:

evidence for overexpression in a subset of human mammary tumors. *Proc Natl Acad Sci U S A*, 1989. **86**(23): 9193–9197.

15. Plowman, G.D., *et al.*, Ligand-specific activation of HER4/p180erbB4, a fourth member of the epidermal growth factor receptor family. *Proc Natl Acad Sci U S A*, 1993. **90**(5): 1746–1750.
16. Plowman, G.D., *et al.*, Molecular cloning and expression of an additional epidermal growth factor receptor-related gene. *Proc Natl Acad Sci U S A*, 1990. **87**(13): 4905–4909.
17. Avraham, R., and Y. Yarden, Feedback regulation of EGFR signalling: decision making by early and delayed loops. *Nat Rev Mol Cell Biol*, 2011. **12**(2): 104–117.
18. Pinkas-Kramarski, R., *et al.*, ErbB tyrosine kinases and the two neuregulin families constitute a ligand-receptor network. *Mol Cell Biol*, 1998. **18**(10): 6090–6101.
19. Pinkas-Kramarski, R., *et al.*, Diversification of Neu differentiation factor and epidermal growth factor signaling by combinatorial receptor interactions. *EMBO J*, 1996. **15**(10): 2452–2467.
20. Riese, D.J., 2nd, *et al.*, The cellular response to neuregulins is governed by complex interactions of the erbB receptor family. *Mol Cell Biol*, 1995. **15**(10): 5770–5776.
21. Tzahar, E., *et al.*, A hierarchical network of interreceptor interactions determines signal transduction by Neu differentiation factor/neuregulin and epidermal growth factor. *Mol Cell Biol*, 1996. **16**(10): 5276–5287.
22. Lee, K.F., *et al.*, Requirement for neuregulin receptor erbB2 in neural and cardiac development. *Nature*, 1995. **378**(6555): 394–398.
23. Lin, W., *et al.*, Aberrant development of motor axons and neuromuscular synapses in erbB2-deficient mice. *Proc Natl Acad Sci U S A*, 2000. **97**(3): 1299–1304.
24. Morris, J.K., *et al.*, Rescue of the cardiac defect in ErbB2 mutant mice reveals essential roles of ErbB2 in peripheral nervous system development. *Neuron*, 1999. **23**(2): 273–283.
25. Park, S.K., *et al.*, The erbB2 gene is required for the development of terminally differentiated spinal cord oligodendrocytes. *J Cell Biol*, 2001. **154**(6): 1245–1258.
26. Meyer, D., and C. Birchmeier, Multiple essential functions of neuregulin in development. *Nature*, 1995. **378**(6555): 386–390.

27. Falls, D.L., Neuregulins: functions, forms, and signaling strategies. *Exp Cell Res*, 2003. **284**(1): 14–30.

28. Ozcelik, C., *et al.*, Conditional mutation of the ErbB2 (HER2) receptor in cardiomyocytes leads to dilated cardiomyopathy. *Proc Natl Acad Sci U S A*, 2002. **99**(13): 8880–8885.

29. Liu, F.F., *et al.*, Heterozygous knockout of neuregulin-1 gene in mice exacerbates doxorubicin-induced heart failure. *Am J Physiol Heart Circ Physiol*, 2005. **289**(2): H660–H666.

30. Bird, B.R., and S.M. Swain, Cardiac toxicity in breast cancer survivors: review of potential cardiac problems. *Clin Cancer Res*, 2008. **14**(1): 14–24.

31. Bargmann, C.I., M.C. Hung, and R.A. Weinberg, Multiple independent activations of the neu oncogene by a point mutation altering the transmembrane domain of p185. *Cell*, 1986. **45**(5): 649–657.

32. Bargmann, C.I., and R.A. Weinberg, Oncogenic activation of the neu-encoded receptor protein by point mutation and deletion. *EMBO J*, 1988. **7**(7): 2043–2052.

33. Muller, W.J., *et al.*, Single-step induction of mammary adenocarcinoma in transgenic mice bearing the activated c-neu oncogene. *Cell*, 1988. **54**(1): 105–115.

34. Guy, C.T., *et al.*, Expression of the neu protooncogene in the mammary epithelium of transgenic mice induces metastatic disease. *Proc Natl Acad Sci U S A*, 1992. **89**(22): 10578–10582.

35. Siegel, P.M., *et al.*, Novel activating mutations in the neu proto-oncogene involved in induction of mammary tumors. *Mol Cell Biol*, 1994. **14**(11): 7068–7077.

36. Andrechek, E.R., *et al.*, Amplification of the neu/erbB-2 oncogene in a mouse model of mammary tumorigenesis. *Proc Natl Acad Sci U S A*, 2000. **97**(7): 3444–3449.

37. Lemoine, N.R., *et al.*, Absence of activating transmembrane mutations in the c-erbB-2 proto-oncogene in human breast cancer. *Oncogene*, 1990. **5**(2): 237–239.

38. Zoll, B., *et al.*, Alterations of the c-erbB2 gene in human breast cancer. *J Cancer Res Clin Oncol*, 1992. **118**(6): 468–473.

39. Chazin, V.R., *et al.*, Transformation mediated by the human HER-2 gene independent of the epidermal growth factor receptor. *Oncogene*, 1992. **7**(9): 1859–1866.

40. Ciardiello, F., *et al.*, Additive effects of c-erbB-2, c-Ha-ras, and transforming growth factor-alpha genes on *in vitro* transformation of human mammary epithelial cells. *Mol Carcinog*, 1992. **6**(1): 43–52.

41. Hudziak, R.M., J. Schlessinger, and A. Ullrich, Increased expression of the putative growth factor receptor p185HER2 causes transformation and tumorigenesis of NIH 3T3 cells. *Proc Natl Acad Sci U S A*, 1987. **84**(20): 7159–7163.

42. Pierce, J.H., *et al.*, Oncogenic potential of erbB-2 in human mammary epithelial cells. *Oncogene*, 1991. **6**(7): 1189–1194.

43. Al-Kuraya, K., *et al.*, HER2, TOP2A, CCND1, EGFR and C-MYC oncogene amplification in colorectal cancer. *J Clin Pathol*, 2007. **60**(7): 768–772.

44. Fleming, G.F., *et al.*, Phase II trial of trastuzumab in women with advanced or recurrent, HER2-positive endometrial carcinoma: a Gynecologic Oncology Group study. *Gynecol Oncol*, 2010. **116**(1): 15–20.

45. Gravalos, C., and A. Jimeno, HER2 in gastric cancer: a new prognostic factor and a novel therapeutic target. *Ann Oncol*, 2008. **19**(9): 1523–1529.

46. Heinmoller, P., *et al.*, HER2 status in non-small cell lung cancer: results from patient screening for enrollment to a phase II study of herceptin. *Clin Cancer Res*, 2003. **9**(14): 5238–5243.

47. Press, M.F., *et al.*, Amplification and overexpression of HER-2/neu in carcinomas of the salivary gland: correlation with poor prognosis. *Cancer Res*, 1994. **54**(21): 5675–5682.

48. Schuell, B., *et al.*, HER 2/neu protein expression in colorectal cancer. *BMC Cancer*, 2006. **6**: 123.

49. Skagias, L., *et al.*, Prognostic impact of HER2/neu protein in urothelial bladder cancer. Survival analysis of 80 cases and an overview of almost 20 years' research. *J BUON*, 2009. **14**(3): 457–462.

50. Slamon, D.J., *et al.*, Human breast cancer: correlation of relapse and survival with amplification of the HER-2/neu oncogene. *Science*, 1987. **235**(4785): 177–182.

51. Slamon, D.J., *et al.*, Studies of the HER-2/neu proto-oncogene in human breast and ovarian cancer. *Science*, 1989. **244**(4905): 707–712.

52. Fontaine, J., *et al.*, Gene amplification and expression of the neu (c-erbB-2) sequence in human mammary carcinoma. *Oncology*, 1988. **45**(5): 360–363.

53. Guerin, M., *et al.*, Overexpression of either c-myc or c-erbB-2/neu proto-oncogenes in human breast carcinomas: correlation with poor prognosis. *Oncogene Res*, 1988. **3**(1): 21–31.
54. Wolff, A.C., *et al.*, American society of clinical oncology/college of American pathologists guideline recommendations for human epidermal growth factor receptor 2 testing in breast cancer. *J Clin Oncol*, 2007. **25**(1): 118–145.
55. Baselga, J., *et al.*, Recombinant humanized anti-HER2 antibody (Herceptin) enhances the antitumor activity of paclitaxel and doxorubicin against HER2/neu overexpressing human breast cancer xenografts. *Cancer Res*, 1998. **58**(13): 2825–2831.
56. Press, M.F., *et al.*, Sensitivity of HER-2/neu antibodies in archival tissue samples: potential source of error in immunohistochemical studies of oncogene expression. *Cancer Res*, 1994. **54**(10): 2771–2777.
57. Press, M.F., *et al.*, Evaluation of HER-2/neu gene amplification and overexpression: comparison of frequently used assay methods in a molecularly characterized cohort of breast cancer specimens. *J Clin Oncol*, 2002. **20**(14): 3095–3105.
58. Pauletti, G., *et al.*, Detection and quantitation of HER-2/neu gene amplification in human breast cancer archival material using fluorescence *in situ* hybridization. *Oncogene*, 1996. **13**(1): 63–72.
59. Kallioniemi, O.P., *et al.*, ERBB2 amplification in breast cancer analyzed by fluorescence in situ hybridization. *Proc Natl Acad Sci U S A*, 1992. **89**(12): 5321–5325.
60. Naber, S.P., *et al.*, Strategies for the analysis of oncogene overexpression. Studies of the neu oncogene in breast carcinoma. *Am J Clin Pathol*, 1990. **94**(2): 125–136.
61. Press, M.F., *et al.*, Diagnostic evaluation of HER-2 as a molecular target: an assessment of accuracy and reproducibility of laboratory testing in large, prospective, randomized clinical trials. *Clin Cancer Res*, 2005. **11**(18): 6598–6607.
62. Dietel, M., *et al.*, Comparison of automated silver enhanced in situ hybridisation (SISH) and fluorescence ISH (FISH) for the validation of HER2 gene status in breast carcinoma according to the guidelines of the American Society of Clinical Oncology and the College of American Pathologists. *Virchows Arch*, 2007. **451**(1): 19–25.
63. Isola, J., *et al.*, Interlaboratory comparison of HER-2 oncogene amplification as detected by chromogenic and fluorescence *in situ* hybridization. *Clin Cancer Res*, 2004. **10**(14): 4793–4798.

64. Persons, D.L., *et al.*, HER-2 fluorescence *in situ* hybridization: results from the survey program of the College of American Pathologists. *Arch Pathol Lab Med*, 2006. **130**(3): 325–331.

65. Varley, J.M., *et al.*, Alterations to either c-erbB-2(neu) or c-myc proto-oncogenes in breast carcinomas correlate with poor short-term prognosis. *Oncogene*, 1987. **1**(4): 423–430.

66. Ross, J.S., *et al.*, The HER-2 receptor and breast cancer: ten years of targeted anti-HER-2 therapy and personalized medicine. *Oncologist*, 2009. **14**(4): 320–368.

67. Sorlie, T., *et al.*, Gene expression patterns of breast carcinomas distinguish tumor subclasses with clinical implications. *Proc Natl Acad Sci U S A*, 2001. **98**(19): 10869–10874.

68. Khleif, S.N., J.H. Doroshow, and W.N. Hait, AACR-FDA-NCI Cancer Biomarkers Collaborative consensus report: advancing the use of biomarkers in cancer drug development. *Clin Cancer Res*, 2010. **16**(13): 3299–3318.

69. Hudziak, R.M., *et al.*, P185HER2 monoclonal antibody has antiproliferative effects in vitro and sensitizes human breast tumor cells to tumor necrosis factor. *Mol Cell Biol*, 1989. **9**(3): 1165–1172.

70. Pegram, M., *et al.*, Inhibitory effects of combinations of HER-2/neu antibody and chemotherapeutic agents used for treatment of human breast cancers. *Oncogene*, 1999. **18**(13): 2241–2251.

71. Pietras, R.J., *et al.*, Antibody to HER-2/neu receptor blocks DNA repair after cisplatin in human breast and ovarian cancer cells. *Oncogene*, 1994. **9**(7): 1829–1838.

72. Sliwkowski, M.X., *et al.*, Nonclinical studies addressing the mechanism of action of trastuzumab (Herceptin). *Semin Oncol*, 1999. **26**(4 Suppl 12): 60–70.

73. Baselga, J., *et al.*, Phase II study of efficacy, safety, and pharmacokinetics of trastuzumab monotherapy administered on a 3-weekly schedule. *J Clin Oncol*, 2005. **23**(10): 2162–2171.

74. Baselga, J., *et al.*, Phase II study of weekly intravenous trastuzumab (Herceptin) in patients with HER2/neu-overexpressing metastatic breast cancer. *Semin Oncol*, 1999. **26**(4 Suppl 12): 78–83.

75. Tokuda, Y., *et al.*, Dose escalation and pharmacokinetic study of a humanized anti-HER2 monoclonal antibody in patients with HER2/neu-overexpressing metastatic breast cancer. *Br J Cancer*, 1999. **81**(8): 1419–1425.

76. Vogel, C.L., *et al.*, Efficacy and safety of trastuzumab as a single agent in first-line treatment of HER2-overexpressing metastatic breast cancer. *J Clin Oncol*, 2002. **20**(3): 719–726.
77. Pegram, M.D., *et al.*, Phase II study of receptor-enhanced chemosensitivity using recombinant humanized anti-p185HER2/neu monoclonal antibody plus cisplatin in patients with HER2/neu-overexpressing metastatic breast cancer refractory to chemotherapy treatment. *J Clin Oncol*, 1998. **16**(8): 2659–2671.
78. Slamon, D.J., *et al.*, Use of chemotherapy plus a monoclonal antibody against HER2 for metastatic breast cancer that overexpresses HER2. *N Engl J Med*, 2001. **344**(11): 783–792.
79. Joensuu, H., *et al.*, Adjuvant docetaxel or vinorelbine with or without trastuzumab for breast cancer. *N Engl J Med*, 2006. **354**(8): 809–820.
80. Slamon, D., *et al.*, Phase III trial comparing AC-T with AC-TH and with TCH in the adjuvant treatment of HER2 positive early breast cancer patients: first interim efficacy analysis. *Breast Cancer Res Treat*, 2005. **94**(suppl. 1): A1045.
81. Piccart-Gebhart, M.J., *et al.*, Trastuzumab after adjuvant chemotherapy in HER2-positive breast cancer. *N Engl J Med*, 2005. **353**(16): 1659–1672.
82. Romond, E.H., *et al.*, Trastuzumab plus adjuvant chemotherapy for operable HER2-positive breast cancer. *N Engl J Med*, 2005. **353**(16): 1673–1684.
83. Smith, I., *et al.*, 2-year follow-up of trastuzumab after adjuvant chemotherapy in HER2-positive breast cancer: a randomised controlled trial. *Lancet*, 2007. **369**(9555): 29–36.
84. Buzdar, A.U., *et al.*, Significantly higher pathologic complete remission rate after neoadjuvant therapy with trastuzumab, paclitaxel, and epirubicin chemotherapy: results of a randomized trial in human epidermal growth factor receptor 2-positive operable breast cancer. *J Clin Oncol*, 2005. **23**(16): 3676–3685.
85. Buzdar, A.U., *et al.*, Neoadjuvant therapy with paclitaxel followed by 5-fluorouracil, epirubicin, and cyclophosphamide chemotherapy and concurrent trastuzumab in human epidermal growth factor receptor 2-positive operable breast cancer: an update of the initial randomized study population and data of additional patients treated with the same regimen. *Clin Cancer Res*, 2007. **13**(1): 228–233.
86. Mandrekar, S.J., and D.J. Sargent, Clinical trial designs for predictive biomarker validation: theoretical considerations and practical challenges. *J Clin Oncol*, 2009. **27**(24): 4027–4034.

87. Mass, R.D., *et al.*, Evaluation of clinical outcomes according to HER2 detection by fluorescence *in situ* hybridization in women with metastatic breast cancer treated with trastuzumab. *Clin Breast Cancer*, 2005. **6**(3): 240–246.

88. Paik, S., C. Kim, and N. Wolmark, HER2 status and benefit from adjuvant trastuzumab in breast cancer. *N Engl J Med*, 2008. **358**(13): 1409–1411.

89. Perez, E.A., *et al.*, HER2 and chromosome 17 effect on patient outcome in the N9831 adjuvant trastuzumab trial. *J Clin Oncol*, 2010. **28**(28): 4307–4315.

90. *A Randomized Phase III Trial of Adjuvant Therapy Comparing Chemotherapy Alone (Six Cycles of Docetaxel Plus Cyclophosphamide or Four Cycles of Doxorubicin Plus Cyclophosphamide Followed by Weekly Paclitaxel) to Chemotherapy Plus Trastuzumab in Women with Node-Positive or High-Risk Node-Negative HER2-Low Invasive Breast Cancer.* 2011 [cited 2011 8-27-2011]; NSABP B47]. Available from: http://www.nsabp.pitt.edu/B47_Information.asp.

91. Blackwell, K.L., *et al.*, Randomized study of Lapatinib alone or in combination with trastuzumab in women with ErbB2-positive, trastuzumab-refractory metastatic breast cancer. *J Clin Oncol*, 2010. **28**(7): 1124–1130.

92. Kaufman, B., *et al.*, Lapatinib monotherapy in patients with HER2-overexpressing relapsed or refractory inflammatory breast cancer: final results and survival of the expanded HER2+ cohort in EGF103009, a phase II study. *Lancet Oncol*, 2009. **10**(6): 581–588.

93. Blackwell, K.L., *et al.*, Single-agent lapatinib for HER2-overexpressing advanced or metastatic breast cancer that progressed on first- or second-line trastuzumab-containing regimens. *Ann Oncol*, 2009. **20**(6): 1026–1031.

94. Burstein, H.J., *et al.*, A phase II study of lapatinib monotherapy in chemotherapy-refractory HER2-positive and HER2-negative advanced or metastatic breast cancer. *Ann Oncol*, 2008. **19**(6): 1068–1074.

95. Cameron, D., *et al.*, A phase III randomized comparison of lapatinib plus capecitabine versus capecitabine alone in women with advanced breast cancer that has progressed on trastuzumab: updated efficacy and biomarker analyses. *Breast Cancer Res Treat*, 2008. **112**(3): 533–543.

96. Di Leo, A., *et al.*, Phase III, double-blind, randomized study comparing lapatinib plus paclitaxel with placebo plus paclitaxel as first-line

treatment for metastatic breast cancer. *J Clin Oncol*, 2008. **26**(34): 5544–5552.

97. Geyer, C.E., *et al.*, Lapatinib plus capecitabine for HER2-positive advanced breast cancer. *N Engl J Med*, 2006. **355**(26): 2733–2743.
98. Gomez, H.L., *et al.*, Efficacy and safety of lapatinib as first-line therapy for ErbB2-amplified locally advanced or metastatic breast cancer. *J Clin Oncol*, 2008. **26**(18): 2999–3005.
99. Johnston, S., *et al.*, Lapatinib combined with letrozole versus letrozole and placebo as first-line therapy for postmenopausal hormone receptor-positive metastatic breast cancer. *J Clin Oncol*, 2009. **27**(33): 5538–5546.
100. Johnston, S., *et al.*, Phase II study of predictive biomarker profiles for response targeting human epidermal growth factor receptor 2 (HER-2) in advanced inflammatory breast cancer with lapatinib monotherapy. *J Clin Oncol*, 2008. **26**(7): 1066–1072.
101. Schwartzberg, L.S., *et al.*, Lapatinib plus letrozole as first-line therapy for HER-2+ hormone receptor-positive metastatic breast cancer. *Oncologist*, 2010. **15**(2): 122–129.
102. Toi, M., *et al.*, Lapatinib monotherapy in patients with relapsed, advanced, or metastatic breast cancer: efficacy, safety, and biomarker results from Japanese patients phase II studies. *Br J Cancer*, 2009. **101**(10): 1676–1682.
103. Press, M.F., *et al.*, HER-2 gene amplification, HER-2 and epidermal growth factor receptor mRNA and protein expression, and lapatinib efficacy in women with metastatic breast cancer. *Clin Cancer Res*, 2008. **14**(23): 7861–7870.
104. Agus, D.B., *et al.*, Targeting ligand-activated ErbB2 signaling inhibits breast and prostate tumor growth. *Cancer Cell*, 2002. **2**(2): 127–137.
105. Franklin, M.C., *et al.*, Insights into ErbB signaling from the structure of the ErbB2-pertuzumab complex. *Cancer Cell*, 2004. **5**(4): 317–328.
106. Jackson, J.G., *et al.*, Blockade of epidermal growth factor- or heregulin-dependent ErbB2 activation with the anti-ErbB2 monoclonal antibody 2C4 has divergent downstream signaling and growth effects. *Cancer Res*, 2004. **64**(7): 2601–2609.
107. Baselga, J., *et al.*, Phase II trial of pertuzumab and trastuzumab in patients with human epidermal growth factor receptor 2-positive metastatic breast cancer that progressed during prior trastuzumab therapy. *J Clin Oncol*, 2010. **28**(7): 1138–1144.

108. Portera, C.C., *et al.*, Cardiac toxicity and efficacy of trastuzumab combined with pertuzumab in patients with [corrected] human epidermal growth factor receptor 2-positive metastatic breast cancer. *Clin Cancer Res*, 2008. **14**(9): 2710–6.

109. Gianni, L., *et al.*, Open-label, phase II, multicenter, randomized study of the efficacy and safety of two dose levels of Pertuzumab, a human epidermal growth factor receptor 2 dimerization inhibitor, in patients with human epidermal growth factor receptor 2-negative metastatic breast cancer. *J Clin Oncol*, 2010. **28**(7): 1131–1137.

110. Muss, H.B., *et al.*, c-erbB-2 expression and response to adjuvant therapy in women with node-positive early breast cancer. *N Engl J Med*, 1994. **330**(18): 1260–1266.

111. Paik, S., *et al.*, erbB-2 and response to doxorubicin in patients with axillary lymph node-positive, hormone receptor-negative breast cancer. *J Natl Cancer Inst*, 1998. **90**(18): 1361–1370.

112. Pritchard, K.I., *et al.*, HER2 and responsiveness of breast cancer to adjuvant chemotherapy. *N Engl J Med*, 2006. **354**(20): 2103–2111.

113. Thor, A.D., *et al.*, erbB-2, p53, and efficacy of adjuvant therapy in lymph node-positive breast cancer. *J Natl Cancer Inst*, 1998. **90**(18): 1346–1360.

114. Levine, M.N., *et al.*, Randomized trial of intensive cyclophosphamide, epirubicin, and fluorouracil chemotherapy compared with cyclophosphamide, methotrexate, and fluorouracil in premenopausal women with node-positive breast cancer. National Cancer Institute of Canada Clinical Trials Group. *J Clin Oncol*, 1998. **16**(8): 2651–2658.

115. Gennari, A., *et al.*, HER2 status and efficacy of adjuvant anthracyclines in early breast cancer: a pooled analysis of randomized trials. *J Natl Cancer Inst*, 2008. **100**(1): 14–20.

116. Pegram, M.D., *et al.*, The effect of HER-2/neu overexpression on chemotherapeutic drug sensitivity in human breast and ovarian cancer cells. *Oncogene*, 1997. **15**(5): 537–547.

117. Bieche, I., *et al.*, Two distinct amplified regions at 17q11-q21 involved in human primary breast cancer. *Cancer Res*, 1996. **56**(17): 3886–3890.

118. Hoare, S.F., *et al.*, Identification of genetic changes associated with drug resistance by reverse *in situ* hybridization. *Br J Cancer*, 1997. **75**(2): 275–282.

119. Keith, W.N., *et al.*, Co-amplification of erbB2, topoisomerase II alpha and retinoic acid receptor alpha genes in breast cancer and allelic loss

at topoisomerase I on chromosome 20. *Eur J Cancer*, 1993. **29A**(10): 1469–1475.

120. Murphy, D.S., *et al.*, Interphase cytogenetic analysis of erbB2 and topoII alpha co-amplification in invasive breast cancer and polysomy of chromosome 17 in ductal carcinoma *in situ*. *Int J Cancer*, 1995. **64**(1): 18–26.

121. Smith, K., *et al.*, Topoisomerase II alpha co-amplification with erbB2 in human primary breast cancer and breast cancer cell lines: relationship to m-AMSA and mitoxantrone sensitivity. *Oncogene*, 1993. **8**(4): 933–938.

122. Tomasetto, C., *et al.*, Identification of four novel human genes amplified and overexpressed in breast carcinoma and localized to the q11–q21.3 region of chromosome 17. *Genomics*, 1995. **28**(3): 367–376.

123. van de Vijver, M., *et al.*, Amplification of the neu (c-erbB-2) oncogene in human mammmary tumors is relatively frequent and is often accompanied by amplification of the linked c-erbA oncogene. *Mol Cell Biol*, 1987. **7**(5): 2019–2023.

124. Press, M.F., *et al.*, Alteration of topoisomerase II-alpha gene in human breast cancer: association with responsiveness to anthracycline-based chemotherapy. *J Clin Oncol*, 2011. **29**(7): 859–867.

125. Knoop, A.S., *et al.*, retrospective analysis of topoisomerase IIa amplifications and deletions as predictive markers in primary breast cancer patients randomly assigned to cyclophosphamide, methotrexate, and fluorouracil or cyclophosphamide, epirubicin, and fluorouracil: Danish Breast Cancer Cooperative Group. *J Clin Oncol*, 2005. **23**(30): 7483–7490.

126. O'Malley, F.P., *et al.*, Topoisomerase II alpha and responsiveness of breast cancer to adjuvant chemotherapy. *J Natl Cancer Inst*, 2009. **101**(9): 644–650.

127. Gudkov, A.V., *et al.*, Isolation of genetic suppressor elements, inducing resistance to topoisomerase II-interactive cytotoxic drugs, from human topoisomerase II cDNA. *Proc Natl Acad Sci U S A*, 1993. **90**(8): 3231–3235.

128. Withoff, S., *et al.*, Selection of a subpopulation with fewer DNA topoisomerase II alpha gene copies in a doxorubicin-resistant cell line panel. *Br J Cancer*, 1996. **74**(4): 502–507.

129. Zhou, Z., *et al.*, Adenovirus-mediated human topoisomerase IIalpha gene transfer increases the sensitivity of etoposide-resistant human breast cancer cells. *Cancer Res*, 1999. **59**(18): 4618–4624.

130. Gunnarsdottir, K.A., *et al.*, CEF is superior to CMF for tumours with TOP2A aberrations: a Subpopulation Treatment Effect Pattern Plot (STEPP) analysis on Danish Breast Cancer Cooperative Group Study 89D. *Breast Cancer Res Treat*, 2010. **123**(1): 163–169.

131. O'Malley, F.P., *et al.*, Topoisomerase II alpha protein and responsiveness of breast cancer to adjuvant chemotherapy with CEF compared to CMF in the NCIC CTG randomized MA.5 adjuvant trial. *Breast Cancer Res Treat*, 2011. **128**(2): 401–409.

132. Hayes, D.F., *et al.*, HER2 and response to paclitaxel in node-positive breast cancer. *N Engl J Med*, 2007. **357**(15): 1496–1506.

133. Konecny, G.E., *et al.*, Her-2/neu gene amplification and response to paclitaxel in patients with metastatic breast cancer. *J Natl Cancer Inst*, 2004. **96**(15): 1141–1151.

134. Henderson, I.C., *et al.*, Improved outcomes from adding sequential Paclitaxel but not from escalating Doxorubicin dose in an adjuvant chemotherapy regimen for patients with node-positive primary breast cancer. *J Clin Oncol*, 2003. **21**(6): 976–983.

135. De Laurentiis, M., *et al.*, Taxane-based combinations as adjuvant chemotherapy of early breast cancer: a meta-analysis of randomized trials. *J Clin Oncol*, 2008. **26**(1): 44–53.

136. Sparano, J.A., *et al.*, Weekly paclitaxel in the adjuvant treatment of breast cancer. *N Engl J Med*, 2008. **358**(16): 1663–1671.

137. Yang, S.X., *et al.*, Akt phosphorylation at Ser473 predicts benefit of paclitaxel chemotherapy in node-positive breast cancer. *J Clin Oncol*, 2010. **28**(18): 2974–2981.

138. Carlomagno, C., *et al.*, c-erb B2 overexpression decreases the benefit of adjuvant tamoxifen in early-stage breast cancer without axillary lymph node metastases. *J Clin Oncol*, 1996. **14**(10): 2702–2708.

139. Arpino, G., *et al.*, HER-2 amplification, HER-1 expression, and tamoxifen response in estrogen receptor-positive metastatic breast cancer: a southwest oncology group study. *Clin Cancer Res*, 2004. **10**(17): 5670–5676.

140. De Laurentiis, M., *et al.*, A meta-analysis on the interaction between HER-2 expression and response to endocrine treatment in advanced breast cancer. *Clin Cancer Res*, 2005. **11**(13): 4741–4748.

141. Houston, S.J., *et al.*, Overexpression of c-erbB2 is an independent marker of resistance to endocrine therapy in advanced breast cancer. *Br J Cancer*, 1999. **79**(7–8): 1220–1226.

142. Tetu, B., and J. Brisson, Prognostic significance of HER-2/neu oncoprotein expression in node-positive breast cancer. The influence of the pattern of immunostaining and adjuvant therapy. *Cancer*, 1994. **73**(9): 2359–2365.

143. Read, L.D., *et al.*, Hormonal modulation of HER-2/neu protooncogene messenger ribonucleic acid and p185 protein expression in human breast cancer cell lines. *Cancer Res*, 1990. **50**(13): 3947–3951.

144. Liu, Y., *et al.*, MCF-7 breast cancer cells overexpressing transfected c-erbB-2 have an *in vitro* growth advantage in estrogen-depleted conditions and reduced estrogen-dependence and tamoxifen-sensitivity *in vivo*. *Breast Cancer Res Treat*, 1995. **34**(2): 97–117.

145. Shou, J., *et al.*, Mechanisms of tamoxifen resistance: increased estrogen receptor-HER2/neu cross-talk in ER/HER2-positive breast cancer. *J Natl Cancer Inst*, 2004. **96**(12): 926–935.

146. Jordan, V.C., and B.W. O'Malley, Selective estrogen-receptor modulators and antihormonal resistance in breast cancer. *J Clin Oncol*, 2007. **25**(36): 5815–5824.

147. Osborne, C.K., *et al.*, Role of the estrogen receptor coactivator AIB1 (SRC-3) and HER-2/neu in tamoxifen resistance in breast cancer. *J Natl Cancer Inst*, 2003. **95**(5): 353–361.

148. Berry, D.A., *et al.*, HER-2/neu and p53 expression versus tamoxifen resistance in estrogen receptor-positive, node-positive breast cancer. *J Clin Oncol*, 2000. **18**(20): 3471–3479.

149. Knoop, A.S., *et al.*, Value of epidermal growth factor receptor, HER2, p53, and steroid receptors in predicting the efficacy of tamoxifen in high-risk postmenopausal breast cancer patients. *J Clin Oncol*, 2001. **19**(14): 3376–3384.

150. Lin, Y.Z., and G.M. Clinton, A soluble protein related to the HER-2 proto-oncogene product is released from human breast carcinoma cells. *Oncogene*, 1991. **6**(4): 639–643.

151. Pupa, S.M., *et al.*, The extracellular domain of the c-erbB-2 oncoprotein is released from tumor cells by proteolytic cleavage. *Oncogene*, 1993. **8**(11): 2917–2923.

152. Zabrecky, J.R., *et al.*, The extracellular domain of p185/neu is released from the surface of human breast carcinoma cells, SK-BR-3. *J Biol Chem*, 1991. **266**(3): 1716–1720.

153. Codony-Servat, J., *et al.*, Cleavage of the HER2 ectodomain is a pervanadate-activable process that is inhibited by the tissue inhibitor of metalloproteases-1 in breast cancer cells. *Cancer Res*, 1999. **59**(6): 1196–1201.

154. Di Fiore, P.P., *et al.*, erbB-2 is a potent oncogene when overexpressed in NIH/3T3 cells. *Science*, 1987. **237**(4811): 178–182.
155. Segatto, O., *et al.*, Different structural alterations upregulate *in vitro* tyrosine kinase activity and transforming potency of the erbB-2 gene. *Mol Cell Biol*, 1988. **8**(12): 5570–5574.
156. Langton, B.C., *et al.*, An antigen immunologically related to the external domain of gp185 is shed from nude mouse tumors overexpressing the c-erbB-2 (HER-2/neu) oncogene. *Cancer Res*, 1991. **51**(10): 2593–2598.
157. Baselga, J., *et al.*, Phase II study of weekly intravenous recombinant humanized anti-p185HER2 monoclonal antibody in patients with HER2/neu-overexpressing metastatic breast cancer. *J Clin Oncol*, 1996. **14**(3): 737–744.
158. Leitzel, K., *et al.*, Elevated serum c-erbB-2 antigen levels and decreased response to hormone therapy of breast cancer. *J Clin Oncol*, 1995. **13**(5): 1129–1135.
159. Yamauchi, H., *et al.*, Prediction of response to antiestrogen therapy in advanced breast cancer patients by pretreatment circulating levels of extracellular domain of the HER-2/c-neu protein. *J Clin Oncol*, 1997. **15**(7): 2518–2525.
160. Brodowicz, T., *et al.*, Soluble HER-2/neu neutralizes biologic effects of anti-HER-2/neu antibody on breast cancer cells *in vitro*. *Int J Cancer*, 1997. **73**(6): 875–879.
161. Harris, L., *et al.*, American society of clinical oncology 2007 update of recommendations for the use of tumor markers in breast cancer. *J Clin Oncol*, 2007. **25**(33): 5287–5312.
162. Leyland-Jones, B., and B.R. Smith, Serum HER2 testing in patients with HER2-positive breast cancer: the death knell tolls. *Lancet Oncol*, 2011. **12**(3): 286–295.
163. Finn, R.S., *et al.*, Prognostic and predictive value of HER2 extracellular domain in metastatic breast cancer treated with lapatinib and paclitaxel in a randomized phase III study. *J Clin Oncol*, 2009. **27**(33): 5552–5558.
164. Gravalos, C., *et al.*, Phase II study of trastuzumab and cisplatin as first-line therapy in patients with HER2-positive advanced gastric or gastroesophageal junction cancer. *Clin Transl Oncol*, 2011. **13**(3): 179–184.
165. Bang, Y.J., *et al.*, Trastuzumab in combination with chemotherapy versus chemotherapy alone for treatment of HER2-positive advanced gastric

or gastro-oesophageal junction cancer (ToGA): a phase 3, open-label, randomised controlled trial. *Lancet*, 2010. **376**(9742): 687–697.

166. Berchuck, A., *et al.*, Overexpression of HER-2/neu is associated with poor survival in advanced epithelial ovarian cancer. *Cancer Res*, 1990. **50**(13): 4087–4091.

167. Fajac, A., *et al.*, c-erbB2 gene amplification and protein expression in ovarian epithelial tumors: evaluation of their respective prognostic significance by multivariate analysis. *Int J Cancer*, 1995. **64**(2): 146–151.

168. Felip, E., *et al.*, Overexpression of c-erbB-2 in epithelial ovarian cancer. Prognostic value and relationship with response to chemotherapy. *Cancer*, 1995. **75**(8): 2147–2152.

169. Garcia-Velasco, A., *et al.*, Prognostic value of hormonal receptors, p53, ki67 and HER2/neu expression in epithelial ovarian carcinoma. *Clin Transl Oncol*, 2008. **10**(6): 367–371.

170. Hogdall, E.V., *et al.*, Distribution of HER-2 overexpression in ovarian carcinoma tissue and its prognostic value in patients with ovarian carcinoma: from the Danish MALOVA Ovarian Cancer Study. *Cancer*, 2003. **98**(1): 66–73.

171. Kupryjanczyk, J., *et al.*, TP53 status determines clinical significance of ERBB2 expression in ovarian cancer. *Br J Cancer*, 2004. **91**(11): 1916–1923.

172. Lassus, H., *et al.*, ERBB2 amplification is superior to protein expression status in predicting patient outcome in serous ovarian carcinoma. *Gynecol Oncol*, 2004. **92**(1): 31–39.

173. Meden, H., *et al.*, Overexpression of the oncogene c-erb B2 in primary ovarian cancer: evaluation of the prognostic value in a Cox proportional hazards multiple regression. *Int J Gynecol Pathol*, 1994. **13**(1): 45–53.

174. Nielsen, J.S., *et al.*, Prognostic significance of p53, Her-2, and EGFR overexpression in borderline and epithelial ovarian cancer. *Int J Gynecol Cancer*, 2004. **14**(6): 1086–1096.

175. Tuefferd, M., *et al.*, HER2 status in ovarian carcinomas: a multicenter GINECO study of 320 patients. *PLoS One*, 2007. **2**(11): e1138.

176. Vermeij, J., *et al.*, Genomic activation of the EGFR and HER2-neu genes in a significant proportion of invasive epithelial ovarian cancers. *BMC Cancer*, 2008. **8**: 3.

177. Verri, E., *et al.*, HER2/neu oncoprotein overexpression in epithelial ovarian cancer: evaluation of its prevalence and prognostic significance. Clinical study. *Oncology*, 2005. **68**(2–3): 154–161.

178. Young, S.R., *et al.*, ERBB2 and chromosome 17 centromere studies of ovarian cancer by fluorescence *in situ* hybridization. *Genes Chromosomes Cancer*, 1996. **16**(2): 130–137.

179. Bookman, M.A., *et al.*, Evaluation of monoclonal humanized anti-HER2 antibody, trastuzumab, in patients with recurrent or refractory ovarian or primary peritoneal carcinoma with overexpression of HER2: a phase II trial of the Gynecologic Oncology Group. *J Clin Oncol*, 2003. **21**(2): 283–290.

180. Chu, Q.S., *et al.*, A phase I and pharmacokinetic study of lapatinib in combination with letrozole in patients with advanced cancer. *Clin Cancer Res*, 2008. **14**(14): 4484–4490.

181. Kimball, K.J., *et al.*, A phase I study of lapatinib in combination with carboplatin in women with platinum sensitive recurrent ovarian carcinoma. *Gynecol Oncol*, 2008. **111**(1): 95–101.

182. Makhija, S., *et al.*, Clinical activity of gemcitabine plus pertuzumab in platinum-resistant ovarian cancer, fallopian tube cancer, or primary peritoneal cancer. *J Clin Oncol*, 2010. **28**(7): 1215–1223.

183. Weroha, S.J., *et al.*, Phase II trial of lapatinib and topotecan (LapTop) in patients with platinum-refractory/resistant ovarian and primary peritoneal carcinoma. *Gynecol Oncol*, 2011. **122**(1): 116–120.

184. Yamamoto, N., *et al.*, Phase I and pharmacokinetic study of HER2-targeted rhuMAb 2C4 (Pertuzumab, RO4368451) in Japanese patients with solid tumors. *Jpn J Clin Oncol*, 2009. **39**(4): 260–266.

185. Hirsch, F.R., *et al.*, Evaluation of HER-2/neu gene amplification and protein expression in non-small cell lung carcinomas. *Br J Cancer*, 2002. **86**(9): 1449–1456.

186. Shi, D., *et al.*, Overexpression of the c-erbB-2/neu-encoded p185 protein in primary lung cancer. *Mol Carcinog*, 1992. **5**(3): 213–218.

187. Clamon, G., *et al.*, Lack of trastuzumab activity in nonsmall cell lung carcinoma with overexpression of erb-B2: 39810: a phase II trial of Cancer and Leukemia Group B. *Cancer*, 2005. **103**(8): 1670–1675.

188. Gatzemeier, U., *et al.*, Randomized phase II trial of gemcitabine-cisplatin with or without trastuzumab in HER2-positive non-small-cell lung cancer. *Ann Oncol*, 2004. **15**(1): 19–27.

189. Herbst, R.S., *et al.*, Efficacy and safety of single-agent pertuzumab, a human epidermal receptor dimerization inhibitor, in patients with non small cell lung cancer. *Clin Cancer Res*, 2007. **13**(20): 6175–6181.

190. Krug, L.M., *et al.*, Randomized phase II study of weekly docetaxel plus trastuzumab versus weekly paclitaxel plus trastuzumab in patients

with previously untreated advanced nonsmall cell lung carcinoma. *Cancer*, 2005. **104**(10): 2149–2155.

191. Langer, C.J., *et al.*, Trastuzumab in the treatment of advanced non-small-cell lung cancer: is there a role? Focus on Eastern Cooperative Oncology Group study 2598. *J Clin Oncol*, 2004. **22**(7): 1180–1187.

192. Lara, P.N., Jr., *et al.*, Trastuzumab plus docetaxel in HER2/neu-positive non-small-cell lung cancer: a California Cancer Consortium screening and phase II trial. *Clin Lung Cancer*, 2004. **5**(4): 231–236.

193. Ross, H.J., *et al.*, Randomized phase II multicenter trial of two schedules of lapatinib as first- or second-line monotherapy in patients with advanced or metastatic non-small cell lung cancer. *Clin Cancer Res*, 2010. **16**(6): 1938–1949.

194. Zinner, R.G., *et al.*, Trastuzumab in combination with cisplatin and gemcitabine in patients with Her2-overexpressing, untreated, advanced non-small cell lung cancer: report of a phase II trial and findings regarding optimal identification of patients with Her2-overexpressing disease. *Lung Cancer*, 2004. **44**(1): 99–110.

195. Lae, M., *et al.*, Assessing HER2 gene amplification as a potential target for therapy in invasive urothelial bladder cancer with a standardized methodology: results in 1005 patients. *Ann Oncol*, 2010. **21**(4): 815–819.

196. Mayr, D., *et al.*, Comprehensive immunohistochemical analysis of Her-2/neu oncoprotein overexpression in breast cancer: HercepTest (Dako) for manual testing and Her-2/neuTest 4B5 (Ventana) for Ventana BenchMark automatic staining system with correlation to results of fluorescence in situ hybridization (FISH). *Virchows Arch*, 2009. **454**(3): 241–248.

197. Press, M.F., *et al.*, HER-2/neu gene amplification characterized by fluorescence in situ hybridization: poor prognosis in node-negative breast carcinomas. *J Clin Oncol*, 1997. **15**(8): 2894–2904.

Chapter 5

Hormone Receptors and Endocrine Therapy in Breast Cancer

Sherry X. Yang
National Clinical Target Validation Laboratory,
Division of Cancer Treatment & Diagnosis,
National Cancer Institute, National Institutes of Health,
Bethesda, MD 20892, USA
Sherry.Yang@nih.gov

Breast cancer is the most commonly diagnosed cancer among females in the western countries and the second leading cause of cancer death in women in the United States. In 2013, an estimated 234,580 women are expected to be diagnosed with new cases of invasive breast cancer (http://www.cancer.org/Cancer/BreastCancer) in the United States. Approximately, two thirds of breast cancer cases are estrogen receptor alpha (ER) and/or progesterone receptor (PgR) positive. Patients with ER-positive breast cancer attain substantial benefit from endocrine therapy compared with those with ER-negative disease.[1,2] Assessment of ER status and PgR status is mandatory in newly diagnosed invasive breast cancer, although the data for PgR is not as extensive as that for ER. This chapter summarizes the evolving

Handbook of Therapeutic Biomarkers in Cancer
Edited by Sherry X. Yang and Janet E. Dancey
Copyright © 2013 Pan Stanford Publishing Pte. Ltd.
ISBN 978-981-4364-65-2 (Hardcover), 978-981-4364-66-9 (eBook)
www.panstanford.com

process leading to ER and PgR accepted as prognostic and predictive biomarkers in breast cancer, including a brief discussion of biology of hormone receptors (HR), current practice on ER/PgR-targeted endocrine therapy, and resistant mechanisms of endocrine therapy in breast cancer. We also discuss diagnostic tests of ER/PgR as well as issues evolved during the process of using ER and PgR as the predictive (therapeutic) biomarkers over several decades. At this time, ER-β, another form of ER, will not be discussed in this chapter.

5.1 Introduction

Endocrine therapy in breast cancer dates back to the late 1800s, when Sir George Beatson performed bilateral oophorectomy on two patients with inoperable breast cancer, and one of them had apparent beneficial effects.[3] His hypothesis about the non-nervous connection between the ovaries and breasts was the reason that he performed the surgeries for his patients. Subsequently, laboratory studies provided further evidence on the connection between estrogen and breast cancer. In 1916, Lathrop and Loeb demonstrated that early ovariectomy significantly reduced the incidence of mammary cancer development in mice.[4] Moreover, Antoine Lacassagne illustrated the relationship between estrogen and development of adenocarcinoma of the breast also using mouse model and suggested that deprivation of estrogen could be a useful treatment of breast cancer.[5] However, it was the identification of ER and discovery of "anti-estrogen" drug tamoxifen in the late 1960s and early 1970s that marked the beginning of a long and tortuous journey of ER and PgR as the prototype molecular targets and, in the meantime, as the therapeutic biomarkers of endocrine therapy in breast cancer.[6–8]

5.2 Biology of Hormone Receptors

ER and PgR are members of a large family of the nuclear hormone receptors that regulate gene expression. ER is a 66 kDa protein composed of five functional domains (A through E), which is encoded by *ESR*1 gene (Fig. 5.1). The N-terminal A/B domain is able to transactivate gene transcription in the absence of ligand estrogen (AF1). The C domain, also known as the DNA-binding domain, binds to estrogen response elements (EREs) in DNA. The D domain is a hinge region that connects the C and E domains. The E domain contains the ligand binding cavity as well as binding sites for coactivator and corepressor proteins. Upon binding of a lipid-soluble hormone 17β-estradiol to the ligand-binding domain (AF2) of the receptors, the hormone-bound receptor undergoes a conformational change that allows dimerization and translocation to the nucleus, and binds to EREs in the promoter regions of target genes.[9] In particular, the transcriptional activity of ER is enhanced by binding of coactivators such as SRC1, SRC2, and AIB1 to the transcriptional complex. Examples of ER-regulated genes are PgR, IGF-1 receptor (IGF1R), cyclin D1, antiapoptotic factor Bcl-2, and vascular endothelial growth receptor.[10–12] Thus, transcriptional activation of ER-responsive genes leads to the growth and survival of normal as well as mammary tumor cells.[13,14]

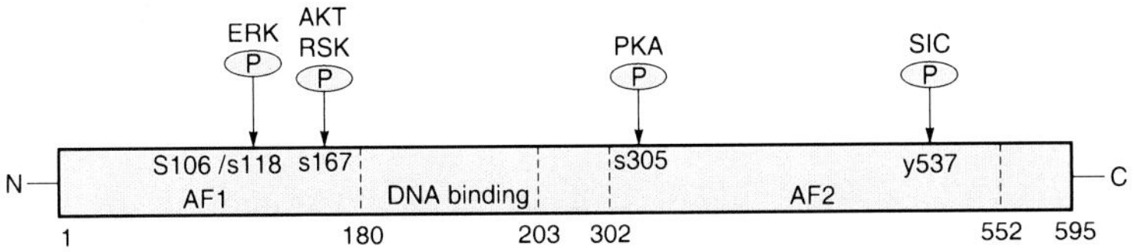

Figure 5.1 The functional domains and phosphorylations sites of estrogen receptor. ERK, extracellular signal-regulated kinases; PKA, protein kinase A; RSK, ribosomal protein S6 kinase.

5.3 ER/PgR as Prognostic and Therapeutic Biomarkers

In the era of personalized medicine, selection of patients who benefit from a drug is of critical importance for treatment success. Perhaps ER, the prototype of molecular target in cancer, is one of the best characterized biomarkers in oncology. Prognostic biomarkers are disease-specific markers and predict disease outcome in the absence of therapeutic interventions. Predictive biomarkers are usually treatment-specific and predict whether a patient will respond to a specific therapy. ER and PgR have been established as both prognostic biomarkers and, in particular, predictive biomarkers in breast cancer, which are discussed in greater detail in the following sections.

5.3.1 ER and PgR as Prognostic Factors in Breast Cancer

ER-positive (ER+) tumors (Fig. 5.2) are relatively well differentiated with less aggressive features compared with ER-negative (ER-) tumors, and comprise up to 75% of the breast cancer cases worldwide.[15,16] Several retrospective population-based studies have demonstrated that hormone receptor status is an independent prognostic indicator of outcome in women with breast cancer. Patients with positive hormone receptors tend to have a better prognosis: better survival overall, within each stage, and in the younger and older age groups.[15,17,18] Breast cancer, recently using cDNA microarray technology, has been classified as luminal A and B, HER2, basal, and normal-like molecular subtypes.[19] Luminal A and B are recognized as two biologically distinct ER-positive subtypes of breast cancer; luminal A tumors have lower proliferation rates and better prognosis than luminal B tumors. The 10-year breast cancer-specific survival was 79% for luminal A, 64% for luminal B,

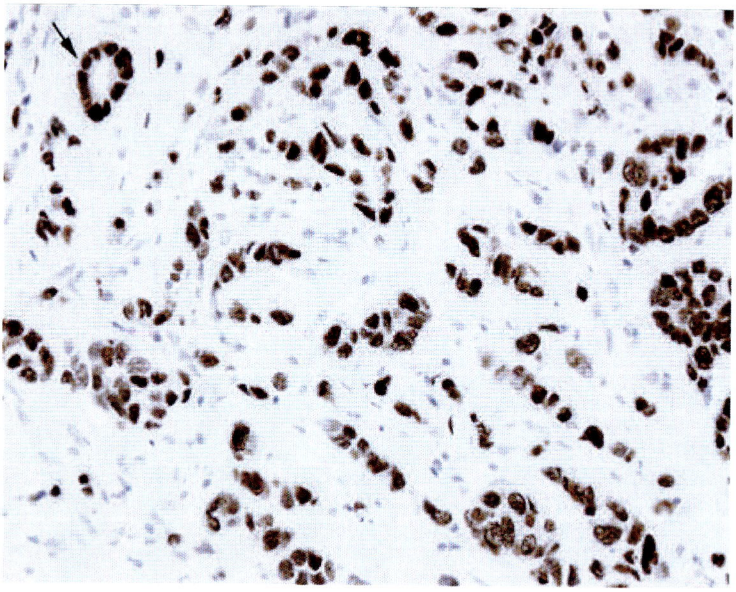

Figure 5.2 Expression of ER in human breast tumor cells and epithelial cells of a benign lesion (arrow) detected by immunohistochemistry using monoclonal antibodies to human ER, clones 1D5 and ER-2–123 (ER PharmDx™ Kit, DAKO).

and 57% for luminal-HER2 subtypes, among patients who received adjuvant tamoxifen therapy.[20] In a study of 1,434 consecutive patients with invasive breast cancer who received breast conserving therapy, 5-year cumulative local recurrence was 0.8% for luminal A, and 2.3% for luminal B.[21] The molecular subtypes defined by the gene expression profiling approach provide understanding of the molecular basis of breast cancer, and improving prognosis in breast cancer.[22]

5.3.2 ER and PgR as Therapeutic Biomarkers in Breast Cancer

Evaluation of hormonal receptor status as predictive/therapeutic biomarkers became possible when pharmacologic agents that bind to ER were developed. Tamoxifen, a failed contraceptive drug that had been shown to have active anti-tumor activity with low toxicities, was approved for hormonal treatment of advanced breast cancer in the United Kingdom in 1973 and in the United States in 1977.[7,8,23,24] It is a selective ER modulator (SERM) and exerts anti-tumor function through its antagonistic effect on breast tissues. Tamoxifen has agonist effect on other tissues such as uterus.[25] The connection between the presence of hormonal receptors and response to tamoxifen has been gradually established. Patients with ER-positive disease derived benefits from tamoxifen while paucity of hormonal receptor was associated with a lack of response.[13,14,26]

In 1998, an Oxford Overview conducted by the Early Breast Cancer Trialists' Collaborative Group (EBCTCG) consolidated the evidence of adjuvant tamoxifen treatment and revealed that tamoxifen improved 10-year survival of women with ER-positive tumors and unknown ER status. The proportional reductions of breast cancer recurrence and mortality appeared to be largely unaffected by other patient characteristics or treatments.[27] The updated meta-analyses, including 56 randomized trials with 48,000 patients, further confirmed the absolute clinical benefit of tamoxifen in women with ER+ breast cancer.[2] Tamoxifen reduced ER+ breast cancer recurrence by 11.8%, and reduced breast cancer mortality by 9.2% (Fig. 5.3). Furthermore, benefit from tamoxifen was independent of age, nodal status and use of chemotherapy. Therefore, the relationship between hormone receptor expression and efficacy of endocrine therapy is firmly established by these convincing clinical data.

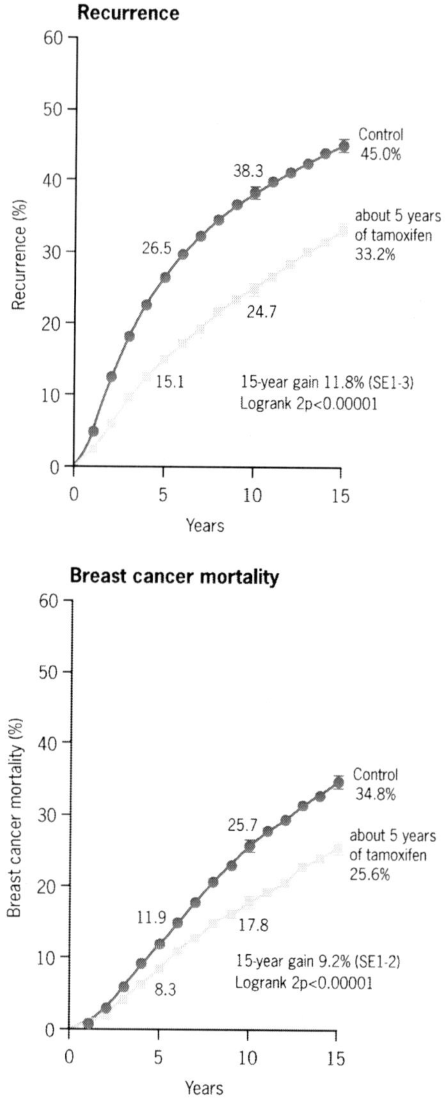

Figure 5.3 About 5 years of tamoxifen versus not in ER-positive (or ER-unknown) disease: 15-year probabilities of recurrence and of breast cancer mortality. Reprinted from *The Lancet* 365, Early Breast Cancer Trialists' Collaborative Group (EBCTCG), Effects of chemotherapy and hormonal therapy for early breast cancer on recurrence and 15-year survival: an overview of the randomised trials, 1687–717, copyright (2005), with permission from Elsevier.

5.4 ER/PgR Targeted Therapy for Breast Cancer

Determination of ER/PgR status is mandatory for all newly diagnosed invasive breast cancer, and on metastatic lesions if they influence the treatment decisions.[28] Immunohistochemistry (IHC) assays are recommended for use in testing ER/PgR contents. American Society of Clinical Oncology/College of American Pathologists Guideline recommends that tumors are defined as ER/PgR positive if there are at least 1% positive tumor nuclei by IHC.[29,30] Endocrine therapy should be given to patients with ER-positive and/or PgR-positive invasive breast cancer regardless of age, lymph node status, or whether or not adjuvant chemotherapy is to be administered.

Currently, major ER-targeted endocrine pharmacological therapies are with the use of SERMs, aromatase inhibitors (AIs), and selective ER down-regulators (SERDs). SERMs such as tamoxifen directly bind to ER and block estrogen from binding to its receptors and alter the molecular conformation of the receptor. This results in preferential recruitment of the corepressor proteins rather than the coactivator complex and, therefore, inhibits estrogen-stimulated tumor growth.[31] AIs reduce the production of estrogen by blocking the aromatization of androstenedione and testosterone—estrogen precursors in peripheral tissues.[32] SERDs that bind to ER with high affinity are pure steroidal antiestrogen with no agonist effects, leading to both down-regulation and degradation of the receptor.[33-35] The FDA-approved ER/PgR-targeted drugs for the treatment of breast cancer are listed in Table 5.1.

Selection for appropriate endocrine therapy depends on the disease state (early versus recurrent or metastatic) and patient endocrine status (premenopausal versus postmenopausal). In premenopausal women, most estrogen is produced in the ovaries in response to pituitary-derived luteinizing and follicle-stimulating hormones. In postmenopausal women, estrogen is produced mainly in peripheral tissues by aromatase conversion of rostenedione and testosterone produced in the adrenal gland. Inhibition of peripheral conversion of estrogen precursors to estrogen in premenopausal patients may lead to a reduced feedback of estrogen to the hypothalamus and pituitary axis and, consequently, stimulation of ovarian estrogen production.[36] Because of this concern, use of AIs should be avoided in premenopausal patients with functional ovaries.

Table 5.1 FDA-approved ER/PgR-targeted drugs for treatment of breast cancer*

Drug	Class/ mechanisms of action	Indications	Dose	Major side effects
Tamoxifen (Nolvadex®)	SERM: -Competitive inhibitors of estrogen binding to ERs, and mixed agonist and antagonist activity, depending on the target tissue -Tamoxifen is metabolized to its active metabolite, endoxifen by cytochrome P450 2D6 (CYP2D6)	**Advanced/Metastatic:** Premenopausal women with metastatic breast cancer **Adjuvant:** Node-positive and node-negative breast cancer in women following mastectomy or segmental mastectomy, axillary dissection, and breast irradiation **DCIS:** Women with DCIS following breast surgery and radiation **Chemoprevention:** Women at high risk for breast cancer (with a 5-year predicted risk of breast cancer ≥ 1.67% calculated by the Gail Model)	20–40 mg daily (20 mg daily or 10–20 mg twice daily)	-Hot flashes -Vaginal discharge -Thromboembolic event -Endometrial cancer

Drug	Class/ mechanisms of action	Indications	Dose	Major side effects
Toremifene (Fareston®)	SERM: -Nonsteroidal triphenylethylene derivative -Binds to ER and may exert estrogenic, antiestrogenic, or both activities, depending upon the duration of treatment, animal species, gender, target organ, or endpoint selected -The antitumor effect is believed to be mainly due to its ability to compete with estrogen for binding sites in the cancer, blocking the growth-stimulating effects of estrogen in the tumor	**Advanced/Metastatic:** Postmenopausal women with estrogen-receptor positive or unknown metastatic breast cancer	60 mg once daily	-Hot flashes -Vaginal discharge -Thromboembolic event -Endometrial cancer

(Continued)

Table 5.1 (*Continued*)

Drug	Class/ mechanisms of action	Indications	Dose	Major side effects
Letrozole (Femara®)	AI: A nonsteroidal competitive inhibitor of the aromatase enzyme system.	**Adjuvant:** - Postmenopausal women with hormone-receptor positive early breast cancer - Extended adjuvant treatment of early breast cancer in postmenopausal women, who have received 5 years of adjuvant tamoxifen therapy **Advanced/Metastatic:** - First-line treatment of postmenopausal women with hormone receptor positive or unknown, locally advanced or metastatic breast cancer. - Advanced breast cancer in postmenopausal women with disease progression following anti-estrogen therapy	2.5 mg once daily	-Decrease in bone mineral density -Hot flushes -Arthralgia, arthritis and myalgia

Drug	Class/ mechanisms of action	Indications	Dose	Major side effects
Anastrozole (Arimidex®)	AI: A selective non-steroidal AI	**Adjuvant:** Postmenopausal women with hormone-receptor positive early breast cancer **Advanced/Metastatic:** - First-line treatment of postmenopausal women with hormone receptor-positive or hormone receptor unknown locally advanced or metastatic - Second-line treatment in postmenopausal women with disease progression following tamoxifen therapy	1 mg once daily	-Decrease in bone mineral density -Fatigue / asthenia - Hot flushes -Arthralgia - Musculoskeletal pain

(*Continued*)

Table 5.1 *(Continued)*

Drug	Class/ mechanisms of action	Indications	Dose	Major side effects
Exemestane (Aromasin®)	AI: An irreversible, steroidal aromatase inactivator	**Adjuvant:** Postmenopausal women with estrogen receptor-positive early breast cancer who have received two to 3 years of tamoxifen and are switched to AROMASIN for completion of a total of five consecutive years of adjuvant hormonal therapy. **Advanced/Metastatic:** Advanced breast cancer in postmenopausal women whose disease has progressed following tamoxifen therapy.	1 mg once daily	-Decrease in bone mineral density -Fatigue / asthenia - Hot flushes - Evaluation in serum liver function test
Fulvestrant (Faslodex®)	SERD: Binding to ER competitively; rapid degradation of the bounded receptor leading to the loss of ER protein	**Advanced/Metastatic:** Hormone receptor-positive metastatic breast cancer in postmenopausal women with disease progression following anti-estrogen therapy	500 mg intra-muscular, on days 1, 15, 29, first month, then 500 mg monthly	-Injection site pain -Nausea - Bone pain -Arthralgia -Musculoskeletal pain -Hot flushes

*Detailed information of the listed drugs can be found in the package usage sheets of each drug.

5.4.1 Adjuvant Endocrine Therapy

After local and regional treatment, some patients eventually develop local recurrence and/or distant metastasis. Adjuvant systemic endocrine and chemotherapy are administered in order to achieve treatment benefit in terms of increasing disease-free survival (DFS) and/or overall survival (OS). The benefit from tamoxifen treatment has been established since the meta-analysis conducted by the EBCTCG through an international collaborative effort.[2,27] In addition, clinical evidence has demonstrated that chemotherapy is less effective in some patients with HR+ and HER2-negative early breast cancer.[37,38] Low-risk patients identified by the recently developed Oncotype DX and MammaPrint assays can get cured with endocrine therapy alone after locoregional therapy.[39,40]

5.4.1.1 Premenopausal patients

Based on the existing evidence, tamoxifen 20 mg daily for 5 years is recommended as adjuvant endocrine therapy for premenopausal women with hormone receptor-positive invasive breast cancer by major practice guidelines. The National Comprehensive Cancer Network (NCCN) recommends that tamoxifen can be used alone if tumors are small (≤0.5 cm) or have micro-invasion with minimal lymph node involvement (≤ 2 mm in axillary node metastasis). For patients with node-positive disease, tamoxifen is given after chemotherapy and radiation therapy. If tumors are >0.5 cm but with minimal lymph node involvement, tamoxifen is indicated regardless of whether chemotherapy is recommended (http://www.nccn.org/professionals/physician_gls/f_guidelines.asp). Tamoxifen is well tolerated in general. The major known side effects include hot flashes, thromboembolic event and endometrial cancer (Table 5.1). Tamoxifen is teratogenic and reduces bone density in premenopausal women.[41] For premenopausal patients taking tamoxifen, pregnancy should be avoided. Bisphosphonates may be considered to prevent bone loss. Because cytochrome P450 2D6 (CYP2D6) is required for conversion of tamoxifen to its active metabolite, endoxifen, patients with low levels of CYP2D6 activity may have reduced production of endoxifen. CYP2D6 activity depends on the genetic variability of patients and co-administration of other drugs that inhibit CYP2D6. Although single nucleotide polymorphisms (SNPs) of CYP2D6 affect the effectiveness of tamoxifen endocrine therapy by retrospective analyses, prospective evidence has yet to be established.[42,43] The clinical impact of co-administration of CYP2D6 inhibitors, e.g. selective serotonin reuptake inhibitors (SSRIs), has not been fully evaluated at this time. Nonetheless, drugs that inhibit CYP2D6 should be used with caution.

In addition to tamoxifen, ovarian suppression/ablation is also an effective adjuvant endocrine therapy for premenopausal women with hormone-receptor positive breast cancer.[2] Surgical oophorectomy and pharmacologic ovarian suppression with luteinizing hormone-releasing hormone (LHRH) agonists are commonly used approaches.[44] While surgical oophorectomy permanently stops ovarian steroid production, pharmacological ovarian function suppression using agents such as the LHRH agents cause reversible and non-permanent ovarian dysfunction. Therefore, suppression of ovarian function may be incomplete or may require significant amount of time in some women. If pharmacological agents are used for this purpose, biochemical verification of ovarian function suppression through laboratory evaluation should be performed. Radiation can also be used to induce ovarian ablation; however, it is less frequently used.[45]

5.4.1.2 Postmenopausal patients

Tamoxifen is approved for adjuvant treatment of node-positive or node-negative breast cancer following total mastectomy or segmental mastectomy, axillary dissection, and/or breast irradiation in postmenopausal women with breast cancer. Three aromatase inhibitors, letrozole (Femara®), anastrozole (Arimidex®), and exemestane (Aromasin®) are also approved by the FDA for the adjuvant treatment of hormone receptor-positive breast cancer in the postmenopausal patient population.

5.4.1.2.1 *Tamoxifen*

Early randomized clinical trials clearly demonstrated that adjuvant tamoxifen treatment for 5 years reduces the risk of disease recurrence and prolongs survival.[46,47] As discussed in Section 5.3.2, the Oxford overviews fully established the effectiveness of tamoxifen as adjuvant therapy. For postmenopausal patients, the indication of adjuvant tamoxifen is the same as that for premenopausal women as mentioned earlier. Side effects of tamoxifen are similar to those of premenopausal patients except that tamoxifen does not reduce bone mineral density in the postmenopausal women.[41]

5.4.1.2.2 *Aromatase inhibitors*

The utility of AIs for the treatment of hormone receptor-positive breast cancer is based on the scientific rationale discussed in Section 5.4. The approved AIs are either nonsteroidal (anastrozole and letrozole) or steroidal (exemestane) aromatase inhibitors. Several randomized clinical trials, and meta-analyses evaluated AIs as adjuvant endocrine therapy for postmenopausal patients with early stage breast cancer (see their indications and usages in Table 5.1).

- Anastrozole

The Anastrozole, Tamoxifen alone or in Combination (ATAC) trial, a three-arm study, compared anastrozole (1 mg daily), or anastrozole and tamoxifen (20 mg daily) with tamoxifen for 5 years in postmenopausal women with ER+ or ER unknown early breast cancer.[48,49] The majority (84%) of patients were hormone receptor positive. Anastrozole compared with tamoxifen significantly improved DFS, increased time to recurrence, reduced distant metastasis and contralateral breast cancer (42% reduction) in HR+ patients. There were no differences in OS for patients taking anastrozle compared with tamoxifen. No benefit was observed for the combination arm relative to tamoxifen arm. Anastrozole caused significantly higher arthralgia and fractures, while tamoxifen was associated with more hot flushes, vaginal bleeding and discharge, endometrial cancer, venous thromboemobolic and deep venous thromboembolic events.[49]

Several studies examined the effect of switching to anastrozole after 2 to 3 years of tamoxifen for a total of 5 years of endocrine therapy.[50,51] A meta-analysis of these studies, which evaluated 4,006 patients, showed that women who switched to anastrozole had significant improvement in DFS, distant recurrence-free survival, event-free survival and OS compared with patients who remained on tamoxifen.[52]

- Letrozole

Breast International Group (BIG) 1–98 trial evaluated the use of tamoxifen alone for 5 years, letrozole alone for 5 years, or tamoxifen for 2 years followed sequentially by letrozole for 3 years or letrozole for 2 years followed by tamoxifen for 3 years. Letrozole for 5 years, compared with tamoxifen for 5 years, significantly improved DFS by an initial analysis of 8,010 postmenopausal women and the updated analysis of 4,922 patients. In both analyses, OS was not statistically significant.[53,54]

MA-17 trial evaluated the effect of letrozole after patients who had received tamoxifen for 4.5 to 6 years. The planned interim analysis with a median follow-up of 2.4 years showed a significant improvement in DFS, favoring letrozole to placebo.[55] The updated analysis confirmed the clinical benefit of the extended therapy with letrozole.[56] More low-grade hot flashes, arthritis, arthralgia, and myalgia were observed in patients on letrozole. It appears that there were more new diagnoses of osteoporosis on letrozole than on placebo (5.8% versus 4.5%), but the difference did not reach statistical significance.[55]

- Exemestane

The Intergroup Exemestane Study (IES) evaluated the effect of exemestane after 2 to 3 years of tamoxifen for a total of 5 years versus tamoxifen for 5 years in 4,724 postmenopausal women with ER-positive or ER unknown breast cancer. After a median follow-up of 30.6 months, a significant improvement in DFS favoring exemestane arm was reported.[57] In a subsequent analysis after a median follow-up of 55.7 months, DFS benefit was confirmed although OS benefit was not statistically significant.[58] Patients on exemestane had more arthralgia, diarrhea, hypertension, fractures, arthritis, musculoskeletal pain, carpal tunnel syndrome, insomnia, and osteoporosis. Patients on tamoxifen had more gynecologic symptoms, thromboembolic disease, endometrial hyperplasia and uterine polyps.

These studies demonstrated that use of an AI in postmenopausal women with hormone receptor-positive early breast cancer reduced risk of recurrence (ipsilateral and contralateral) and distant metastasis compared with tamoxifen alone. These benefits were observed when AIs are used as initial therapy (anastrozole or letrozole), sequential therapy (anastrozole, letrozole or exemestane), or extended therapy (letrozole). However, optimal choice of using AI endocrine therapy for postmenopausal women is not well established. Based on the existing clinical evidence, it is recommended that an AI should be used as initial adjuvant endocrine therapy for 5 years, or AI for 2–3 years followed by tamoxifen, or tamoxifen 2–3 years followed by an AI, or as extended therapy for postmenopausal women with hormone receptor-positive early breast cancer. Tamoxifen alone for 5 years is recommended for patients who decline or who have a contraindication to AIs (http://www.nccn.org/professionals/physician_gls/f_guidelines.asp).[59,60]

The side effect profiles of AIs are different from those of tamoxifen (Table 5.1). The major side effects of AIs are bone loss and musculoskeletal symptoms, including joint pain or stiffness and myalgia as described earlier in this chapter. There might be a slight increase in the risk of ischemic cardiovascular disease with AIs compared with tamoxifen.[61] A recent meta-analysis on seven randomized trials, including 30,032 patients, found that longer duration of AIs, compared with tamoxifen for 5 years, is significantly associated with an increased risk of cardiovascular disease and bone fracture but decreased odds of venous thrombosis and endometrial cancer. Despite a significant improvement in DFS, the cumulative toxicity from the upfront use of AIs has been attributed to the lack of OS benefit. The data suggest that a switch from tamoxifen to AIs reduces toxicity and is likely the best balance between efficacy and toxicity.[62]

5.4.2 Endocrine Therapy for Recurrent and Metastatic Disease

There are three classes of ER/PgR-targeted agents to choose from for treatment of hormonal receptor-positive recurrent or metastatic breast cancer. These are SERMs (tamoxifen and toremifene), AIs (letrozole, anastrozole and exemestane) and SERDs (fulvestrant). The initial endocrine treatment options in premenopausal women are tamoxifen (or toremifene), or ovarian function suppression with oophorectomy or gonadotropin releasing hormone (GnRH) agonists. Although AIs reduce estrogen production effectively for the reasons discussed in Section 5.4, they are not recommended for use in premenopausal women without a permanent suppression of ovarian function.

For postmenopausal women with advanced disease, either tamoxifen or an AI is an appropriate option.[63–65] A meta-analysis of first-line randomized studies comparing any type of AIs versus standard treatment with tamoxifen in advanced breast cancer demonstrated a slight but statistically significant survival benefit, with 11% relative hazard reduction. Treatment with AIs is associated with superior response rates, TTP, and overall survival compared with first-line tamoxifen.[66]

The pure anti-estrogen drug, fulvestrant (Faslodex®), as mentioned earlier, competes with estrogen for the ER, effectively blocks ER dimerization and DNA binding, increases ER turnover, and inhibits nuclear uptake of the receptor.[33,34] It is approved for postmenopausal patients with disease progression following anti-estrogen therapy. Fulvestrant is an active agent in patients who are refractory to tamoxifen or an AI.[67–69] Fulvastrant given at 500 mg every 14 days for the first month followed by every 28 days thereafter, significantly increased progression-free survival compared with 250 mg every 28 days.[70] When fulvestrant was compared with tamoxifen as the first-line treatment for metastatic disease, no differences in response rates, TTP and treatment tolerability were observed.[71] One advantage of fulvestrant is the monthly intramuscular injection instead of daily intake.

In general, multiple lines of endocrine therapies can be used for patients with ER+/PgR+ advanced disease before the initiation of systemic chemotherapy. The objective response rates of the first line therapy are higher (21 to 33%, 33%, 32%, and 46% for tamoxifen, anastrozole, letrozole and exemestane, respectively) than those of the second line treatment (10.3–12%, 19%, and 23% for anastrozole, letrozole and exemestane, respectively).[64,66,72–76] A meta-analysis of 37 trials, including 11,403 women with advanced breast cancer, showed a significant survival benefit for treatment with an AI over other endocrine therapies.[63]

The selection of treatment choices also takes into consideration of the toxicity profiles (Table 5.1), tumor burden/location, and history of response to hormonal treatment. Considerations are also given for a third or fourth-line endocrine therapy using progestins, androgens or estrogens for selected patients. The various endocrine therapies differ somewhat in terms of efficacy and substantially vary with regard to tolerability. Because the toxicities of endocrine therapy are usually tolerable, some physicians would prefer exhaust all endocrine treatment options before considering chemotherapy.

Based on the current clinical data, AIs are recommended as the treatment of choice if a patient has relapsed while receiving adjuvant tamoxifen or has not received any adjuvant endocrine treatment or has relapsed more than 1 year after discontinuing adjuvant tamoxifen or an AI. If a patient has relapsed during or within 12 months of receiving an AI in the adjuvant setting, a SERM such as tamoxifen or toremifene, or fulvestrant would be appropriate as the first-line treatment options (Breast Cancer, http://www.nccn.org/professionals/physician_gls/f_guidelines.asp).

5.4.3 Neoadjuvant Endocrine Therapy

Neoadjuvant endocrine therapy is still largely considered under experimental clinical trial settings. In early studies, neoadjuvant tamoxifen yielded clinical response rate of 30% or higher in older women with breast cancer, who were too frail to undergo other treatment.[77,78] Subsequent randomized studies comparing tamoxifen alone and surgery followed by tamoxifen in elderly patients with operable breast cancer showed that while surgery is better for local disease control, OS was similar in the two treatment groups.[79–81] Several studies compared AIs with tamoxifen in younger and healthier postmenopausal women with HR+ bulky disease. These studies and a subsequent meta-analysis concluded that AIs were more effective than tamoxifen for improvement in the rate of breast cancer conservation surgeries.[82–84]

Several studies evaluated factors that predict long-term outcome after neoadjuvant endocrine therapy. Unlike neoadjuvant chemotherapy in which pathological complete response (pCR) is a strong predictor for survival, neoadjuvant endocrine therapy resulted in low pCR rate and pCR was not associated with survival.[85] Recent studies identified pathological tumor size, pathological node status, and Ki67 and ER status in the surgical resection specimen as having independent prognostic values for relapse or death after recurrence post endocrine neoadjuvant therapy.[86,87] As the molecular heterogeneity of breast cancer is increasingly recognized, neoadjuvant therapy provides a perfect clinical setting to assess molecular factors that predict endocrine treatment response.[88]

5.5 Endocrine Therapy Resistance

Although the benefit of endocrine therapy in HR+ breast cancer is clearly established, only 50 to 60% ER+ and/or PgR+ tumors respond to endocrine therapy.[27] Tumors with low histological grade and with a low level of proliferative activity are likely to be sensitive to endocrine therapy.[89] These observations are consistent with the scientific rationale that breast tumors relying on hormones for growth likely respond to endocrine therapy. Low levels of ER or PgR expression and positive HER2 status have been associated with less responsiveness to endocrine therapy.

All patients who responded to endocrine therapy for the treatment of an advanced disease will eventually become resistant. Up to 40% of patients relapse and develop tamoxifen-resistant disease among those who received adjuvant tamoxifen therapy.[90] A lack of ER expression has been attributed to both *de novo* and intrinsic resistance. Studies have shown that epigenetic alterations through CpG island hypermethylation appear to be, at least in part, responsible for the lack of ER expression.[91–94] Many theories, most of them using preclinical models, have been tested to elucidate the acquired resistance mechanisms to endocrine therapy. Loss of ER expression occurs in about 15% to 20% of breast cancer and loss of function by gene mutation represents <1% of ER-positive tumors.[95–97] The other mechanisms of *de novo* and acquired resistances include (i) continued ER signaling in the presence of ER antagonists; (ii) use of non-ER pathways that circumvent the reliance on ER signaling. The molecular cross talk between ER and receptor tyrosine kinase signaling of epidermal growth factor receptor (EGFR), HER2 and IGF1R leading to the activation of mitogen-activated kinase (MAPK) and PI3K pathways contributes to the resistance to endocrine therapy.[98–100] In the metastatic breast cancer setting, targeting HER2 with trastuzumab or lapatinib in combination with endocrine therapy demonstrated an improvement in PFS, compared with endocrine therapy alone.[101,102] However, targeting the IGF pathway with octreotide, a somatostatin analog, in combination with tamoxifen did not demonstrate significant clinical activity compared with tamoxifen alone in patients with HR-positive or -unknown breast tumors in the adjuvant setting.[103] Due to the complexity of endocrine resistant mechanisms, no factor thus far related to endocrine resistance, has been approved or recommended for use in the clinic at this point.[28]

Using an unbiased genome-wide functional screen, Medes-Pereira *et al.* identified multiple genes that affect sensitivity or resistance to tamoxifen.[104] Those include genes whose silencing cause tamoxifen resistance, which include BAP1, CLPP, GPRC5D, NAE1, NF1, NIPBL, NSD1, RAD21, RARG, SMC3, and UBA3. This study also identified a set of genes whose silencing lead to sensitivity to tamoxifen (C10 or f72, C15 or f55/NUT, EDF1, ING5, KRAS, NOC3L, PPP1R5B, RRAS2, TMPRSS2, and TPM4). The genome-wide approach could help to evaluate the molecular determinants of sensitivity and resistance to endocrine therapy although clinical implications of the identified genes have yet to be tested and validated.

5.6 Diagnostic Tests for ER and PgR

In the early 1980s, the need for establishing uniformed assays was recognized and recommended for examination through large randomized trials conducted by the oncology cooperative groups.[105,106] Over the past several decades, several issues on the evaluation of hormone receptor status have been identified and evolved. These include assay formats and methodologies, and determinations of positive receptor status.

5.6.1 Methods for Evaluation of ER/PgR Expression

Since the 1970s, several assays with different biochemical principles, analytic sensitivity, and analytic precision have been used for ER/PgR evaluation. The initial assays developed were ligand-binding assays (LBAs), which had significant limitations due to the technical complexity, requirement for fresh-frozen tissues, high variability, and non-specific to tumor cells.[107] Nonetheless, earlier randomized trials used LBAs led to our understanding of the relationship between the presence of ER and response to endocrine therapy. In the late 1980s, immuno-assays using polyclonal and monoclonal antibodies to detect ER emerged. The enzyme immunoassay (EIA) was developed and used for quantification of ER from the cytosol extracts of fresh tumor tissues. The amount of steroid hormone receptors detected by EIA was highly correlated with that measured through LBAs, which EIA was less labor-intensive.[108] IHC was subsequently developed and evolved to allow detection of ER in formalin-fixed and paraffin-embedded tissues.[109,110] It is the use of IHC method to detect the hormone receptors in the malignant nuclei that has led to routine and convenient evaluation of ER/PgR a practical reality.[109]

With different techniques evaluating hormonal receptor status, one of the critical issues is the assignment of cutoff values for establishing receptor positivity. These are usually arbitrarily chosen in clinical validation studies. For LBAs, the cutoff points were varied from ≥ 3 to ≥ 20 fmol/mg cytosol protein in early studies.[16,111,112] The clinical useful cutoff for ER was accepted as ≥ 10 fmol/mg cytosol protein, which is specified in the FDA approved commercial LBA assay kits.[111] As EIA was also a cytosolic protein-based assay, the cutoff for EIA was determined as ≥ 15 fmol/mg cytosol protein, which was equivalent to ≥ 10 fmol/mg cytosol protein measured by LBAs. Starting from the mid-1990s, IHC has been the test predominantly used and replaced LBAs and EIA. Clinical validations of IHC testing were based on several studies comparing the concordance of IHC with LBAs. One of the most important studies was the analysis of nearly 2000 women, in which a semi-quantitative scoring system, known as the Allred' scores, was used. Allred scoring takes into consideration of the proportion of positive cells (scored on a scale of 0–5) and staining intensity (on a scale of 0–3). The proportion and intensity are then summed to produce total scores of 0 or 2 through 8. A score of 0–2 is regarded as negative while 3–8 as positive. High concordance between IHC and LBAs was established.[16,113,114] Importantly, IHC is proved to be superior to the LBAs for assessing ER status as it is easier, safer, and less expensive. It has equivalent or better ability to predict response to adjuvant endocrine therapy.

5.6.2 Immunohistochemical Testing for ER and PgR

As few as 1% ER labeling of invasive carcinoma nuclei has been shown to predict clinical benefit from endocrine therapy.[16] Because endocrine therapy is highly effective and relatively non-toxic, one would never want to deny a patient the potential benefits of this type of treatment. Further standardization of ER and PgR testing (especially for reducing false negative results) becomes an important issue. To improve the accuracy of IHC testing for ER and PgR in breast cancer, the American Society of Clinical Oncology/College of American Pathologists issued guideline recommendations in 2010.[30,115]

The guideline addressed many factors influencing the detection accuracy of ER and PgR, and issued optimal algorithm for ER/PgR testing, which ensures accurate and reproducible assay performance. The testing laboratory is required to document the proof of initial assay validation, ongoing internal QA procedure, participation in external proficiency testing, and biennial accreditation by a valid accrediting agency. Preanalytical standardization procedure includes the time from tumor removal to fixation of ≤ 1 hour, use of 10% neutral buffered formalin (NBF) as the fixative, and the duration of fixation from ≥ 6 hours to ≤ 72 hours, and storage of slides of ≤ 6 weeks. The procedure for analytic standardization comprises the use of ER/PgR antibodies that have been clinically validated, and optimal testing conditions with acceptable recommended positive and negative controls on each batch run. Clones 1D5 and 6F11 are the most widely used and clinically validated mouse monoclonal antibodies for IHC use (Fig. 5.2). It is recommended that ER and PgR assays are considered positive if there are at least 1% positive tumor nuclei in the sample on testing in the presence of expected reactivity of internal (normal epithelial elements) and external controls.[115]

5.7 Conclusions

The prognostic and predictive values of HRs in breast cancer treatment are established through numerous studies over several decades. The evolution of hormonal receptor serving as predictive biomarkers has historical value for future biomarkers studies in breast cancer and other cancer types. Robust clinical validation of the molecular factors that predict sensitivity and resistance of endocrine therapy identified in preclinical and small sample size clinical studies should be a focus in future investigations. In turn, more patients could experience response and benefit from endocrine therapy.

References

1. Fisher, B., *et al.* Influence of tumor estrogen and progesterone receptor levels on the response to tamoxifen and chemotherapy in primary breast cancer. *J Clin Oncol* **1**, 227–241 (1983).
2. Early Breast Cancer Trialists' Collaborative Group (EBCTCG) Effects of chemotherapy and hormonal therapy for early breast cancer on recurrence and 15-year survival: an overview of the randomised trials. *Lancet* **365**, 1687–1717 (2005).
3. Beatson, G.T. On the treatment of inoperable cases of carcinoma of the mamma: suggestions for a new method of treatment, with illustrative cases *The Lancet* **148**, 162–165 (1896).

4. Lathrop, A.E., and Loeb, L. Further investigations on the origin of tumors in mice. III. On the part played by internal secretion in the spontaneous development of tumors. *J Cancer Res* **1**, 1–19 (1916).
5. Lacassagne, A. The Relation Between Hormones and Cancer. *Can Med Assoc J* **37**, 112–117 (1937).
6. De Sombre, E.R., Puca, G.A., and Jensen, E.V. Purification of an estrophilic protein from calf uterus. *Proc Natl Acad Sci U S A* **64**, 148–154 (1969).
7. Cole, M.P., Jones, C.T., and Todd, I.D. A new anti-oestrogenic agent in late breast cancer. An early clinical appraisal of ICI46474. *Br J Cancer* **25**, 270–275 (1971).
8. Jordan, V.C. Tamoxifen: a most unlikely pioneering medicine. *Nat Rev Drug Discov* **2**, 205–213 (2003).
9. Bjornstrom, L., and Sjoberg, M. Mechanisms of estrogen receptor signaling: convergence of genomic and nongenomic actions on target genes. *Mol Endocrinol* **19**, 833–842 (2005).
10. Lee, A.V., Cui, X., and Oesterreich, S. Cross-talk among estrogen receptor, epidermal growth factor, and insulin-like growth factor signaling in breast cancer. *Clin Cancer Res* **7**, 4429s–4435s; discussion 4411s–4412s (2001).
11. Schiff, R., *et al.* Cross-talk between estrogen receptor and growth factor pathways as a molecular target for overcoming endocrine resistance. *Clin Cancer Res* **10**, 331S–336S (2004).
12. Arpino, G., Wiechmann, L., Osborne, C.K., and Schiff, R. Crosstalk between the estrogen receptor and the HER tyrosine kinase receptor family: molecular mechanism and clinical implications for endocrine therapy resistance. *Endocr Rev* **29**, 217–233 (2008).
13. McGuire, W.L. Estrogen receptors in human breast cancer. *J Clin Invest* **52**, 73 77 (1973).
14. McGuire, W.L., Carbone, P.P. Sears, M.E., *et al.* Estrogen receptors in human breast cancer: An overview. in *Estrogen Receptors in Human Breast Cancer* (eds. McGuire, W.L., Carbone, P.P. & Vollmer, E.P.) 1–8 (New York, NY, Raven Press, 1974).
15. Anderson, W.F., Chatterjee, N., Ershler, W.B., and Brawley, O.W. Estrogen receptor breast cancer phenotypes in the Surveillance, Epidemiology, and End Results database. *Breast Cancer Res Treat* **76**, 27–36 (2002).
16. Harvey, J.M., Clark, G.M., Osborne, C.K., and Allred, D.C. Estrogen receptor status by immunohistochemistry is superior to the ligand-binding assay for predicting response to adjuvant endocrine therapy in breast cancer. *J Clin Oncol* **17**, 1474–1481 (1999).

17. Murphy, L.C., and Watson, P. Steroid receptors in human breast tumorigenesis and breast cancer progression. *Biomed Pharmacother* **56**, 65–77 (2002).

18. Dunnwald, L.K., Rossing, M.A., and Li, C.I. Hormone receptor status, tumor characteristics, and prognosis: a prospective cohort of breast cancer patients. *Breast Cancer Res* **9**, R6 (2007).

19. Perou, C.M., *et al.* Molecular portraits of human breast tumours. *Nature* **406**, 747–752 (2000).

20. Cheang, M.C., *et al.* Ki67 index, HER2 status, and prognosis of patients with luminal B breast cancer. *J Natl Cancer Inst* **101**, 736–750 (2009).

21. Arvold, N.D., *et al.* Age, breast cancer subtype approximation, and local recurrence after breast-conserving therapy. *J Clin Oncol* **29**, 3885–3891 (2011).

22. Hu, Z., *et al.* The molecular portraits of breast tumors are conserved across microarray platforms. *BMC Genomics* **7**, 96 (2006).

23. Ingle, J.N., *et al.* Randomized clinical trial of diethylstilbestrol versus tamoxifen in postmenopausal women with advanced breast cancer. *N Engl J Med* **304**, 16–21 (1981).

24. Fisher, B., *et al.* Treatment of primary breast cancer with chemotherapy and tamoxifen. *N Engl J Med* **305**, 1–6 (1981).

25. Gallo, M.A., and Kaufman, D. Antagonistic and agonistic effects of tamoxifen: significance in human cancer. *Semin Oncol* **24**, S1-71-S1–80 (1997).

26. Osborne, C.K., and McGuire, W.L. Therapy for cancer of the breast. Current status of steroid hormone receptors. *West J Med* **130**, 401–407 (1979).

27. Tamoxifen for early breast cancer: an overview of the randomised trials. Early Breast Cancer Trialists' Collaborative Group. *Lancet* **351**, 1451–1467 (1998).

28. Harris, L., *et al.* American Society of Clinical Oncology 2007 update of recommendations for the use of tumor markers in breast cancer. *J Clin Oncol* **25**, 5287–5312 (2007).

29. Hammond, M.E., *et al.* American Society of Clinical Oncology/College of American Pathologists guideline recommendations for immunohistochemical testing of estrogen and progesterone receptors in breast cancer (unabridged version). *Arch Pathol Lab Med* **134**, e48–e72 (2010).

30. Hammond, M.E., *et al.* American Society of Clinical Oncology/College Of American Pathologists guideline recommendations for

immunohistochemical testing of estrogen and progesterone receptors in breast cancer. *J Clin Oncol* **28**, 2784–2795 (2010).

31. Osborne, C.K. Tamoxifen in the treatment of breast cancer. *N Engl J Med* **339**, 1609–1618 (1998).
32. Chumsri, S., Howes, T., Bao, T., Sabnis, G., and Brodie, A. Aromatase, aromatase inhibitors, and breast cancer. *J Steroid Biochem Mol Biol* **125**, 13–22 (2011).
33. DeFriend, D.J., *et al.* Effects of 4-hydroxytamoxifen and a novel pure antioestrogen (ICI 182780) on the clonogenic growth of human breast cancer cells *in vitro*. *Br J Cancer* **70**, 204–211 (1994).
34. Dauvois, S., White, R., and Parker, M.G. The antiestrogen ICI 182780 disrupts estrogen receptor nucleocytoplasmic shuttling. *J Cell Sci* **106**, 1377–1388 (1993).
35. Wakeling, A.E., Dukes, M., and Bowler, J. A potent specific pure antiestrogen with clinical potential. *Cancer Res* **51**, 3867–3873 (1991).
36. Miller, W.R. Biological rationale for endocrine therapy in breast cancer. *Best Pract Res Clin Endocrinol Metab* **18**, 1–32 (2004).
37. Berry, D.A., *et al.* Estrogen-receptor status and outcomes of modern chemotherapy for patients with node-positive breast cancer. *JAMA* **295**, 1658–1667 (2006).
38. Hayes, D.F., *et al.* HER2 and response to paclitaxel in node-positive breast cancer. *N Engl J Med* **357**, 1496–1506 (2007).
39. Paik, S., *et al.* A multigene assay to predict recurrence of tamoxifen-treated, node-negative breast cancer. *N Engl J Med* **351**, 2817–2826 (2004).
40. van 't Veer, L.J., *et al.* Gene expression profiling predicts clinical outcome of breast cancer. *Nature* **415**, 530–536 (2002).
41. Powles, T.J., Hickish, T., Kanis, J.A., Tidy, A., and Ashley, S. Effect of tamoxifen on bone mineral density measured by dual-energy x-ray absorptiometry in healthy premenopausal and postmenopausal women. *J Clin Oncol* **14**, 78–84 (1996).
42. Dezentje, V.O., *et al.* Effect of concomitant CYP2D6 inhibitor use and tamoxifen adherence on breast cancer recurrence in early-stage breast cancer. *J Clin Oncol* **28**, 2423–2429 (2010).
43. Madlensky, L., *et al.* Tamoxifen metabolite concentrations, CYP2D6 genotype, and breast cancer outcomes. *Clin Pharmacol Ther* **89**, 718–725 (2011).

44. Cuzick, J., *et al.* Use of luteinising-hormone-releasing hormone agonists as adjuvant treatment in premenopausal patients with hormone-receptor-positive breast cancer: a meta-analysis of individual patient data from randomised adjuvant trials. *Lancet* **369**, 1711–1723 (2007).
45. Hughes, L.L., *et al.* Efficacy of radiotherapy for ovarian ablation: results of a breast intergroup study. *Cancer* **101**, 969–972 (2004).
46. Adjuvant tamoxifen in the management of operable breast cancer: the Scottish Trial. Report from the Breast Cancer Trials Committee, Scottish Cancer Trials Office (MRC), Edinburgh. *Lancet* **2**, 171–175 (1987).
47. Fisher, B., *et al.* A randomized clinical trial evaluating tamoxifen in the treatment of patients with node-negative breast cancer who have estrogen-receptor-positive tumors. *N Engl J Med* **320**, 479–484 (1989).
48. Baum, M., *et al.* Anastrozole alone or in combination with tamoxifen versus tamoxifen alone for adjuvant treatment of postmenopausal women with early breast cancer: first results of the ATAC randomised trial. *Lancet* **359**, 2131–2139 (2002).
49. Howell, A., *et al.* Results of the ATAC (Arimidex, Tamoxifen, Alone or in Combination) trial after completion of 5 years' adjuvant treatment for breast cancer. *Lancet* **365**, 60–62 (2005).
50. Jakesz, R., *et al.* Switching of postmenopausal women with endocrine-responsive early breast cancer to anastrozole after 2 years' adjuvant tamoxifen: combined results of ABCSG trial 8 and ARNO 95 trial. *Lancet* **366**, 455–462 (2005).
51. Boccardo, F., *et al.* Switching to an aromatase inhibitor provides mortality benefit in early breast carcinoma: pooled analysis of 2 consecutive trials. *Cancer* **109**, 1060–1067 (2007).
52. Jonat, W., *et al.* Effectiveness of switching from adjuvant tamoxifen to anastrozole in postmenopausal women with hormone-sensitive early-stage breast cancer: a meta-analysis. *Lancet Oncol* **7**, 991–996 (2006).
53. Thurlimann, B., *et al.* A comparison of letrozole and tamoxifen in postmenopausal women with early breast cancer. *N Engl J Med* **353**, 2747–2757 (2005).
54. Coates, A.S., *et al.* Five years of letrozole compared with tamoxifen as initial adjuvant therapy for postmenopausal women with endocrine-responsive early breast cancer: update of study BIG 1–98. *J Clin Oncol* **25**, 486–492 (2007).

55. Goss, P.E., *et al.* A randomized trial of letrozole in postmenopausal women after five years of tamoxifen therapy for early-stage breast cancer. *N Engl J Med* **349**, 1793–1802 (2003).

56. Goss, P.E., *et al.* Randomized trial of letrozole following tamoxifen as extended adjuvant therapy in receptor-positive breast cancer: updated findings from NCIC CTG MA.17. *J Natl Cancer Inst* **97**, 1262–1271 (2005).

57. Coombes, R.C., *et al.* A randomized trial of exemestane after two to three years of tamoxifen therapy in postmenopausal women with primary breast cancer. *N Engl J Med* **350**, 1081–1092 (2004).

58. Coombes, R.C., *et al.* Survival and safety of exemestane versus tamoxifen after 2–3 years' tamoxifen treatment (Intergroup Exemestane Study): a randomised controlled trial. *Lancet* **369**, 559–570 (2007).

59. Winer, E.P., *et al.* American Society of Clinical Oncology technology assessment on the use of aromatase inhibitors as adjuvant therapy for women with hormone receptor-positive breast cancer: status report 2002. *J Clin Oncol* **20**, 3317–3327 (2002).

60. Winer, E.P. *et al.* American Society of Clinical Oncology technology assessment on the use of aromatase inhibitors as adjuvant therapy for postmenopausal women with hormone receptor-positive breast cancer: status report 2004. *J Clin Oncol* **23**, 619–629 (2005).

61. Cuppone, F., *et al.* Do adjuvant aromatase inhibitors increase the cardiovascular risk in postmenopausal women with early breast cancer? Meta-analysis of randomized trials. *Cancer* **112**, 260–267 (2008).

62. Amir, E., Seruga, B., Niraula, S., Carlsson, L., and Ocana, A. Toxicity of adjuvant endocrine therapy in postmenopausal breast cancer patients: a systematic review and meta-analysis. *J Natl Cancer Inst* **103**, 1299–1309 (2011).

63. Gibson, L., Lawrence, D., Dawson, C., and Bliss, J. Aromatase inhibitors for treatment of advanced breast cancer in postmenopausal women. *Cochrane Database Syst Rev*, CD003370 (2009).

64. Bonneterre, J., *et al.* Anastrozole versus tamoxifen as first-line therapy for advanced breast cancer in 668 postmenopausal women: results of the Tamoxifen or Arimidex Randomized Group Efficacy and Tolerability study. *J Clin Oncol* **18**, 3748–3757 (2000).

65. Paridaens, R.J., *et al.* Phase III study comparing exemestane with tamoxifen as first-line hormonal treatment of metastatic breast cancer in postmenopausal women: the European Organisation for Research

and Treatment of Cancer Breast Cancer Cooperative Group. *J Clin Oncol* **26**, 4883–4890 (2008).

66. Mauri, D., Pavlidis, N., Polyzos, N.P., and Ioannidis, J.P. Survival with aromatase inhibitors and inactivators versus standard hormonal therapy in advanced breast cancer: meta-analysis. *J Natl Cancer Inst* **98**, 1285–1291 (2006).
67. Ingle, J.N., *et al.* Fulvestrant in women with advanced breast cancer after progression on prior aromatase inhibitor therapy: North Central Cancer Treatment Group Trial N0032. *J Clin Oncol* **24**, 1052–1056 (2006).
68. Perey, L., *et al.* Clinical benefit of fulvestrant in postmenopausal women with advanced breast cancer and primary or acquired resistance to aromatase inhibitors: final results of phase II Swiss Group for Clinical Cancer Research Trial (SAKK 21/00). *Ann Oncol* **18**, 64–69 (2007).
69. Chia, S., *et al.* Double-blind, randomized placebo controlled trial of fulvestrant compared with exemestane after prior nonsteroidal aromatase inhibitor therapy in postmenopausal women with hormone receptor-positive, advanced breast cancer: results from EFECT. *J Clin Oncol* **26**, 1664–1670 (2008).
70. Di Leo, A., *et al.* Results of the CONFIRM phase III trial comparing fulvestrant 250 mg with fulvestrant 500 mg in postmenopausal women with estrogen receptor-positive advanced breast cancer. *J Clin Oncol* **28**, 4594–4600 (2010).
71. Howell, A., *et al.* Comparison of fulvestrant versus tamoxifen for the treatment of advanced breast cancer in postmenopausal women previously untreated with endocrine therapy: a multinational, double-blind, randomized trial. *J Clin Oncol* **22**, 1605–1613 (2004).
72. Milla-Santos, A., *et al.* Anastrozole versus tamoxifen as first-line therapy in postmenopausal patients with hormone-dependent advanced breast cancer: a prospective, randomized, phase III study. *Am J Clin Oncol* **26**, 317–322 (2003).
73. Mouridsen, H., *et al.* Phase III study of letrozole versus tamoxifen as first-line therapy of advanced breast cancer in postmenopausal women: analysis of survival and update of efficacy from the International Letrozole Breast Cancer Group. *J Clin Oncol* **21**, 2101–2109 (2003).
74. Buzdar, A.U., *et al.* A phase III trial comparing anastrozole (1 and 10 milligrams), a potent and selective aromatase inhibitor, with megestrol acetate in postmenopausal women with advanced breast carcinoma. Arimidex Study Group. *Cancer* **79**, 730–739 (1997).

75. Dombernowsky, P., *et al.* Letrozole, a new oral aromatase inhibitor for advanced breast cancer: double-blind randomized trial showing a dose effect and improved efficacy and tolerability compared with megestrol acetate. *J Clin Oncol* **16**, 453–461 (1998).

76. Kvinnsland, S., *et al.* High activity and tolerability demonstrated for exemestane in postmenopausal women with metastatic breast cancer who had previously failed on tamoxifen treatment. *Eur J Cancer* **36**, 976–982 (2000).

77. Horobin, J.M., Preece, P.E., Dewar, J.A., Wood, R.A., and Cuschieri, A. Long-term follow-up of elderly patients with locoregional breast cancer treated with tamoxifen only. *Br J Surg* **78**, 213–217 (1991).

78. Bergman, L., van Dongen, J.A., van Ooijen, B., and van Leeuwen, F.E. Should tamoxifen be a primary treatment choice for elderly breast cancer patients with locoregional disease? *Breast Cancer Res Treat* **34**, 77–83 (1995).

79. Mustacchi, G., *et al.* Tamoxifen alone versus adjuvant tamoxifen for operable breast cancer of the elderly: long-term results of the phase III randomized controlled multicenter GRETA trial. *Ann Oncol* **14**, 414–420 (2003).

80. Hind, D., Wyld, L., Beverley, C.B., and Reed, M.W. Surgery versus primary endocrine therapy for operable primary breast cancer in elderly women (70 years plus). *Cochrane Database Syst Rev*, CD004272 (2006).

81. Willsher, P.C., Robertson, J.F., Chan, S.Y., Jackson, L., and Blamey, R.W. Locally advanced breast cancer: early results of a randomised trial of multimodal therapy versus initial hormone therapy. *Eur J Cancer* **33**, 45–49 (1997).

82. Eiermann, W., *et al.* Preoperative treatment of postmenopausal breast cancer patients with letrozole: A randomized double-blind multicenter study. *Ann Oncol* **12**, 1527–1532 (2001).

83. Cataliotti, L., *et al.* Comparison of anastrozole versus tamoxifen as preoperative therapy in postmenopausal women with hormone receptor-positive breast cancer: the Pre-Operative "Arimidex" Compared to Tamoxifen (PROACT) trial. *Cancer* **106**, 2095–2103 (2006).

84. Seo, J.H., Kim, Y.H., and Kim, J.S. Meta-analysis of pre-operative aromatase inhibitor versus tamoxifen in postmenopausal woman with hormone receptor-positive breast cancer. *Cancer Chemother Pharmacol* **63**, 261–266 (2009).

85. Buchholz, T.A., *et al.* Factors predictive of outcome in patients with breast cancer refractory to neoadjuvant chemotherapy. *Cancer J* **7**, 413–420 (2001).

86. Dowsett, M., *et al.* Prognostic value of Ki67 expression after short-term presurgical endocrine therapy for primary breast cancer. *J Natl Cancer Inst* **99**, 167–170 (2007).

87. Ellis, M.J., *et al.* Outcome prediction for estrogen receptor-positive breast cancer based on postneoadjuvant endocrine therapy tumor characteristics. *J Natl Cancer Inst* **100**, 1380–1388 (2008).

88. Chia, Y.H., Ellis, M.J., and Ma, C.X. Neoadjuvant endocrine therapy in primary breast cancer: indications and use as a research tool. *Br J Cancer* **103**, 759–764 (2010).

89. Rastelli, F., and Crispino, S. Factors predictive of response to hormone therapy in breast cancer. *Tumori* **94**, 370–383 (2008).

90. Ring, A., and Dowsett, M. Mechanisms of tamoxifen resistance. *Endocr Relat Cancer* **11**, 643–658 (2004).

91. Ottaviano, Y.L., *et al.* Methylation of the estrogen receptor gene CpG island marks loss of estrogen receptor expression in human breast cancer cells. *Cancer Res* **54**, 2552–2555 (1994).

92. Yang, X., Yan, L., and Davidson, N.E. DNA methylation in breast cancer. *Endocr Relat Cancer* **8**, 115–127 (2001).

93. Lapidus, R.G., *et al.* Methylation of estrogen and progesterone receptor gene 5′ CpG islands correlates with lack of estrogen and progesterone receptor gene expression in breast tumors. *Clin Cancer Res* **2**, 805–810 (1996).

94. Yang, X., *et al.* Synergistic activation of functional estrogen receptor (ER)-alpha by DNA methyltransferase and histone deacetylase inhibition in human ER-alpha-negative breast cancer cells. *Cancer Res* **61**, 7025–7029 (2001).

95. Gutierrez, M.C., *et al.* Molecular changes in tamoxifen-resistant breast cancer: relationship between estrogen receptor, HER-2, and p38 mitogen-activated protein kinase. *J Clin Oncol* **23**, 2469–2476 (2005).

96. Herynk, M.H., and Fuqua, S.A. Estrogen receptor mutations in human disease. *Endocr Rev* **25**, 869–898 (2004).

97. Musgrove, E.A., and Sutherland, R.L. Biological determinants of endocrine resistance in breast cancer. *Nat Rev Cancer* **9**, 631–643 (2009).

98. Osborne, C.K., and Schiff, R. Mechanisms of endocrine resistance in breast cancer. *Annu Rev Med* **62**, 233–247 (2011).

99. Perez-Tenorio, G., and Stal, O. Activation of AKT/PKB in breast cancer predicts a worse outcome among endocrine treated patients. *Br J Cancer* **86**, 540–545 (2002).

100. Hurtado, A., *et al.* Regulation of ERBB2 by oestrogen receptor-PAX2 determines response to tamoxifen. *Nature* **456**, 663–666 (2008).

101. Johnston, S., *et al.* Lapatinib combined with letrozole versus letrozole and placebo as first-line therapy for postmenopausal hormone receptor-positive metastatic breast cancer. *J Clin Oncol* **27**, 5538–5546 (2009).

102. Kaufman, B., *et al.* Trastuzumab plus anastrozole versus anastrozole alone for the treatment of postmenopausal women with human epidermal growth factor receptor 2-positive, hormone receptor-positive metastatic breast cancer: results from the randomized phase III TAnDEM study. *J Clin Oncol* **27**, 5529–5537 (2009).

103. Pritchard, K.I., *et al.* Randomized trial of tamoxifen versus combined tamoxifen and octreotide LAR therapy in the adjuvant treatment of early-stage breast cancer in postmenopausal women: NCIC CTG MA.14. *J Clin Oncol* **29**, 3869–3876 (2011).

104. Mendes-Pereira, A.M., *et al.* Breast Cancer Special Feature: Genome-wide functional screen identifies a compendium of genes affecting sensitivity to tamoxifen. *Proc Natl Acad Sci U S A* (2012).

105. Wittliff, J.L., Fisher, B., and Durant, J.R. Establishment of uniformity in steroid receptor analyses used in cooperative clinical trials of breast cancer treatment. *Recent Results Cancer Res* **71**, 198–206 (1980).

106. Wittliff, J.L., Brown, A.M., and Fisher, B. Establishment of uniformity in steroid receptor determinations for protocol B-09 of the National Surgical Adjuvant Breast Project in *Estrogen receptor assays in breast cancer: laboratory discrepancies and quality assurance* (eds. Sarfaty, G.A., Nash, A.R. & Keightley, D.D.) (New York (NY): Masson Publishing USA, 1981).

107. Hull, D.F., 3rd, *et al.* Multiple estrogen receptor assays in human breast cancer. *Cancer Res* **43**, 413–416 (1983).

108. Pasic, R., Djulbegovic, B., and Wittliff, J.L. Comparison of sex steroid receptor determinations in human breast cancer by enzyme immunoassay and radioligand binding. *J Clin Lab Anal* **4**, 430–436 (1990).

109. King, W.J., and Greene, G.L. Monoclonal antibodies localize oestrogen receptor in the nuclei of target cells. *Nature* **307**, 745–747 (1984).

110. Greene, G.L., Sobel, N.B., King, W.J., and Jensen, E.V. Immunochemical studies of estrogen receptors. *J Steroid Biochem* **20**, 51–56 (1984).

111. McGuire, W.L. Quantitation of estrogen receptor in mammary carcinoma. *Methods Enzymol* **36**, 248–254 (1975).

112. McCarty, K.S., Jr., Miller, L.S., Cox, E.B., Konrath, J., and McCarty, K.S., Sr. Estrogen receptor analyses. Correlation of biochemical and immunohistochemical methods using monoclonal antireceptor antibodies. *Arch Pathol Lab Med* **109**, 716–721 (1985).

113. Elledge, R.M., *et al.* Estrogen receptor (ER) and progesterone receptor (PgR), by ligand-binding assay compared with ER, PgR and pS2, by immuno-histochemistry in predicting response to tamoxifen in metastatic breast cancer: a Southwest Oncology Group Study. *Int J Cancer* **89**, 111–117 (2000).

114. Barnes, D.M., Harris, W.H., Smith, P., Millis, R.R., and Rubens, R.D. Immunohistochemical determination of oestrogen receptor: comparison of different methods of assessment of staining and correlation with clinical outcome of breast cancer patients. *Br J Cancer* **74**, 1445–1451 (1996).

115. Hammond, M.E., *et al.* American Society of Clinical Oncology/College of American Pathologists guideline recommendations for immunohistochemical testing of estrogen and progesterone receptors in breast cancer. *Arch Pathol Lab Med* **134**, 907–922 (2010).

Chapter 6

Predictive Biomarkers for Epidermal Growth Factor Receptor Agents in Non-Small Cell Lung Cancer

John Hilton, Penelope A. Bradbury, and Janet Dancey
NCIC Clinical Trials Group, Queen's University,
10 Stuart Street, Kingston, ON K7L 3N6, Canada
jdancey@ctg.queensu.ca

6.1 Introduction

Lung cancer is the leading cause of cancer-related death and despite efforts to improve treatment options, the overall 5-year survival still remains poor at 15%.[1] As in other malignancies, novel systemic therapies for the treatment of lung cancer have shifted from cytotoxic chemotherapy to molecularly targeted agents. To date, the most successful novel targeted agents for the treatment of non-small cell lung cancer (NSCLC) have inhibited the epidermal growth factor receptor (EGFR) pathway. Although EGFR inhibitors have entered clinical practice for the treatment of advanced NSCLC, the

Handbook of Therapeutic Biomarkers in Cancer
Edited by Sherry X. Yang and Janet E. Dancey
Copyright © 2013 Pan Stanford Publishing Pte. Ltd.
ISBN 978-981-4364-65-2 (Hardcover), 978-981-4364-66-9 (eBook)
www.panstanford.com

benefits of these agents in unselected patients with advanced NSCLC have been modest, driving efforts to identify biomarkers to define subpopulations of patients with NSCLC most likely to benefit from these agents. This chapter will review the status of this predictive biomarker research for the EGFR inhibitors erlotinib, gefitinib, and cetuximab, which are currently in clinical use for the treatment of NSCLC.

6.2 The Epidermal Growth Factor Receptor Family

The epidermal growth factor receptor (EGFR) family consists of four members: EFGR, Human epidermal growth factor receptor-2 (HER2), HER3, and HER4. All members of the EGFR family contain an extracellular ligand binding region, a membrane spanning region and cytoplasmic region that possesses tyrosine kinase activity. The binding of a ligand to the receptor leads to the formation of either homo- or heterodimers between the members of the epidermal growth factor receptor family and activation of tyrosine kinase activity.[2] This leads to the binding of adenosine triphosphate (ATP) and phosphorylation of the cytoplasmic component of the kinase. This phosphorylation event allows adaptor proteins to interact with the receptor and initiate the downstream signaling pathways[2] (see Fig. 6.1). Without ligand, the EGFR receptor is found in a closed conformation with the dimerization interface blocked. Unlike EGFR, HER2 has a different extracellular region and has a fixed conformation, which results in permanent exposure of the dimerization domain.[3] In addition, HER2 is unique among the receptor family in that it binds none of the potential EGF ligands. As a result, it appears that its primary role in the pathway is to form heterodimers with the other receptors.[4] HER3 also plays a distinct role in the HER family signaling network. Although it is kinase inactive and therefore incapable of initiating downstream signaling pathways on its own, HER3 can dimerize with other receptors, particularly HER2, for potent cellular signaling. Evidence is growing to show that the transactivating nature of HER3 is an essential aspect of oncogenic function of the related receptors HER1/EGFR and HER2.

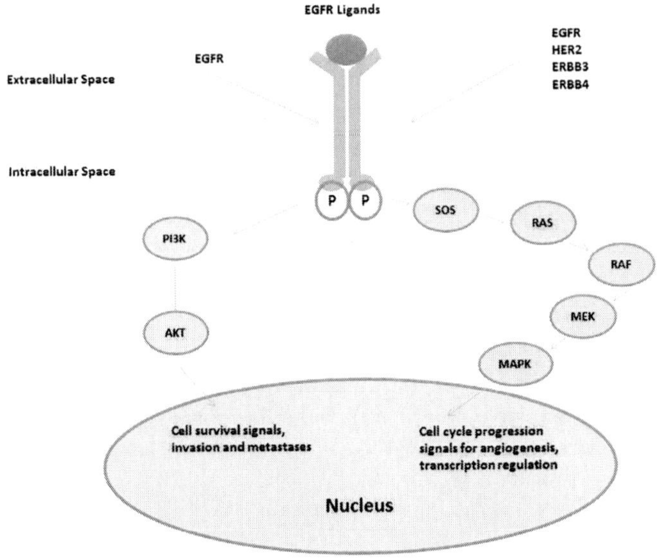

Figure 6.1 The EGFR Pathway. EGFR signaling commences with the binding of ligand (e.g. EGF, TGF-α) to the EGFR receptor. This leads to dimerization, with one part being EGFR and the other part being one of EGFR, HER2, ErbB3 or ErbB4. Dimerization leads to the activation of the intracellular tyrosine kinase domains through autophosphorylation. Through docking proteins, both the RAS/RAF/MAPK pathways and PI3K/Akt pathways are activated. Activation of these pathways leads to cell cycle progression, increased cell division, increased cell survival and increased probability of metastasis. EGFR also activates the STAT protein family in the cytosol. This leads to translocation into the nucleus and further transcription regulation.[2]

6.3 Signal Transduction Pathways Controlled by the Activation of EGFR

A schematic of the signal transduction pathways under the control of EGFR is shown in (Fig. 6.1). Overall, two major pathways are involved: the RAS-RAF-MAPK (Ras) pathway and the phosphoinositol-3-phosphate-Akt (PI3K) pathway. For the Ras pathway, activated EGFR recruits Ras and initiates activation through the exchange of guanine diphosphate (GDP) to guanine triphosphate (GTP).[5] This leads to signal transduction and transcription of genes controlling cycle–cycle progression and cellular proliferation.[6] Unlike the Ras pathway, EGFR activation of the PI3K pathway requires only the presence of HER3 and is not dependent on adaptor proteins.[2] Interaction with PI3K leads to phosphorylation of phosphatidylinositol 4,5-diphosphate, resulting in phosphatidylinositol 3,4,5-triphosphate. This molecule then activates of Akt and other downstream effector pathways controlling cell proliferation and survival.[7] This results in increased cell growth, apoptosis resistance, increased tumor invasion, and cell migration.[8]

In addition to the above pathways, EGFR has the ability to regulate signal transducers and activators of transcription (STAT) and sarcoma kinase (Src) pathways through the Janus kinase (JAK). STAT proteins have been implicated in oncogenesis and tumor progression while Src can cause increased EGFR signaling and resistance to EGFR-targeted therapy.[9,10]

6.4 EGFR Inhibitors for the Management of NSCLC

Two major classes of EGFR-targeted therapies that have entered clinical practice are the anti-EFGR monoclonal antibodies (MoAbs), cetuximab (Bristol-Myers Squibb) and panitumumab (Amgen), and EGFR-specific tyrosine kinase inhibitors (TKIs), erlotinib (OSI Pharmaceuticals) and gefitinib (AstraZeneca). The MoAbs bind to the extracellular domain of EGFR when it is in the inactive configuration and compete with EGF receptor binding by occluding the ligand-binding region. By doing so, receptor dimerization and activation of the tyrosine kinase domain is prevented.[11] Anti-EGFR MoAbs recognize EGFR exclusively and represent a highly selective treatment. In contrast, TKIs compete reversibly with ATP for binding to the intracellular tyrosine kinase catalytic domain. By inhibiting the binding of ATP, the EGFR TKIs prevent phosphorylation of the cytoplasmic kinase domain and the initiation of downstream signaling. Unlike the MoAbs, the TKIs have the ability to inhibit other growth factor receptor tyrosine kinases, potentially leading to a greater range of side-effects with their usage.[12]

EGFR inhibitors have been evaluated in phase III randomized clinical trials in the curative and palliative setting (Table 6.1); however, currently, these agents have only been shown to prolong survival for patients with advanced, incurable NSCLC. Trials have demonstrated that the combination of EGFR inhibitors and cytotoxic chemotherapy in patients with advanced, incurable disease have shown no survival benefit over cytotoxic chemotherapy alone in the first-line setting.[13,14,15,16] Trials assessing these agents in the adjuvant setting (RADIANT) are currently ongoing and have not yet reported.[17]

The first phase III clinical trial to demonstrate an overall survival benefit from an EGFR TKI was NCIC Clinical Trials Group (CTG) BR21 clinical trial.[18] This trial evaluated erlotinib versus placebo in patients with NSCLC after one or two lines of chemotherapy for advanced disease. Patients randomized to the erlotinib arm had a superior overall survival (OS) and quality of life benefit, compared with patients on the placebo arm (median OS 6.7 months vs. 4.7 months; HR = 0.70, p = <0.001). The Iressa NSCLC Trial Evaluating REsponse and Survival versus Taxotere (INTEREST) was a phase III clinical trial with a non-inferiority design that evaluated gefitinib versus standard second-line cytotoxic chemotherapy docetaxel. The trial met its non-inferiority endpoint.[19] The Iressa Pan-Asian Study (IPASS) was a second non-inferiority trial in the first-line metastatic setting, evaluated gefitinib and standard first-line chemotherapy with carboplatin and paclitaxel in patients of Asian descent, with adenocarcinoma histology and never smoking/former light smoker status. This design was based on the observation that the patients with these clinical features appeared to respond better to EGFR TKIs.[13,14,15,16] The trial met its non-inferiority endpoint, and has become a pivotal trial for the validation of the EGFR activating mutations as a predictive biomarker.[20]

Two phase III trials have assessed cetuximab in combination with platinum-doublets. The FLEX (first-line Erbitux) trial compared the addition of cetuximab with cisplatin and vinorelbine in the first-line setting for patients with advanced NSCLC. The addition of cetuximab resulted in a statistically significant 1.1 month overall survival benefit and a superior response rate (36% vs. 29%).[21] The BMS-099 trial assessed whether cetuximab was beneficial in combination with carboplatin and a taxane. The primary endpoint of the trial was progression-free survival (PFS). No statistically significant benefit in terms of the primary endpoint was detected with the trial. There was a 1.3 month benefit on OS detected, although this was not statistically significant.[22]

Table 6.1 Selected phase III trials for EGFR inhibitors in NSCLC

Trial	Agent	# of participants	Enrollment criteria	Comparison arms	Results
ISEL[29]	Gefitinib	1692	Advanced NSCLC in second- or third-line setting	Gefitinib vs. placebo	Primary endpoint of OS was not met (HR = 0.89, $p = 0.087$, 5.6 months vs. 5.1 months)
INTEREST[19]	Gefitinib	1466	Advanced NSCLC in second-line setting	Gefitinib vs. docetaxel	Non-inferiority of gefitinib for OS was met (HR = 1.02, 96% CI 0·905–1·150, 7.6 months vs. 8.0 months)
IPASS[20,30]	Gefitinib	1217	Advanced NSCLC in first-line setting	Gefitinib vs. carboplatin/ paclitaxel	Gefitinib demonstrated a superior PFS in *EGFR* mutated patients (HR = 0.48, $p < 0.001$). Gefitinib was inferior in *EGFR* wild-type patients (HR = 2.85, $p < 0.001$). No OS benefit was identified in final analysis between the two arms in *EGFR* mutated patients.
BR21[18]	Erlotinib	731	Advanced NSCLC in second- and third-line setting	Erlotinib vs. placebo	Primary endpoint of OS was met (HR = 0.70, $p = <0.001$, 6.7 months vs. 4.7 months)

Trial	Agent	# of participants	Enrollment criteria	Comparison arms	Results
OPTIMAL[31]	Erlotinib	165	Advanced NSCLC in first-line setting in patients with activating EGFR mutations	Erlotinib vs. carboplatin/gemcitabine	Primary endpoint of PFS benefit was met for erlotinib (HR = 0.16, $p < 0.0001$, 13.1 months vs. 4.6 months)
FLEX[21]	Cetuximab	1125	Advanced NSCLC in first-line setting	Cisplatin/vinorelbine plus cetuximab vs. cisplatin/vinorelbine alone	Primary endpoint of OS was met (HR = 0·871, $p = 0·044$, 11.3 months vs. 10.1 months)
BMS-099[22]	Cetuximab	676	Advanced NSCLC in first-line setting	Carboplatin/taxane plus cetuximab vs. carboplatin/taxane alone	Primary endpoint of OS was not met (HR = 0.89, $p = 0.169$, 9.69 months vs. 8.38 months)

Despite the positive findings across trials in the first, second, and third-line setting, it was evident that the survival benefit from these agents in unselected populations was modest (see Table 6.1); however, it was also apparent that some patients enjoyed a rapid and long-lasting response. As a result, efforts were initiated to identify predictive biomarkers for the EGFR-directed agents. The markers most consistently associated with benefit are mutations in *EGFR*. Additional markers that have been evaluated include *EGFR* gene copy number, receptor and ligand expression and mutations in downstream components such as the *RAS* gene. A summary of the known biomarkers in NSCLC and their methods of detection are presented in Tables 6.2 and 6.3, respectively.

Table 6.2 Evidence from clinical trials for predictive and prognostic factors in NSCLC

Biomarker	Predictive for EGFR TKIs	Predictive for anti-EGFR MoAbs	Prognostic	Predictive for toxicity
Activating *EGFR* mutations	Phase III evidence provided by IPASS and OPTIMAL[20,31]	FLEX[21]	BR21 and TRIBUTE[35,36]	Unknown
EGFR Amplification	BR21 and ISEL[33,48,49,50]	FLEX[21]	Limited retrospective data[61,62]	Unknown
***KRAS* Mutations**	BR21, INTEREST and SATURN[43,73,74]	FLEX and BMS099[57,70]	Retrospective work and phase III adjuvant study data (BR10)[67,68]	Unknown
TGF-α	BR21[79]	Unknown	Retrospective work from BR21[79]	Unknown
Amphiregulin	BR21[79]	Unknown	BR21[79]	Unknown
Short *EGFR* Intron 1 repeats	Uncertain	Unknown	Uncertain	Retrospective evidence[86]

Table 6.3 Methods for detecting biomarkers

Biomarker	Method for detection
***EGFR* Activating Mutations**	One step reverse transcription/PCR followed by PCR and DNA sequencing via the Sanger Method (termination sequencing through the utilization of di-deoxynucleic acids)[23] Direct DNA sequencing from tumor samples via the Sanger method following nested PCR amplification[27,33]
EGFR Protein Overexpression	IHC with >10% staining considered positive[33]
***EGFR* Gene Overexpression**	FISH probes designed for chromosome 7p12 and centromeric region[33,48]
***Kras* Mutations**	PCR amplification of Kras gene from extracted tumor cells followed by DNA sequencing via the Sanger Method[23]
EGFR Ligand Evaluation	ELISA testing[79]
***EGFR* Polymorphisms**	Direct DNA sequencing via the Sanger method[82,88]

Abbreviations: PCR, polymerase chain reaction; IHC = immunohistochemistry; FISH = fluorescence *in situ* hybridization; ELISA, enzyme-linked immunosorbent assay.

6.5 Activating EGFR Receptor Mutations

The presence of activating *EGFR* receptor mutations have proven to be an important predictor of benefit for patients receiving EGFR TKIs and is the first biomarker for a molecularly targeted therapy that has been widely adopted into clinical practice for patients with NSCLC. *EGFR* mutations are reported in approximately 10% of North American and European patients with NSCLC but in 30–50% of patients of east Asian descent. Activating mutations lead to hyperactivation of the kinase domain, resulting in sustained activation of the downstream pathways regulated by EGFR. Most activating mutations involve exons 18–21, which encode the ATP binding pocket of the tyrosine kinase domain.[23] EGFR mutations have been divided into three major classes. Class I mutations involve in-frame deletions of exon 19 at amino acid positions 747 and 749, respectively. Class II mutations involve single base pair substitutions, creating single amino acid changes. The most common Class II mutation is L858R, where arginine is substituted for leucine. Class III mutations involve in-frame duplications or insertions in exon 20.[24] Comparisons of mutant receptors with that of the wild-type demonstrate a threefold increase in amount of activated receptor in the presence of EGF ligand and a prolonged period of activation (15 minutes versus 3 hours).[25] In addition, when compared with wild-type receptor, it appears that Class I and L858R mutations results in a protein that has a higher Km for ATP and a lower Ki for erlotinib.[26] This suggests that these mutant receptors preferentially bind its inhibitor compared with its activator.

The discovery of activating *EGFR* mutations came from retrospective evaluation of clinical trials evaluating EGFR TKIs. Initial phase III trials were conducted in combination with standard first-line chemotherapy.[13,14,15,16] These trials in total demonstrated that there was no additional benefit with the addition of EGFR TKIs to standard first-line chemotherapy; however, subgroup analyses of the trials suggested patients with adenocarcinomas and no smoking history were more likely to respond to EGFR TKIs. This observation was supported by two phase II trials evaluating gefitinib following first-line platinum-based chemotherapy. Both of these trials demonstrated that gefitinib had anti-tumor activity in this setting, with significant responses seen in patients meeting the above clinical profile. In the evaluation of clinical responders in the phase II trials, it was reported that mutations to the ATP binding pocket of EGFR resulted in a significant response to EGFR TKIs and that the frequency of *EGFR* mutations were much higher in samples of patients of east Asian descent than those from North America.[25,27,28] This finding was supported by data from the ISEL (IRESSA Survival Evaluation in Lung cancer) trial, a phase III randomized trial of gefitinib versus best supportive care, which failed to demonstrate either a median or 1-year survival benefit in the second- or third-line setting patients with advanced NSCLC; a pre-planned subgroup analysis showed a significant benefit only in patients of Asian origin and non-smokers, the clinical profile identified in previous work.[29]

Based on this preliminary work, the IPASS study was initiated to see whether chemotherapy or gefitinib was the superior treatment for this select population. As it was demonstrated that activating *EGFR* mutations occurred at a higher frequency among patients of Asian heritage with adenocarcinoma histology with a never/light smoking history, this clinical profile was required for participation in the trial. The trial compared gefitinib with platinum chemotherapy (carboplatin/paclitaxel) in the first-line setting with the objective of demonstrating that gefitinib was non-inferior to platinum-based doublet chemotherapy. In addition, the trial evaluated *EGFR* mutation status for the participants and was successful in doing so in approximately a third of the trial participants. IPASS demonstrated that the *EGFR* mutation status of a patient's tumor was a strong predictor of benefit from gefitinib. Pre-planned sub-group analyses showed that PFS was significantly longer for gefitinib in patients with *EGFR* mutation positive tumors. In contrast, doublet chemotherapy was superior in patients with *EGFR* mutation negative tumors.[20] In the trial's final analysis, IPASS failed to demonstrate an overall survival advantage between the two arms in patients with activating *EGFR* mutations, although this is believed to be due to the use of subsequent lines of therapy, which would have included EGFR TKIs in the group initially receiving cytotoxic chemotherapy.[30] The OPTIMAL trial randomized Asian patients with advanced NSLCLC who were known to have activating *EGFR* mutations to either erlotinib or carboplatin/gemcitabine in the first-line setting. Patients receiving erlotinib had a significantly improved PFS (13.1 months vs. 4.6 months; HR = 0.16, p = <0.0001).[31] This trial has confirmed that activating *EGFR* mutations are predictive for benefit from EGFR TKIs in the first-line setting. The molecular profiling of patients with advanced NSCLC for *EGFR* mutations to enable selection of first-line therapy with either chemotherapy (*EGFR* mutation negative) or gefitinib (*EGFR* mutation positive) has now become standard of care in many jurisdictions.

The role of *EGFR* mutation status after first-line therapy is less clear. In the second-line setting, the INTEREST study demonstrated that patients with activating *EGFR* mutations had a significantly longer progression-free survival (PFS) with gefitinib than with docetaxel (7.0 vs. 4.1 months; HR = 0.16; p = 0.001); however, no overall survival advantage was demonstrated for patients with *EGFR* mutation positive tumors.[32] In the BR21 study, a phase III randomized trial, demonstrated that erlotinib increased median survival by approximately 2 months compared with best supportive care alone in the second- or third-line setting. The evaluation of tissue samples from patients enrolled in BR21 failed to demonstrate a statistically significant association between activating mutations and benefit from erlotinib.[33] There are a number of possibilities for this discrepancy. One may be the limited statistical power of the analysis; only 27% of the patients enrolled in the study were able to have tissue evaluated for mutation status, which is compounded by the rarity of EGFR mutations in non-Asian populations. The third issue is whether activating *EGFR* mutations remain equally predictive outside of the first-line setting. Currently, erlotinib is indicated as a possible treatment in the second- and third-line setting for treatment of advanced NSCLC regardless of activating *EGFR* mutation status.

Although it has been demonstrated that activating *EGFR* mutations are predictive for EGFR TKIs, it is uncertain whether this also applies for the anti-EGFR MoAb cetuximab. Cetuximab was evaluated in FLEX and BMS-099, two phase III clinical trials. Unlike the work in EGFR TKIs, these trials focused on EGFR overexpression as the biomarker of interest. At the current time, there is no evidence that *EGFR* activating mutations predicts for response to cetuximab.[34]

There is accruing evidence that EGFR mutations are also prognostic of an improved outcome. In the placebo arm of BR21, patients with activating *EGFR* mutations demonstrated a longer median survival time than patients with wild-type variants.[35] This possibility is supported by a subset analysis of the TRIBUTE study, a phase III trial of erlotinib plus chemotherapy versus chemotherapy alone. This analysis suggested that there were significantly longer survival times in patients with *EGFR* mutations compared with those who had wild-type *EGFR* when treated with chemotherapy alone.[36] In contrast, mutational status may not affect the outcome of patients with early-stage NSCLC. A Japanese study of patients with early-stage lung cancer who had undergone surgical resection NSCLC suggested that *EGFR* activating mutations did not affect survival.[37]

6.6 Biomarkers for Acquired Resistance to EGFR TKIs

Although some patients experience long-lasting remissions from EGFR TKIs, it is not curative and acquired resistance ultimately develops. The most common mutation of acquired resistance is the threonine to methionine mutation at codon 790 (T790M) of the EGFR, which occurs in 50% of all patients with acquired drug resistance to EGFR TKIs).[24,38] *In vitro* experiments have shown that the mutation reduces the inhibitory effects of erlotinib and gefitinib on the EGFR receptor.[39,40] Two mechanisms of resistance have been proposed. The first is that the mutation results in an alteration in the ATP binding pocket, which prevents the binding of reversible EGFR TKIs through steric interactions.[39,40] The second involves an increase in the affinity of the kinase domain for ATP, therefore reducing the ability of gefitinib or erlotinib to competitively inhibit the binding of ATP.[41]

Although T790M is believed to be acquired through selective pressure during treatment, the utilization of a highly specific assay has shown that low levels of T790M can be detected even in pretreated samples. In this report, Maheswaran and colleagues demonstrated that the presence of T790M prior to treatment resulted in a significantly shorter PFS compared with those patients where the mutation was not detectable.[42] This suggests that T790M may be a useful pretreatment biomarker for predicting benefit to EGFR TKIs. Other resistance mutations have been identified, such as aspartic acid to tyrosine at codon 261, which weakens the ability of the TKIs to bind to the EGFR receptor.[43] Unfortunately, identification of other resistance biomarkers in clinical practice has been challenging due to the lack of tumor tissue acquired from patients at the time of progressive disease.

6.7 EGFR Gene Amplification and Increased Protein Levels

Overexpression EGFR mRNA and protein has been implicated in tumor development in the lung since the 1980s, when it was first noted that abnormalities in the expression levels of EGFR occurred in NSCLC. Laboratory research demonstrated that NSCLC cell lines were found to overexpress the EGFR receptor compared with small cell lung cancer (SCLC) cell lines and that there was increasing receptor expression with advancing stage of disease.[44,45] Retrospective analyses report EGFR protein overexpression in 62% of the NSCLC, more frequently in adenocarcinomas, and that gene copy number correlated with protein expression.[46,47] These observations have provided the basis for evaluating *EGFR* gene amplification and protein expression as a biomarker for benefit from anti-EGFR therapy.

A high *EGFR* gene copy number, through either gene amplification or high polysomy, may predict for response from EGFR TKIs. Cappuzzo and colleagues demonstrated that patients with *EGFR* overexpression detected by fluorescence *in situ* hybridization (FISH) had a higher objective response rate (36% vs. 3%), a superior time to progression (9 months vs. 2.5 months) and median overall survival (19 months vs. 7 months) with gefitinib than those who were FISH negative.[48] In subset analysis of BR21, it was found that patients who were FISH positive demonstrated a superior response rate and an overall survival benefit with erlotinib in both univariate and multivariate analysis.[33,49] The ISEL trial also demonstrated a survival benefit for patients with a high *EGFR* gene copy number with benefits even in those with poor prognostic factors, such as smokers and patients with non-adenocarcinoma histology.[50] A preplanned subgroup analysis of the INTEREST trial evaluated patients with increased *EGFR* copy number. In this selected population, gefitinib did not demonstrate superiority to docetaxel.[51] A study of Japanese patients who were randomized to either gefitinib or docetaxel reported similar conclusions with regards to increased *EGFR* copy number as INTEREST.[52]

Assessing for overexpression of EGFR protein through immunohistochemistry (IHC) has also been evaluated. Tumors overexpressing EGFR through IHC (defined as at least 10% of cells positive for EGFR protein expression on immunohistochemistry) was found to correlate with response rate in the BR21 study, and survival in patients with EGFR positive and unknown tumors in univariate analysis.[33] In contrast, the ISEL study found that IHC positive patients did have a survival benefit with gefitinib compared with IHC negative patients.[50] Erlotinib is currently indicated for patients with advanced NSCLC with EGFR positive or unknown status and has been subsequently been utilized as an eligibility criterion in a currently active trial of adjuvant erlotinib either through FISH or IHC.[53,54]

It is possible that the combination of the two techniques may offer the best method for predicting benefit to TKI treatment. Hirsch and colleagues found that patients who were both IHC and FISH positive for EGFR overexpression had a significantly better response rate, progression free and overall survival compared with patients who were negative via both techniques.[55] This suggests that both IHC and FISH are independently predictive of benefit and when utilized together represent a significant biomarker for benefit from TKIs.

Overexpression of EGFR protein has also been explored as a predictive factor for cetuximab. Two phase III trials have provided some insight whether IHC methods are capable of evaluating EGFR protein overexpression as a predictive factor. The FLEX trial, which demonstrated a statistically significant survival benefit, required that participants have EGFR overexpression detected through IHC for inclusion in the trial.[21] In contrast to FLEX, the BMS-099 trial did not require pre-selection based on EGFR expression status and reported no statistically significant benefit.[56] A subsequent retrospective evaluation of the BMS-099 trial found no association between EGFR overexpression and benefit from therapy.[57] The Southwest Oncology Group explored the use of FISH as a predictive factor for cetuximab. In a phase II trial, patients who were FISH-positive had a superior progression-free survival and overall survival with the use of cetuximab compared with patients who were FISH-negative.[58] Although not conclusive, these results suggest that EGFR overexpression may predict for benefit from cetuximab therapy in patients with advanced NSCLC.

Overexpression HER2 protein with or without overexpression of EGFR may also be predictive for patients with NSCLC. Approximately 18–33% of NSCLC patients will demonstrate HER2 overexpression via IHC and overexpression correlates with a poorer prognosis.[59] One study of patients found that patients who were both *EGFR* and *HER2* positive via FISH analysis had a superior response rate, time to progression and overall survival when treated with gefitinib.[60]

EGFR overexpression as a prognostic factor for patients with NSCLC has been evaluated. In a 2004 study, Selvaggi and colleagues demonstrated a significant correlation between EGFR overexpression and a worse overall survival in patients with completely resected stage I-IIIA NSCLC.[61] In contrast, Kanematsu and colleagues showed that a higher degree of EGFR phosphorylation, rather than receptor overexpression, was what correlated with a shorter time to progression and overall survival.[62] Both studies evaluated only a small number of patients, and the significance of the findings is uncertain.

6.8 *KRAS* Mutations and Anti-EGFR Therapy

Activating mutations to the oncogene *KRAS* occurs in approximately 30% of adenocarcinomas and 5% of squamous cell carcinomas of the lung.[63] *KRAS* activating mutations are associated with a history of tobacco use with its frequency varying among different ethnic groups. For example, its frequency is high among African-Americans and low among Asians after exposure to either cigarette smoke or environmental carcinogens.[64] RAS is a critical molecule in the oncogenesis pathway as overexpression and activation of EGFR leads to hyperactivation of Ras pathway. In addition, activation of RAS leads to transcriptional activation of the gene responsible for transforming growth factor α (TGF-α), which is a ligand for EGFR; this can lead to persistent activation of the EGFR pathway and the PI3K/Akt survival pathways.[65]

The potential prognostic and predictive implications of *KRAS* mutation status has been evaluated in numerous studies, however, many have been in studies of small sample size.[66] A meta-analysis of 5216 patients over 43 studies has suggested that activating *KRAS* mutations are a negative prognostic factor for survival (HR = 1.35), particularly in patients with adenocarcinomas.[67] *KRAS* mutation status was prospectively evaluated in 93% of patients in NCIC CTG BR10 trial of post operative chemotherapy in patients with stage IB or II NSCLC; no difference in survival between patients with wild-type *KRAS* versus mutant *KRAS* enrolled in the observation arm.[68] The prognostic significance of *KRAS* mutations remains uncertain.

KRAS mutations have proven to be a clinically important predictive factor for EGFR MoAbs in colorectal cancer[69]; however, this is uncertain in NSCLC. Both the FLEX and BMS-099 (phase III randomized trials evaluating the addition of cetuximab to platinum-based chemotherapy in the first-line setting) evaluated markers to identify predictors of benefit. With regards to *KRAS* mutations, neither the FLEX trial or BMS-099 trials found that *KRAS* mutations were predictive for cetuximab efficacy when combined with first-line chemotherapy in advanced NSCLC.[57,70]

Two meta-analyses have been performed evaluating activating *KRAS* mutations and response to EGFR TKIs. The first analyzed 17 studies for a total of 165 *KRAS* mutant patients while the second evaluated 22 studies for a total of 231 *KRAS* mutant patients. Both studies found a lack of response to TKIs in patients with *KRAS* mutations, although this did not result in a significant difference in PFS or OS.[71,72] In a multivariate analysis, the NCIC BR21 study found that *KRAS* mutations were not predictive of a differential treatment effect.[49] The INTEREST trial found no differences between treatments in OS, PFS or response rate according to *KRAS* mutation status in patients who received gefitinib.[73] The Sequential Tarceva in Unresectable NSCLC (SATURN) trial, which evaluated maintenance erlotinib in patients without progression of disease after 4 cycles of first-line cytotoxic chemotherapy, found no differences in PFS between wild-type and mutant *KRAS* patients.[74] *KRAS* mutation status in the ISEL trial could not be evaluated. It should be noted that most of these evaluations were performed on only a fraction of trial participants due to a lack of tissue availability. At the current time, there is no evidence suggesting that *KRAS* mutations affect EGFR TKI benefit in NSCLC.

6.9 EGFR Ligands

Ligands that target the epidermal growth factor receptor family consist of three main groups. The first group binds specifically to EGFR and consists of EGF, TGF-α, and amphiregulin. The second group has specificity for both EGFR and HER4 and includes betacellulin and heparin-binding EGF as members. The third group binds HER3 and HER4 and is composed of the neuregulins.[75]

Relative to the evaluation of other EGFR biomarkers, there is a paucity of data on the prognostic and predictive implications of the EGFR ligands. Within the EGF ligand family, TGF-α and amphiregulin have been studied more extensively, either as circulating ligands within the blood or within the tumor. However, most work has been undertaken in single cohorts of patients with inconsistent results reported.[76–78] Currently, the only published study to evaluate the prognostic and predictive significance of circulating ligands within the context of a NSCLC clinical trial containing a placebo arm is a retrospective analysis of the BR21 clinical trial of erlotinib. Baseline plasma was collected from 565 patients of the randomized 731 patients, which was analyzed for TGF-α and amphiregulin levels. In multivariate analysis, high amphiregulin levels (defined as ≥ 10pg/mL) were prognostic for a poorer outcome compared with patients with low amphiregulin, but did not predict for benefit from erlotinib. Conversely, high TGF-α (defined as >12 pg/mL) was predictive of lack of benefit from erlotinib but was not prognostic.[79] These results are exploratory and require validation. Furthermore, the prognostic and predictive significance of EGF ligands in patients defined by other EGFR biomarkers is unclear. For example, in a small series of patients with EGFR mutation positive NSCLC, Masago and colleagues reported that five of the six patients who had primary resistance to gefitinib had TGF-α and amphiregulin positive serum.[80] This observation requires further evaluation, but as subsequent phase III trials prospectively include ligand analyses, the role of ligand expression may be better elucidated.

6.10 Polymorphism Studies and Anti-EGFR Therapy

Functional germline polymorphisms within the *EGFR* gene, pathway or related genes including the multidrug transporter *ABCG2* have been evaluated for their association with outcome and toxicity from EGFR active agents.[81] To date, the majority of published analyses have utilized a candidate gene approach focusing on polymorphisms for which there is evidence of functional change, primarily within the *EGFR* gene.

Intron 1 within the *EGFR* gene contains a *CA* sequence repeat 1 varying from 14 to 22 *CA* dinucleotide repeats. Shorter repeats are associated with higher transcription rates and increased *EGFR* gene expression compared with the longer sequences.[82,83] However, the reported association between sequence repeat length and toxicity or outcome from EGFR TKIs has been inconsistent across studies. In a study of 70 patients treated with EGFR TKIs, Nie *et al.* reported an association between shorter *CA* repeats (defined as ≤16 repeats) and superior response and survival than patients with >16 *CA* repeats.[84] Similar results were demonstrated in a study of 92 gefitinib treated patients, with patients carrying shorter *CA* repeats enjoying a superior PFS than patients with longer repeats.[85] Huang *et al.* described an association with shorter *CA* repeat and the presence of skin rash.[86] However, this was not the finding when evaluated using 242 samples collected as part of the NCIC CTG BR.21 clinical trial in which patients with at least one short allele (<16 repeats) randomized to the placebo had an improved PFS compared with those patients that had homozygous long alleles, whereas no association for patients randomized to the erlotinib arm.[87] This may indicate that *intron 1 CA* repeats has prognostic but not predictive significance.

Other *EGFR* polymorphisms of interest include *-216G/T* and *-191C/A*, which are located in the gene promoter at transcription factor binding sites including SP1.[88] Again, studies that have evaluated the association between these genetic polymorphisms and outcome have reported inconsistent results[89,90]; however, these were not shown to be predictive of benefit from erlotinib when evaluated retrospectively as part of the BR.21 clinical trial.[87]

While germline polymorphisms may contribute to the understanding of variation in benefit and toxicity experienced by patients on EGFR TKIs, results are inconsistent and currently exploratory in nature. Studies conducted in small cohorts of patients may have contributed to the inconsistent findings and further studies are needed to define the role of germline variation and outcome and toxicity of EGFR inhibitory agents.

6.11 Circulating Tumor Cells in NSCLC Biomarker Research

Successful evaluation of tumor biomarkers requires adequate tissue sampling in order to perform the required analysis. Unfortunately, tissue acquisition has been limited in NSCLC. Frequently, the initial diagnosis of advanced disease is generally made utilizing fine-needle aspiration or bronchial washing methods; unfortunately, these techniques make it more difficult to acquire an adequate sample for molecular analysis. For example, fewer than 50% of patients in the BR21, INTEREST and IPASS clinical trials were evaluated for

EGFR mutations status, although this low number may be partially due to patient refusal to consent for tissue analysis, which was optional.[20,32,33] In addition, validation and identification of predictive biomarkers appears to be most successful when tissue acquisition occurs shortly before initiation of the experimental therapy as this minimizes the opportunity for tumors to undergo further genetic mutation that may influence the outcome of therapy.[91] However, performing numerous invasive procedures strictly for the purposes of research poses ethical, cost and logistic challenges in clinical trials. Circulating tumor cells (CTCs), which are malignant cells detected in the bloodstream, represents an opportunity to repeatedly acquire tumor cells in a minimally invasive manner and at the time of treatment decisions.

One of the limitations to the use of CTCs in biomarker research is their acquisition. In general, patients with metastatic cancer will have a CTC frequency of one per billion blood cells, thus purification of CTCs is necessary before any analysis can be performed.[92] Immunomagnetic bead-based and microfluidic chip-based immunologic techniques, two available methods for isolating CTCs, have a purification rate of 0.1% and 1–80%, respectively.[93,94] These purification techniques, and their further refinement, represent an important step in the development of this field.

Although the number of CTCs have been shown to be prognostic in a number of tumor sites, their uses in NSCLC are still very preliminary. Nevertheless, it has been successfully demonstrated that CTCs can be utilized to identify resistance mutations to EGFR TKIs in patients receiving these medications.[42] This work is suggestive that CTCs can play an important role in biomarker research in the future.

6.12 Conclusions

Although first-generation EGFR inhibitors have demonstrated an advance in the treatment of advanced NSCLC, most of the benefits of treatment appear to occur in appropriately selected patients; when used in unselected patients, overall benefits are modest. Biomarkers have the potential to select patients most likely to benefit from therapy, avoiding those who are unlikely to respond, and potentially improving the cost effectiveness of treatment.[87] Some success has been realized in the first-line palliative setting with the validation of *EGFR* mutations as a predictive biomarker for gefitinib; however, many questions remain unanswered. It is apparent that biomarker discovery and validation has been hampered by the limited availability of tissue within clinical trials. In the future, it is hoped that the incorporation of biomarker discovery in parallel with the drug development process with emerging technologies such as the collection of CTCs from blood samples will assist the discovery and validation of biomarkers for the next generation of trials of EGFR active agents in NSCLC. However, despite challenges and unanswered questions, the EGFR inhibitors in lung cancer have highlighted the heterogeneity of NSCLC as a disease that can no longer be treated uniformly.

References

1. American Cancer Society. Cancer Facts & Figures 2013. Atlanta: American Cancer Society, 2013.
2. Yarden Y, Sliwkowski MX. Untangling the ErbB signaling network. *Nat Rev Mol Cell Biol* 2, 127–137 (2001).
3. Garrett TP, McKern MN, Lou M, *et al.* The crystal structure of a truncated ErbB2 ectodomain reveals an active conformation, poised to interact with other ErbB receptors. *Mol Cell* 11, 495–505 (2003).
4. Graus-Porta D, Beerli RR, Daly JM, *et al.* ErbB-2, the preferred heterodimerization partner of all ErbB receptors, is a mediator of lateral signaling. *EMBO J* 16, 1647–1655 (1997).
5. Lurje G, Lenz HJ. EGFR signaling and drug discovery. *Oncology* 77, 400–410 (2009).
6. Hill CS, Treisman R. Transcriptional regulation by extracellular signals: mechanisms and specificity. *Cell* 80, 199–211 (1995).
7. Vivanco I, Sawyers CL. The phosphatidylinositol 3-kinase AKT pathway in human cancer. *Nat Rev Cancer* 2, 489–501 (2002).
8. Stokoe D, Stephens LR, Copeland T, *et al.* Dual role of phosphatidylinositol-3-4-5-triphosphate in the activation of protein kinase B. *Science* 277, 567–570 (1997).
9. Yu H, Jove R. The STATs of cancer—new molecular targets come of age. *Nat Rev Cancer* 4, 97–105 (2004).
10. Leu TH, Maa MC. Functional implication of the interaction between EGF receptor and c-src. *Front Biosci* 8, s28–s38 (2003).

11. Li S, Schmitz KR, Jeffrey PD, *et al.* Structural basis for inhibition of the epidermal growth factor receptor by cetuximab. *Cancer Cell* 7, 201–311 (2005).
12. *Ciardiello F, Tortora G. EGFR Antagonists in Cancer Treatment. NEJM* 358, 1160–1174 (2008).
13. Giaccone G, Herbst RS, Manegold C, *et al.* Gefitinib in combination with gemcitabine and cisplatin in advanced non-small cell lung cancer: a phase III trial—INTACT1. *JCO* 22, 777–784 (2004).
14. Herbst RS, Giaccone G, Shiller JH, *et al.* Gefitinib in combination with paclitaxel and carboplatin in advanced non-small-cell lung cancer: a phase III trial—INTACT 2. *JCO* 22, 785–794 (2004).
15. Herbst RS, Prager D, Hermann R, *et al.* TRIBUTE: a phase III trial of erlotinib hydrochloride (OSI-774) combined with carboplatin and paclitaxel chemotherapy in advanced non-small-cell cancer. *JCO* 23, 5892–5899 (2005).
16. Gatzemeier U, Pluzanska A, Szczesna A, *et al.* Phase III study of erlotinib in combination with cisplatin and gemcitabine in advanced non-small-cell lung cancer: the Tarceva Lung Cancer Investigation Trial. *JCO* 25, 1545–1552 (2007).
17. http://clinicaltrials.gov/show/NCT00373425.
18. Shepherd FA, Rodrigues Pereira J, Ciuleanu T, *et al.* Erlotinib in previously treated non-small-cell lung cancer. *NEJM* 353, 123–32.
19. Kim ES, Hirsh V, Mok T, *et al.* Gefitinib versus docetaxel in previously treated non-small-cell lung cancer (INTEREST): a randomised phase III trial. *Lancet* 372, 1809–1818 (2008).
20. Mok TS, *et al.* Gefitinib or carboplatin-paclitaxel in pulmonary adenocarcinoma. *NEJM* 361, 947–957 (2009).
21. Pirker R, Pereira JR, Szczesna A, *et al.* Cetuximab plus chemotherapy in patients with advanced non-small-cell lung cancer (FLEX): an open-label randomised phase III trial. *Lancet* 373, 1525–1531 (2009).
22. Lynch TJ, Patel T, Dreisbach L, *et al.* Cetuximab and first-line taxane/carboplatin chemotherapy in advanced non-small-cell lung cancer: results of the randomized multicenter phase III trial BMS099. *JCO* 28, 911–917 (2010).
23. Kosaka T, Yatabe Y, Endoh H, *et al.* Mutations of the epidermal growth factor receptor gene in lung cancer: biological and clinical implications. *Cancer Res* 64, 8919–8923 (2004).
24. Gazdar AF. Activating and resistance mutations of EGFR in non-small-cell lung cancer: role in clinical response to EGFR tyrosine kinase inhibitors. *Oncogene* 28, s24–s31 (2009).

25. Lynch TJ, Bell DW, Sordella R, *et al.* Activating mutations in the epidermal growth factor receptor underlying responsiveness of non-small cell lung cancer to gefitinib. *NEJM* 350, 2129–2139 (2004).

26. Carey KJ, Garton AJ, Romero MS, *et al.* Kinetic analysis of epidermal growth factor receptor somatic mutant proteins shows increased sensitivity to the epidermal growth factor receptor tyrosine kinase inhibitor, erlotinib. *Cancer Res* 66, 8163–8171 (2006).

27. Paez JG, Janne PA, Lee JC, *et al.* EGFR mutations in lung cancer: correlation with clinical response to gefitinib therapy. *Science* 304, 1497–1500 (2004).

28. Pao W, Miller V, Zakowski M, *et al.* EGF receptor gene mutations are common in lung cancers from "never smokers" and are associated with sensitivity of tumors to gefitinib and erlotinib. *Proc Natl Acad Sci USA* 101, 13306–13311 (2004).

29. Thatcher N, Chang A, Parikh P, *et al.* Gefitinib plus best supportive care in previously treated patients with refractory advanced non-small cell lung cancer: results from a randomized, placebo-controlled, multicentre study (Iressa Survival Evaluation in Lung Cancer). *Lancet* 366, 1527–1537 (2005).

30. Yang C-H, Fukuoka M, Mok TS, *et al.* Final overall survival results from a phase III randomized, open-label, first-line study of gefitinib V carboplatin/paclitaxel in clinically selected patients with advanced non-small cell lung cancer in Asia. 35th ESMO Congress. Abstract LBA2.

31. Wu Y-L, Zhou C, Chen G, *et al.* first biomarker analyses from a phase iii, randomised, open-label, first-line study of erlotinib versus carboplatin (CBDCA) plus gemcitabine (GEM) in chinese patients (pts) with advanced non-small-cell lung cancer (NSCLC) with EGFR activating mutations (OPTIMAL, CTONG 0802) [Abstract] *Ann Oncol* 21s, LBA 14 (2010).

32. Douillard J, Hirsh V, Mok TS, *et al.* Molecular and clinical subgroup analyses from a phase III trial comparing gefitinib with docetaxel in previously treated non-small cell lung cancer (INTEREST) [abstract] *JCO* 26s, 8001 (2008).

33. Tsao MS, Chang KT, Wu LH, *et al.* Erlotinib in lung cancer—molecular and clinical predictors of outcome. *NEJM* 353, 133–144 (2005).

34. Morgensztern D, Gavidan R. Is there a role for cetuximab in non-small cell lung cancer. *Clin Cancer Res* 13, 4602s–4605s (2007).

35. Shepherd F, Tsao M. Unraveling the mystery of prognostic and predictive factors from epidermal growth factor receptor therapy. *JCO* 27, 1219–1220 (2006).

36. Eberhard DA, Johnson BE, Amler LC, *et al.* Mutations in the epidermal growth factor receptor and in KRAS are predictive and prognostic indicators in patients with non-small-cell lung cancer treated with chemotherapy alone and in combination with erlotinib. *JCO* 23, 5900–5909 (2005).
37. Kosaka T, Yatabe Y, Endoh H, *et al.* Mutations of the epidermal growth factor receptor gene in lung cancer: biological and clinical implications. *Cancer Res* 64, 8919–8923 (2004).
38. Kosaka T, Yatabe Y, Endoh H, *et al.* Mutations of the epidermal growth factor receptor gene in lung cancer: biological and clinical implications. *Cancer Res* 64, 8919–8923 (2006).
39. Kobayashi S, Boggon TJ, Dayaram T, *et al.* EGFR mutation and resistance of non-small cell lung cancer to gefitinib. *NEJM* 352, 786–792 (2005).
40. Pao W, Wang TY, Riely GJ, *et al.* Acquired resistance of lung adenocarcinomas to gefitinib and erlotinib is associated with a second mutation in the EGFR kinase domain. *PLoS Med* 2, e73 (2005).
41. Yun C-H, Mengwasser KE, Toms AV, *et al.* The T790M mutation in EGFR kinase causes drug resistance by increasing the affinity for ATP. *Proc Natl Acad Sci U S A* 105, 2070–2075 (2008).
42. Maheswaran S, Sequist LV, Nagrath S, *et al.* Detection of mutations in EGFR in circulating lung cancer cells. *NEJM* 359, 1–12 (2008).
43. Balak MN, Gong Y, Riely GJ, *et al.* Novel D761Y and common secondary T790M mutations in epidermal growth factor receptor-mutant lung adenocarcinomas with acquired resistance to kinase inhibitors. *Clin Cancer Res* 12, 6494–6501 (2006).
44. Haeder M, Rotsch M, Bepler G, *et al.* Epidermal growth factor receptor expression in human lung cancer cell lines. *Cancer Res* 48, 1132–1136 (1988).
45. Veale D, Ashcroft T, Marsh C, *et al.* Epidermal growth factor receptors in non-small cell lung cancer. *Br J Cancer* 55, 513–516 (1987).
46. Hirsch FR, Varella-Garcia M, Bunn Jr PA, *et al.* Epidermal growth factor receptor in non-small cell lung carcinomas: correlation between gene copy number and protein expression and impact on prognosis. *JCO* 21, 3798–3807 (2008).
47. Hirsch FR, Varella-Garcia M, Cappuzzo F. Predictive value of EGFR and HER2 overexpression in advanced non-small-cell lung cancer. *Oncogene* 28, s32–s37 (2009).
48. Cappuzzo F, Hirsch FR, Rossi E, *et al.* Epidermal growth factor receptor gene and protein and gefitinib sensitivity and non small cell lung cancer. *JNCI* 97, 643–655 (2005).

49. Zhu C-Q, da Cunha Santos G, Ding K, *et al.* Role of KRAS and EGFR as biomarkers for response to erlotinib in National Cancer Institute of Canada Clinical Trials Group study BR21. *JCO* 26, 4268–4275 (2008).

50. Hirsch FR, Varella-Garcia M, Bunn Jr PA, *et al.* Molecular predictors of outcome with gefitinib in a phase III placebo-controlled study in advanced non-small-cell lung cancer. *JCO* 24, 5034–5042 (2006).

51. Kim ES, Hirsh V, Mok T, *et al.* Gefitinib versus docetaxel in previously treated non-small-cell lung cancer (INTEREST): a randomised phase III trial. *Lancet* 372, 1809–1818 (2008).

52. Niho S, Ichinose Y, Tamura T, *et al.* Results of a randomized phase III study to compare the overall survival of gefitinib versus docetaxel in Japanese patients with non-small cell lung cancer who failed one or two chemotherapy regimens. *JCO* 25, 387s [abstract] (2007).

53. http://www.rochecanada.com/portal/servlet/staticfilesServlet?type=data&communityId=re753001&id=static/attachedfile/re7300002/re77300002/AttachedFile_09853.pdf.

54. Cancer Research UK: http://www.cancerhelp.org.uk/trials/a-trial-looking-at-erlotinib-after-treatment-for-non-small-cell-lung-cancer.

55. Hirsch FR, Varella-Garcia M, Cappuzzo F, *et al.* Combination of EGFR gene copy number and protein expression predicts outcome for advanced non-small-cell l ung cancer patients treated with gefitinib. *Ann Oncol* 18, 752–760 (2007).

56. Lynch TJ, Patel T, Dreisbach L, *et al.* Cetuximab and first-line taxane/carboplatin chemotherapy in advanced non-small-cell lung cancer: results of the randomized multicenter phase III trial BMS099. *JCO* 28, 911–917 (2010).

57. Khambata-Ford S, Harbison CT, Hart LL, *et al.* Analysis of potential predictive markers of cetuximab benefit in BMS099, a phase III study of cetuximab and first-line taxane/carboplatin in advanced non-small-cell lung cancer. *JCO* 28, 918–27 (2010).

58. Hirsch FR, Herbst RS, Olsen C, *et al.* Increased EGFR gene copy number detected by fluorescent in situ hybridization predicts outcome in non-small-cell lung cancer patients treated with cetuximab and chemotherapy. *JCO* 26, 3351–3357 (2008).

59. Hirsch FR, Franklin WA, Veve R, *et al.* HER2/neu expression in malignant lung tumors. *Semin Oncol* 29, 51–58 (2002).

60. Cappuzzo F, Varella-Garcia M, Shigematsu H, *et al.* Increased HER2 gene copy number is associated with response to gefitinib therapy in epidermal growth factor receptor-positive non-small cell lung cancer patients. *JCO* 23, 5007–5018 (2005).

61. Selvaggi G, Novello S, Torri V, *et al.* Epidermal growth factor receptor overexpression correlates with a poor prognosis in completely resected non-small-cell lung cancer. *Ann Oncol* 15, 28–32 (2004).

62. Kanematsu T, Yano S, Uehara H, *et al.* Phosphorylation, but not overexpression, of epidermal growth factor receptor is associated with poor prognosis of non-small cell lung cancer patients. *Oncol Res* 13,289–98 (2003).

63. Jemal A, Siegel R, Ward E, *et al.* Cancer statistics, 2008. *CA Cancer J Clin* 58, 71–96 (2008).

64. Hunt JD, Stimas A, Martin JE, *et al.* Differences in KRAS mutation spectrum in lung cancer cases between African Americans and Caucasians after occupational or environmental exposure to known carcinogens. *Cancer Epidemiol Biomarkers Prev* 11, 1405–1412 (2002).

65. Bardeesy N, Kim M, Xu J, *et al.* Role of epidermal growth factor receptor signaling in RAS driven melanoma. *Mol Cell Biol* 25, 4176–4188 (2005).

66. Roberts PJ, Stinchcombe TE, Channing J, *et al.* Personalized medicine in non-small-cell lung cancer: is KRAS a useful marker in selecting patients for epidermal growth factor receptor-targeted therapy? *JCO eprint* Oct 4th, (2010).

67. Mascaux C, Iannino N, Martin B, *et al.* The role of RAS oncogene in survival of patients with lung cancer: a systemic review of the literature with meta-analysis. *Br J Cancer* 92, 131–139 (2005).

68. Winton T, Livingston R, Johnson D, *et al.* Vinorelbine plus cisplatin vs. observation in resected non-small-cell lung cancer. *NEJM* 352, 2589–2597 (2005).

69. Karapetis CS, Khambata-Ford S, Jonker DJ, *et al.* KRAS mutations and benefit from cetuximab in advanced colorectal cancer. *NEJM* 359, 1757–1765 (2008).

70. O'Byrne, KJ, Bondarenko, I, Barrios, C, *et al.* Molecular and clinical predictors of outcome for cetuximab in non-small cell lung cancer (NSCLC): Data from the FLEX study. *JCO* 27:408s (Abstract #8007) (2009).

71. Linardou H, Dahabreh IJ, Kanaloupiti D, *et al.* Assessment of somatic k-RAS mutations as a mechanism associated with resistance to EGFR-targeted agents: a systematic review and meta-analysis of studies in advanced non-small-cell lung cancer and metastatic colorectal cancer. *Lancet Oncol* 9, 962–972 (2008).

72. Mao C, Qiu LX, Liao RY, *et al.* KRAS mutations and resistance to EGFR-TKIs treatment in patients with non-small cell lung cancer: a meta-analysis of 22 studies. *Lung Cancer* 69, 272–278 (2010).

73. Douillard JY, Shepherd FA, Hirsh V, *et al.* Molecular predictors of outcome with gefitinib and docetaxel in previously treated non-small-cell lung cancer: Data from the randomized phase III INTEREST trial. *JCO* 28, 744–752 (2010).

74. Brugger W, Triller N, Blasinska-Morawiec M, *et al.* Biomarker analyses from the phase III placebo-controlled SATURN study of maintenance erlotinib following first-line chemotherapy for advanced NSCLC. *JCO* 27, 411s (abstract 8020) (2009).

75. Hynes NE, Lane HA. ERBB Receptors and cancer: the complexity of targeted inhibitors. *Nate Rev Cancer* 5, 341–354 (2005).

76. Ishikawa N, Daigo Y, Takano A, *et al.* Increases of amphiregulin and transforming growth factor-alpha in serum as predictors of poor response to gefitinib among patients with advanced non-small cell lung cancers. *Cancer Res* 65, 9176–9184 (2005).

77. Fontanini G, De Laurentiis M, Vignati S, *et al.* Evaluation of epidermal growth factor-related growth factors and receptors and of neoangiogenesis in completely resected stage I-IIIA non-small cel lung cancer: Amphiregulin and microvessel count are independent prognostic indicators of survival. *Clin Cancer Res* 4, 241–249 (1998).

78. Tateishi M, Ishida T, Mitsudomi T, *et al.* Immunohistochemical evidence of autocrine growth factors in adenocarcinoma of the human lung. *Cancer Res* 50, 7077–7080 (1990).

79. Addison CL, Ding K, Zhao H, *et al.* Plasma transforming growth factor α and amphiregulin protein levels in National Cancer Institute of Canada Clinical Trials Group BR.21. *J Clin Oncol* 28: 5247–5256 (2010).

80. Masago K, Fujita S, Hatachi Y, *et al.* Clinical significance of pretreatment serum amphiregulin and transforming growth factor-alpha, and an epidermal growth factor receptor somatic mutation in patients with advanced non-squamous, non-small cell lung cancer. *Cancer Sci* 99, 2295–2301 (2008).

81. Suk Heist R, Christiani D. EGFR-targeted therapies in lung cancer: predictors of response and toxicity. *Pharmacogenomics* 10, 59–68 (2009).

82. Gebhart F, Zanker KS, Brandt B. Modulation of epidermal growth factor receptor gene transcription by a polymorphic dinucleotide repeat in intron 1. *J Biol Chem* 274, 13176–13180 (1999).

83. Etienne-Grimaldi MC, Pereira S, Magné N, *et al.* Analysis of the dinucleotide repeat polymorphism in the epidermal growth factor receptor (EGFR) gene in head and neck cancer patients. *Ann Oncol* 16, 934–941 (2005).

84. Nie Q, Wang Z, Zhang GC, *et al.* The epidermal growth factor receptor intron1 (CA)*n* microsatellite polymorphism is a potential predictor of treatment outcome in patients with advanced lung cancer treated with Gefitinib. *Eur J Pharmacol* 570, 175–181 (2007).

85. Liu G, Gurubhagavatula S, Zhou W, *et al.* Epidermal growth factor receptor polymorphisms and clinical outcomes in non-small-cell lung cancer patients treated with gefitinib. *Pharmacogenomics* 8, 129–138 (2008).

86. Huang C-L, Yang C-H, Yeh K-H, *et al. EGFR* intron 1 dinucleotide repeat polymorphism is associated with the occurrence of skin rash with gefitinib treatment. *Lung Cancer* 64, 346–351 (2009).

87. Liu G, Cheng D, Le Maitre A, *et al.* EGFR and ABCG2 polymorphisms as prognostic and predictive markers in the NCIC CTG BR.21 trial of single-agent erlotinib in advanced non-small cell lung cancer (NSCLC). [Abstract 7538] JCO 28s (2010).

88. Liu W, Innocenti F, Wu MH, *et al.* A functional common polymorphism in a Sp1 recognition site of the epidermal growth factor receptor gene promoter. *Cancer Res* 65, 46–53 (2005).

89. Gregorc V, Hidalgo M, Spreafico A, *et al.* Germline polymorphisms in EGFR and survival in patients with lung cancer receiving gefitinib. *Clin Pharmacol Ther* 83, 477–484 (2008).

90. Cusatis G, Gregorc V, Li J, *et al.* Pharmacogenetics of ABCG2 and adverse reactions to gefitinib. *JNCI* 98, 1739–1742 (2006).

91. Hodgson DR, Wellings R, Orr MCM, *et al.* Circulating tumor-derived predictive biomarkers in oncology. *Drug Discov Today* 15, 98–101 (2010).

92. Maheswaran S Haber DA. Circulating tumor cells: a window into cancer biology and metastasis. *Curr Opin Genet Dev* 20, 96–99 (2010).

93. Allard WJ, Matera J, Miller MC, *et al.* Tumor cells circulate in the peripheral blood of all major carcinomas but not in healthy subjects or patients with nonmalignant diseases. *Clin Cancer Res* 10, 6897–6904 (2004).

94. Nagrath S, Sequist LV, Maheswaran S, *et al.* Isolation of rare circulating tumour cells in cancer patients by microchip technology. *Nature* 450, 1235–1239 (2007).

Chapter 7

Markers of Sensitivity and Resistance to EGFR Inhibitors in Colorectal Cancer

Jose G. Monzon and Janet Dancey
NCIC Clinical Trials Group, Queen's University,
10 Stuart Street, Kingston, ON K7L 3N6, Canada

7.1 Introduction

Colorectal cancer (CRC) is the third most frequently diagnosed cancer, and the fourth leading cause of cancer deaths worldwide (608,000 deaths/year) [1]. When detected in its early stages colorectal cancer is curable with surgery and, in some high-risk cases, adjuvant chemotherapy. However, once colorectal cancer has spread to distant sites, patients are rarely curable and have a median survival of 18 to 21 months [2].

Colorectal cancer progresses in a stepwise fashion. It starts as a benign adenomatous polyp, developing into an advanced adenoma with high-grade dysplasia and ultimately an invasive malignancy. These steps are associated with distinct molecular changes and altered cellular signaling pathways that ultimately drive the

Handbook of Therapeutic Biomarkers in Cancer
Edited by Sherry X. Yang and Janet E. Dancey
Copyright © 2013 Pan Stanford Publishing Pte. Ltd.
ISBN 978-981-4364-65-2 (Hardcover), 978-981-4364-66-9 (eBook)
www.panstanford.com

progression of the disease [3]. Understanding these molecular changes have clinical ramifications in disease prognosis, treatment development, and predicting response and resistance to therapy.

One such cellular signaling cascade that plays a major role in colorectal carcinogenesis, is the epidermal growth factor receptor (EGFR) pathway [4]. EGFR was one of the first potential candidates proposed for targeted cancer treatment. Among several newly developed EGFR targeting agents, two monoclonal antibodies (moAbs), cetuximab and panitumumab, are now approved in the treatment of metastatic colorectal cancer (mCRC). Activation of the EGFR, confers sensitivity to EGFR inhibitors, while alterations to the components of the intracellular signaling downstream of the EGFR have been reported to confer resistance to anti-EGFR antibody treatments. This chapter focuses on the EGFR pathway, the concomitant derangements found in colorectal cancers, and how some of these alterations affect sensitivity and resistance to these targeted treatments.

7.2 The Epidermal Growth Factor Receptor (EGFR) Pathway and Colorectal Cancer

For normal growth and development, a cell must receive and interpret extracellular signals from the environment and transmit them intracellularly, such that the cell can respond appropriately. It does so via networks of signaling cascades involving transmembrane and intracellular protein kinases. Malfunctioning of these signaling mechanisms can lead to cancer, as they are important for controlling cellular growth, invasion, apoptosis, and angiogenesis [5]. As a prototypical model, the EGFR signaling cascade is functionally activated in the majority of colorectal cancers.

The EGFR is one of four members of the EGFR/HER (**H**uman **E**pidermal Growth Factor **R**eceptor) family of receptor tyrosine kinases (RTKs) (reviewed in [6]). There are three steps required for the activation of EGFR dependent intracellular signaling (See Fig. 7.1). The first step requires the binding of specific ligands to the extracellular portion of the EGFR. Seven ligands have been identified that selectively bind the EGFR: EGF, transforming growth factor alpha (TGFa), heparin-binding EGF-like growth factor (HBEGF), amphiregulin (AREG), betacellulin (BTC), epiregulin (EREG) and epigen (EPGM) (reviewed in [7]). Second, upon ligand binding, the receptor forms homodimers with itself or heterodimers with another family RTK member. Dimerization activates the kinase domain on the intracellular portion of the receptor resulting in ATP-dependent autophosphorylation of specific tyrosine residues. Receptor autophosphorylation leads to the activation of two intracellular signaling pathways, namely the RAS/RAF/MAPK (**Ra**t **S**arcoma/**R**apidly **A**ccelerated **F**ibrosarcoma/**M**itogen **A**ctivated **P**rotein **K**inase) and PI3K/AKT (**P**hosphatidyl**i**nositol **3**-**K**inase/AKT is named after a viral oncogene) signaling pathways (reviewed in [4]). Normally, these pathways play a role in organogenesis, however when abnormally activated can lead to malignant proliferation, inhibition of apoptosis, invasion, and angiogenesis.

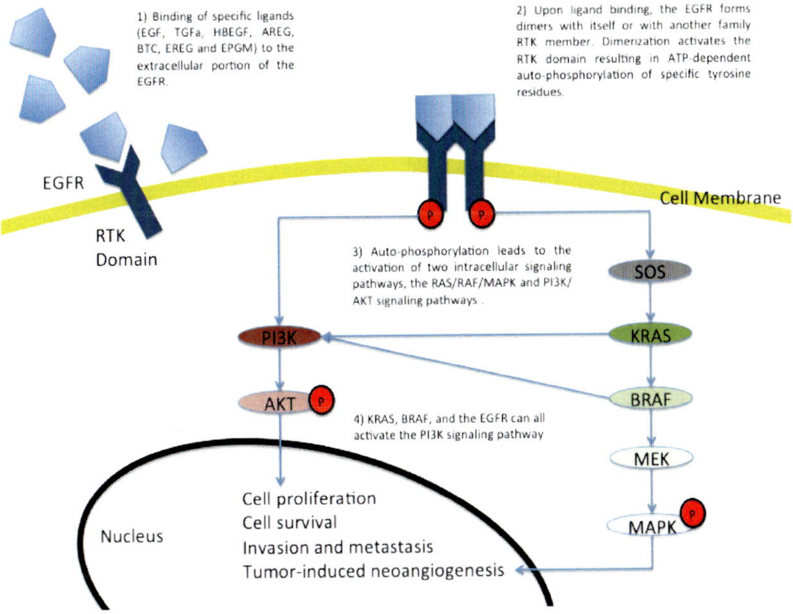

Figure 7.1 EGFR intracellular signaling leads to activation of the RAS/RAF/MAPK and PI3K/AKT signaling pathways.

Although the *EGFR* gene is rarely mutated in colorectal tumors [8, 9], somatic mutations of genes encoding for effectors of the EGFR pathway are common. The majority of activating mutations in colorectal cancer occur in the *KRAS, BRAF* and *PIK3CA* (catalytic subunit of PI3K protein), occurring in 32–37%, 10–17% and 15% of cases, respectively, based on large population based studies [10–18].

7.2.1 RAS/RAF/MAPK Pathway

RAS is a 21 kDa GDP/GTP-binding protein (**G**uanosine **Dip**hosphate/**G**uanosine **Trip**hosphate) that is localized at the inner surface of the plasma membrane. RAS resides primarily in its inactive state bound to GDP. RAS is triggered by upstream signals transmitted by an activated EGFR, and stimulated to release GDP and exchange it with GTP, via guanine nucleotide exchange factors (GEFs) [19]. RAS activation is reversed by the GTPase activating proteins (GAPs) that stimulate the GTPase activity of RAS leading to the hydrolysis of bound GTP [20]. Phosphorylation of effector molecules leads to the activation of downstream RAF and PI3K pathways.

RAS is a well-known proto-oncogene mutated in several human cancers including colorectal cancers. There are 3 RAS isoforms encoded by the *KRAS, NRAS*, and *HRAS* genes. The vast majority of mutations identified in colorectal cancers occur in the KRAS isoform, the human homolog of the Kirsten rat sarcoma-2 virus oncogene [17]. The most commonly affected amino acid residues are G12, G13, and more rarely Q61 [17]. *NRAS* mutations are much less frequent in colorectal cancer, accounting for less than 5% of mutations [21], while *HRAS* mutations have not been identified [17]. The majority of these mutations result in prolonged RAS activation by impairing the GTPase domain of the RAS protein and conferring resistance to the GAPs, ultimately leading to the constitutive activation of the downstream growth promoting pathways [12].

Once activated RAS recruits the serine/threonine kinase BRAF. Upon activation BRAF sequentially activates MEK (mitogen activated ERK kinase) and ERK (extracellular signal-regulated kinase) downstream effectors of the MAPK pathway. More than 30 different *BRAF* mutations have been identified [17], but the *BRAF* V600E mutation constitutes about 90% of all activating *BRAF* mutations and is observed in 10–15% of colorectal tumors [22]. In colonic cell lines harboring the V600E mutation, increased cellular proliferation and inhibition of apoptosis is observed [23]. *BRAF* and *RAS* activating mutations are often mutually exclusive events suggesting only one mutation within the same pathway is sufficient for signaling activation and denotes the importance of the activation of this pathway in cancer pathogenesis [12, 18].

7.2.2 PI3K/AKT Pathway

In addition to KRAS and BRAF, the EGFR activates the PI3K signaling pathway. Activating mutations in the *PIK3CA* p110 subunit (PI3K Catalytic subunit p110) can deregulate this latter pathway. The *PIK3CA* gene encodes the catalytic subunit p110 of PI3K that can be activated directly by phosphorylation of the EGFR tyrosine kinase domain or the interaction with RAS proteins. This *PIK3CA* gene is mutated in 15–30% of colorectal cancers, with frequent mutations found in exons 1, 2, 9, and 20 These mutations lead to activation of AKT signaling in the absence of growth factors *in vitro* [24–26].

Less common genetic alterations that may activate AKT signaling and substitute for *PIK3CA* mutations include amplification of *insulin receptor substrate 2 (IRS2)*, loss of PTEN expression, or coamplification of *AKT* and *PAK4* [24]. The latter two alterations are downstream mediators of PI3K signaling, while IRS2 is an upstream activator of PI3K. Of these alterations, the best characterized is the inactivation of the tumor suppressor gene *PTEN*, which encodes a phosphatase that dephosphorylates a lipid second messenger, phosphatidyllinositol (3,4,5) triphosphate (PIP3) to phosphatidylinositol (4,5) biphosphate (PIP2). PIP3 is the primary product of PI3K activity and its production antagonizes PI3K function and activates downstream signaling components, most notably the protein kinase AKT. Hence, the dephosphorylation of PIP3 is important because it results in the inhibition of the AKT signaling pathway. Loss of PTEN activity and subsequent loss of inhibition of AKT signaling is sufficient to cause a cancer phenotype by increasing cellular proliferation and reducing cell death. The *PTEN* gene may be inactivated via somatic mutations, allelic losses, and epigenetic hypermethylation [27–29].

7.3 EGFR Inhibitors Used in Metastatic Colorectal Cancer (mCRC)

As noted in the above sections, several derangements in the EGFR pathway have been identified in colorectal cancers. Given this phenomenon, several therapeutic agents have been developed targeting this pathway and the EGFR. Two broad classes of EGFR targeted agents have been tested in colorectal cancer patients (See Table 7.1), namely anti-EGFR monoclonal antibodies (moAbs) and small-molecule EGFR tyrosine kinase inhibitors (TKIs). Although these classes differ in several aspects, the main dissimilarity is in the mechanism of interference with EGFR activation. For instance, the anti-EGFR monoclonal antibodies bind to the extracellular portion of the receptor, preventing ligand binding and receptor dimerization, while the TKIs bind the intracellular tyrosine kinase domain of the EGFR, competing with ATP and inhibiting receptor autophosphorylation [6].

Although, both anti-EGFR moAbs and EGFR TKIs have been evaluated in clinical trial in patients with mCRC (Phase 3 data summarized in Table 7.2), only the moAbs, cetuximab and panitumumab, have demonstrated clinical benefit and have been approved for use in patients with mCRC. Cetuximab is a human mouse chimeric immunoglobulin G1 antibody. The efficacy of cetuximab was first demonstrated in combination with irinotecan in chemotherapy-resistant mCRC patients expressing EGFR by immunohistochemistry (IHC), then as monotherapy in chemotherapy refractory patients [30–33]. Recently, based on a trial demonstrating the superiority of cetuximab and multi-agent chemotherapy versus chemotherapy alone resulted in the approval of cetuximab use in the first-line treatment of patients with mCRC [34].

Panitumumab is a fully human anti-EGFR monoclonal IgG2 antibody also approved for use in patients with mCRC. A randomized phase 3 clinical trial demonstrated that panitumumab improved progression-free survival compared with the best supportive care in patients with mCRC previously treated with chemotherapy [35]. Most recently, two phase 3 trials demonstrated the efficacy of adding panitumumab to two different chemotherapy backbones (FOLFOX, 5-flourouracil and oxaliplatin based [36] and FOLFIRI, 5 fluorouracil and irinotecan based chemotherapy protocols [37]) in the first and second line chemotherapy settings.

Neovascularization, or the development of new blood vessels is a necessary prerequisite for tumor growth. An important pathway controlling neovascularization is the vascular endothelial growth factor signaling pathway. Inhibition of this pathway by a humanized moAb targeting VEGF, namely bevacizumab, has been developed and approved in the first line treatment of patients with mCRC.

Table 7.1 Representative EGFR/HER1-targeted antibodies and TKIs tested in mCRC

Agent (company)	Molecule	Specificity	Characteristics	Status
Antibodies				
Cetuximab/IMC-225 (ImClone, Bristol-Myers Squibb)	Chimeric IgG1 mAb	EGFR-ECD	-Intravenously administered -Binds exclusively to EGFR -Binds extracellular domain of EGFR	FDA-Approved (CRC, SCCHN)
Panitumumab/ABX-EGF (Amgen, Abgenix)	Fully human IgG2 mAb	EGFR-ECD		FDA-Approved (CRC)
Matuzumab/EMD 72000 (EMD Pharmaceuticals, Merck KGaA	Humanized mAb	EGFR-ECD		II/III
MDX-447 (Medarex)	Bispecific mAb	EGFR, CD64		I/II
Reversible TKIs				
Gefitinib (AstraZeneca)	Anilinoquinazoline	HER1	-Orally administered -May bind other EGFR family members -Binds intracellular domain of EGFR	FDA-Approved (NSCLC)
Erlotinib (Genentech, OSIP, Roche)	Anilinoquinazoline	HER1		FDA-Approved (NSCLC, pancreatic cancer)
Irreversible TKIs				
Pelitinib/EKB-569 (Wyeth)	3-Cyanoquinolone	HER1/2	-Orally administered -May bind other EGFR family members -Binds intracellular domain of EGFR	I/II
BIBW 2992 (Boehringer Ingelheim)	NA	HER1/2		II

Abbreviations: CD64, cluster of differentiation 64; mCRC, metastatic colorectal cancer; ECD, extracellular domain; EGFR/HER1, epidermal growth factor receptor 1/human epidermal growth factor receptor 1; FDA, Food and Drug Administration; HER2, human epidermal growth factor receptor 2; mAbs, monoclonal antibodies; NSCLC, non-small cell lung cancer; TKI, tyrosine kinase inhibitor.

Table 7.2 Phase III trials of anti-EGFR mAbs in metastatic CRC

Study/Reference	Arms	N	ORR	mPFS (mo)		mOS (mo)	
First Line Therapy							
CRYSTAL Van Cutsem *et al.*, 2009 [34]	Cetuximab +FOLFIRI	599	46.9	8.9	HR = 0.851 (0.726–0.998); *P* = 0.0479	19.9	HR = 0.93 (0.81–1.07); *P* = 0.31
	FOLFIRI	599	38.7 [*P* = 0.004]	8		18.6	
OPUS Bokemeyer *et al.*, 2009 [81]	Cetuximab +FOLFOX4	169	46	7.2	HR = 0.931 (0.705–1.23); *P* = 0.6170	NR	
	FOLFOX4	168	36 [*P* = 0.064]	7.2		NR	
COIN Maughan TS *et al.*, 2010 [137]	Cetuximab +XELOX/ FOLFOX		64	8.6	HR = 0.96 (0.84–1.09); *P* = 0.60	17.9	HR= 1.04 (0.90–1.20); *P* = 0.68
	XELOX/FOLFOX		57 [*P* = 0.049]	8.6		17.0	
PRIME Douillard J, *et al.*, 2010 [36]	Panitumumab + FOLFOX	593	55	9.6	HR = 0.80 (0.66–0.97); *P* = 0.02	23.9	HR = 0.83 (0.67–1.02) *P* = 0.72
	FOLFOX	590	48 [*P* = 0.068]	8.0		19.7	
Second Line Therapy							

Study/Reference	Arms	N	ORR	mPFS (mo)		mOS (mo)	
Peeters M *et al.*, 2010 [37]	Panitumumab + FOLFIRI	591	35	5.9	HR = 0.73 (0.59–0.90); *P* = 0.004	14.5	HR = 0.85 (0.70–1.04); *P* = 0.12
	FOLFIRI	595	10	3.9		12.5	
EPIC Sobrero *et al.*, 2008 [30]	Cetuximab +irinotecan	648	16.4	4	HR = 0.692 (0.617–0.776); *P*≤0.0001	10.7	HR = 0.975 (0.875–0.114); *P* = 0.71
	Irinotecan	650	4.2 [*P*<0.0001]	2.6		10	
Salvage Monotherapy							
CO17 Jonker *et al.*, 2007 [32]	Cetuximab	287	NR	1.9	HR = 0.68 (0.57–0.8); *P*<0.001	6.1	HR = 0.77 (0.64–0.92); *P* = 0.005
	BSC	285	NR	1.8		4.6	
Van Cutsem *et al.*, 2007 [35]	Panitumumab	231	10	1.8	HR = 0.54 (0.44–0.66); *P*<0.0001	NR	
	BSC	232	0	1.7		NR	
Salvage Therapy Cetuximab and Chemotherapy							
BOND Cunningham *et al.*, 2004 [31] 2004	Cetuximab+Irinotecan	218	22.9			8.6	
	Cetuximab	111	10.8			6.9	

(*Continued*)

Table 7.2 *(Continued)*

Study/Reference	Arms	N	ORR	mPFS (mo)		mOS (mo)	
Anti-EGFR mAbs added to chemotherapy-bevacizumab backbone							
CAIRO2 Tol *et al.*, 2008 [43]	Cetuximab + CAPOX/ bevacizumab	368	44	9.6	HR = 1.12 (1.03-1.45); $P = 0.018$	20.3	HR = 1.15 (0.93-1.43); $P = 0.21$
	CAPOX bevacizumab	368	44 [$P = 0.88$]	10.7		20.4	
PACCE Hecht *et al.*, 2009 [44]	Panitumumab + oxaliplatin-chemotherapy/ Bevacizumab	413	46	10	HR = 1.27 (1.06-1.52)	19.4	HR = 1.43 (1.11-1.83)
	oxaliplatin-chemotherapy/ bevacizumab	410	48	11.4		24.5	
	Panitumumab +irinotecan-chemotherapy/ Bevacizumab	115	43	10.1	HR = 1.19 (1.79-1.79)	20.7	HR = 1.42 (0.77-2.62)
	irinotecan-chemotherapy/ Bevacizumab	115	40	11.7		20.5	

Abbreviations: BSC, best supportive care; FOLFIRI, leucovorin, fluorouracil, and irinotecan FOLFOX, leucovorin, fluorouracil, and oxaliplatin; HR, hazard ratio; mCRC, metastatic colorectal cancer; mo, month mOS, median overall survival; mPFS, median progression free survival; NR, not reported; ORR, overall response rate; yr, year.

Several trials have demonstrated that bevacizumab in combination with standard chemotherapy improved response rates, progression free survival (PFS) and overall survival (OS) when compared with chemotherapy alone in the treatment of patients with mCRC [38–42]. In hopes of improving outcomes in patients with mCRC, researchers have attempted dual inhibition of both the VEGF and EGFR pathways by combining bevacizumab and chemotherapy with either cetuximab or panitumumab [43–45]. Unfortunately, these trials did not demonstrate any clinical benefit with the addition of an anti-EGFR moAb and may have been a detriment to survival (See Table 7.2) [43, 44].

In contrast to result with moAbs, phase 2 trials using the EGFR TKIs, erlotinib and gefitinib, have been disappointing as monotherapy did not demonstrate activity when tested on patients with previously treated mCRC [46, 47]. There is some phase 2 data suggesting that these agents may be useful when combined with chemotherapy [48–50], however in the absence of completed phase 3 trials demonstrating clinical benefit, these agents remain experimental in the treatment of patients with mCRC.

7.4 Determinants of Sensitivity and Resistance to EGFR Targeting moAbs

The benefit of cetuximab and panitumumab in the treatment of patients with mCRC has been clearly demonstrated, albeit with only modest objective response rates of approximately 10% and improvements of survival of 3–4 months [31, 35]. These response and survival rates improve, by combining these anti-EGFR moAbs with multi-agent chemotherapy and using these regimens in treatment naïve patients with mCRC [34]. To further improve the treatment of patients with mCRC it is imperative to identify predictive markers of response or resistance to cetuximab and panitumumab. These markers will aide in identifying patients who may benefit from treatment, while avoiding the unnecessary toxicities and costs associated with these drugs in those who may not.

A predictive marker is not necessarily the same as a prognostic marker. A predictive marker gives information about the effect of a therapeutic intervention, whilst a prognostic marker provides information about a patients overall cancer outcome, regardless of therapy. An ideal predictive marker is simple, reproducible, correlates with a clinical endpoint, and is available prior to initiation of treatment. Both clinical and cellular alterations have emerged as potential predictive markers for anti-EGFR moAb therapies. However, as with most predictive tests, none have been ideal at predicting response or resistance to anti-EGFR moAbs. This is, in part, due to the complexities in identifying appropriate markers for testing and the scoring of these markers. EGFR inhibitor induced skin rash is a clinical feature that has been evaluated as a potential predictive marker. The cellular features that alter DNA, microRNA, and/or protein in the EGFR pathway from colorectal cancers have also been evaluated for predictive potential.

7.4.1 Clinical Features

7.4.1.1 EGFR inhibitor induced-skin rash

The toxicity profile of the anti-EGFR moAbs excludes several of the severe side effects commonly associated with cytotoxic chemotherapy, such as hematological suppression. However, the use of these agents have been hampered by a constellation of dermatological toxicities experienced by more than 90% of patients, notably, papulopustular rash, xerosis, pruritis, and paronychia [51]. The patho-physiology of the cetuximab-induced rash remains poorly understood, but is likely due to the predicted downstream perturbations of EGFR signaling that have been demonstrated in studies of skin biopsies, such as phosphorylated EGFR, phosphorylated AKT, Ki67, and p27 in patients treated with anti-EGFR moAbs [52–54]. Though rarely life threatening, these skin effects impact quality of life, increase risk for infection, and may lead to dose modification or treatment discontinuation, thus potentially affecting the overall clinical benefits of therapy.

A significant positive correlation between cutaneous toxicity and rates of response, PFS, and OS has been noted in all trials of cetuximab [34], or panitumumab [35] in patients with advanced colorectal cancer. Based on the association of cetuximab and rash, the EVEREST trial was launched [55]. Patients with irinotecan-refractory EGFR-expressing CRC were randomized to either a standard dose of cetuximab or the dose escalation arm, which increased the cetuximab dose up to 500 mg/m^2 if the patient did not experience more than Grade 1 rash by week 3. Preliminary results from this trial suggests that cetuximab dose escalation is safe and increases response rates in patients without a rash or those who develop only a mild skin reaction within 3 weeks of starting therapy. The response rate among patients with less than or equal to a grade 1 skin toxicity who were randomized to dose escalation was 30% versus 13% in the control group who only continued standard dose therapy [55]. This trend was not statistically significant and requires more mature data from a larger sample population.

Using rash as an early clinical marker of benefit from anti-EGFR moAbs does have some limitations. Rash has occurred in patients without any apparent benefit from anti-EGFR moAbs, and conversely clinical benefit has occurred in patients without rash. Some suspect that rash may be a surrogate marker for a genetic profile that predisposes individuals to rash and also modulate the response to anti-EGFR moAbs. Also, rash could merely be a prognostic marker of intact immunity or merely the result of treatment bias; as patients who have intact immune systems or are treated long enough to develop a rash may have prolonged survival [56]. Although, a clear association between signaling inhibition in skin and antitumor response has not been found, rash seems to correlate with drug exposure, concentration, and likely reflects effectiveness.

7.4.2 Potential Predictive Genetic Alterations of the EGFR Pathway in Patients with mCRC

7.4.2.1 KRAS mutations

Activating mutations in the *KRAS* gene, which result in EGFR independent activation of the MAPK pathway and possibly PI3K pathway through direct KRAS activation of PI3K, are found in approximately 40% patients with CRC [57]. These mutations typically occur in codons 12 (79%; exon 2), 13 (17.6%; exon 2) or 61 (exon 3) [57] and do not seem to have any prognostic value in patients with mCRC. However, in patients with mCRC mutations in *KRAS* are the strongest negative predictive marker for benefit for anti-EGFR moAbs as single agents and in combination with chemotherapy (see Section 7.4.2.3).

7.4.2.2 KRAS mutation detection

The presence of KRAS mutations in colorectal tumor tissue can be detected by several different molecular methods (see Table 7.3).

Table 7.3 Methods used for *KRAS* mutation testing

Method	Sensitivity	Advantages	Disadvantages
Sequencing Methods			
Sanger Sequencing	10–30%	-Gold standard -Detects most small mutations (single nucleotide, changes, insertions and deletions)	-Not quantitative -Less sensitive -Risk of contamination
Pyrosequencing	5–10%	-Detects most small mutations (single nucleotide, changes, insertions and deletions) -Can sequence short pieces of DNA from fixed tissue more readily	-Only sequences short reading lengths -Risk of contamination
PCR Based			
Allele Specific Real-Time PCR/ Amplification refractory mutation System (TheraScreen)	1–5%	-Fast -Highly sensitive -Therascreen available as commercial kit and closed system, so decreased risk for contamination	-Detects only known mutations -Requires specifically designed probes Several reactions, so more reagents and template DNA required
Post-PCR Fluorescent melting curvc analysis with specific probes/ High Resolution Melting Analysis	5–10%	-Screens for mutations prior to sequencing	-Requires sequencing for verification -Expensive -Requires expensive equipment
PCR Clamping	1%	-Rapid -Closed system, so decreased risk for contamination	-Requires sequencing for verification -No control reaction
Restriction Fragment Length Polymorphisms	1%	-Inexpensive -No specialized equipment required	-Not quantitative -Requires sequencing for verification

Commonly used methods to evaluate samples for KRAS testing are DNA sequencing and real time polymerase reaction (PCR). DNA sequencing is considered the gold standard for the detection of mutations and is currently the most commonly used method for *KRAS* testing. DNA sequencing can identify most small mutations in the analyzed *KRAS* gene segment. The analytic sensitivity of this method is reported to be 10 to 30%; i.e. at least 10–30% of the template DNA needs to be mutant for this method to detect the mutant sequence [58, 59]. To improve the low sensitivity and the expense of direct sequencing, other methods have been employed that are based on real-time PCR based techniques.

Real time PCR assays are an excellent option for clinical testing for *KRAS* mutations, because they are rapid, simple, and have improved sensitivity. In real time PCR, the KRAS amplification product, or amplicon, is detected by fluorescent probes or DNA dyes as nucleotides are being incorporated into the synthesized DNA strand. For KRAS mutation testing, two general designs are employed with real-time PCR, either allele specific probes or post-PCR melting curve analysis.

The TheraScreen™ KRAS testing kit combines allele specific probes with real time PCR (also called amplification refractory mutation systems: ARMS combined with real time PCR). This kit utilizes PCR primers, which have complementary sequences to wild-type and mutant templates. These primers amplify only specific complementary templates and can discriminate among templates that differ by a single nucleotide. The TheraScreen has a sensitivity of 1–5%, and hence can detect the presence of KRAS mutant alleles that have low allelic concentration [60, 61]. Moreover, since only the mutant allele is amplified and the primer sequence is known, direct sequencing is not required to confirm the mutant sequence, making this system both efficient and practical. A drawback to this system is that only known mutations can be tested and hence novel unknown mutations may be missed. Also, separate reactions are required for each mutation tested, thus requiring more reagents and DNA template.

Real time PCR with post-PCR melting curve analysis, detects the disrupted affinity of two DNA chains caused by a mutation. A single strand of DNA containing a mutation has an altered binding affinity when bound to either a complementary mutant (MT) strand or wild-type (WT) strand, such that they dissociate at a different temperature compared with the WT double stranded sequence. Hence, real time PCR is performed using probes complementary to the WT sequence and flanking the mutation site (codons 12 and 13). Following PCR, the amplicon melting temperature is determined, defined as the temperature at which 50% of the amplicons have dissociated into single stranded DNA. Based on the different binding affinities of the WT/WT, WT/MT, and MT/MT double strands, different melting temperatures are detected. This method demonstrates a high sensitivity for the detection of KRAS mutations that was similar to direct sequencing [62]. However, if an altered melting temperature is detected (suggesting a mutant sequence), the exact nucleotide change is still not identified, and hence direct sequencing is still required.

To improve sensitivity for KRAS mutation detection, the PCR-clamp approach has been developed as another variation on real-time PCR melting curve analysis. This method uses mutation specific and WT specific primers, except the WT specific primer is bound to a peptide. This WT peptide/nucleic acid probe suppresses amplification of the wild-type KRAS sequence, preferentially amplifying mutant template if present. By doing so, this technique has a reported sensitivity of less than 1% [63]. Again, only known mutations can be identified and since the probes are mutation specific, no direct sequencing is required.

7.4.2.3 Specimen selection for KRAS mutation testing

As mentioned in Section 7.3, the benefit of anti-EGFR moAbs has been demonstrated in patients with mCRC. Hence, in this setting it would make sense that molecular markers of response should be tested from metastatic tissue as opposed to the primary lesion as this would likely reflect the pathway changes that would render a tumor resistant or sensitive to the applied targeted agent. The problem with using metastatic tissue for testing is that tissue is usually limited or often not available. However, if there was no discordance between the primary and metastatic sites in regards to *KRAS* mutation status, then either tissue sample could be tested. Studies comparing *KRAS* status in primary and metastatic tumors are limited and have had inconsistent results [64–67]; however, two studies have recently attempted to address this issue. First, in a study of 48 patients with mCRC, DNA sequencing revealed an overall concordance of *KRAS* mutational status in 92% of cases when comparing primary and metastatic lesions from the same patient, suggesting that evaluation of *KRAS* status can be performed in either the primary tumor or metastatic sites [68]. Similarly, a recent study demonstrated a concordance of 90% between primary and distant metastases for *KRAS* mutations detected by direct sequencing and pyro-sequencing techniques [69]. Based on the high concordance rate, guidelines have been formulated suggesting that primary lesions, if metastatic tissue is not available, be tested for *KRAS* mutations prior to utilizing an anti-EGFR moAb. However, it remains unclear whether there should be attempts to determine the *KRAS* status in the untested primary or metastatic tissue, if the tested sample is *KRAS* WT. Also, whether the *KRAS* discordance affects its predictive role remains unknown.

7.4.2.4 Prognostic significance of KRAS mutation status

There have been discrepant reports on the prognostic significance of *KRAS* mutations with some suggestion that KRAS mutations may be prognostic in the earlier stages of colorectal cancer but not in the metastatic setting. Earlier studies have demonstrated that *KRAS* mutations portend a poorer prognosis in patients with non-metastatic stages of colorectal cancer [70–72]. Some of these earlier smaller studies may have been subject to publication bias, as more recent analyses from the CALGB 89803 [73] and the PETACC-3 [74] trials demonstrate that KRAS mutations are not prognostic for patients with stage 2 or 3 colorectal cancer treated with adjuvant 5-fluoruracil based chemotherapy. Finally, a trial investigating the role of cetuximab in chemotherapy pre-treated patients with mCRC did not demonstrate a prognostic effect for KRAS status in those patients randomized to receive best supportive care [75]. Although debatable, there is evidence to suggest that there may be a prognostic role for patients with earlier stages of colorectal cancer and none for those with mCRC.

7.4.2.5 Predictive significance of KRAS mutation status

Despite *KRAS* mutations having questionable prognostic value in patients with mCRC, it is certainly predictive of treatment benefit in this setting. The presence of *KRAS* mutations correlates with lack of objective response rate and improvement in both progression free and overall survival in patients with mCRC who received anti-EGFR moAbs. Interestingly, in a study of 30 patients with mCRC, all of the *KRAS* mutations were harbored by patients who did not respond to cetuximab (13/19), while no mutations were identified in the patients who did respond (0/11) [76]. Since then, several retrospective studies, have demonstrated that *KRAS* mutations are significantly associated with lack of response or benefit to cetuximab or panitumumab in patients with chemotherapy refractory mCRC [75, 77–79].

Two phase 3 randomized control studies investigated the role of either monotherapy cetuximab [31] or monotherapy panitumumab [35] compared with best supportive care (BSC) in chemotherapy refractory patients with mCRC. Both panitumumab and cetuximab studies demonstrated improvements in PFS (HR of 0.54 and 0.68, respectively), however only cetuximab demonstrated an improvement in OS compared with BSC (HR 0.77). This latter negative result was likely because patients initially assigned to the BSC arm were allowed to receive panitumumab treatment upon progression, while those in the cetuximab study were not allowed to crossover.

Retrospectively conducted subgroup analyses looking at *KRAS* mutational status in both of the aforementioned studies were also conducted (see Table 7.4). *KRAS* mutations were significantly associated with lack of response to panitumumab alone (0% for *KRAS* mutated patients vs. 17% for *KRAS WT* [80]) and cetuximab alone (1.2% for *KRAS* mutated patients vs. 12.8% for *KRAS* WT [75]). Moreover, PFS in *KRAS*-mutated patients receiving BSC compared with panitumumab (7.3 and 7.4 weeks, respectively) or with cetuximab (1.8 months for both the cetuximab and BSC groups) were not statistically different. The treatment effect of panitumumab or cetuximab on PFS in the *KRAS* WT group (HR 0.45 for panitumumab and 0.40 for cetuximab) was significantly greater than in the mutant group (HR 0.99 for both panitumumab and cetuximab). In patients with *KRAS* WT tumors, treatment with cetuximab as compared with supportive care alone significantly improved OS (median, 9.5 vs. 4.8 months; HR 0.55). As mentioned above, the OS benefit associated with cetuximab treatment can be explained by the fact that crossover was not allowed in this study unlike in the phase III study with panitumumab. Combined, these studies concluded that patients with a colorectal tumor harboring a *KRAS* mutation did not benefit from anti-EGFR moAbs, whereas patients with a tumor bearing *KRAS* WT did benefit. The mutation status of the *KRAS* gene did not influence survival among patients treated with BSC alone, (as noted above) suggesting no prognostic disadvantage.

Table 7.4 Clinical efficacy with anti-EGFR mAb in mCRC according to KRAS status (randomized trials with single agent or combinations)

Study	Treatment (# pts in trial/# pts with marker analysis)	Method of *KRAS* testing	*KRAS* status	N (%) on treatment arms	mPFS	OS	ORR (%)
Amado *et al.*, 2008 [80]	Panitumumab vs. BSC (463/427)	Allele specific real time PCR	WT	124 vs. 119	12.3 vs. 7.3 wks	NR	17 vs. 0
			MT	84 vs. 100	7.4 vs. 7.3 wks	NR	0 vs. 0
Karapetis *et al.*, 2008 [75]	Cetuximab vs. BSC (573/394)	PCR and Sequencing	WT	117 vs. 81	3.7 vs. 1.9 mo	9.5 vs. 4.8 mo	12.8 vs. 0
			MT	113 vs. 83	1.8 vs. 1.8 mo	4.5 vs. 4.6 mo	1.2 vs. 0
CRYSTAL [92]	Cetuximab + FOLFIRI vs. FOLFIRI (1198/540)	qPCR	WT	172 vs. 176	9.9 vs. 8.7 mo. (HR = 0.68, *P* = 0.017)	NR	59.3 vs. 43.2 (*P* = 0.0025;)
			MT	105 vs. 87	7.6 vs. 8.1 mo (HR = 1.07, *P* = 0.47)	NR	36.2 vs. 40.2 (*P* = 0.46;)
OPUS [81]	Cetuximab + FOLFOX vs. FOLFOX (337/233)	Allele specific real time PCR	WT	61 vs. 73	7.7 vs. 7.2 mo (HR = 0.57, *P* = 0.016)	NR	61 vs. 37 (*P* = 0.011)
			MT	52 vs. 47	5.5 vs. 8.6 mo (HR = 1.83, *P* = 0.019)	NR	33 vs. 49 (*P* = 0.106)

(*Continued*)

Table 7.4 (*Continued*)

Study	**Treatment (# pts in trial/# pts with marker analysis)**	**Method of *KRAS* testing**	***KRAS* status**	**N (%) on treatment arms**	**mPFS**	**OS**	**ORR (%)**
PRIME Douillard J *et al.*, 2010 [36]	Panitumumab and FOLFOX vs. FOLFOX	Allele specific real time PCR	WT	325 vs. 331	9.6 vs. 8.0 mo (HR = 0.80, *P* = 0.02)	23.9 vs. 19.7 mo (HR = 0.83, *P* = 0.72)	55 vs. 48
			MT	221 vs. 219	7.3 vs. 8.8 mo (HR = 1.29; *P* = 0.02)	15.5 vs. 19.3 mo (HR = 1.24, *P* = 0.07)	40 vs. 40
Peeters M *et al.*, 2010 [37]	Panitumumab + FOLFIRI vs. FOLFIRI	Allele specific real time PCR	WT	303 vs. 294	5.9 vs. 3.9 mo (HR = 0.73, *P* = 0.004)	14.5 vs. 12.5 mo (HR = 0.85, *P* = 0.12)	35 vs. 10
			MT	238 vs. 248	5.0 vs. 4.0 (HR_0.85; *P* = 0.14)	11.8 vs. 11.1 mo (HR = 0.94)	13 vs. 14

Study	Treatment (# pts in trial/# pts with marker analysis)	Method of *KRAS* testing	*KRAS* status	N (%) on treatment arms	mPFS	OS	ORR (%)
CAIRO2 [43]	Cetuximab + CAPOX/ bevacizumab vs. CAPOX/ bevacizumab (755/520)	Sequencing and real time PCR	WT	158 vs. 156	10.5 vs. 10.6 mo ($P = 0.30$)	21.8 vs. 22.4 mo ($P = 0.64$;)	61.4 vs. 50.0 ($P = 0.06$;)
			MT	98 vs. 108	8.1 mo vs. 12.5 mo ($P = 0.003$;)	17.2 mo vs. 24.9 ($P = 0.03$;)	45.9 vs. 59.2 ($P = 0.03$;)
PACCE (interim analysis [44])	Panitumumab + irinotecan/ bevacizumab vs. irinotecan/ bevacizumab (230/200)	Allele specific real time PCR	WT	57 vs. 58	NR	NR	47 vs. 54 (HR = 1.42)
			MT	46 vs. 39	NR	NR	30 vs. 38 (HR = 0.59)

Abbreviations: CI, confidence interval; EGFR, epidermal growth factor receptor; mAb, monoclonal antibody; mCRC, metastatic colorectal cancer; mo, months; mPFS, median progression-free survival; MT, mutant; NR, not reported; ORR, overall response rate; OS, overall survival; pts, patients; qPCR, quantitative polymerase chain reaction; wks, weeks; WT, wild-type.

Resistance to anti-EGFR moAbs in *KRAS* mutated CRC was not overcome by the combination with chemotherapy in the setting of first- and second-line treatment (see Table 7.4). Phase 3 trials of second or third line chemotherapy and moAbs have also demonstrated benefit with the addition of and EGFR inhibitor is limited to patients who harbor *KRAS* wild-type tumors. Based on the above findings, panitumumab and cetuximab are approved in Europe and North America, only for patients with mCRC and *KRAS* WT tumors, in the first or second-line second as monotherapy or in combination with standard chemotherapy.

Interestingly, a recent study suggests that not all *KRAS* mutations are resistant to the effects of cetuximab. Among patients with chemotherapy refractory CRC, a particular *KRAS* mutation in codon 13, namely p.G13D, was associated with longer overall and PFS when compared with other *KRAS* mutated tumors [82]. This was only an observational study and evaluation of cetuximab therapy in these patients with tumors harboring the p.G13D mutation in prospective randomized trials is warranted.

Finally, predicting response to an anti-EGFR moAbs based on *KRAS* mutation status is not perfect. Although a positive mutation status is a highly specific test for non-response (almost 95% of *KRAS* mutant tumors do not respond), it is only moderately sensitive For instance, only 40–60% of patients with WT *KRAS* colorectal cancers respond to an anti-EGFR moAb. This suggests that there may be other components in the EGFR pathway that may determine resistance and predict for response to treatment.

7.4.2.6 BRAF mutations in patients with mCRC

The *BRAF* gene encodes a serine threonine protein kinase that acts as a downstream effector of KRAS signaling [83]. The *BRAF* gene is composed of 18 exons, and the most common activating mutation occurs in exon 15, a thymine to adenine transversion, leading to a valine to glutamate substitution at codon 600, denoted V600E [84]. The encoded mutant protein has more than a 10-fold increase in kinase activity [85]. The V600E *BRAF* activating mutation, which occur mutually exclusive with *KRAS* mutations, are found in about 10% of mCRCs [86, 87].

7.4.2.7 BRAF mutation detection in patients with mCRC

To detect the *BRAF* mutations similar methods to *KRAS* mutation screening have been employed. These include direct sequencing, real-time PCR with melting curve analysis and allele specific PCR. Although sequencing is the gold standard for testing, real-time PCR-based techniques have the added benefit of being simpler, quicker, cheaper, and less error prone. Moreover, the sensitivity and specificity of real-time PCR for detecting the V600E *BRAF* mutation were both 100%, using direct sequencing as the standard assay [88]. Despite these benefits of employing a PCR based approach, unknown or rare mutations in the *BRAF* gene can only be identify or confirmed using direct sequencing.

7.4.2.8 Specimen selection for BRAF mutation testing

Although a high concordance rate is observed between primary tumor and metastasis in the *KRAS* gene status, only limited data on the concordance rate for *BRAF* exists and estimates are variable. Two studies report high concordance rates of 94–100% [89, 90], while one retrospective study reported a concordance rate of only 38.5% [91]. Hence, when the primary tumor harbors the V600E mutation, the *BRAF* mutational status of the metastatic lesion may need to be verified. Whether a change of the *BRAF* mutational status from primary tumor to metastatic site alters prognosis or predictive potential remains to be determined.

7.4.2.9 Prognostic and predictive role of BRAF mutations

In several studies, the V600E *BRAF* mutation has been implicated as a marker for poor prognosis in all stages of CRC, independent of the administration of anti-EGFR moAbs [74, 92, 93]. Preliminary data also suggests that *BRAF* gene mutation may predict for resistance to therapy. Di Nicolantonio *et al.* reported analyses on 113 tumors from cetuximab- or panitumumab-treated mCRC patients [94]. The *BRAF* V600E mutation was detected in 11 of 79 patients who had wild-type *KRAS*. None of the patients with *BRAF*-mutated tumors responded to treatment, whereas none of the responders carried *BRAF* mutations. Moreover, patients with *BRAF*-mutated tumors had significantly shorter PFS and OS than wild-type patients. Since then, several independent studies have consistently demonstrated that the V600E *BRAF* mutation predicts for anti-EGFR moAbs resistance and decreased survival in the subgroup of chemotherapy refractory patients with mCRC and WT *KRAS* tumors [95–97]. The above data suggests that *BRAF* testing could be employed in those patients with *KRAS* WT tumors, to further delineate resistance to anti-EGFR moAbs. However, a preliminary update from the CRYSTAL study, a phase 3 trial randomizing first line treatment with irinotecan with or without cetuximab in patients with mCRC, contradicted this hypothesis. These results suggested that *BRAF* mutations were associated with a poor prognosis overall, but did not appear to be a strong predictive biomarker for lack of benefit to cetuximab in *KRAS* WT patients with mCRC. In this analysis, treatment with fluorouracil and irinotecan (FOLFIRI) plus cetuximab in *KRAS* WT or *KRAS* WT/*BRAF* WT patients provided a median overall survival of 23.5 months and of 25.1 months, respectively. However, for *KRAS* WT patients who had *BRAF* mutations the median overall survival was just 14.1 months. Among the 59 patients with WT *KRAS*, *BRAF* mutant tumors, there was no difference in PFS, response rate, and OS with the addition of cetuximab to FOLFIRI compared with FOLFIRI alone. Although this data, demonstrates a prognostic as opposed to a predictive role for *BRAF* mutations status, the data is immature and the sample may be too small to be conclusive [92]. Although promising as a predictive marker for resistance to anti-EGFR moAbs, due to the potential prognostic impact of *BRAF* mutations, randomized controlled trials are required before it can be implemented as such into clinical practice.

7.4.2.10 KRAS Let-7 single nucleotide polymorphism

A unique mechanism of KRAS activation has recently been hypothesized that may predict for response to anti-EGFR moAbs. MicroRNAs (miRNAs) are a class of small RNAs that play a role in gene regulation in the cell. Binding of miRNA to messenger RNA (mRNA) recruits and activates an RNA-induced silencing complex to the target message [98].

Members of the *Let-7 (Lethal-7)* family of miRNAs have been found to bind to specific regions in the 3′ untranslated region (3′UTR) of the *KRAS* mRNA, namely the *Let-7* complementary site (LCS), resulting in down regulation of transcript translation [99]. A single nucleotide polymorphism (SNP) has been described and characterized within the LCS of the *KRAS* mRNA [100]. This SNP is denoted LCS6 (for SNP rs61764370; rs stands for reference SNP) and consists of a thymidine (T) to guanine (G) base change. The LCS6 nucleotide change alters the binding capability of Let-7 to the *KRAS* mRNA and this loss of transcript silencing results in *KRAS* over expression in *in vitro* models [100].

Graziano *et al.* recently demonstrated that the LCS6 G allele had a worse OS and PFS than the T/T genotype carriers, in 130 patients with mCRC treated with salvage cetuximab/irinotecan and *KRAS* WT and *BRAF* WT tumors [101]. Around the same time, another study, contradicted these results, publishing that the WT LCS6 genotype (T/T) had a worse objective response rate (ORR 9%) compared with patients with the T/G or G/G genotype (42%; P = 0.02) [102]. Although there was a trend to improved PFS and OS in those patients with a G allele, these results were not statistically significant. Possible reasons for the conflicting results are that *KRAS* exon 61, known to harbor activating mutations, and *BRAF* were not sequenced in the latter study. The non-responders to cetuximab among the LCS6 T/T genotype patients could be partly attributed to the presence of unidentified *KRAS* and *BRAF* mutations. This could also explain the differences in survival analyses. These studies were both retrospective studies and ultimately require larger prospective controlled studies utilizing other accepted predictive markers before the LCS6 polymorphism can be accepted as a predictive marker of anti-EGFR moAbs efficacy.

7.4.3 Genetic Mutations Affecting the EGFR Gene

7.4.3.1 Somatic EGFR gene mutations

Somatic *EGFR* gene mutations are found in approximately 5–15% of unselected white patients and in 25–35% of unselected Asian patients with non-small cell lung cancer (NSCLC) [103, 104]. Also, the *EGFR* gene has somatic mutations in 30–50% of patients with glioblastoma (GBM) [9]. However, these mutations in the *EGFR* gene rarely occur in CRC. When DNA from 293 colorectal cancers were screened for *EGFR* mutations within the tyrosine kinase domain only one mutation was identified [9]. Moroni *et al.* also detected only one mutation (3.2%) among 31 patients with metastatic colorectal cancer who had responded to cetuximab [8]. The patient harboring this mutation achieved stable disease for 24 weeks with cetuximab and chemotherapy treatment. This mutation was a heterozygous missense mutation in exon 21 (Gly857Arg) affecting a residue located within the activation loop of the EGFR catalytic domain. Although this patient demonstrated benefit from cetuximab, due to the rarity of this and *EGFR* mutations in general, it is difficult to determine the true impact of this mutation on predicting response to ant-EGFR moAbs in patients with mCRC. As a consequence, *EGFR* mutations are unlikely to be a predictive marker for response to anti-EGFR moABs agents in this cancer site.

7.4.3.2 EGFR gene copy number

Variations in gene copy number (GCN) abrogate normal cellular functions and can be hallmarks of tumorigenesis. The mechanisms best described underlying increased *EGFR* GCN are gene amplification and chromosome 7 (where the *EGFR* gene resides) polysomy [8, 105, 106]. Increased *EGFR* GCN has been observed in as low as 2.2% and as high as 31% of tumors obtained from patients with mCRC [8, 106–109]. There have been several studies trying to address whether variations in *EGFR* GCN predicts for response to anti-EGFR moABs in patients with advanced CRC [8, 106–109]. Delineating this premise has proven more than difficult, in part due to the difficulties in interpreting and standardizing the assays used to measure GCN.

7.4.3.3 Measuring EGFR gene copy number

GCN may be assessed by PCR or *in situ* hybridization (ISH) based techniques. Of these, ISH is more accurate or less prone to false negative results, due to normal cell contamination. However, techniques and scoring have been variable, likely resulting in discordant conclusions regarding the prognostic and predictive value of GCN in this setting [8, 107, 110, 111].

To avoid the problems of DNA contamination from normal cells, FISH (fluorescent *in situ* hybridization) and CISH (chromogenic *in situ* hybridization) assays can be utilized as they are performed on whole cell populations, with only the cancerous cells being scored and the normal adjacent cells acting as controls. FISH and CISH analyses use fluorescently or chromogenic probes, respectively, that bind specific sequences of DNA, in this case the *EGFR* gene. By counting the number of fluorescently or chromogenically tagged chromosomes or interphase chromatin spreads within nuclei, the copy number of the *EGFR* gene can be tabulated. Counting the *EGFR* signal from several nuclei and averaging the number of signals per nucleus is how GCN is measured. Normally, the GCN is two per nucleus, representing the diploid nature of cells and the alleles inherited from either parent. Variations of GCN, either gains or losses, can be detected via these methods.

As mentioned above, there are two copies of the *EGFR* gene in normal cells and anything above two would be considered abnormal on a cellular basis. Although abnormal, this may not be the threshold level for which gene copy number (a continuous variable) would predict for benefit to anti-EGFR moAbs, in fact, predictive thresholds for GCN have varied from 2.47 and above, to greater than and equal to 6 (see Table 7.5) [8, 76, 105, 106, 108, 112]. This varied threshold is due to the variable ratios of disomy versus polysomy or amplification within nuclei from the same tumor specimen, representing the inhomogeneous *EGFR* FISH pattern of mCRC [106, 113]. Moreover, gene amplification representing high levels of genomic gains is more readily identifiable by FISH and CISH, compared with subtle degrees of chromosomal gains (as with polysomy). Both of these reasons make it difficult to score and evaluate GCN with FISH and CISH.

7.4.3.4 Specimen selection for EGFR gene copy number

Statistically significant concordance between primary colorectal tumors and their metastases with regard to *EGFR* gene copy number has been demonstrated by FISH analyses [105, 108]. Capuzzo *et al.*, reported that in only one out of 21 patients was a different *EGFR* gene score encountered from primary lesion and distant metastatic site, as measured by FISH [105]. Although not conclusive, it suggests that FISH analyses for *EGFR* GCN can be performed on either primary or metastatic tissue.

7.4.3.5 Prognostic value of EGFR gene copy number

It was initially thought that increased *EGFR* GCN might affect patient prognosis, possibly relating to CRC biology, as opposed to being a predictor of the therapeutic benefits with cetuximab. This is based on an earlier phase II study that evaluated the antitumor activity of cetuximab, in patients with mCRC refractory to standard chemotherapy [111]. No relationships between *EGFR* GCN, response and PFS were observed in this study, but OS did appear to be related to GCN. However, in this study the *EGFR* GCN was measured with a quantitative PCR assay and may have been prone to false negative

Table 7.5 EGFR FISH results and response rates in anti-EGFR mAb clinical trials

Study	Anti-EGFR mAb	GCN Threshold; methodology	Tumor samples (*N*)	Analysis	*P*	*P*-value
Moroni *et al.*, 2005 [138]	Cetuximab or Panitumumab	≥3; FISH	31	EGFR FISH + tumors (n)/ response (n)	8 /10 responders 1 /21 non-responders	$P < 0.0001$
Lievre *et al.*, 2006 [76]	Cetuximab	≥6; CISH	30	EGFR CISH + tumors/response (n)	3/3 responder 0/27 non-responders	$P = 0.04$
Frattini *et al.*, 2007 [112]	Cetuximab	≥3 EGFR/CEP7; FISH	27	EGFR FISH + tumors (n)/ response (n)	10 /10 responders 2/17 non-responders	$P < 0.05$
Sartore-Bianchi *et al.*, 2007	Panitumumab	≥2.47; FISH	30	Responders (n)/ EGFR FISH status (n)	6/20 EGFR FISH+ 0/38 EGFR FISH-	$P = 0.0009$
Cappuzzo *et al.*, 2008 [105]	Cetuximab	≥2.92; FISH	85	Responders (n)/ EGFR FISH status (n)	14 /43 EGFR FISH+ 1 /42 EGFR FISH-	$P = 0.0001$
Personeni *et al.*, 2008 [108]	Cetuximab	≥2.83; FISH	NR	According to ROC curve analysis based on objective response, a significantly longer overall survival in patients harboring increased mean EGFR GCN		$P = 0.037$

Abbreviations: EGFR, epidermal growth factor receptor; FISH, fluorescent *in situ* hybridization; CISH, chromogenic *in situ* hybridization; mAB, monoclonal antibody; NR, not reported; ROC, receiver operating characteristic.

results as mentioned above. Contrary to this result, in a larger randomized study investigating panitumumab plus best supportive care (BSC) versus BSC alone in patients with mCRC refractory to standard chemotherapy, the BSC arm did not demonstrate a correlation between *EGFR* GCN or chromosome 7 polysomy status (both measured by FISH analyses) and PFS [106]. This suggests that these parameters are not prognostic in patients with mCRC.

7.4.3.6 Predictive role of EGFR gene copy number

Using FISH analyses, Moroni *et al.* were the first to report that patients with mCRC with increased *EGFR* copy numbers are more likely to respond to cetuximab or panitumumab [8]. Several studies using FISH or CISH have since corroborated this data, and demonstrated that this association was independent of *KRAS* mutation status [106, 108]. This increased *EGFR* GCN does not seem to translate into increased expression of the corresponding proteins [114, 115]. Hence, it is unclear how and why the copy number of the EGFR receptor is correlated with improved response. Also, it was soon apparent, that only a fraction of tumors with an increased gene copy number respond to treatment [109].

Compared with patients with increased *EGFR* GCN, patients with normal GCN exhibit lower response rates to anti-EGFR moAbs. In patients treated with panitumumab, a mean *EGFR* GCN of less than 2.5 per nucleus or less than 40% of tumor cells displaying chromosome 7 polysomy within the tumor predicted for shorter progression-free survival ($P = 0.039$ and $P = 0.029$, respectively) and overall survival ($P = 0.015$ and $P = 0.014$, respectively). Similarly, the presence of chromosome disomy 7 has been investigated as a predictor for resistance to anti-EGFR moAbs. Chromosome 7 homogeneous disomy, the most frequent FISH pattern in mCRC with decreased *EGFR* GCN, is more easy to detect than an increase in *EGFR* copy number and may be more amendable to interpretation and reproducibility by FISH analyses [116]. In fact, Li *et al.* demonstrated that patients with *EGFR* FISH pattern of chromosome 7 disomy have a low chance to benefit from cetuximab based therapy [117].

It seems that elevations in *EGFR* GCN, as measured by FISH or CISH, may predict for benefit with treatment with anti-EGFR moABs. Conversely, normal GCN may predict for resistance to treatment. However, based on the retrospective and limited number of cases in these studies, the *EGFR* GCN has not been recommended as a clinical biomarker for the selection of mCRC patients responsive to anti-EGFR moAbs until larger prospective studies suggest otherwise (see Table 7.5).

7.4.3.7 PIK3CA mutations

PIK3CA gene encodes the p110 alpha kinase domain of PI3K. The encoded lipid kinase regulates signaling pathways downstream of the EGFR [25]. The PIK3CA gene is mutated in 20% of CRCs with mutations frequently occurring in exon 9 and 20 [24]. In preclinical models, these mutations result in activation of the EGFR pathway and are oncogenic in cellular models. In various colon cancers cell lines, activating *PIK3CA* mutations appear to confer resistance to cetuximab [118].

7.4.3.8 Measuring PIK3CA mutations

To detect these mutations, studies have typically employed direct sequencing of exons 9 and 20 or mass spectrometry [90, 119–121]. The latter involves PCR amplification of the region containing the mutation of interest, an extension reaction to generate allele-specific DNA products, and chip-based mass spectrometry for separation and analysis of the DNA. A single base extension reaction was performed using extension primers that hybridize immediately adjacent to the mutation. The extension reaction generates diagnostic products that, based on their unique mass values, allow discrimination between two alleles. This method is very sensitive, but only known mutations can be tested and it sometimes requires direct sequencing for verification.

7.4.3.9 Specimen selection for PIK3CA mutation testing

Currently, only one study has addressed the potential heterogeneity of the PIK3CA mutation status among primary tumors and metastatic tissue from the same patient [90]. In this study, there was no significant discordance between primary tumor and distant metastasis with only 1 case of 20 showing a PIK3CA mutation in the primary tumor but not in the distant metastasis. Hence, this suggests that PIK3CA mutation testing of primary tumors as well as distant metastases is appropriate to determine the PIK3CA mutation status.

7.4.3.10 Predictive role of PIK3CA mutation testing

Activating mutations in PIK3CA may predict clinical response to anti-EGFR moAbs. There have been several conflicting reports on the predictive role of *PIK3CA* mutations [8, 122, 123]. This discrepancy arises because different mutations identified in the *PIK3CA* gene disrupt different functional domains of the protein, and hence may not confer the same sensitivity to anti-EGFR moABs. For instance, the two most common mutations, within exons 9 and 20, encode regions of the protein that are RAS dependent and independent, respectively. In fact, in the first study attempting to distinguish the predictive role of these different *PIK3CA* mutations in a large chemotherapy refractory cohort of mCRC patients did demonstrate that exon 20 mutations had a low response rate to cetuximab and chemotherapy, while the mutations in exon 9 did not [119]. Prior studies attempting to address the predictive role of *PIK3CA* mutations did not divide their populations based on specific mutations. Also, the magnitude of the predictive effect for *PIK3CA* mutations is low relative to that observed for *KRAS* and *BRAF* mutations and these mutations can co-occur with *KRAS, BRAF* and *PTEN* mutations making the interpretation of the independent effect of *PIK3CA* mutations difficult [119]. Therefore, large patient cohorts will be required before drawing any definitive conclusions.

7.4.4 Potential Predictive Alterations in Gene Expression of the EGFR Pathway

7.4.4.1 EGFR protein expression

There are several ways to detect EGFR protein expression, including immunohistochemistry, enzyme-link immunosorbent assay (ELISA), and western blotting [124]. By far, the most commonly used test for EGFR protein expression is IHC. The process of IHC staining exploits the specific interaction between an antibody and a particular antigen. Fixed cells are exposed to, in this case, an anti-EGFR antibody. Visualizing the antibody-antigen interaction can be accomplished in a number of ways, but most commonly an antibody is conjugated to an enzyme that can catalyze a color-producing reaction. The staining results are measured semi-quantitatively on a scale of 0, 1+, 2+, and 3+. The score is based on staining intensity and percentage of tumor cells stained.

Based on the high specificity of cetuximab for EGFR, it was originally assumed that expression of EGFR, as detected by IHC, would be a predictor of benefit to the agent. As a result, the early clinical trials testing cetuximab (and panitumumab) were limited to those patients whose tumors expressed EGFR as detected by IHC. Contrary to this assumption, subsequent studies demonstrated that objective responses could be observed in patients who had EGFR negative tumors as determined by IHC and treated with cetuximab or panitumumab [125]. Moreover, most studies have failed to demonstrate any relationship between level of EGFR expression and the clinical activity of anti-EGFR drugs. There are several hypotheses as to the reasons behind the benefit observed in patients with EGFR negative tumors, as detected by IHC, from anti-EGFR moAbs. First, there have been several technical reasons proposed causing the inconsistent results. This includes, variability in EGFR staining based on processing and handling of tumor tissue samples, such as fixation methods and prolonged storage time. Also, the variability may arise from the use of different antibodies, targeting different epitopes as compared with those targeted by the anti-EGFR moAbs. Finally, false-negative results may be encountered due to the partially subjective scoring of IHC. Based on these inconsistencies, IHC testing for EGFR is inadequate at identifying patients who would and would not benefit from anti-EGFR moAbs.

7.4.4.2 EGFR ligands: amphiregulin and epiregulin

Both amphiregulin (AREG) and epiregulin (EREG) are alternative ligands for EGFR. Epiregulin is known to bind more weakly to EGFR than EGF but is more potent than EGF and results in prolonged receptor activation [126]. It was hypothesized that elevated expression of these ligands may stimulate an autocrine loop through the EGFR. As a result, this may identify those cancer cells that are dependent on EGFR signaling and therefore indicate sensitivity to anti-EGFR agents that block receptor to ligand binding.

A recent study by Khambata-Ford *et al.* demonstrated that in chemotherapy refractory patients treated with cetuximab monotherapy, tumors that had high gene expression levels of the EGFR ligands, AREG and EREG, and WT *K-RAS* were more likely to have disease control on anti-EGFR treatment. This study employed a global genomic approach, using microarray analyses of 95 colorectal cancer liver metastases, to identify markers that were associated with disease control in patients treated with cetuximab [78]. Moreover, patients with high AREG and EREG expression, as detected by ELISA, had a longer PFS compared with those with a low expression (EREG: median of 103.5 vs. 57 days; and AREG: median of 115.5 vs. 57 days). This study was soon corroborated in an independent cohort of patients with mCRC testing for ligand mRNA in primary tumor samples [127]. A chemotherapy refractory cohort was accrued, but this time treated with irinotecan and cetuximab as opposed to monotherapy. Again, increased expression of AREG and EREG in *KRAS* WT tumors predicted for response, PFS and OS.

Although promising as an additional test to predict for benefit from anti-EGFR moAbs in patients with mCRC and *KRAS* WT tumors, further validation of the above ligand data is required in prospective controlled trials, as well as in trials involving anti-EGFR moAbs in earlier lines of treatment and in combination with chemotherapy. Moreover, the techniques to detect and score the levels of amphiregulin and epiregulin expression must be standardized before this can be applied to clinical practice.

7.4.4.3 PTEN loss of expression

PTEN (phosphatase and tensin homologue gene) encodes an enzyme that when functioning properly, signals cells to stop dividing and induces apoptosis when necessary. PTEN inactivation leads to uncontrolled signaling via PIP3 accumulation and subsequent AKT hyperphosphorylation, resulting in loss of responsiveness to apoptotic stimuli of cancer cells [128]. PTEN loss has been observed in different sporadic tumors, including colorectal cancers [129]. Preclinical data has demonstrated the importance of a functioning PTEN/PI3K/AKT pathway in determining the sensitivity of CRC cell lines to cetuximab [118].

7.4.4.4 Measuring PTEN expression

PTEN expression is most often measured with immunohistochemistry (IHC). Although, several antibodies exist for the detection of PTEN with IHC, none have been determined to be superior over another (see Table 7.6). Similar to EGFR, the IHC for PTEN is subject to some of the same criticisms. To date, there has been no standardization in the IHC assay in regards to tissue handling, optimized storage duration, and semi-subjective scoring algorithms.

Table 7.6 Loss of PTEN Expression results and clinical outcomes in anti-EGFR mAb clinical trials

Study	Anti-EGFR mAb	Tumor samples	PTEN status	Clinical outcome (95% CI)	*P*-value
Frattini *et al.*, 2007 [112]	Cetuximab	27	Loss	0/11 responses	$P < 0.001$
			Expressed	10/16 responses	
Laurent-Puig *et al.*, 2009 [97]	Cetuximab	116 *KRAS* WT	Loss	10/22 responses, 11.8 months (9.1-17.9) median OS	NS for response, but $P = 0.013$ for OS
			Expressed	41//89 responses 16.2 months (13.9-20.7) median OS	
Razis *et al.*, 2009 [130]	Cetuximab	72	Loss	5/10 responses 9.44 months TTP	$P = 0.54$ for TTP
			Expressed	19/62 responses 6.39 months TTP	

Abbreviations: EGFR, epidermal growth factor receptor; OS, overall survival; TTP, time to progression; NS, non-significant.

7.4.4.5 Predictive role of loss of PTEN expression

The loss of PTEN protein expression occurs in 30% of colorectal tumors and may be an additional determinant of PI3K/AKT pathway activation responsible for resistance to anti-EGFR moABs [112]. The first study suggesting that loss of PTEN had a predictive role was in a small cohort of 27 chemotherapy refractory patients with mCRC treated with cetuximab and irinotecan [112]. This study demonstrated that in 0 of the 11 patients with loss of PTEN expression benefited from the treatment with cetuximab and irinotecan, while a response was observed in 10 out of 16 patients with intact PTEN expression. The six patients with intact PTEN expression who did not benefit from cetuximab had *KRAS* mutation in four of the cases. Some studies have corroborated this data [69, 121], but conflicting reports [97, 130] have emerged making it difficult to interpret the predictive efficacy of loss of PTEN expression. For instance, one study demonstrated discordance between primary tumors and loss of PTEN expression and that only loss of PTEN in the metastatic lesions were predictive of response to cetuximab and irinotecan [69]. Another reported on the lack of PTEN gene amplification by FISH as predictive to cetuximab response, while the expression of PTEN as measured by IHC did not have this capability [130]. Differences in IHC for pre-analytic and analytic variables, and scoring may be why studies have conflicting results.

7.5 Future Directions

Consistent problems have arisen when attempting to measure the effect of predictive markers for anti-EGFR moAbs. The first difficulty encountered is that some of these genetic alterations are rare and/or the magnitude of the effect on predicting resistance or sensitivity is small to moderate. Requiring large prospective randomized cohorts to accurately examine the effects of these molecular changes, although ideal, is understandably difficult. If retrospective studies are to be undertaken, multiple known predictive markers should be tested for each tumor sample. By doing so, this limits the problem of falsely labeling a marker as predictive, when in fact the true predictive marker was merely not tested. Moreover, testing several markers in combination may increase the magnitude of predictive efficacy. Recently a published report indicated that when expression of PTEN and mutations of *KRAS*, *BRAF*, and *PIK3CA* are concomitantly ascertained, up to 70% of patients with mCRC unlikely to respond to anti-EGFR therapies could be identified [131]. Finally, by studying several prospective markers at once, it may identify other components or pathways involved in cancer and help design logical targeted treatment protocols that can be administered sequentially or concurrently.

Second, only *KRAS* and *BRAF* have commercially available tests that have been standardized. As of yet, the other molecular markers of the EGFR pathway do not have standard ways of detecting, measuring, and scoring the alteration of interest, ultimately making it difficult to interpret the biological effects, clinical outcomes, and comparisons across studies. Further complicating the picture, is the fact that for some of the molecular changes occurs late in cancer progression and might not be concordant from primary to metastatic tissue.

Third, all of these studies discussed within this chapter focus on tumor alterations to identify these markers of resistance and sensitivity. Certainly, there are host factors, such as genomic DNA from non-tumor tissue that may also dictate response to anti-EGFR moAbs. For instance, several independent studies have recently found that Fc gamma receptor [132], COX-2 [133], and EGFR germline polymorphisms [134], are associated with clinical outcome in mCRC patients treated with single-agent cetuximab independent of *KRAS* status. Other potential sites to test for surrogate markers for response would be circulating tumor cells and lymph node metastases.

Finally, as of yet, few studies have addressed the molecular changes that underlie the acquired resistance to anti-EGFR moAbs in mCRC. Inevitably, those patients who initially responded to anti-EGFR treatments ultimately progress and succumb to the disease. Most of the studies that have investigated acquired resistance in NSCLC [135] or are merely hypothesis generating *in vitro* and mouse models [136]. Several novel agents are in development, targeting the EGFR and other signaling pathways involved in carcinogenesis. These agents could be implemented purposefully as salvage treatment if the mechanisms of acquired resistance to anti-EGFR moAbs was better understood.

As per Bayesian theory, the ultimate goal of all predictive tests is to refine the pretest probability to the point where one can confidently make a recommendation based on the post-test probability of benefit to treatment. This is an old paradigm, now applied to the novel targeted agents used in the treatment of mCRC with the hopes of identifying predictive markers that will identify patients who respond to treatment and avoid costly and toxic agents in those who do not. For now, only *KRAS* WT patients with mCRC are approved for cetuximab and panitumumab treatment based on the positive response observed in these patients. Additional dependable and easily measured biomarkers are needed to improve the identification of patients who will benefit from anti-EGFR moAbs.

References

1. GLOBOCAN 2008 (IARC) Section of Cancer Information accessed 25/2/2013 http://globocan.iarc.fr/factsheets/populations/factsheet.asp?uno=900.
2. Poston GJ, Figueras J, Giuliante F, *et al.* Urgent need for a new staging system in advanced colorectal cancer. *J Clin Oncol* 2008; 26: 4828–4833. Published on 20080818. doi:10.1200/JCO.2008.17.6453.
3. Markowitz SD, Bertagnolli MM. Molecular origins of cancer: Molecular basis of colorectal cancer. *N Engl J Med* 2009; 361: 2449–2460. doi:10.1056/NEJMra0804588.
4. Lievre A, Blons H, Laurent-Puig P. Oncogenic mutations as predictive factors in colorectal cancer. *Oncogene* 2010; 29: 3033–3043. Published on 20100412. doi:10.1038/onc.2010.89.
5. Parsons DW, Wang TL, Samuels Y, *et al.* Colorectal cancer: mutations in a signalling pathway. *Nature* 2005; 436: 792. doi:10.1038/436792a.
6. Ciardiello F, Tortora G. EGFR antagonists in cancer treatment. *N Engl J Med* 2008; 358: 1160–1174. doi:10.1056/NEJMra0707704.
7. Schneider MR, Wolf E. The epidermal growth factor receptor ligands at a glance. *J Cell Physiol* 2009; 218: 460–466. doi:10.1002/jcp.21635.
8. Moroni M, Veronese S, Benvenuti S, *et al.* Gene copy number for epidermal growth factor receptor (EGFR) and clinical response to antiEGFR treatment in colorectal cancer: a cohort study. *Lancet Oncol* 2005; 6: 279–286. doi:10.1016/S1470-2045(05)70102-9.
9. Barber TD, Vogelstein B, Kinzler KW, Velculescu VE. Somatic mutations of EGFR in colorectal cancers and glioblastomas. *N Engl J Med* 2004; 351: 2883. doi:10.1056/NEJM200412303512724.
10. Samowitz WS, Albertsen H, Herrick J, *et al.* Evaluation of a large, population-based sample supports a CpG island methylator phenotype in colon cancer. *Gastroenterology* 2005; 129: 837–845. doi:10.1053/j.gastro.2005.06.020.
11. Samowitz WS, Curtin K, Schaffer D, *et al.* Relationship of Ki-ras mutations in colon cancers to tumor location, stage, and survival: a population-based study. *Cancer Epidemiol Biomarkers Prev* 2000; 9: 1193–1197.

12. Rajagopalan H, Bardelli A, Lengauer C, *et al.* Tumorigenesis: RAF/RAS oncogenes and mismatch-repair status. *Nature* 2002; 418: 934. doi:10.1038/418934a.

13. Nosho K, Irahara N, Shima K, *et al.* Comprehensive biostatistical analysis of CpG island methylator phenotype in colorectal cancer using a large population-based sample. *PLoS One* 2008; 3: e3698. Published on 20081112. doi:10.1371/journal.pone.0003698.

14. Nosho K, Kawasaki T, Ohnishi M, *et al.* PIK3CA mutation in colorectal cancer: relationship with genetic and epigenetic alterations. *Neoplasia* 2008; 10: 534–541.

15. Brink M, de Goeij AF, Weijenberg MP, *et al.* K-ras oncogene mutations in sporadic colorectal cancer in The Netherlands Cohort Study. *Carcinogenesis* 2003; 24: 703–710.

16. de Vogel S, Weijenberg MP, Herman JG, *et al.* MGMT and MLH1 promoter methylation versus APC, KRAS and BRAF gene mutations in colorectal cancer: indications for distinct pathways and sequence of events. *Ann Oncol* 2009; 20: 1216–1222. Published on 20090122. doi:10.1093/annonc/mdn782.

17. Bamford S, Dawson E, Forbes S, *et al.* The COSMIC (Catalogue of Somatic Mutations in Cancer) database and website. *Br J Cancer* 2004; 91: 355–358. doi:10.1038/sj.bjc.6601894.

18. Barault L, Veyrie N, Jooste V, *et al.* Mutations in the RAS-MAPK, PI(3)K (phosphatidylinositol-3-OH kinase) signaling network correlate with poor survival in a population-based series of colon cancers. *Int J Cancer* 2008; 122: 2255–2259. doi:10.1002/ijc.23388.

19. Vetter IR, Wittinghofer A. The guanine nucleotide-binding switch in three dimensions. *Science* 2001; 294: 1299–1304. doi:10.1126/science.1062023.

20. Bos JL. Ras Oncogenes in Human Cancer: a Review. *Cancer Res* 1989; 49: 4682–4689.

21. Lea IA, Jackson MA, Dunnick JK. Genetic pathways to colorectal cancer. *Mutat Res* 2009; 670: 96–98. Published on 20090701. doi:10.1016/j.mrfmmm.2009.06.011.

22. Wan PT, Garnett MJ, Roe SM, *et al.* Mechanism of activation of the RAF-ERK signaling pathway by oncogenic mutations of B-RAF. *Cell* 2004; 116: 855–867.

23. Preto A, Figueiredo J, Velho S, *et al.* BRAF provides proliferation and survival signals in MSI colorectal carcinoma cells displaying BRAF(V600E) but not KRAS mutations. *J Pathol* 2008; 214: 320–327. doi:10.1002/path.2295.

24. Samuels Y, Wang Z, Bardelli A, *et al.* High frequency of mutations of the PIK3CA gene in human cancers. *Science* 2004; 304: 554. Published on 20040311. doi:10.1126/science.1096502.

25. Ikenoue T, Kanai F, Hikiba Y, *et al.* Functional analysis of PIK3CA gene mutations in human colorectal cancer. *Cancer Res* 2005; 65: 4562–4567. doi:10.1158/0008-5472.CAN-04-4114.

26. Kang S, Bader AG, Vogt PK. Phosphatidylinositol 3-kinase mutations identified in human cancer are oncogenic. *Proc Natl Acad Sci U S A* 2005; 102: 802–807. Published on 20050112. doi:10.1073/pnas.0408864102.

27. Wang ZJ, Taylor F, Churchman M, *et al.* Genetic pathways of colorectal carcinogenesis rarely involve the PTEN and LKB1 genes outside the inherited hamartoma syndromes. *Am J Pathol* 1998; 153: 363–366.

28. Shin KH, Park YJ, Park JG. PTEN gene mutations in colorectal cancers displaying microsatellite instability. *Cancer Lett* 2001; 174: 189–194.

29. Guanti G, Resta N, Simone C, *et al.* Involvement of PTEN mutations in the genetic pathways of colorectal cancerogenesis. *Hum Mol Genet* 2000; 9: 283–287.

30. Sobrero AF, Maurel J, Fehrenbacher L, *et al.* EPIC: phase III trial of cetuximab plus irinotecan after fluoropyrimidine and oxaliplatin failure in patients with metastatic colorectal cancer. *J Clin Oncol* 2008; 26: 2311–2319. Published on 20080407. doi:10.1200/JCO.2007.13.1193.

31. Cunningham D, Humblet Y, Siena S, *et al.* Cetuximab monotherapy and cetuximab plus irinotecan in irinotecan-refractory metastatic colorectal cancer. *N Engl J Med* 2004; 351: 337–345. doi:10.1056/NEJMoa033025.

32. Jonker DJ, O'Callaghan CJ, Karapetis CS, *et al.* Cetuximab for the treatment of colorectal cancer. *N Engl J Med* 2007; 357: 2040–2048. doi:10.1056/NEJMoa071834.

33. Saltz LB, Meropol NJ, Loehrer PJ S, *et al.* Phase II trial of cetuximab in patients with refractory colorectal cancer that expresses the epidermal growth factor receptor. *J Clin Oncol* 2004; 22: 1201–1208. Published on 20040301. doi:10.1200/JCO.2004.10.182.

34. Van Cutsem E, Kohne CH, Hitre E, *et al.* Cetuximab and chemotherapy as initial treatment for metastatic colorectal cancer. *N Engl J Med* 2009; 360: 1408–1417. doi:10.1056/NEJMoa0805019.

35. Van Cutsem E, Peeters M, Siena S, *et al.* Open-label phase III trial of panitumumab plus best supportive care compared with best supportive care alone in patients with chemotherapy-refractory metastatic

colorectal cancer. *J Clin Oncol* 2007; 25: 1658–1664. doi:10.1200/JCO.2006.08.1620.

36. Douillard JY, Siena S, Cassidy J, *et al.* Randomized, Phase III Trial of Panitumumab With Infusional Fluorouracil, Leucovorin, and Oxaliplatin (FOLFOX4) Versus FOLFOX4 Alone As First-Line Treatment in Patients With Previously Untreated Metastatic Colorectal Cancer: The PRIME Study. *J Clin Oncol* 2010; 28: 4697–4705. Published on 20101004. doi:10.1200/JCO.2009.27.4860.
37. Peeters M, Price TJ, Cervantes A, *et al.* Randomized phase III study of panitumumab with Fluorouracil, Leucovorin, and Irinotecan (FOLFIRI) compared with FOLFIRI alone as second-line treatment in patients with metastatic colorectal cancer. *J Clin Oncol* 2010; 28: 4706–4713. Published on 20101004. doi:10.1200/JCO.2009.27.6055.
38. Giantonio BJ, Catalano PJ, Meropol NJ, *et al.* Bevacizumab in combination with oxaliplatin, fluorouracil, and leucovorin (FOLFOX4) for previously treated metastatic colorectal cancer: results from the Eastern Cooperative Oncology Group Study E3200. *J Clin Oncol* 2007; 25: 1539–1544. doi:10.1200/JCO.2006.09.6305.
39. Saltz LB, Clarke S, Diaz-Rubio E, *et al.* Bevacizumab in combination with oxaliplatin-based chemotherapy as first-line therapy in metastatic colorectal cancer: a randomized phase III study. *J Clin Oncol* 2008; 26: 2013–2019. doi:10.1200/JCO.2007.14.9930.
40. Van Cutsem E, Rivera F, Berry S, *et al.* Safety and efficacy of first-line bevacizumab with FOLFOX, XELOX, FOLFIRI and fluoropyrimidines in metastatic colorectal cancer: the BEAT study. *Ann Oncol* 2009; 20: 1842–1847. Published on 20090430. doi:10.1093/annonc/mdp233.
41. Fuchs CS, Marshall J, Barrueco J. Randomized, controlled trial of irinotecan plus infusional, bolus, or oral fluoropyrimidines in first-line treatment of metastatic colorectal cancer: updated results from the BICC-C study. *J Clin Oncol* 2008; 26: 689–690. doi:10.1200/JCO.2007.15.5390.
42. Hurwitz H, Fehrenbacher L, Novotny W, *et al.* Bevacizumab plus irinotecan, fluorouracil, and leucovorin for metastatic colorectal cancer. *N Engl J Med* 2004; 350: 2335–2342. doi:10.1056/NEJMoa032691.
43. Tol J, Koopman M, Cats A, *et al.* Chemotherapy, bevacizumab, and cetuximab in metastatic colorectal cancer. *N Engl J Med* 2009; 360: 563–572. doi:10.1056/NEJMoa0808268.
44. Hecht JR, Mitchell E, Chidiac T, *et al.* A randomized phase IIIB trial of chemotherapy, bevacizumab, and panitumumab compared with chemotherapy and bevacizumab alone for metastatic colorectal cancer.

J Clin Oncol 2009; 27: 672–680. Published on 20081229. doi:10.1200/JCO.2008.19.8135.

45. Saltz LB, Lenz HJ, Kindler HL, *et al.* Randomized phase II trial of cetuximab, bevacizumab, and irinotecan compared with cetuximab and bevacizumab alone in irinotecan-refractory colorectal cancer: the BOND-2 study. *J Clin Oncol* 2007; 25: 4557–4561. Published on 20070917. doi:10.1200/JCO.2007.12.0949.
46. Rothenberg ML, LaFleur B, Levy DE, *et al.* Randomized phase II trial of the clinical and biological effects of two dose levels of gefitinib in patients with recurrent colorectal adenocarcinoma. *J Clin Oncol* 2005; 23: 9265–9274. doi:10.1200/JCO.2005.03.0536.
47. Townsley CA, Major P, Siu LL, *et al.* Phase II study of erlotinib (OSI-774) in patients with metastatic colorectal cancer. *Br J Cancer* 2006; 94: 1136–1143. doi:10.1038/sj.bjc.6603055.
48. Zampino MG, Magni E, Massacesi C, *et al.* First clinical experience of orally active epidermal growth factor receptor inhibitor combined with simplified FOLFOX6 as first-line treatment for metastatic colorectal cancer. *Cancer* 2007; 110: 752–758. doi:10.1002/cncr.22851.
49. Kuo T, Cho CD, Halsey J, *et al.* Phase II study of gefitinib, fluorouracil, leucovorin, and oxaliplatin therapy in previously treated patients with metastatic colorectal cancer. *J Clin Oncol* 2005; 23: 5613–5619. doi:10.1200/JCO.2005.08.359.
50. Meyerhardt JA, Zhu AX, Enzinger PC, *et al.* Phase II study of capecitabine, oxaliplatin, and erlotinib in previously treated patients with metastastic colorectal cancer. *J Clin Oncol* 2006; 24: 1892–1897. doi:10.1200/JCO.2005.05.3728.
51. Hynes NE, Lane HA. ERBB receptors and cancer: the complexity of targeted inhibitors. *Nat Rev Cancer* 2005; 5: 341–354. doi:10.1038/nrc1609.
52. Albanell J, Rojo F, Averbuch S, *et al.* Pharmacodynamic studies of the epidermal growth factor receptor inhibitor ZD1839 in skin from cancer patients: histopathologic and molecular consequences of receptor inhibition. *J Clin Oncol* 2002; 20: 110–124.
53. Baselga J, Rischin D, Ranson M, *et al.* Phase I safety, pharmacokinetic, and pharmacodynamic trial of ZD1839, a selective oral epidermal growth factor receptor tyrosine kinase inhibitor, in patients with five selected solid tumor types. *J Clin Oncol* 2002; 20: 4292–4302.
54. Vanhoefer U, Tewes M, Rojo F, *et al.* Phase I study of the humanized antiepidermal growth factor receptor monoclonal antibody EMD72000 in patients with advanced solid tumors that express the epidermal

growth factor receptor. *J Clin Oncol* 2004; 22: 175–184. doi:10.1200/JCO.2004.05.114.

55. Tejpar S, Peeters M, Humblet Y, *et al.* Phase I/II study of cetuximab dose-escalation in patients with metastatic colorectal cancer (mCRC) with no or slight skin reactions on cetuximab standard dose treatment (EVEREST): Pharmacokinetic (PK), Pharmacodynamic (PD) and efficacy data. *J Clin Oncol* 2007; 25: 4037, (June 20 supplement).
56. Perez-Soler R, Saltz L. Cutaneous adverse effects with HER1/EGFR-targeted agents: is there a silver lining? *J Clin Oncol* 2005; 23: 5235–5246. doi:10.1200/JCO.2005.00.6916.
57. Bos JL, Fearon ER, Hamilton SR, *et al.* Prevalence of ras gene mutations in human colorectal cancers. *Nature* 1987; 327: 293–297. doi:10.1038/327293a0.
58. Li J, Wang L, Mamon H, *et al.* Replacing PCR with COLD-PCR enriches variant DNA sequences and redefines the sensitivity of genetic testing. *Nat Med* 2008; 14: 579–584. Published on 20080413. doi:10.1038/nm1708.
59. Ogino S, Kawasaki T, Brahmandam M, *et al.* Sensitive sequencing method for KRAS mutation detection by Pyrosequencing. *J Mol Diagn* 2005; 7: 413–421.
60. Clayton SJ, Scott FM, Walker J, *et al.* K-ras point mutation detection in lung cancer: comparison of two approaches to somatic mutation detection using ARMS allele-specific amplification. *Clin Chem* 2000; 46: 1929–1938.
61. Soulieres D, Greer W, Magliocco AM, *et al.* KRAS mutation testing in the treatment of metastatic colorectal cancer with anti-EGFR therapies. *Curr Oncol* 2010; 17 Suppl 1: S31–S40.
62. Nikiforova MN, Lynch RA, Biddinger PW, *et al.* RAS point mutations and PAX8-PPAR gamma rearrangement in thyroid tumors: evidence for distinct molecular pathways in thyroid follicular carcinoma. *J Clin Endocrinol Metab* 2003; 88: 2318–2326.
63. Dabritz J, Hanfler J, Preston R, *et al.* Detection of Ki-ras mutations in tissue and plasma samples of patients with pancreatic cancer using PNA-mediated PCR clamping and hybridisation probes. *Br J Cancer* 2005; 92: 405–412. doi:10.1038/sj.bjc.6602319.
64. Al-Mulla F, Going JJ, Sowden ET, *et al.* Heterogeneity of mutant versus wild-type Ki-ras in primary and metastatic colorectal carcinomas, and association of codon-12 valine with early mortality. *J Pathol* 1998; 185: 130–138. doi:2-M.

65. Oudejans JJ, Slebos RJ, Zoetmulder FA, *et al.* Differential activation of ras genes by point mutation in human colon cancer with metastases to either lung or liver. *Int J Cancer* 1991; 49: 875–879.

66. Suchy B, Zietz C, Rabes HM. K-ras point mutations in human colorectal carcinomas: relation to aneuploidy and metastasis. *Int J Cancer* 1992; 52: 30–33.

67. Tortola S, Steinert R, Hantschick M, *et al.* Discordance between K-ras mutations in bone marrow micrometastases and the primary tumor in colorectal cancer. *J Clin Oncol* 2001; 19: 2837–2843.

68. Artale S, Sartore-Bianchi A, Veronese SM, *et al.* Mutations of KRAS and BRAF in primary and matched metastatic sites of colorectal cancer. *J Clin Oncol* 2008; 26: 4217–4219. doi:10.1200/JCO.2008.18.7286.

69. Loupakis F, Pollina L, Stasi I, *et al.* PTEN expression and KRAS mutations on primary tumors and metastases in the prediction of benefit from cetuximab plus irinotecan for patients with metastatic colorectal cancer. *J Clin Oncol* 2009; 27: 2622–2629. Published on 20090427. doi:10.1200/JCO.2008.20.2796.

70. Ahnen DJ, Feigl P, Quan G, *et al.* Ki-ras mutation and p53 overexpression predict the clinical behavior of colorectal cancer: a Southwest Oncology Group study. *Cancer Res* 1998; 58: 1149–1158.

71. Andreyev HJ, Norman AR, Cunningham D, *et al.* Kirsten ras mutations in patients with colorectal cancer: the 'RASCAL II' study. *Br J Cancer* 2001; 85: 692–696. doi:10.1054/bjoc.2001.1964.

72. Belly RT, Rosenblatt JD, Steinmann M, *et al.* Detection of mutated K12-ras in histologically negative lymph nodes as an indicator of poor prognosis in stage II colorectal cancer. *Clin Colorectal Cancer* 2001; 1: 110–116.

73. Ogino S, Meyerhardt JA, Irahara N, *et al.* KRAS mutation in stage III colon cancer and clinical outcome following intergroup trial CALGB 89803. *Clin Cancer Res* 2009; 15: 7322–7329. Published on 20091124. doi:10.1158/1078-0432.CCR-09-1570.

74. Roth AD, Tejpar S, Delorenzi M, *et al.* Prognostic role of KRAS and BRAF in stage II and III resected colon cancer: results of the translational study on the PETACC-3, EORTC 40993, SAKK 60–00 trial. *J Clin Oncol* 2010; 28: 466–474. Published on 20091214. doi:10.1200/JCO.2009.23.3452.

75. Karapetis CS, Khambata-Ford S, Jonker DJ, *et al.* K-ras mutations and benefit from cetuximab in advanced colorectal cancer. *N Engl J Med* 2008; 359: 1757–1765. doi:10.1056/NEJMoa0804385.

76. Lievre A, Bachet JB, Le Corre D, *et al.* KRAS mutation status is predictive of response to cetuximab therapy in colorectal cancer. *Cancer Res* 2006; 66: 3992–3995. doi:10.1158/0008-5472.CAN-06-0191.

77. Lievre A, Bachet JB, Boige V, *et al.* KRAS mutations as an independent prognostic factor in patients with advanced colorectal cancer treated with cetuximab. *J Clin Oncol* 2008; 26: 374–379. doi:10.1200/JCO.2007.12.5906.

78. Khambata-Ford S, Garrett CR, Meropol NJ, *et al.* Expression of epiregulin and amphiregulin and K-ras mutation status predict disease control in metastatic colorectal cancer patients treated with cetuximab. *J Clin Oncol* 2007; 25: 3230–3237. doi:10.1200/JCO.2006.10.5437.

79. Benvenuti S, Sartore-Bianchi A, Di Nicolantonio F, *et al.* Oncogenic activation of the RAS/RAF signaling pathway impairs the response of metastatic colorectal cancers to anti-epidermal growth factor receptor antibody therapies. *Cancer Res* 2007; 67: 2643–2648. doi:10.1158/0008-5472.CAN-06-4158.

80. Amado RG, Wolf M, Peeters M, *et al.* Wild-type KRAS is required for panitumumab efficacy in patients with metastatic colorectal cancer. *J Clin Oncol* 2008; 26: 1626–1634. Published on 20080303. doi:10.1200/JCO.2007.14.7116.

81. Bokemeyer C, Bondarenko I, Makhson A, *et al.* Fluorouracil, leucovorin, and oxaliplatin with and without cetuximab in the first-line treatment of metastatic colorectal cancer. *J Clin Oncol* 2009; 27: 663–671. Published on 20081229. doi:10.1200/JCO.2008.20.8397.

82. De Roock W, Jonker DJ, Di Nicolantonio F, *et al.* Association of KRAS p.G13D mutation with outcome in patients with chemotherapy-refractory metastatic colorectal cancer treated with cetuximab. *JAMA* 2010; 304: 1812–1820. doi:10.1001/jama.2010.1535.

83. Raman M, Chen W, Cobb MH. Differential regulation and properties of MAPKs. *Oncogene* 2007; 26: 3100–3112. doi:10.1038/sj.onc.1210392.

84. Davies H, Bignell GR, Cox C, *et al.* Mutations of the BRAF gene in human cancer. *Nature* 2002; 417: 949–954. Published on 20020609. doi:10.1038/nature00766.

85. Pollock PM, Harper UL, Hansen KS, *et al.* High frequency of BRAF mutations in nevi. *Nat Genet* 2003; 33: 19–20. Published on 20021125. doi:10.1038/ng1054.

86. Tol J, Nagtegaal ID, Punt CJ. BRAF mutation in metastatic colorectal cancer. *N Engl J Med* 2009; 361: 98–99. doi:10.1056/NEJMc0904160.

87. Domingo E, Laiho P, Ollikainen M, *et al.* BRAF screening as a low-cost effective strategy for simplifying HNPCC genetic testing. *J Med Genet* 2004; 41: 664–668. doi:10.1136/jmg.2004.020651.

88. Benlloch S, Paya A, Alenda C, *et al.* Detection of BRAF V600E mutation in colorectal cancer: comparison of automatic sequencing and real-time chemistry methodology. *J Mol Diagn* 2006; 8: 540–543.

89. Italiano A, Hostein I, Soubeyran I, *et al.* KRAS and BRAF mutational status in primary colorectal tumors and related metastatic sites: biological and clinical implications. *Ann Surg Oncol* 2010; 17: 1429–1434. Published on 20100105. doi:10.1245/s10434-009-0864-z.

90. Baldus SE, Schaefer KL, Engers R, *et al.* Prevalence and heterogeneity of KRAS, BRAF, and PIK3CA mutations in primary colorectal adenocarcinomas and their corresponding metastases. *Clin Cancer Res* 2010; 16: 790–799. Published on 20100126. doi:10.1158/1078-0432. CCR-09-2446.

91. Santini D, Spoto C, Loupakis F, *et al.* High concordance of BRAF status between primary colorectal tumours and related metastatic sites: implications for clinical practice. *Ann Oncol* 2010; 21: 1565. doi:10.1093/annonc/mdq318.

92. Van Cutsem E, Dicato M, Arber N, *et al.* Molecular markers and biological targeted therapies in metastatic colorectal cancer: expert opinion and recommendations derived from the 11th ESMO/World Congress on Gastrointestinal Cancer, Barcelona, 2009. *Ann Oncol* 2010; 21 Suppl 6: vi1-10. doi:10.1093/annonc/mdq273.

93. Ogino S, Nosho K, Kirkner GJ, *et al.* CpG island methylator phenotype, microsatellite instability, BRAF mutation and clinical outcome in colon cancer. *Gut* 2009; 58: 90–96. Published on 20081002. doi:10.1136/gut.2008.155473.

94. Di Nicolantonio F, Martini M, Molinari F, *et al.* Wild-type BRAF is required for response to panitumumab or cetuximab in metastatic colorectal cancer. *J Clin Oncol* 2008; 26: 5705–5712. Published on 20081110. doi:10.1200/JCO.2008.18.0786.

95. Souglakos J, Philips J, Wang R, *et al.* Prognostic and predictive value of common mutations for treatment response and survival in patients with metastatic colorectal cancer. *Br J Cancer* 2009; 101: 465–472. Published on 20090714. doi:10.1038/sj.bjc.6605164.

96. Loupakis F, Ruzzo A, Cremolini C, *et al.* KRAS codon 61, 146 and BRAF mutations predict resistance to cetuximab plus irinotecan in KRAS codon 12 and 13 wild-type metastatic colorectal cancer. *Br J*

Cancer 2009; 101: 715–721. Published on 20090714. doi:10.1038/sj.bjc.6605177.

97. Laurent-Puig P, Cayre A, Manceau G, *et al.* Analysis of PTEN, BRAF, and EGFR status in determining benefit from cetuximab therapy in wild-type KRAS metastatic colon cancer. *J Clin Oncol* 2009; 27: 5924–5930. Published on 20091102. doi:10.1200/JCO.2008.21.6796.
98. Sethupathy P, Collins FS. MicroRNA target site polymorphisms and human disease. *Trends Genet* 2008; 24: 489–497. Published on 20080906. doi:10.1016/j.tig.2008.07.004.
99. Johnson SM, Grosshans H, Shingara J, *et al.* RAS is regulated by the let-7 microRNA family. *Cell* 2005; 120: 635–647. doi:10.1016/j.cell.2005.01.014.
100. Chin LJ, Ratner E, Leng S, *et al.* A SNP in a let-7 microRNA complementary site in the KRAS 3′ untranslated region increases non-small cell lung cancer risk. *Cancer Res* 2008; 68: 8535–8540. doi:10.1158/0008-5472.CAN-08–2129.
101. Graziano F, Canestrari E, Loupakis F, *et al.* Genetic modulation of the Let-7 microRNA binding to KRAS 3'-untranslated region and survival of metastatic colorectal cancer patients treated with salvage cetuximab-irinotecan. *Pharmacogenomics J* 2010; 10: 458–464. Published on 20100223. doi:10.1038/tpj.2010.9.
102. Zhang W, Winder T, Ning Y, *et al.* A let-7 microRNA-binding site polymorphism in 3'-untranslated region of KRAS gene predicts response in wild-type KRAS patients with metastatic colorectal cancer treated with cetuximab monotherapy. *Ann Oncol* 2010 Published on 20100705. doi:10.1093/annonc/mdq315.
103. Paez JG, Janne PA, Lee JC, *et al.* EGFR mutations in lung cancer: correlation with clinical response to gefitinib therapy. *Science* 2004; 304: 1497–1500. Published on 20040429. doi:10.1126/science.1099314.
104. Lynch TJ, Bell DW, Sordella R, *et al.* Activating mutations in the epidermal growth factor receptor underlying responsiveness of non-small-cell lung cancer to gefitinib. *N Engl J Med* 2004; 350: 2129–2139. Published on 20040429. doi:10.1056/NEJMoa040938.
105. Cappuzzo F, Finocchiaro G, Rossi E, *et al.* EGFR FISH assay predicts for response to cetuximab in chemotherapy refractory colorectal cancer patients. *Ann Oncol* 2008; 19: 717–723. Published on 20071031. doi:10.1093/annonc/mdm492.
106. Sartore-Bianchi A, Moroni M, Veronese S, *et al.* Epidermal growth factor receptor gene copy number and clinical outcome of metastatic

colorectal cancer treated with panitumumab. *J Clin Oncol* 2007; 25: 3238–3245. doi:10.1200/JCO.2007.11.5956.

107. Bengala C, Bettelli S, Bertolini F, *et al.* Prognostic role of EGFR gene copy number and KRAS mutation in patients with locally advanced rectal cancer treated with preoperative chemoradiotherapy. *Br J Cancer* 2010; 103: 1019–1024. Published on 20100914. doi:10.1038/sj.bjc.6605853.

108. Personeni N, Fieuws S, Piessevaux H, *et al.* Clinical usefulness of EGFR gene copy number as a predictive marker in colorectal cancer patients treated with cetuximab: a fluorescent *in situ* hybridization study. *Clin Cancer Res* 2008; 14: 5869–5876. doi:10.1158/1078-0432.CCR-08-0449.

109. Italiano A, Follana P, Caroli FX, *et al.* Cetuximab shows activity in colorectal cancer patients with tumors for which FISH analysis does not detect an increase in EGFR gene copy number. *Ann Surg Oncol* 2008; 15: 649–654. Published on 20071107. doi:10.1245/s10434-007-9667-2.

110. De Roock W, Piessevaux H, De Schutter J, *et al.* KRAS wild-type state predicts survival and is associated to early radiological response in metastatic colorectal cancer treated with cetuximab. *Ann Oncol* 2008; 19: 508–515. Published on 20071112. doi:10.1093/annonc/mdm496.

111. Lenz HJ, Van Cutsem E, Khambata-Ford S, *et al.* Multicenter phase II and translational study of cetuximab in metastatic colorectal carcinoma refractory to irinotecan, oxaliplatin, and fluoropyrimidines. *J Clin Oncol* 2006; 24: 4914–4921. doi:10.1200/JCO.2006.06.7595.

112. Frattini M, Saletti P, Romagnani E, *et al.* PTEN loss of expression predicts cetuximab efficacy in metastatic colorectal cancer patients. *Br J Cancer* 2007; 97: 1139–1145. Published on 20071016. doi:10.1038/sj.bjc.6604009.

113. Ooi A, Takehana T, Li X, *et al.* Protein overexpression and gene amplification of HER-2 and EGFR in colorectal cancers: an immunohistochemical and fluorescent *in situ* hybridization study. *Mod Pathol* 2004; 17: 895–904. doi:10.1038/modpathol.3800137.

114. Shia J, Klimstra DS, Li AR, *et al.* Epidermal growth factor receptor expression and gene amplification in colorectal carcinoma: an immunohistochemical and chromogenic *in situ* hybridization study. *Mod Pathol* 2005; 18: 1350–1356. doi:10.1038/modpathol.3800417.

115. Sauer T, Guren MG, Noren T, Dueland S. Demonstration of EGFR gene copy loss in colorectal carcinomas by fluorescence *in situ* hybridization

(FISH): a surrogate marker for sensitivity to specific anti-EGFR therapy? *Histopathology* 2005; 47: 560–564. doi:10.1111/j.1365-2559.2005.02252.x.

116. Moroni M, Sartore-Bianchi A, Veronese S, Siena S. EGFR FISH in colorectal cancer: what is the current reality? *Lancet Oncol* 2008; 9: 402–403. doi:10.1016/S1470-2045(08)70109-8.

117. Li YH, Wang F, Shen L, *et al.* EGFR FISH pattern of chromosome 7 disomy predicts resistance to cetuximab in KRAS wild-type metastatic colorectal cancer patients. *Clin Cancer Res* 2010 Published on 20100930. doi:10.1158/1078-0432.CCR-10-0208.

118. Jhawer M, Goel S, Wilson AJ, *et al.* PIK3CA mutation/PTEN expression status predicts response of colon cancer cells to the epidermal growth factor receptor inhibitor cetuximab. *Cancer Res* 2008; 68: 1953–1961. doi:10.1158/0008-5472.CAN-07-5659.

119. De Roock W, Claes B, Bernasconi D, *et al.* Effects of KRAS, BRAF, NRAS, and PIK3CA mutations on the efficacy of cetuximab plus chemotherapy in chemotherapy-refractory metastatic colorectal cancer: a retrospective consortium analysis. *Lancet Oncol* 2010; 11: 753–762. Published on 20100708. doi:10.1016/S1470-2045(10)70130-3.

120. Prenen H, De Schutter J, Jacobs B, *et al.* PIK3CA mutations are not a major determinant of resistance to the epidermal growth factor receptor inhibitor cetuximab in metastatic colorectal cancer. *Clin Cancer Res* 2009; 15: 3184–3188. Published on 20090414. doi:10.1158/1078-0432.CCR-08-2961.

121. Sartore-Bianchi A, Martini M, Molinari F, *et al.* PIK3CA mutations in colorectal cancer are associated with clinical resistance to EGFR-targeted monoclonal antibodies. *Cancer Res* 2009; 69: 1851–1857. Published on 20090217. doi:10.1158/0008-5472.CAN-08-2466.

122. Perrone F, Lampis A, Orsenigo M, *et al.* PI3KCA/PTEN deregulation contributes to impaired responses to cetuximab in metastatic colorectal cancer patients. *Ann Oncol* 2009; 20: 84–90. Published on 20080731. doi:10.1093/annonc/mdn541.

123. Niendorf KB, Goggins W, Yang G, *et al.* MELPREDICT: a logistic regression model to estimate CDKN2A carrier probability. *J Med Genet* 2006; 43: 501–506. doi:10.1136/jmg.2005.032441.

124. Nannini M, Pantaleo MA, Paterini P, *et al.* Molecular detection of epidermal growth factor receptor (EGFr) in colorectal cancer: does it still make sense? *Colorectal Dis* 2010 Published on 20100112. doi:10.1111/j.1463-1318.2010.02212.x.

125. Chung KY, Shia J, Kemeny NE, *et al.* Cetuximab shows activity in colorectal cancer patients with tumors that do not express the epidermal growth factor receptor by immunohistochemistry. *J Clin Oncol* 2005; 23: 1803–1810. Published on 20050127. doi:10.1200/JCO.2005.08.037.

126. Shelly M, Pinkas-Kramarski R, Guarino BC, *et al.* Epiregulin is a potent pan-ErbB ligand that preferentially activates heterodimeric receptor complexes. *J Biol Chem* 1998; 273: 10496–10505.

127. Jacobs B, De Roock W, Piessevaux H, *et al.* Amphiregulin and epiregulin mRNA expression in primary tumors predicts outcome in metastatic colorectal cancer treated with cetuximab. *J Clin Oncol* 2009; 27: 5068–5074. Published on 20090908. doi:10.1200/JCO.2008.21.3744.

128. Di Cristofano A, Pandolfi PP. The multiple roles of PTEN in tumor suppression. *Cell* 2000; 100: 387–390.

129. Zhou XP, Loukola A, Salovaara R, *et al.* PTEN mutational spectra, expression levels, and subcellular localization in microsatellite stable and unstable colorectal cancers. *Am J Pathol* 2002; 161: 439–447.

130. Razis E, Briasoulis E, Vrettou E, *et al.* Potential value of PTEN in predicting cetuximab response in colorectal cancer: an exploratory study. *BMC Cancer* 2008; 8: 234. Published on 20080813. doi:10.1186/1471-2407-8–234.

131. Sartore-Bianchi A, Di Nicolantonio F, Nichelatti M, *et al.* Multi-determinants analysis of molecular alterations for predicting clinical benefit to EGFR-targeted monoclonal antibodies in colorectal cancer. *PLoS One* 2009; 4: e7287. Published on 20091002. doi:10.1371/journal.pone.0007287.

132. Bibeau F, Lopez-Crapez E, Di Fiore F, *et al.* Impact of Fc{gamma}RIIa-Fc{gamma}RIIIa polymorphisms and KRAS mutations on the clinical outcome of patients with metastatic colorectal cancer treated with cetuximab plus irinotecan. *J Clin Oncol* 2009; 27: 1122–1129. Published on 20090121. doi:10.1200/JCO.2008.18.0463.

133. Lurje G, Nagashima F, Zhang W, *et al.* Polymorphisms in cyclooxygenase-2 and epidermal growth factor receptor are associated with progression-free survival independent of K-ras in metastatic colorectal cancer patients treated with single-agent cetuximab. *Clin Cancer Res* 2008; 14: 7884–7895. doi:10.1158/1078-0432.CCR-07–5165.

134. Zhang W, Gordon M, Schultheis AM, *et al.* FCGR2A and FCGR3A polymorphisms associated with clinical outcome of epidermal growth factor receptor expressing metastatic colorectal cancer patients

treated with single-agent cetuximab. *J Clin Oncol* 2007; 25: 3712–3718. doi:10.1200/JCO.2006.08.8021.

135. Gazdar AF. Activating and resistance mutations of EGFR in non-small-cell lung cancer: role in clinical response to EGFR tyrosine kinase inhibitors. *Oncogene* 2009; 28 Suppl 1: S24–S31. doi:10.1038/onc.2009.198.

136. Li C, Iida M, Dunn EF, *et al.* Nuclear EGFR contributes to acquired resistance to cetuximab. *Oncogene* 2009; 28: 3801–3813. Published on 20090817. doi:10.1038/onc.2009.234.

137. Maughan TS, Adams R, Smith CG, *et al.* Identification of potentially responsive subsets when cetuximab is added to oxaliplatin-fluoropyrimidine chemotherapy (CT) in first-line advanced colorectal cancer (aCRC): Mature results of the MRC COIN trial. *J Clin Oncol* 28:15s, 2010 (suppl; abstr 3502) 2010.

138. Moroni M, Sartore-Bianchi A, Benvenuti S, *et al.* Somatic mutation of EGFR catalytic domain and treatment with gefitinib in colorectal cancer. *Ann Oncol* 2005; 16: 1848–1849. Published on 20050712. doi:10.1093/annonc/mdi356.

Chapter 8

Targeting BCR-ABL for Molecular Therapy of Chronic Myelogenous Leukemia

Shamudheen Rafiyath,[a] Guoqing Wei,[a,b] and Delong Liu[a]
[a]*Division of Hematology and Oncology,*
New York Medical College and Westchester Medical Center,
Valhalla, NY 10595, USA
[b]*Bone Marrow Transplantation Center, The First Affiliated University Hospital,*
Zhejiang University School of Medicine, Hangzhou 310003, China
delong_liu@nymc.edu

Chronic myeloid leukemia (CML) accounts for approximately 20% of all adult leukemia in the United States. Progression of CML is generally described as a three-phase process, beginning in a mostly asymptomatic chronic phase (CP), progressing to an intermediate accelerated phase (AP) and followed by a usually terminal blast phase (BP). Left untreated, CML usually progresses from CP to BP over a period of 3 to 5 years, with survival in the blastic phase measured in months.

Handbook of Therapeutic Biomarkers in Cancer
Edited by Sherry X. Yang and Janet E. Dancey
Copyright © 2013 Pan Stanford Publishing Pte. Ltd.
ISBN 978-981-4364-65-2 (Hardcover), 978-981-4364-66-9 (eBook)
www.panstanford.com

Treatment of CML patients in the last century was based predominantly on radiotherapy, busulfan, hydroxyurea and interferon-alfa. Allogeneic hematopoietic stem cell transplantation is the only curative therapy currently for CML. A series of discoveries led to the recognition that the BCR-ABL protein, which results from a balanced, reciprocal translocation involving chromosomes 9 and 22, has a central role in the pathogenesis of CML. Important mile stones in our understanding of the biological basis of the disease were the discovery of the Philadelphia (Ph) chromosome in 1960, the characterization of the breakpoint cluster region on chromosome 22 in 1984 and the demonstration of the BCR-ABL fusion gene in 1986 [1]. The BCR-ABL Protein (now known as BCR-ABL1) functions as a constitutively activated tyrosine kinase. This knowledge laid the foundation for the preclinical work and led to the development of imatinib mesylate (formerly ST1571), which inhibits the kinase activity of BCR-ABL1 protein. While most patients benefit from imatinib treatment, a substantial number either are initially refractory (primary resistance) or develop resistance during the course of treatment (acquired resistance). Several new BCR-ABL tyrosine kinase inhibitors have substantial activity in patients with imatinib resistance and are now being compared with imatinib for newly diagnosed patients. Integration of the various treatment modalities and determining optimal therapy or response to therapy is evolving rapidly.

8.1 Pathogenesis

The Philadelphia chromosome is a shortened chromosome 22(22q-), resulted from balanced reciprocal translocation between the long arms of chromosomes 9 and 22, (t 9, 22) (q34; q11) [2,3]. The (9; 22) translocation resulted in the formation of BCR-ABL chimeric gene and protein, which played the central role in the pathogenesis of CML [1, 4]. The (9; 22) translocation transposes the ABL (Abelson) proto-oncogene from chromosome 9 into a relatively small 5.8 kb genomic region on chromosome 22 named the breakpoint cluster region (BCR) [5]. The break point is almost always occur upstream of the second axon (a2) resulting in the translocation of all except exon 1 of ABL. Depending on the precise location of the break point in the BCR gene, two different chimeric genes can be formed. The breaks between b2 and b3 (also known as e13) can produce a b2a2 fusion messenger RNA and breaks between b3 and b4 (e14) produces a b3a2 fusion messenger RNA [1]. Although the BCR-ABL protein encoded by the b2a2 mRNA is 25 amino acids larger than the BCR-ABL protein encoded by the b3a2 mRNA, both are referred to as p210 BCR-ABL and both have similar prognosis. The Philadelphia chromosome can also be found in 25–50% of the adult acute lymphoblastic leukemia (ALL) and rare cases of acute myeloid leukemia (AML). One third of the Philadelphia chromosome present in the ALL is indistinguishable from the Philadelphia chromosome in the CML. In two thirds of the patients the break point in the chromosome 22 is between e1 and e2 (minor break point cluster region (m-BCR) resulting in a protein of 190 Kd (p190). The p190 transcript is rarely found in CML. An uncommon Ph chromosome has also been described with an e19a2 BCR-ABL mRNA product that encodes a 230-kD protein (p230). The genomic breakpoint of the BCR gene at e19 is referred to as the micro-breakpoint cluster region (μ-BCR) and is rarely found in chronic neutrophilic leukemia.

8.2 Structure of BCR-ABL

ABL is a non-receptor tyrosine kinase present in many tissues. It is distributed in the nucleus and the cytoplasm and shuttle between the two compartments. It serves a complex role as a cellular module that integrates signals that influences decisions in regard to cell cycle and apoptosis [6,7].

The BCR protein is ubiquitously expressed [8] (Fig. 8.1). The N-terminal exon encodes a serine—threonine kinase. The center of the molecule contains a region with ABL-like and pleckstrin homology (PH) homology domains that stimulate the exchanges of guanidine triphosphate (GTP) for guanidine diphosphate (GDP) on Rho guanidine exchange factors, which in turn may activate transcription factors such as NF-kappaB [9,10]. BCR can be phosphorylated on several tyrosine residues like tyrosine 177, which binds Grb-2, which is an adaptor molecule involved in the activation of Ras pathway [11–13].

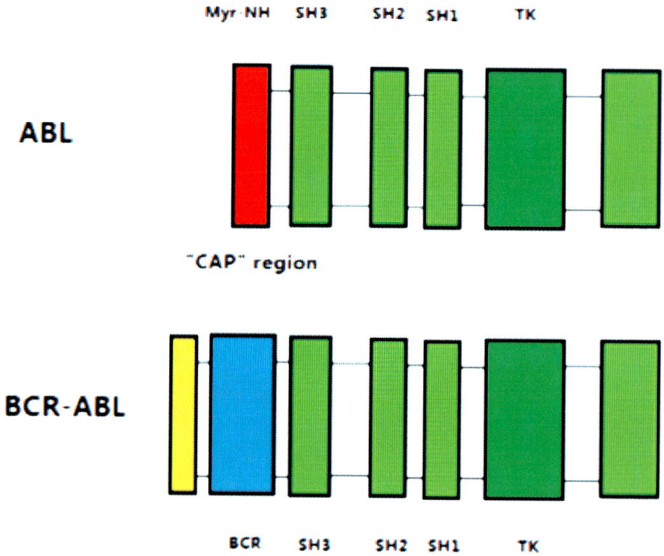

Figure 8.1 The structure of BCR-ABL. The BCR-ABL oncoprotein is a hybrid containing functional domains from the N-terminal end of BCR and the C-terminal end of ABL. The NH2 *terminus* in ABL1 and BCR-ABL kinases is the 'Cap' region. The ABL1b contains a myristate site Myr-NH) at the extreme end of the amino terminal segment. It binds to the kinase domain and keeps the SH2-SH3 in the inhibitory state. In the BCR-ABL protein the lack of 'Cap' region and a dimerization of domain encoded by the first exon of BCR are responsible for constitutive activation of the ABL SH1 domain, resulting in uncontrolled signal transduction and an abnormal cellular phenotype.

8.3 Mechanism of CML

The tyrosine kinase encoded by the SRC homology1 (SH1) domain of the ABL component of BCR-ABL plays the most important role for the oncogenic transformation. The other important regions in the ABL portion are protein-interaction SRC homology 2 (SH2) and the C-terminal actin-binding domains. On the BCR portions, the coiled-coil motif encoded by the first BCR exon is responsible for dimerization of the Oncoprotein; a tyrosine at position 177 is crucial for binding of adaptor proteins such as growth factor receptor-bound protein 2; and N-terminal phosphothreonine residues are required for interaction with SH2-containing proteins including ABL itself.

8.4 Essential Features of BCR-ABL

ABL tyrosine kinase is tightly regulated under physiologic conditions. The SH3 domains appear to play major role in this inhibitory process because it is deletions or position alterations activates the kinase [14,15]. The fusion of BCR sequences 5′ of the ABL SH3 domain abrogates the physiologic suppression of the kinase. This might be the consequences of homodimer formation. A host of substrates can be tyrosine phosphorylated by BCR-ABL. With autophosphorylation, there is a marked increase of phosphotyrosine on BCR-ABL itself. This creates additional binding sites for the SH2 domains of other proteins.

The four major different mechanisms for the malignant transformation by BCR-ABL are as follows:

1. **Altered adhesion to stromal cells and extracellular matrix.** BCR-ABL exhibit negative adhesion to bone marrow stromal cells and extracellular matrix. Adhesion to stroma negatively regulates cell proliferation and CML cells escapes this regulation [16].
2. **Activation of mitogenic signaling.**
   (i) Several link between BCR-ABL and Ras have been defined. Autophosphorylation of tyrosine 177 provides a docking site for adapter molecule Grb-2. Grb-2 binds to Sos Protein and stabilizes Ras in its active GTP-bound form.

   (ii) JAK-STAT pathway. Phosphorylation of stat transcription factors (stat 1 and stat 5) has been reported in several BCR-ABL positive cell lines and in primary CML cells; stat 5 appears to contribute to malignant transformation [17].

   (i) PI3 kinase pathway. PI3 kinase activity is required for the malignant transformation of CML cells. BCR-ABL forms multimeric complexes with the PI3 kinase, *cbl*, and adaptor molecules *crk* and *crkl*, in which PI3 kinase is activated [17,18].

   (ii) Myc Pathway. Overexpression of *myc* has been demonstrated in many human malignancies. Depending on the cellular contact myc may constitute a proliferative or an apoptotic signal [19,20].

3. **Inhibition of apoptosis.** Multiple signals initiated by BCR-ABL have proliferative and anti-apoptotic potential. The BCR-ABL shifts the balance toward the inhibition of apoptosis while simultaneously providing a proliferative stimulus [21,22].
4. **Degradation of inhibitory proteins.** The BCR-ABL induces the proteosome mediated degradation of Abi-1 and Abi-2 proteins which have inhibitory function, another mechanism by which BCR-ABL induces cellular transformation [23].

8.5 Targeted Therapies of Chronic Myeloid Leukemia

8.5.1 First-Generation Tyrosine Kinase Inhibitors

8.5.1.1 Imatinib mesylate

Imatinib is a 2-phenylaminopyrimidine compound [24]. The drug was first evaluated in the second line treatment for CML following interferon failure. In a phase II trial, 454 of the 532 patients had CP CML. At 18 months 60% of patients had a major cytogenetic response (MCyR) and 41% had a complete cytogenetic response (CCyR), with no Ph-positive cells in metaphase seen in the bone marrow. At 60 months, 67% had achieved MCyR and 57% had achieved CCyR [25]. Followed by the success of the phase II study, a phase III trial, International Randomized study of Interferon and imatinib (IRIS trial) was designed to compare prospectively the administration of imatinib 400 mg with the interferon-alpha and cytarabine in previously untreated CML-CP. The study recruited 1106 patients and the preliminary data were published in 2003.

In the IRIS trial [26], the CCyR at 18 months was 76% for patients in the imatinib group and 14% for those in the interferon-cytarabine group. The rate of freedom from progression to AP or BP was 97% and 92%, respectively. The 5-year follow-up study of the IRIS trial showed the estimated cumulative CCyR was 87%, progression-free survival (PFS) was 83%, and overall survival (OS) was 89% in the imatinib group.

Seven-year follow-up data have been recently reported [27]. The OS is 86% and the event free survival (EFS) is 81% at 7 years. Seven percent of the patients progressed to the accelerated or the blastic phase. CCyR was 82% and 83% of these patients maintain that response.

Recently, 8-year follow-up of IRIS study, evaluating long-term efficacy and safety of imatinib [28]. At the 8-year cut-off, 55% of patients remained on IM study treatment, and 45% had discontinued treatment due to adverse events/safety(6%), unsatisfactory therapeutic outcome (16%), stem cell transplant (SCT) (3%), death (3%) or other reasons (17% for withdrawal or lack of renewal of consent and miscellaneous). No new safety issues were identified in a long-term analysis of serious adverse events. Estimated EFS at 8 years was 81% and freedom from progression to AP/BP was 92%. Estimated OS was 85% at 8 years, and 93% when only CML-related deaths and those prior to SCT were considered. A Stable CCyR was achieved by patients who had minor to partial cytogenetic response (CyR) (>0–65% Ph+ metaphases) at 3 months and those with partial CyR (PCyR; >0–35% Ph+ metaphases) at 6 and 12 months compared with patients with less than CCyR at 18 months.

In the TOPS trial newly diagnosed patients across all risk categories received either 400 or 800 mg imatinib as front-line therapy [29]. At 12 months, differences in major molecular response (MMR) and CCyR rates were not statistically significant (MMR, 46% vs. 40% and CCyR, 70% vs. 66%. MMR occurred faster among patients randomly assigned to imatinib 800 mg/d, who had higher rates of MMR at 3 and 6 months compared with those in the imatinib 400 mg/d arm CCyR also occurred faster in the 800 mg/d arm (CCyR at 6 months, 57% vs. 45%.) The improved time to achieve MMR with higher dose was more pronounced in patients with high-risk Sokal scores (41% vs. 26%). The most common adverse events were edema, gastrointestinal problems, and rash, and all were more common in patients in the 800 mg/d arm. Grades 3 to 4 hematologic toxicity also occurred more frequently in patients receiving imatinib 800 mg/d. In the Central European Leukemia Study Group (CELSG) trial, the ISTAHIT study, the previously treated patients received imatinib 800 mg versus 400 mg in the first 6 months of treatment followed by 400 mg/day as maintenance therapy [30]. Compared with the standard-dose, high-dose imatinib led to higher rates of major and CCyR at both 3 months (major: 21% versus 37%, $P = 0.01$; complete: 6% versus 25%, $P < 0.001$) and 6 months (major: 34% versus 54%, $P = 0.009$; complete: 20% versus 44%, $P < 0.001$). At 12 months, the rates of major CyR (the primary end-point) were comparable between the two arms (57% versus 59%). CyR rates were higher in patients without dose reduction in the high-dose arm (61%) than in the patients with no dose reduction in the standard-dose arm (36%). Whether the faster achievement of MMR will translate into improved long-term outcome remains to be determined. A randomized trial is currently ongoing to address whether the standard dose should be changed.

8.5.1.2 Monitoring

All patients have to be evaluated with cytogenetic analysis before the start of therapy, and a baseline quantitative PCR analysis. In the United States, the guidelines are to repeat bone marrow at 6, 12 and 18 months from the initiation of therapy. The follow-up of bone marrow can be necessary once the patient achieved CCyR. Quantitative RT-PCR should be tested every 3 months when a patient is responding to treatment or every 3–6 months after CCyR (Table 8.1). FISH may be used if RQ-PCR for BCR-ABL1 is not available. The most sensitive available test for low levels of leukemia is to measure BCR-ABL1 transcript numbers in the blood or marrow using a real-time quantitative reverse transcriptase PCR (RQ-PCR) [31,32].

8.5.1.3 Toxicity

Myelosuppression is the most common grade 3–4 adverse events. Treatment should be held for grade 3–4 neutropenia and thrombocytopenia and can be restarted when the count recovers. Dose reductions should be considered when the recovery of neutropenia and platelets takes longer than two weeks. However the myelosuppression is very uncommon after initial 1–2 months of treatment. The other side effects are nausea, diarrhea, peripheral or periorbital edema, muscle cramps, rash, weight gain, fatigue etc. These are very often minor events and respond to early intervention. Nausea infrequently needs antiemetics (prochlorperazine, promethazine, or other agents) (Table 8.3) [33].

8.5.1.4 Imatinib resistance

8.5.1.4.1 *Primary resistance*

The primary hematological resistance is failure to achieve hematologic remission within 3–6 months of initiation of treatment,

Table 8.1 Definitions of responses based on European Leukemia Net, 2010 update

Hematologic Complete Hematologic Response (CHR)	**Platelet count <450 × 109/L WBC count-<10 × 109/L Differential: no immature granulocytes, basophils <5%, non-palpable spleen**	**Check at diagnosis, then every 15 days until CHR has been achieved and confirmed, then at least every 3 months or as required**
Cytogenetic Complete Cytogenetic Response (CCyR)	No ph+ metaphases	Check at diagnosis, at 3 months and at 6 months, then every 6 months until a CCyR has been achieved and confirmed, then every 12 months if regular molecular monitoring cannot be assured.
Partial Cytogenetic Response (PCyR)	1–35% ph+ metaphases	Check always for occurrences of treatment failure
Minor Cytogenetic Response	36–65% ph+ metaphases	
Minimal Cytogenetic Response	66–95% ph+ metaphases	
None	>95% ph+ metaphases	
Molecular Complete Molecular Response (CMR)	Undetectable BCR-ABL mRNA transcripts by real time quantitative or nested PCR in two consecutive blood samples of adequate quality (sensitivity >104)	RT- Q-PCR every 3 months, until MMR has been achieved and confirmed, then at least every 6 month
Major Molecular Response (MMR)	Ratio of BCR-ABL to ABL (or other housekeeping genes) ≤ 0.1% on the international scale.	Mutational analysis: In occurrences of suboptimal response or failure.

Source: 1. Baccarani M, Cortes J, Pane F, Niederwieser D. (2009) Chronic myeloid leukemia. An update of concepts and management Recommendations of the EuropeanLeukemiaNet. *J Clin Oncol*, 27, pp. 6041–6051.

2. Baccarani M, Saglio G, Goldman J, Hochhaus A. (2006) Evolving concepts in the management of chronic myeloid leukemia: recommendations from an expert panel on behalf of the European LeukemiaNet. *Blood*, 108, pp. 1809–1820.

whereas primary cytogenetic resistance is failure to achieve any level of cytogenetic response at 6 months, MCyR at 12 months or CCyR at 18 months. Approximately 15 to 20% of patients with newly diagnosed CML do not have a CCyR to imatinib. In imatinib resistance it is important to confirm that the patient is taking imatinib as prescribed. A serum level may be checked to determine patient's adherence to therapy. However there are not data at present to support whether change of therapy based on plasma imatinib levels will affect treatment outcomes. The levels of human organic cation transporter-1 (HOCT1) have been reported as major predictor of response to imatinib. Most patients with suboptimal response to imatinib have low HOCT1 activity [34]. The cellular uptake of dasatinib or nilotinib seems to be independent of HOCT1 expression.

8.5.1.4.2 *Secondary resistance*

The most common mechanism for secondary resistance is the presence of point mutations within the ABL kinase domain of BCR/ABL. Such mutations inhibit the ability of imatinib to bind to BCR-ABL by corrupting the binding sites or preventing the kinase domain from assuming the inactive conformation required for imatinib binding [35]. Point mutations develop in approximately 35% to 70% of patients displaying resistance to imatinib, either spontaneously or through the evolutionary pressure of imatinib [36,37].

The majority of resistance-conferring mutations fall within four regions of the kinase domain: the ATP-binding loop (P-loop) of the ABL kinase domain, the contact site, the SH2 binding site (activation loop), and the catalytic domain [38]. Approximately 85% of all imatinib-resistant mutations are associated with amino acid substitutions at just seven residues (P-loop: M244V, G250E, Y253F/H and E255K/V; contact site: T315I; and catalytic domain: M351T and F359V) [39]. The most frequently mutated region of BCRABL is the P-loop, accounting for 36–48% of all mutations [36,37].

In various biological assays, P-loop mutants Y253F and E255K exhibit an increased transformation potency relative to unmutated BCR-ABL. Overall, the relative transformation potencies of various mutations were found to be as follows: Y253F > E255K > native BCR-ABL ≥ T315I > H396P > M351T. Transformation potency also correlated with intrinsic BCR-ABL kinase activity in this study.

The presence of T3151 mutation confers the highest resistance to imatinib and other second-generation tyrosine kinase inhibitors.

8.5.1.5 Mutation analysis

It is important to do mutation analysis for patients who failed imatinib or have accelerated or blast phase CML. In Gruppo Italiano Malattie Ematologiche dell'Adulto (GIMEMA) study, 297 patients were evaluated for mutations. Among them, mutations were found in 27% with CP patients, 52% of AP patients, and 75% of myeloid BC, and 83% lymphoid BC/ Ph+ ALL BCR-ABL mutant can be grouped based on imatinib sensitivity: sensitive (IC50 ≤ 1000 nM); intermediately sensitive (IC50 ≤ 3000 nM; i.e., M244V, G250E, Q252H, F317L and E355G); and insensitive (IC50 > 3000 nM; i.e., Y253F/H, E255K/V and T315I). The various mutations occur at different frequencies and confer different levels of imatinib resistance [40]. The sensitivity of many of these mutations to imatinib has been studied *in vitro*. The P-loop mutants are 70- to 100-fold less sensitive to imatinib compared with native BCR-ABL. The presence of these mutations also has been associated with poor prognosis for patients receiving imatinib. Prior studies showed increased frequency of death rates, progression of disease to AP and worsening of OS associated with P-loop mutations [36,37,41,42]. Ernst *et al.* found that P-loop mutations were detectable months before the development of resistance in patients taking imatinib [43]. Because the appearance of such mutations seems to indicate impending relapse and resistance to imatinib, earlier detection may provide clinical benefit for patients by prompting earlier reconsideration of therapeutic interventions.

8.5.2 Second-Generation Tyrosine Kinase Inhibitors

8.5.2.1 Dasatinib

Dasatinib is a potent, orally active, dual BCR-ABL/SRC family kinase inhibitor [44]. Dasatinib is structurally unrelated to imatinib and can bind both the inactive and active configurations of BCR-ABL. Due to structural differences from imatinib and nilotinib, dasatinib is active against most of the imatinib related mutations that lead to resistance. Dasatinib binds multiple conformations of BCR-ABL [27,45], unlike imatinib and nilotinib [28,29,46]. The ability to bind both active and inactive conformations of BCR-ABL may explain its potent activity against most of the known imatinib-resistant kinase domain mutations, with the exception of T315I [30,47]. Dasatinib is also more potent than imatinib, with 325-fold greater *in vitro* activity against unmutated BCR-ABL [46]. The sensitivity of BCR-ABL mutants to dasatinib can be classified as sensitive (IC50 ≤ 5 nM), intermediately sensitive (IC50 = 5 to 11 nM; i.e., E255K/V and F317L) and insensitive (IC50 > 11 nM; i.e., T315I). T315I, a contact point mutation, is insensitive to all currently approved BCR-ABL inhibitors [27,28,30–32,45–49]. P-loop mutated BCR-ABL is generally sensitive or intermediately sensitive to dasatinib, with IC50 values falling in the range of 1 to 11 nM [10,35].

The initial approval of dasatinib was based on the data from the START program (SRC/ABL Tyrosine kinase inhibition Activity: Research trials of dasatinib). In the START-C trial dasatinib was evaluated in patients with CML-CP who were previously treated with imatinib or intolerant to imatinib [23,50]. Dasatinib 70 mg twice daily was associated with a high rate of durable cytogenetic response in imatinib-resistant CML-CP or patients who were intolerable to imatinib. After 24 months of treatment 62% of the patients achieved MCyR, 53% achieved CCyR and the 47% achieved MMR. At 24 months the PFS was 75% with the imatinib-resistant group and 94% in the imatinib-intolerant group. The OS was 92% and 100% in imatinib-resistant and imatinib-intolerant group, respectively.

The first approved dose of dasatinib was 70 mg twice daily for all stages of CML. The label has been updated after the open label dose optimization study, in which patients were randomized (1:1:1:1) to receive 100 mg once daily, 50 mg twice daily, 140 mg once daily or 70 mg twice daily [51,52]. The patients on the 100 mg arm showed equivalent efficacy compared with the 70 mg twice daily dosage arm and demonstrated fewer grade 3 and 4 adverse effects (AEs; 30% vs. 48%) with the 100 mg dosage. Most significantly the 100 mg dose was associated with less pleural effusions (7% vs. 16%). The patient's adherence to the medication was superior with the 100 mg dosage. The recommended starting dose for imatinib-resistant CML-CP or imatinib-intolerant patients is 100 mg once daily.

Mueller *et al.* [33,53] assessed the response of dasatinib in imatinib-resistant patients based on the baseline mutational status. The patient s with p-loop mutations had a CCyR of 43% and 42% on all other mutations except those with T315I and F317L mutations. In this study no patients with T315I mutation achieved CCyR and only one out of 14 (7%) patients with F317L achieved CCyR. The patients from START-C trial also evaluated based on the mutational status and the results were similar to this trial. Among all baseline mutations F 317L and T315I mutations have been insensitive to dasatinib and other available TKIs.

8.5.2.1.1 *Dasatinib vs. imatinib as first-line therapy for the CML-CP*

The US Food and Drug Administration (FDA) has approved dasatinib (Sprycel, Bristol-Myers Squibb) 100 mg once daily for the treatment of adult patients with newly diagnosed Philadelphia chromosome-positive (Ph+) CML in CP based on results from the Dasatinib versus Imatinib Study in Treatment-Naïve CP-CML Patients (DASISION) open-label, phase 3 trial [54].

In the DASISION study the rate of confirmed CCyR and MMR by 12 months was significantly higher among patients receiving dasatinib than among patients receiving imatinib (77% vs. 66% p = 0.007 and 52% vs. 34% p < 0.0001). Responses were achieved more quickly with the dasatinib than with the imatinib. The rates of CCyR by 3, 6, and 9 months after the initiation of dasatinib were 54%, 73% and 78%, respectively and the rates after imatinib initiation were 31%, 59% and 67%, respectively. The rates of MMR by 3, 6, and 9 months after the initiation of dasatinib were 8%, 27% and 39%, respectively and the rates after imatinib initiation were 0.4%%, 8% and 18%, respectively. Progression to accelerated or blastic phase of CML occurred in 5 of 259 patients who were receiving dasatinib (1.9%, all blastic phase) and in 9 of 260 patients who were receiving imatinib (3.5%, all blastic phase. At 12 months the estimated rates of PFS and OS were similar for dasatinib and imatinib group (PFS; 96% VS 97% and OS: 97% VS 99%).

8.5.2.1.2 *Safety and tolerability*

The grade 3 or 4 neutropenia occurred with similar frequency among the dasatinib and imatinib groups (21% and 20%), the rate of thrombocytopenia was 19% and 10% among the dasatinib and imatinib group, respectively. Fluid retention of all grades occurred more frequently with imatinib than with dasatinib (42% vs. 19%). Pleural effusion was reported only in the dasatinib group (10%); all events were grade 1 or grade 2 three patients discontinued treatment because of the grade 2 pleural effusion (Table 8.3). The frequency of pleural effusions was lower than the occurrence of pleural effusions with the dasatinib 70 mg twice daily dosage. Gastrointestinal and other bleeding events occurred in 5% of patients in each group. The median change in the QTc interval from baseline was 3.0 msec in the dasatinib group and 8.2 msec in the imatinib group. The overall rates of discontinuation of therapy because of toxic effects of the drug were 5% with the dasatinib group and 4% with the imatinib group [54].

The follow-up period in the study is not long enough to draw the meaningful conclusion of differences in the survival benefit between the treatment groups. The early data of higher rates of CCyR and MMR with dasatinib and early evidence of reduced rates of progression to more aggressive phases of CML suggest that the upfront use of dasatinib therapy may improve the long-term outcome in patients with treatment naïve CML-CP.

8.5.2.2 Nilotinib

Nilotinib (AMN107, Novartis) is an, orally active, aminopyrimidine-derivative tyrosine kinase inhibitor that is more potent against CML cells *in vitro* than is imatinib. Like imatinib, nilotinib functions through competitive inhibition at the ATP-binding site of BCR-ABL, leading to the inhibition of tyrosine phosphorylation of proteins that are involved in the intracellular signal transduction that BCR-ABL mediates. Nilotinib has a higher binding affinity and selectivity for the ABL kinase than imatinib. Nilotinib is 10 to 50 times more potent than imatinib in inhibiting the proliferation and autophosphorylation of wild-type BCR-ABL cell lines and most of the BCR-ABL mutants, except the T315I mutant [49, 55,56].

In the phase II study nilotinib 400 mg twice daily (800 mg/ day was administered to all patients with CML-CP following imatinib resistance or intolerance based on safety, tolerability and pharmacokinetic data from the phase 1 study [57]. At 6 months of treatment rates of MCyR and CCyR rates were 48% and 31%, respectively. Among patients who achieved a MCyR 96% continued treatment without progression or death for at least 6 months. Eleven percent of patients discontinued treatment because of disease progression [58].

A baseline mutation analysis was available on 182 patients in the study. After 6 months of therapy, MCyR was achieved in 42%

of patients and CCyR was achieved in 23% of patients with baseline mutations, versus in 51% and 35 of patients without baseline mutations. Major CyR and Complete Hematologic Response (CHR) were observed across all BCR-ABL genotypes, with the exception of the T315I mutation identified in 4 (2.2%) of 182 patients, and the E255V and E274K mutation each identified in 1 (0.6%) of 182 patients

FDA has approved nilotinib (Tasigna, Novartis) as a first-line treatment of chronic-phase Philadelphia chromosome-positive CML (Ph+ CML). Approval was based on results from ENESTnd (Evaluating Nilotinib Efficacy and Safety in Clinical trials of Newly Diagnosed Ph+ CML Patients), in which nilotinib was found to be superior to imatinib in achieving MMR and CCyR [58]. In this study the newly diagnosed CML patients were randomly assigned in a 1:1:1 ratio to receive nilotinib 300 mg twice daily or nilotinib 400 mg twice daily or imatinib 400 mg once daily. At 12 months rates of MMR were significantly higher in patients receiving 300 mg of nilotinib (44%) or 400 mg of nilotinib (43%) twice daily than in the imatinib arm (22%). Rates of CCyR achieved by 12 months were also significantly higher for nilotinib vs. imatinib (78–80% vs. 65%; $P < 0.001$), and CCyR and MMR occurred faster in the nilotinib arms. After a median 14 months of treatment, fewer nilotinib-treated patients had progressed to AP/BP phase compared with imatinib-treated patients (<1% vs. 4%; $P \leq 0.01$ in an analysis of time to progression). Similar to DASISION, no patient who had a MMR had progression to AP/BP.

8.5.2.2.1 *Safety and tolerability of nilotinib*

Grade 3 or 4 neutropenia were more frequent in imatinib group (20%) than in both nilotinib groups (12% and 10%) and thrombocytopenia was more frequent in the nilotinib group (22% vs. 9%). Elevations of any grades in levels of alanine aminotransferase, aspartate aminotransferase, and bilirubin were more frequently observed in both nilotinib groups than in the imatinib group (Table 8.3).

Dose reductions or interruptions occurred in 59% of patients receiving 300 mg of nilotinib, in 66% receiving 400 mg of nilotinib and in 52% receiving imatinib. No patient in any of the study groups had a QTc more than 500 msec. No decrease from baseline in the mean left ventricular ejection fraction was observed at any time during the study. Discontinuations because of adverse events occurred in 5% of patients receiving 300 mg of nilotinib, 9% of those receiving 400 mg of nilotinib, and 7% of those receiving imatinib.

The data from ENESTnd study were supported by other studies: (i) a single-arm phase 2 trial of nilotinib 400 mg BID performed by MDACC ($n = 61$); and (ii) a second single-arm phase 2 trial of nilotinib 400 mg BID performed by the Italian GIMEMA group ($n = 73$) [59].

8.5.3 Management of Resistance to First-Line TKIs

Based on the FDA approval of nilotinib and dasatinib the guidelines have also included nilotinib or dasatinib for newly diagnosed patients. Many physicians are currently comfortable in using imatinib as the first-line therapy. Dasatinib and nilotinib are effective in patients with imatinib-resistant or -intolerant CP CML. Dose escalation of imatinib to 800 mg daily has been shown to overcome some of the resistance but the duration of response is short lived, especially on patients who never achieved CCyR. Dose escalation is still an option for patients who developed cytogenetic relapse or having incomplete cytogenetic remission (Table 8.2) [60–62].

Table 8.2 Recommendations based on European Leukemia Net, 2010 update

Time	Optimal response	Suboptimal response	Failure	Warnings
Diagnosis	N/A	N/A		High risk CCA/Ph+
3 months	CHR, at least minor CyR	No CyR	Less than CHR	N/A
6 months	At least PCyR	Less than PCyR	Less than PCyR	N/A
12 months	CCyR	PCyR	Less than PCyR	Less than MMR
18 months	MMR	Less than PCyR	Less than CCyR	N/A
Any time during treatment	Stable or improving MMR	Loss of MMR mutations	Loss of CHR, loss of CCyR, mutations, CCA/ph+	Increase in transcript levels

Time	Optimal response	Suboptimal response	Failure	Warnings
Treatment recommendations Imatinib 400 mg daily on diagnosis	Continue imatinib 400 mg daily	Continue imatinib 400 mg or test high dose or dasatinib or nilotinib	Dasatinib or Nilotinib Allo HSCT in patients who have experienced progression to AP/BP and in who carry T315I mutations	N/A

Dasatinib and imatinib failure and accelerated and blast phase CML		
Accelerated and blastic phase	Patients who are TKI naive	Allo HSCT, preceded by imatinib 600 or 800 mg, dasatinib or nilotinib, in case of imatinib-resistant mutations
Accelerated and blastic phase after 2nd line TKI	Patients with prior treatment of imatinib	Allo HSCT preceded by dasatinib or nilotinib
2nd line TKIs	Suboptimal responses	Continue dasatinib or nilotinib with an option for allo HSCT in patients with warning features(i.e. prior hematological resistance to imatinib mutations
2nd line TKIs	Failure	Allo HSCT, preceded by dasatinib or nilotinib

Source: 1. Baccarani M, Cortes J, Pane F, Niederwieser D. (2009) Chronic myeloid leukemia. An update of concepts and management Recommendations of the EuropeanLeukemiaNet. *J Clin Oncol*, 27, pp. 6041–6051.

2. Baccarani M, Saglio G, Goldman J, Hochhaus A. (2006) Evolving concepts in the management of chronic myeloid leukemia: recommendations from an expert panel on behalf of the European LeukemiaNet. *Blood*, 108, pp. 1809–1820.

The failure to first-line TKIs is an indication to identify the patient's mutation status. The mutation T315I is resistant to all the three approved TKIs. Allogeneic HSCT remains a potentially curative option for T315I mutated patients. Several clinical trials with newer agents like aurora kinase inhibitors (danusertib, ponatinib) and omacetaxine mepesuccinate are currently undergoing. The P-loop mutations Y253F and E255V/K are resistant to nilotinib and imatinib, and are sensitive or less resistant to dasatinib therapy. The mutations 317L and V299L are resistant to dasatinib and sensitive to nilotinib and imatinib (Table 8.3).

Table 8.3 Major differences between the front-line tyrosine kinase inhibitors

Drug	Imatinib	Dasatinib	Nilotinib
Mechanism of action	Tyrosine kinase inhibitor, single target BCR-ABL Not inhibit SRC kinase family	multi-targeted tyrosine kinase inhibitor Including BCR-ABL, SRC family of kinases,	Tyrosine kinase inhibitor inhibits BCR-ABL Not inhibit SRC kinase family
Binding	inactive form of BCR-ABL	Both the inactive and active form BCR-ABL	Inactive form only
Potency	Potency less than dasatinib and nilotinib	325-fold greater potency for wild type BCR-ABL.	30 to 50-fold greater potency than imatinib
Dose	400 mg PO q daily-CML-CP 600 mg PO q daily-CML-AP	100 mg PO Q daily	400 mg PO BID
Major side effects	Thrombocytopenia, neutropenia, fluid retention (>10%)	Thrombocytopenia, neutropenia, pleural effusions, fluid retention (>10%)	Thrombocytopenia, neutropenia, peripheral edema, QTc Prolongation Elevated LFTS, lipase
		Mutations and sensitivity	
Wild type	Sensitive	Sensitive	Sensitive
M244V	Intermediate	Sensitive	Sensitive
G250E	Intermediate	Sensitive	Sensitive
Q252H	Intermediate	Sensitive	Sensitive
Y253F	Insensitive	Sensitive	Intermediate
Y253H	Insensitive	Sensitive	Intermediate
E255K	Insensitive	Intermediate	Intermediate

E255V	Insensitive	Intermediate	Intermediate
T315I mutation	Highly resistant	Highly resistant	Highly resistant
Multidrug-resistant protein-1(MDR-1)	Substrate for MDR-1 protein(efflux protein)	Not a substrate for MDR-1 protein	Substrate for MDR-1 protein
Timing	Take with meal	Take with or without meals	Take 1 hour before or 2 hours after a meal

Source: Refs. 27, 40, 46, 49, 54, 55, and 56; Soverini S, Hochhaus A, Nicolini FE, Gruber F. (2011) BCR-ABL kinase domain mutation analysis in chronic myeloid leukemia patients treated with tyrosine kinase inhibitors: recommendations from an expert panel on behalf of European LeukemiaNet. *Blood*, 118, pp. 1208–1215.

8.5.3.1 Bosutinib

Bosutinib (SKI-606) is a dual SRC/ABL TKI with 200-fold greater potency than imatinib against BCR-ABL in biochemical assays [63]. Bosutinib (SKI-606; Wyeth) has potent antiproliferative activity against imatinib-sensitive and -resistant BCR-ABL-positive cell lines, including the Y253F, E255K and D276G mutants, but not the T315I mutant. Like dasatinib, bosutinib is able to bind to both inactive and intermediate conformations of BCR-ABL [64]. Compared with dasatinib, bosutinib has a more favorable toxicity profile because; bosutinib does not significantly inhibit Kit or PDGFR [64].

Early results from Phase II studies have demonstrated its efficacy and safety [65]. Data were reported for 88 patients who were previously treated with imatinib and dasatinib. Among the 88 patients 36 were dasatinib resistant, and 52 were dasatinib-intolerant patients, of which 32 of 36 dasatinib-resistant and 29 of 52 dasatinib-intolerant patients were imatinib resistant. Response rates with bosutinib were comparable to those seen in trials of dasatinib and nilotinib in the second-line setting, including CCyR in 50% and MMR in 52% of evaluated patients, of which 32% were complete. At 24 months, rates of progression-free and OS were 80% and 95%, respectively. Twenty different mutations were found at baseline in 45 (45%) of 99 patients. The CHR rate was 78% in patients with mutations and 89% in patients without mutations, and the MCyR rate was 60% in patients with mutations and 54% in Patients without mutations.

Bosutinib toxicities were similar to those associated with currently available TKIs. Nineteen percent of patients discontinued bosutinib therapy because of toxicity. The most common adverse events involved the gastrointestinal tract (diarrhea, nausea, and vomiting) and usually were grade 1 or 2 in severity. The most common grade 3 or 4 hematological abnormalities were thrombocytopenia (23%), neutropenia (14%) and anemia (9%).

Gambacorti-Passerini *et al.* [66] reported the efficacy of bosutinib in 134 imatinib-resistant or -intolerant patients with advanced leukemia, including 63 patients with accelerated-phase CML and 48 patients with blast-phase CML. A CHR was observed in 61% of patients with AP CML, 32% patients with blast-phase CML. MCyR was achieved in 48% of patients with CML-AP and 52% of patients with CML-BP. Twenty percent of the patients required dose reduction. The median daily dose was 477 mg. nine of ten patients with the T315I mutation were resistant to bosutinib. Overall, preliminary data from this phase 1/2 trial indicate that bosutinib is an active agent for patients with CP-CML who have failed on prior imatinib treatment, with an acceptable toxicity profile. The phase II data also showed activity with various BCR-ABL mutations (except T315I) as well as in imatinib-resistant or -intolerant patients with CML-AP and CML-BP patients.

The data from the phase II trials for bosutinib appear promising. The randomized phase 3 trial of bosutinib vs. imatinib for first-line treatment for newly diagnosed CML (NCT00574873) is still ongoing [67].

8.5.3.2 Bafetinib (INNO-406)

Bafetinib, formerly known as INNO-406, is an oral dual inhibitor of BCR-ABL and Lyn kinases. INNO-406 is up to 55 times more potent than imatinib in BCR-ABL–positive cell lines [68] INNO-406 inhibited the growth of cells with numerous BCR-ABL mutants, including the F317L mutant, but not the T315I mutant [24]. Unlike other second-generation TKIs, INNO-406 inhibits Lyn kinase but has no or limited activity against the other SFK. Since over-expression of Lyn kinase has been implicated in BCR-ABL independent resistance [68], INNO-406 may have further importance in imatinib-resistant CML.

In a Phase I dose escalation study [69] in 56 patients with Ph+ CML or ALL and imatinib resistance or intolerance, six (11%) of patients with CML in the CP experienced a MCyR to bafetinib. No responses were observed in patients with CML in the AP or BP or Ph+ ALL. Transaminase elevations and thrombocytopenia were dose-limiting toxicities at a bafetinib dosage of 480 mg twice daily. A dosage of 240 mg twice daily was recommended for phase II studies.

8.6 Future Horizons in the Treatment of CML

8.6.1 T315I Kinase Inhibitors

The substitution of the amino acid threonine with isoleucine at position 315 of the ABL protein was the first mutation to be detected in patients with imatinib-resistant CML. The T315I mutant can be detected in 4% to 19% of patients with imatinib resistant. The T315I point mutation confers resistance to imatinib, dasatinib, nilotinib, and bosutinib [37]. Several TKIs that are active against the T315I-mutated form of BCR-ABL are being developed.

8.6.2 Aurora Kinase Inhibitor

Aurora kinases are overexpressed in many cancers and are essential for the regulation of mitotic processes during cell division. MK-0457, a potent inhibitor of BCR-ABL and aurora kinases, was the first agent to show clinical activity against the T315I mutation; however, development of this drug was halted due to cardiac toxicity [49, 70].

Danusertib, formerly known as PHA-739358, also exhibited antiproliferative and proapoptotic activity against CML cell lines and BCR-ABL mutants, including the T315I mutant [70]. Hematologic and cytogenetic responses to PHA-739358 were observed in CML patients harboring the T315I mutant. The preliminary reports of phase 1 study [71] in 23 patients with advanced leukemia and resistance or intolerance to imatinib, a second generation TKI, or both were reported. Fifteen of the 23 patients had T315I mutations and a response occurred in 43% of the patients and an acceptable tolerability profile was observed. A phase II study is currently in progress.

Ponatinib (AP24534) is a multi-targeted BCR-ABL/SRC kinase inhibitor with potent *in vitro* activity against all tested mutants of BCR-ABL, including T315I, and clinical activity has been reported in patients with a T315I mutation [72–74]. All 7 (100%) evaluable patients with chronic-phase CML and the T315I mutation had a CHR, 4 (57%) patients had a CCyR, and 1 (14%) patient had a MCyR. Of 11 evaluable patients with CML in the AP or BP or Ph+ ALL, 3 out of 8 patients with T315I had a MHR and 1 patient had a CCyR. Further clinical studies of ponatinib are ongoing, most notably a single-arm phase 2 study in patients with CML or Ph+ ALL who either are resistant or intolerant to either dasatinib or nilotinib, or who harbor the T315I mutation (Ponatinib Ph+ ALL and CML evaluation [PACE]; NCT01207440).

Switch pocket kinase inhibitors, such as DCC-2036 and DCC-2157, target the sites involved in controlling the conformation of BCR-ABL, which ultimately controls the activity state of the kinase. These agents are active against cells expressing a variety of BCR-ABL mutations, including T135I [75, 76]. A phase 1 study of DCC-2036 in patients with T315I or failure on two different TKIs is under way (NCT00827138).

8.6.3 Omacetaxine Mepesuccinate

Omacetaxine (previously homoharringtonine) is a naturally occurring alkaloid derived from evergreen trees that induces apoptosis in leukemic cells, including those harboring the T315I mutation. It is the first agent in a new class of drugs, cetaxines, with a mechanism of action independent of tyrosine kinase inhibition. The drug binds to and inhibits translation of short-lived oncoproteins that are up-regulated in leukemic cells. The safety and efficacy of omacetaxine mepesuccinate in 90 patients with CML in the chronic, accelerated and blast phase and a confirmed BCR-ABL T315I mutations and imatinib resistance were reported [77]. A 1.25 mg/m^2 dose of the drug was given by subcutaneously (S.C.) injection twice daily as induction therapy for 14 days every 28 days until a hematological response was observed, followed by maintenance therapy using 1.25 mg/m^2 twice daily for 7 days every 28 days. All patients failed prior therapy, and 79% failed two or more prior TKIs. The presence of baseline T315I mutation was confirmed in all patients. Data were available for 66 patients (40 CP, 16 AP, and 10 BP). The median follow-up was

6.4 months. In the CP patients, CHR was achieved in 26 patients and maintained in 8 patients (8 patients entered the study in CHR) for an overall CHR rate of 85%; the median duration of CHR was 7.7+ months. Overall CyR was achieved in 27.5% and the median duration of MCyR was 6+ months. The MMR was achieved in 15% of patients. Interestingly a reduction of baseline T315I mutated clone occurred in 56.7% of CP. Toxicity: Grade 3/4 related events occurred in 45 of 66 (68%) patients. The most commonly reported events were thrombocytopenia (58%), anemia (36%) and neutropenia (33%). Grade 3/4 non-hematologic toxicities were uncommon with no events occurring in >5% of patients and infection (3%) the most common event. Treatment delays occurred in approximately 50% of the patients with median duration of approximately 12 days for all disease phases and cycles (CP=12, AP=10, and BP=12 days). The safety data from two long-term phase II studies of omacetaxine in 170 patients with CML in the CP, AP, or BP who were resistant to intolerant of TKIs were reported. Grade 3 or 4 thrombocytopenia, neutropenia, and anemia occurred in 49%, 34% and 30% of the 170 patients, respectively.

8.7 Conclusions and Future Directions

Imatinib has been the standard therapy for CML for the past decade due to its remarkable activity and mild toxicity. Despite the high response rate, imatinib has not been proven as a curative option for the CML patients. The second generation TKIs such as nilotinib and dasatinib are effective in imatinib-resistant and -intolerant patients.

Based on the data from DASISION trial and ENESTnd trials, dasatinib and nilotinib were recommended for the first-line treatment of CML. There were no head to head trial comparing the efficacy of nilotinib and dasatinib. However there are some concerns about the side effect profiles of both drugs. In the dasatinib groups there were mildly increased incidence of pleural effusion and have more CML related deaths than the imatinib group. In the nilotinib group there have been increased lipase, liver enzymes, and glucose levels. QTc prolongation from nilotinib creates concerns about the toxicity in the long-term use. Allogeneic HSCT is an alternative option for patients who progressed to BP or patients with T315I mutations. Ponatinib (AP24534) showed promising results against T315I mutation and F317L mutations in early studies.

Acknowledgment

SR and GW contributed equally to this article. This study was partly supported by the Research Fund for the Doctoral Program of Higher Education of China (GW, No. J20070747), by the International Scholar Exchange Foundation (GW) and by New York Medical College Blood Diseases Fund (DL). Dr. Guoqing Wei is a recipient of CAHON Scholar Award.

References

1. Deininger, M. W., Goldman, J. M., and Melo, J. V. (2000) The molecular biology of chronic myeloid leukemia, *Blood*, 96, pp. 3343–3356.
2. Nowell, P., and Hungerford, D. (1960) A minute chromosome in human chronic granulocytic leukemia (abstract), *Science*, 132, pp. 1497.
3. Rowley, J. D. (1973) Letter: A new consistent chromosomal abnormality in chronic myelogenous leukaemia identified by quinacrine fluorescence and Giemsa staining, *Nature*, 243, pp. 290–293.
4. de Klein, A., van Kessel, A. G., Grosveld, G., Bartram, C. R., Hagemeijer, A., Bootsma, D., Spurr, N. K., Heisterkamp, N., Groffen, J., and Stephenson, J. R. (1982) A cellular oncogene is translocated to the Philadelphia chromosome in chronic myelocytic leukaemia, *Nature*, 300, pp. 765–767.
5. Groffen, J., Stephenson, J. R., Heisterkamp, N., de Klein, A., Bartram, C. R., and Grosveld, G. (1984) Philadelphia chromosomal breakpoints are clustered within a limited region, Bcr, on chromosome 22, *Cell*, 36, pp. 93–99.
6. Kipreos, E. T., and Wang, J. Y. (1990) Differential phosphorylation of c-ABL in cell cycle determined by cdc2 kinase and phosphatase activity, *Science*, 248, pp. 217–220.
7. Sawyers, C. L., McLaughlin, J., Goga, A., Havlik, M., and Witte, O. (1994) The nuclear tyrosine kinase c-ABL negatively regulates cell growth, *Cell*, 77, pp. 121–131.
8. Laneuville, P. (1995) ABL tyrosine protein kinase, *Semin. Immunol.*, 7, pp. 255–266.
9. Denhardt, D. T. (1996) Signal-transducing protein phosphorylation cascades mediated by Ras/Rho proteins in the mammalian cell: the potential for multiplex signalling, *Biochem. J.*, 318 (Pt 3), pp. 729–747.

10. Montaner, S., Perona, R., Saniger, L., and Lacal, J. C. (1998) Multiple signalling pathways lead to the activation of the nuclear factor kappaB by the Rho family of GTPases, *J. Biol. Chem.*, 273, pp. 12779–12785.

11. Diekmann, D., Nobes, C. D., Burbelo, P. D., Abo, A., and Hall, A. (1995) Rac GTPase interacts with GAPs and target proteins through multiple effector sites, *EMBO. J.*, 14, pp. 5297–5305.

12. Ma, G., Lu, D., Wu, Y., Liu, J., and Arlinghaus, R. B. (1997) BCR phosphorylated on tyrosine 177 binds Grb2, *Oncogene*, 14, pp. 2367–2372.

13. Wu, Y., Liu, J., and Arlinghaus, R. B. (1998) Requirement of two specific tyrosine residues for the catalytic activity of BCR serine/threonine kinase, *Oncogene*, 16, pp. 141–146.

14. Mayer, B. J., and Baltimore, D. (1994) Mutagenic analysis of the roles of SH2 and SH3 domains in regulation of the ABL tyrosine kinase, *Mol. Cell. Biol.*, 14, pp. 2883–2894.

15. Van Etten, R. A., Jackson, P., and Baltimore, D. (1989) The mouse type IV c-ABL gene product is a nuclear protein, and activation of transforming ability is associated with cytoplasmic localization, *Cell*, 58, pp. 669–678.

16. Gordon, M. Y., Dowding, C. R., Riley, G. P., Goldman, J. M., and Greaves, M. F. (1987) Altered adhesive interactions with marrow stroma of haematopoietic progenitor cells in chronic myeloid leukaemia, *Nature*, 328, pp. 342–344.

17. de Groot, R. P., Raaijmakers, J. A., Lammers, J. W., Jove, R., and Koenderman, L. (1999) STAT5 activation by BCR-ABL contributes to transformation of K562 leukemia cells, *Blood*, 94, pp. 1108–1112.

18. Skorski, T., Kanakaraj, P., Nieborowska-Skorska, M., Ratajczak, M. Z., Wen, S. C., Zon, G., Gewirtz, A. M., Perussia, B., and Calabretta, B. (1995) Phosphatidylinositol-3 kinase activity is regulated by BCR/ABL and is required for the growth of Philadelphia chromosome-positive cells, *Blood*, 86, pp. 726–736.

19. Bissonnette, R. P., Echeverri, F., Mahboubi, A., and Green, D. R. (1992) Apoptotic cell death induced by c-myc is inhibited by bcl-2, *Nature*, 359, pp. 552–554.

20. Evan, G. I., Wyllie, A. H., Gilbert, C. S., Littlewood, T. D., Land, H., Brooks, M., Waters, C. M., Penn, L. Z., and Hancock, D. C. (1992) Induction of apoptosis in fibroblasts by c-myc protein, *Cell*, 69, pp. 119–128.

21. Bedi, A., Zehnbauer, B. A., Barber, J. P., Sharkis, S. J., and Jones, R. J. (1994) Inhibition of apoptosis by BCR-ABL in chronic myeloid leukemia, *Blood*, 83, pp. 2038–2044.

22. Cortez, D., Kadlec, L., and Pendergast, A. M. (1995) Structural and signaling requirements for BCR-ABL-mediated transformation and inhibition of apoptosis, *Mol. Cell. Biol.*, 15, pp. 5531–5541.

23. Dai, Z., Quackenbush, R. C., Courtney, K. D., Grove, M., Cortez, D., Reuther, G. W., and Pendergast, A. M. (1998) Oncogenic ABL and SRC tyrosine kinases elicit the ubiquitin-dependent degradation of target proteins through a Ras-independent pathway, *Genes. Dev.*, 12, pp. 1415–1424.

24. Goldman, J. M. (2007) How I treat chronic myeloid leukemia in the imatinib era, *Blood*, 110, pp. 2828–2837.

25. Kantarjian, H., Sawyers, C., Hochhaus, A., Guilhot, F., Schiffer, C., Gambacorti-Passerini, C., Niederwieser, D., Resta, D., Capdeville, R., Zoellner, U., Talpaz, M., Druker, B., Goldman, J., O'Brien, S. G., Russell, N., Fischer, T., Ottmann, O., Cony-Makhoul, P., Facon, T., Stone, R., Miller, C., Tallman, M., Brown, R., Schuster, M., Loughran, T., Gratwohl, A., Mandelli, F., Saglio, G., Lazzarino, M., Russo, D., Baccarani, M., and Morra, E. (2002) Hematologic and cytogenetic responses to imatinib mesylate in chronic myelogenous leukemia, *N. Engl. J. Med.*, 346, pp. 645–652.

26. O'Brien, S. G., Guilhot, F., Larson, R. A., Gathmann, I., Baccarani, M., Cervantes, F., Cornelissen, J. J., Fischer, T., Hochhaus, A., Hughes, T., Lechner, K., Nielsen, J. L., Rousselot, P., Reiffers, J., Saglio, G., Shepherd, J., Simonsson, B., Gratwohl, A., Goldman, J. M., Kantarjian, H., Taylor, K., Verhoef, G., Bolton, A. E., Capdeville, R., and Druker, B. J. (2003) Imatinib compared with interferon and low-dose cytarabine for newly diagnosed chronic-phase chronic myeloid leukemia, *N. Engl. J. Med.*, 348, pp. 994–1004.

27. O'Brien, S., Guilhot, F., Goldman, J., Hochhaus, A., Hughes, T., Radich, J., Rudoltz, M., Filian, J., Gathmann, I., Druker, B., and Larson, R. (2008) International randomized study of interferon versus STI571 (IRIS) 7-year ollow-up: sustained survival, low rate of transformation and increased rate of major molecular response (MMR) in patients (pts) with newly diagnosed chronic myeloid leukemia in chronic phase (CMLCP) treated with imatinib (IM), *Blood (ASH Annual Meeting Abstracts)*, 112, Abstract 186.

28. Deininger, M., O'Brien, G., and Guilhot, F. (2009) International Randomized Study of Interferon vs STI571 (IRIS) 8-year follow up: sustained survival and low risk for progression or events in patients with newly diagnosed chronic myeloid leukemia in chronic phase (CML-CP) treated with imatinib, *Blood (ASH Annual Meeting Abstracts)*, 114, abstract 1126.

29. Cortes, J., Baccarani, M., Guilhot, F., Druker, B., Branford, S., Kim, D., Pane, F., Rudoltz, M., Yu, R., Collins, L., Krahnke, T., Radich, J., and Hughes, T. (2008) A phase III, randomized, open-label study of 400 Mg versus 800 Mg of imatinib mesylate (IM) in patients (pts) with newly diagnosed, previously untreated chronic myeloidLeukemia in chronic phase (CML-CP) using molecular endpoints: 1-year results of TOPS (tyrosine kinase inhibitor optimization and selectivity) study, *Blood (ASH Annual Meeting Abstracts) 112*, Abstract 335.

30. Petzer, A., Wolf, D., Fong, D., Lion, T., Dyagil, I., Masliak, Z., Boskovic, D., Griskevicius, L., Lejniece, S., Goranov, S., Gercheva, L., Stojanovic, A., Peytchev, D., Tzvetkov, N., Griniute, R., Oucheva, R., Ulmer, H., Kwakkelstein, M., Fincato, G., and Gastl, G. (2008) CELSG CML 11 "ISTAHIT" phase III study—planned interim analysis: high doses of imatinib mesylate (800 mg/day) significantly improve rates of major and complete cytogenetic remissions (MCR, CCR) in pretreated Ph+/BCR-ABL+ CML patients in chronic phase., *Blood*, 112, Abstract 1112.

31. Hughes, T., and Branford, S. (2003) Molecular monitoring of chronic myeloid leukemia, *Semin. Hematol.*, 40, pp. 62–68.

32. Kaeda, J., Chase, A., and Goldman, J. M. (2002) Cytogenetic and molecular monitoring of residual disease in chronic myeloid leukaemia, *Acta. Haematol.*, 107, pp. 64–75.

33. Schiffer, C. A. (2007) BCR-ABL tyrosine kinase inhibitors for chronic myelogenous leukemia, *N. Engl. J. Med.*, 357, pp. 258–265.

34. Thomas, J., Wang, L., Clark, R. E., and Pirmohamed, M. (2004) Active transport of imatinib into and out of cells: implications for drug resistance, *Blood*, 104, pp. 3739–3745.

35. O'Hare, T., Eide, C. A., and Deininger, M. W. (2007) BCR-ABL kinase domain mutations, drug resistance, and the road to a cure for chronic myeloid leukemia, *Blood*, 110, pp. 2242–2249.

36. Branford, S., Rudzki, Z., Walsh, S., Parkinson, I., Grigg, A., Szer, J., Taylor, K., Herrmann, R., Seymour, J. F., Arthur, C., Joske, D., Lynch, K., and Hughes, T. (2003) Detection of BCR-ABL mutations in patients with CML treated with imatinib is virtually always accompanied by clinical resistance, and mutations in the ATP phosphate-binding loop (P-loop) are associated with a poor prognosis, *Blood*, 102, pp. 276–283.

37. Jabbour, E., Kantarjian, H., Jones, D., Talpaz, M., Bekele, N., O'Brien, S., Zhou, X., Luthra, R., Garcia-Manero, G., Giles, F., Rios, M. B., Verstovsek, S., and Cortes, J. (2006) Frequency and clinical significance of BCR-ABL mutations in patients with chronic myeloid leukemia treated with imatinib mesylate, *Leukemia,* 20, pp. 1767–1773.

38. Hughes, T., Deininger, M., Hochhaus, A., Branford, S., Radich, J., Kaeda, J., Baccarani, M., Cortes, J., Cross, N. C., Druker, B. J., Gabert, J., Grimwade, D., Hehlmann, R., Kamel-Reid, S., Lipton, J. H., Longtine, J., Martinelli, G., Saglio, G., Soverini, S., Stock, W., and Goldman, J. M. (2006) Monitoring CML patients responding to treatment with tyrosine kinase inhibitors: review and recommendations for harmonizing current methodology for detecting BCR-ABL transcripts and kinase domain mutations and for expressing results, *Blood*, 108, pp. 28–37.

39. Soverini, S., Colarossi, S., Gnani, A., Rosti, G., Castagnetti, F., Poerio, A., Iacobucci, I., Amabile, M., Abruzzese, E., Orlandi, E., Radaelli, F., Ciccone, F., Tiribelli, M., di Lorenzo, R., Caracciolo, C., Izzo, B., Pane, F., Saglio, G., Baccarani, M., and Martinelli, G. (2006) Contribution of ABL kinase domain mutations to imatinib resistance in different subsets of Philadelphia-positive patients: by the GIMEMA working party on chronic myeloid leukemia, *Clin. Cancer. Res.*, 12, pp. 7374–7379.

40. Deininger, M., Buchdunger, E., and Druker, B. J. (2005) The development of imatinib as a therapeutic agent for chronic myeloid leukemia, *Blood*, 105, pp. 2640–2653.

41. Nicolini, F. E., Corm, S., Le, Q. H., Sorel, N., Hayette, S., Bories, D., Leguay, T., Roy, L., Giraudier, S., Tulliez, M., Facon, T., Mahon, F. X., Cayuela, J. M., Rousselot, P., Michallet, M., Preudhomme, C., Guilhot, F., and Roche-Lestienne, C. (2006) Mutation status and clinical outcome of 89 imatinib mesylate-resistant chronic myelogenous leukemia patients: a retrospective analysis from the French intergroup of CML (Fi(phi)-LMC GROUP), *Leukemia*, 20, pp. 1061–1066.

42. Soverini, S., Martinelli, G., Rosti, G., Bassi, S., Amabile, M., Poerio, A., Giannini, B., Trabacchi, E., Castagnetti, F., Testoni, N., Luatti, S., de Vivo, A., Cilloni, D., Izzo, B., Fava, M., Abruzzese, E., Alberti, D., Pane, F., Saglio, G., and Baccarani, M. (2005) ABL mutations in late chronic phase chronic myeloid leukemia patients with up-front cytogenetic resistance to imatinib are associated with a greater likelihood of progression to blast crisis and shorter survival: a study by the GIMEMA Working Party on Chronic Myeloid Leukemia, *J. Clin. Oncol.*, 23, pp. 4100–4109.

43. Ernst, T., Erben, P., Muller, M. C., Paschka, P., Schenk, T., Hoffmann, J., Kreil, S., La Rosee, P., Hehlmann, R., and Hochhaus, A. (2008) Dynamics of BCR-ABL mutated clones prior to hematologic or cytogenetic resistance to imatinib, *Haematologica*, 93, pp. 186–192.

44. Lombardo, L. J., Lee, F. Y., Chen, P., Norris, D., Barrish, J. C., Behnia, K., Castaneda, S., Cornelius, L. A., Das, J., Doweyko, A. M., Fairchild, C., Hunt, J. T., Inigo, I., Johnston, K., Kamath, A., Kan, D., Klei, H., Marathe,

P., Pang, S., Peterson, R., Pitt, S., Schieven, G. L., Schmidt, R. J., Tokarski, J., Wen, M. L., Wityak, J., and Borzilleri, R. M. (2004) Discovery of N-(2-chloro-6-methyl-phenyl)-2-(6-(4-(2-hydroxyethyl)-piperazin-1-yl)-2-methylpyrimidin-4-ylamino)thiazole-5-carboxamide (BMS-354825), a dual SRC/ABL kinase inhibitor with potent antitumor activity in preclinical assays, *J. Med. Chem.*, 47, pp. 6658–6661.

45. O'Hare, T., Walters, D. K., Stoffregen, E. P., Jia, T., Manley, P. W., Mestan, J., Cowan-Jacob, S. W., Lee, F. Y., Heinrich, M. C., Deininger, M. W., and Druker, B. J. (2005) *In vitro* activity of BCR-ABL inhibitors AMN107 and BMS-354825 against clinically relevant imatinib-resistant ABL kinase domain mutants, *Cancer. Res*, 65, pp. 4500–4505.

46. Tokarski, J. S., Newitt, J. A., Chang, C. Y., Cheng, J. D., Wittekind, M., Kiefer, S. E., Kish, K., Lee, F. Y., Borzillerri, R., Lombardo, L. J., Xie, D., Zhang, Y., and Klei, H. E. (2006) The structure of Dasatinib (BMS-354825) bound to activated ABL kinase domain elucidates its inhibitory activity against imatinib-resistant ABL mutants, *Cancer. Res*, 66, pp. 5790–5797.

47. Shah, N. P., Tran, C., Lee, F. Y., Chen, P., Norris, D., and Sawyers, C. L. (2004) Overriding imatinib resistance with a novel ABL kinase inhibitor, *Science*, 305, pp. 399–401.

48. La Rosee, P., Corbin, A. S., Stoffregen, E. P., Deininger, M. W., and Druker, B. J. (2002) Activity of the BCR-ABL kinase inhibitor PD180970 against clinically relevant BCR-ABL isoforms that cause resistance to imatinib mesylate (Gleevec, STI571), *Cancer. Res.*, 62, pp. 7149–7153.

49. Weisberg, E., Manley, P. W., Breitenstein, W., Bruggen, J., Cowan-Jacob, S. W., Ray, A., Huntly, B., Fabbro, D., Fendrich, G., Hall-Meyers, E., Kung, A. L., Mestan, J., Daley, G. Q., Callahan, L., Catley, L., Cavazza, C., Azam, M., Neuberg, D., Wright, R. D., Gilliland, D. G., and Griffin, J. D. (2005) Characterization of AMN107, a selective inhibitor of native and mutant BCR-ABL, *Cancer. Cell.*, 7, pp. 129–141.

50. Hochhaus, A., Kantarjian, H. M., Baccarani, M., Lipton, J. H., Apperley, J. F., Druker, B. J., Facon, T., Goldberg, S. L., Cervantes, F., Niederwieser, D., Silver, R. T., Stone, R. M., Hughes, T. P., Muller, M. C., Ezzeddine, R., Countouriotis, A. M., and Shah, N. P. (2007) Dasatinib induces notable hematologic and cytogenetic responses in chronic-phase chronic myeloid leukemia after failure of imatinib therapy, *Blood*, 109, pp. 2303–2309.

51. Druker, B. J., Guilhot, F., O'Brien, S. G., Gathmann, I., Kantarjian, H., Gattermann, N., Deininger, M. W., Silver, R. T., Goldman, J. M., Stone, R. M., Cervantes, F., Hochhaus, A., Powell, B. L., Gabrilove, J. L., Rousselot, P., Reiffers, J., Cornelissen, J. J., Hughes, T., Agis, H., Fischer, T., Verhoef,

G., Shepherd, J., Saglio, G., Gratwohl, A., Nielsen, J. L., Radich, J. P., Simonsson, B., Taylor, K., Baccarani, M., So, C., Letvak, L., and Larson, R. A. (2006) Five-year follow-up of patients receiving imatinib for chronic myeloid leukemia, *N. Engl. J. Med.*, 355, pp. 2408–2417.

52. Shah, N. P., Kantarjian, H. M., Kim, D. W., Rea, D., Dorlhiac-Llacer, P. E., Milone, J. H., Vela-Ojeda, J., Silver, R. T., Khoury, H. J., Charbonnier, A., Khoroshko, N., Paquette, R. L., Deininger, M., Collins, R. H., Otero, I., Hughes, T., Bleickardt, E., Strauss, L., Francis, S., and Hochhaus, A. (2008) Intermittent target inhibition with dasatinib 100 mg once daily preserves efficacy and improves tolerability in imatinib-resistant and -intolerant chronic-phase chronic myeloid leukemia, *J. Clin. Oncol.*, 26, pp. 3204–3212.

53. Mueller, M., Branford, S., Radich, J., Shah, N., Erben, P., and Ernst, T. (2007) Efficacy of dasatinib in chronic phase chronic myelogenous leukemia patients after imatinib failure according to baseline BCR-ABL mutations, *Haematologica (EHA Annual Meeting Abstracts)*, 92, pp. 127–127. Abstract 0356.

54. Kantarjian, H., Shah, N. P., Hochhaus, A., Cortes, J., Shah, S., Ayala, M., Moiraghi, B., Shen, Z., Mayer, J., Pasquini, R., Nakamae, H., Huguet, F., Boque, C., Chuah, C., Bleickardt, E., Bradley-Garelik, M. B., Zhu, C., Szatrowski, T., Shapiro, D., and Baccarani, M. (2010) Dasatinib versus imatinib in newly diagnosed chronic-phase chronic myeloid leukemia, *N. Engl. J. Med.*, 362, pp. 2260–2270.

55. Golemovic, M., Verstovsek, S., Giles, F., Cortes, J., Manshouri, T., Manley, P. W., Mestan, J., Dugan, M., Alland, L., Griffin, J. D., Arlinghaus, R. B., Sun, T., Kantarjian, H., and Beran, M. (2005) AMN107, a novel aminopyrimidine inhibitor of BCR-ABL, has *in vitro* activity against imatinib-resistant chronic myeloid leukemia, *Clin. Cancer. Res.*, 11, pp. 4941–4947.

56. Verstovsek, S., Golemovic, M., Kantarjian, H., Manshouri, T., Estrov, Z., Manley, P., Sun, T., Arlinghaus, R. B., Alland, L., Dugan, M., Cortes, J., Giles, F., and Beran, M. (2005) AMN107, a novel aminopyrimidine inhibitor of p190 BCR-ABL activation and of *in vitro* proliferation of Philadelphia-positive acute lymphoblastic leukemia cells, *Cancer*, 104, pp. 1230–1236.

57. Kantarjian, H., Giles, F., Wunderle, L., Bhalla, K., O'Brien, S., Wassmann, B., Tanaka, C., Manley, P., Rae, P., Mietlowski, W., Bochinski, K., Hochhaus, A., Griffin, J. D., Hoelzer, D., Albitar, M., Dugan, M., Cortes, J., Alland, L., and Ottmann, O. G. (2006) Nilotinib in imatinib-resistant CML and Philadelphia chromosome-positive ALL, *N. Engl. J. Med.*, 354, pp. 2542–2551.

58. Saglio, G., Kim, D. W., Issaragrisil, S., le Coutre, P., Etienne, G., Lobo, C., Pasquini, R., Clark, R. E., Hochhaus, A., Hughes, T. P., Gallagher, N., Hoenekopp, A., Dong, M., Haque, A., Larson, R. A., and Kantarjian, H. M. (2010) Nilotinib versus imatinib for newly diagnosed chronic myeloid leukemia, *N. Engl. J. Med.*, 362, pp. 2251–2259.

59. Rosti, G., Castagnetti, F., Palandri, F., Pane, F., Martinelli, G., Saglio, G., and Baccarani, M. (2008) Nilotinib 800 mg daily as first line treatment of chronic myeloid leukemia in early chronic phase: results of a phase II trial of the GIMEMA CML Working Party, *J. Clin. Oncol.*, 26, (May 20 suppl; Abstract 7054).

60. Kantarjian, H. M., Larson, R. A., Guilhot, F., O'Brien, S. G., Mone, M., Rudoltz, M., Krahnke, T., Cortes, J., and Druker, B. J. (2009) Efficacy of imatinib dose escalation in patients with chronic myeloid leukemia in chronic phase, *Cancer*, 115, pp. 551–560.

61. Kantarjian, H. M., Talpaz, M., O'Brien, S., Giles, F., Garcia-Manero, G., Faderl, S., Thomas, D., Shan, J., Rios, M. B., and Cortes, J. (2003) Dose escalation of imatinib mesylate can overcome resistance to standard-dose therapy in patients with chronic myelogenous leukemia, *Blood*, 101, pp. 473–475.

62. Marin, D., Goldman, J. M., Olavarria, E., and Apperley, J. F. (2003) Transient benefit only from increasing the imatinib dose in CML patients who do not achieve complete cytogenetic remissions on conventional doses, *Blood*, 102, pp. 2702–2703; author reply 2703–2704.

63. Cortes, J., Kantarjian, H., Baccarani, M., Brummendorf, T., Liu, D., Ossenkoppele, G., Volkert, A., Hewes, B., Moore, L., Zacharchuk, C., and Gambacorti, C. (2006) A phase 1/2 study of SKI-606, a dual inhibitor of Src and ABL kinases, in adult patients with Philadelphia chromosome positive (Ph+) chronic myelogenous leukemia (CML) or acute lymphocytic leukemia (ALL) relapsed, refractory or intolerant of imatinib, *Blood (ASH Annual Meeting Abstracts)*, 108, Abstract 168.

64. Puttini, M., Coluccia, A. M., Boschelli, F., Cleris, L., Marchesi, E., Donella-Deana, A., Ahmed, S., Redaelli, S., Piazza, R., Magistroni, V., Andreoni, F., Scapozza, L., Formelli, F., and Gambacorti-Passerini, C. (2006) *In vitro* and *in vivo* activity of SKI-606, a novel SRC-ABL inhibitor, against imatinib-resistant BCR-ABL+ neoplastic cells, *Cancer. Res.*, 66, pp. 11314–11322.

65. Cortes, J., Kantarjian, H., Brummendorf, T., Khoury, H., Kim, D., Turkina, A., Volkert, A., Wang, J., Arkin, S., and Gambacorti-Passerini, C. (2010) Safety and efficacy of bosutinib (SKI-606) in patients (pts) with chronic

phase (CP) chronic myeloid leukemia (CML) following resistance or intolerance to imatinib (IM), *J. Clin. Oncol. 28 (Suppl.)*, abstract 6502.

66. Gambacorti-Passerini, C., Cortes, J., and Khoury, H. (2010) Safety and efficacy of bosutinib in patients with AP and BP CML and Ph+ ALL following resistance/intolerance to imatinib and other TKIs: update from study SKI-200, *J. Clin. Oncol. (American Society of Clinical Oncology 46th Annual Meeting) 28 (15 suppl)*, abstract 6509.

67. Database., U. S. N. I. o. H. C. g. Compare bosutinib to imatinib in subjects with newly diagnosed chronic phase Philadelphia chromosome positive CML.

68. Kimura, S., Naito, H., Segawa, H., Kuroda, J., Yuasa, T., Sato, K., Yokota, A., Kamitsuji, Y., Kawata, E., Ashihara, E., Nakaya, Y., Naruoka, H., Wakayama, T., Nasu, K., Asaki, T., Niwa, T., Hirabayashi, K., and Maekawa, T. (2005) NS-187, a potent and selective dual BCR-ABL/Lyn tyrosine kinase inhibitor, is a novel agent for imatinib-resistant leukemia, *Blood*, 106, pp. 3948–3954.

69. Kantarjian, H., le Coutre, P., Cortes, J., Pinilla-Ibarz, J., Nagler, A., Hochhaus, A., Kimura, S., and Ottmann, O. (2010) Phase 1 study of INNO-406, a dual ABL/Lyn kinase inhibitor, in Philadelphia chromosome-positive leukemias after imatinib resistance or intolerance, *Cancer*, 116, pp. 2665–2672.

70. Gontarewicz, A., Balabanov, S., Keller, G., Colombo, R., Graziano, A., Pesenti, E., Benten, D., Bokemeyer, C., Fiedler, W., Moll, J., and Brummendorf, T. H. (2008) Simultaneous targeting of Aurora kinases and BCR-ABL kinase by the small molecule inhibitor PHA-739358 is effective against imatinib-resistant BCR-ABL mutations including T315I, *Blood*, 111, pp. 4355–4364.

71. Cortes-Franco, J., Dombret, H., and Schafhausen, P. (2009) Danusertib hydrochloride (PHA-739358), a multi-kinase aurora inhibitor, elicits clinical benefit in advanced chronic myeloid leukemia and Philadelphia chromosome positive acute lymphoblastic leukemia, *Blood (ASH Annual Meeting Abstracts)* **114**, abstract 864.

72. Cortes, J., Talpaz, M., Deininger, M., Shah, N., Flinn, I., Mauro, M., O'Hare, T., Spinos, N., Hu, S., Berk, L., Narasimhan, N., Rivera, V., Clackson, T., Haluska, F., and Kantarjian, H. (2009) A phase 1 trial of oral AP24534 in patients with refractory chronic myeloid leukemia and other hematologic malignancies: first results of safety and clinical activity against T315I and resistant mutations, *Blood (ASH Annual Meeting Abstracts)* 114 *(Suppl.)*, 267 (abstract 643).

73. O'Hare, T., Shakespeare, W. C., Zhu, X., Eide, C. A., Rivera, V. M., Wang, F., Adrian, L. T., Zhou, T., Huang, W. S., Xu, Q., Metcalf, C. A., 3rd, Tyner, J. W., Loriaux, M. M., Corbin, A. S., Wardwell, S., Ning, Y., Keats, J. A., Wang, Y., Sundaramoorthi, R., Thomas, M., Zhou, D., Snodgrass, J., Commodore, L., Sawyer, T. K., Dalgarno, D. C., Deininger, M. W., Druker, B. J., and Clackson, T. (2009) AP24534, a pan-BCR-ABL inhibitor for chronic myeloid leukemia, potently inhibits the T315I mutant and overcomes mutation-based resistance, *Cancer. Cell*, 16, pp. 401–412.

74. Rivera, V., Xu, Q., Wang, F., Snodgrass, J., O'Hare, T., Corbin, A., Keats, J., Lamore, S., Ning, Y., Wardwell, S., Russian, K., Broudy, M., Shakespeare, W., Druker, B., Iuliucci, J., and Clackson, T. (2007) Potent Antitumor Activity of AP24534, an Orally Active Inhibitor of BCR-ABL Variants Including T315I, in *In Vitro* and *In Vivo* Models of Chronic Myeloid Leukemia (CML), *Blood (ASH Annual Meeting Abstracts) 110 (Suppl.)*, abstract 1032.

75. Van Etten, R., Chan, W., Zaleskas, V., Evangelista, P., Lazarides, K., Peng C, L., S., Wise, S., Petillo, P., and Flynn, D. (2007) DCC-2036: A Novel Switch Pocket Inhibitor of ABL Tyrosine Kinase with Therapeutic Efficacy Against BCR-ABL T315I *In Vitro* and in a CML Mouse Model, *Blood (ASH Annual Meeting Abstracts)* 110 *(Suppl.)*, abstract 463.

76. Van Etten, R., Chan, W., Zaleskas, V., Walz, C., Evangelista, P., Lazarides, K., Betancur, M., Wise, S., Petillo, P., and Flynn, D. (2008) Switch pocket inhibitors of the ABL Tyrosine Kinase: Distinct kinome inhibition profiles and *in Vivo* efficacy in mouse models of CML and B-lymphoblastic leukemia induced by BCR-ABL T315I, *Blood (ASH Annual Meeting Abstracts) 112 (Suppl.)*, abstract 576.

77. Cortes-Franco., J., Khoury., H. J., Nicolini., F. E., Corm., S., Lipton., J. H., Jones., D., Hochhaus., A., Craig., A. R., Benichou., A.-C., Humphriss., E., and Kantarjian., H. (2009) Safety and efficacy of subcutaneous-administered omacetaxine mepesuccinate in imatinib-resistant chronic myeloid leukemia (CML) patients who harbor the BCR-ABL T315I mutation—results of An ongoing multicenter phase 2/3 study, *Blood (ASH Annual Meeting Abstracts, 114(Suppl.)*, abstract 644.

Chapter 9

Gastrointestinal Stromal Tumors: From Molecular Pathogenesis to Therapy

Joaquina Baranda,[a] Rashna Madan,[b] and Andrew K. Godwin[b]
[a]*Department of Internal Medicine and*
[b]*Department of Pathology and Laboratory Medicine,*
University of Kansas Medical Center, Kansas City, KS 66160 USA
AGodwin@kumc.edu

9.1 Introduction

Gastrointestinal stromal tumors (GISTs), the archetypal and most common mesenchymal neoplasms of the digestive tract, occur most regularly in stomach (60%) and small bowel (25%) but may occur at any site in the gut or even elsewhere in the abdomen [1–3]. They classically manifest in older adults as single, well-circumscribed masses that are composed (in decreasing frequency) of spindle cell histology, epithelioid histology or a combination of those types. GISTs demonstrate widely ranging clinical behavior, which currently may be predicted by a risk stratification system that combines their location as well as two morphologic features, their size and mitotic rate [3]. Typically small (2 cm or less) GISTs with low mitotic counts

Handbook of Therapeutic Biomarkers in Cancer
Edited by Sherry X. Yang and Janet E. Dancey
Copyright © 2013 Pan Stanford Publishing Pte. Ltd.
ISBN 978-981-4364-65-2 (Hardcover), 978-981-4364-66-9 (eBook)
www.panstanford.com

behave very indolently while larger mitotically active GISTs may act more aggressively. In the latter scenario, they may metastasize, usually to the abdominal cavity and liver and can even be fatal.

GISTs are a great example of how science can impact outcomes in cancer patients. The discovery by Hirota in 1998 that GIST contained mutations in *c-KIT* was a breakthrough for advancing the biology and treatment of GIST [4]. From a molecular perspective, these tumors are characterized by activating mutations, primarily in the genes for tyrosine kinase receptors (TKRs): most often in *KIT* [5], the normal cellular homologue of the viral oncoprotein v-Kit (v-Kit, Hardy Zuckerman 4 feline sarcoma viral oncogene homologue) [6, 7] and less commonly in platelet derived growth factor receptor alpha (*PDGFRA*) [8]. These mutations account for the vast majority of GISTs: *KIT* mutations are present in approximately 80% of GISTs and *PDGFRA* mutations in roughly 7% of the tumors [4, 8–16]. A recent protein kinase that been added to this list is BRAF (also known as v-Raf murine sarcoma viral oncogene homolog B1). Mutations in this gene, encoding for a serine-threonine kinase, have been identified in 4–13% of GISTs that lack the aforementioned mutations [17–20].

GISTs are believed to arise from the Interstitial Cells of Cajal (ICC) [21], the pacemaker cells of the gastrointestinal tract, which are located around the myenteric plexus in the muscularis propria of the GI wall or from interstitial mesenchymal precursor stem cells [22] that also have the potential of giving rise to cells in the omentum and peritoneal surfaces. This similarity between GISTs and ICCs is further borne out by the expression of the KIT protein (also called CD117) in non-neoplastic ICC and most GISTs. ICC also require the KIT kinase and its ligand stem cell factor (SCF) for their development [4, 22–26].

GISTs are generally chemotherapy-resistant compared with mesenchymal tumors outside the GI tract [27]. The identification of fundamental hallmark of the biology of GIST resulted in the rapid transformation of this treatment paradigm. Small molecule tyrosine kinase inhibitors have been developed and found to be effective in the treatment of GIST resulting in dramatic improvement in the outcome of patients with this disease. The median survival of patients with advanced GIST on chemotherapy is about 15 months compared with 58 months on imatinib mesylate (also known as Gleevec™) [28], an oral 2-phenylaminopyrimidine derivative that works as a selective inhibitor against mutant forms of type III tyrosine kinases such as KIT, PDGFRA, and BCR/ABL [29].

The true incidence of GIST is unclear. A report based on the SEER (Surveillance, Epidemiology, and End results) data of the National Cancer Institute in 1995 is believed to be an underestimation, as many cases were not captured in the SEER registries [30]. A 13-center cumulative tumor registry SEER database query from 1992 to 2002 determined a 25-fold age-adjusted increase in incidence of GIST, from 0.028 per 100,000 in 1992 to 0.688 per 100,000 in 2002. This increase was primarily because of the reclassification of "smooth-muscle tumors" as GISTs and also represented a 50% increase in population- and age-adjusted gastrointestinal mesenchymal tumor diagnosis since 1992 [31]. The report also suggested that the observed dramatic improvement in survival of patients with GIST coincided with the Food and Drug Administration (FDA) approval of imatinib in the treatment of advanced GIST [31].

9.2 Molecular Pathogenesis of GIST

9.2.1 Mutations in RTKS: KIT

c-KIT, like PDGFRA, belongs to the superfamily of subclass III RTKs. The genes for both *KIT* and *PDGFRA* are located at chromosome 4q12–13 [32]. Structurally KIT is comprised of an extracellular domain (with five immunoglobulin-like repeats), a transmembrane domain, a juxtamembrane domain and two cytoplasmic kinase domains (Fig. 9.1). Other members of this family include receptors for macrophage-colony stimulating factor macrophage colony-stimulating factor (M-CSF) and fetal liver tyrosine kinase 3 (FLT3).

In the non-neoplastic state, KIT is triggered upon binding to SCF, undergoes dimerization, followed by autophosphorylation on tyrosine residues and activation of downstream effectors which affect several cellular processes such as proliferation, apoptosis, adhesion and chemotaxis [32–37]. The juxtamembrane domain is auto-inhibitory in the absence of ligand binding [38]. c-KIT is critical for embryonic development and is also instrumental in several processes including gametogenesis, hematopoiesis, melanogenesis, mast cell growth and development and ICC development [39, 40]. Correspondingly, gain-of-function *KIT* mutations may be detected in seminomas, acute myeloid leukemias, melanomas and systemic mastocytosis.

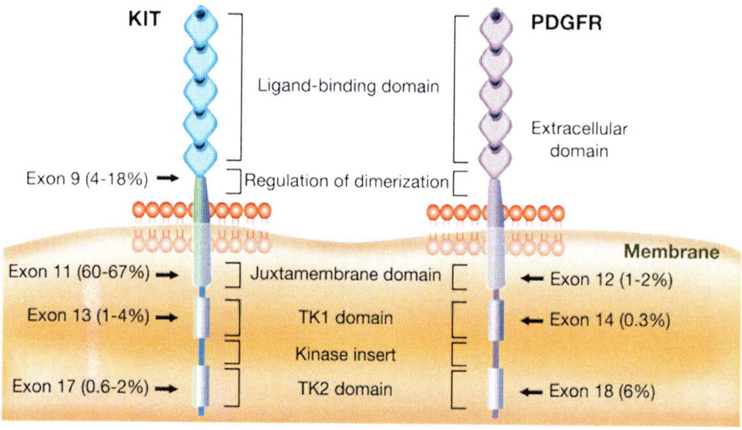

Figure 9.1 Schematic structure of KIT and PDGFRA (adapted from [45]).

In the mutated state, KIT is activated even in the absence of its ligand and its downstream effectors including the phosphatidylinositol 3-kinase (PI3K) and the mitogen-activated kinase (MAPK) pathways affect the cell cycle, translation, metabolism and apoptosis [41] and upon exposure to imatinib show rapid downregulation of these pathways (Fig. 9.2).

Many studies have now evaluated clinical GIST samples and confirmed that up to 80% of GISTs harbor a *KIT* gene mutation (Fig. 9.1) [4, 9, 10, 13, 42–45]. Based on these initial screens, it was determined that the majority of mutations were found in a limited portion of the gene, i.e., exons 9, 11, 13, and 17. These exons encode for a small region of the extracellular domain (exon 9), the juxtamembrane domain (exon 11), and the split kinase domains, I and II (exons 13 and 17), respectively. In approximately two-third of GISTs, *KIT* mutations occur within exon 11 sequences. Mutations at this site release KIT from auto-inhibition and result in constitutive activation [13]. Generally mutations involving exon 11 are responsive to therapy with the tyrosine kinase inhibitor imatinib mesylate (discussed in Section 9.4.4.1). Varying types of mutations may occur. Some studies have shown deletions to characterize more adverse acting tumors than substitutions [46]. A small number of exon 11 mutations are internal tandem duplications and the corresponding tumors tend to be gastric in location and have more favorable behavior [47, 48].

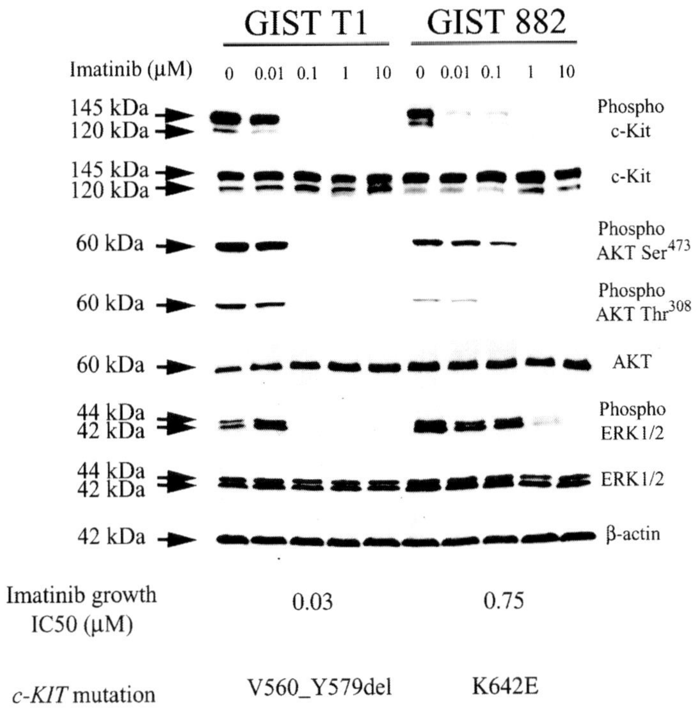

Figure 9.2 Responses of GIST cells to imatinib treatment. Immunoblot assays of GIST T1 and 882 cells treated with imatinib for 6 hours at indicated concentration. Same amount of whole cell protein extracts from each sample were subjected to immunoblot assays using specific anti phospho-KIT and KIT antibodies (adapted from [143] and Frolov and Godwin, unpublished data).

Mutations involving exon 9, which codes for the extracellular domain of KIT, are described in 10–15% of GISTs [1, 42]. Most of these mutations involve tandem duplications of six nucleotides encoding thc amino acids Alanine and Tyrosine (A502-Y503). These tumors have a predilection for the small intestine (unlike exon 11 mutations, which largely do not target a particular location) and tend to an adverse clinical course [47].

A minority of *KIT* mutations (approximately 0.6 to 4%) occur in exons 13 and 17, which code for the split kinase domains (Fig. 9.1)

[49]. Substitutions are the type of mutation usually found and these tumors (especially with exon 17 mutations) have been recently reported to be a little more frequent in intestinal GISTs [50]. Exon 17 mutations are concerning for resistance to imatinib, while untreated exon 13 mutant tumors tend to be responsive [51].

Typically GISTs have one allele of the *KIT* gene mutated; however, approximately 10% of tumors have alteration of the other allele, which may result in a malignant clinical course [15, 42, 52]. Parenthetically, it is worth mentioning that the KIT protein can be detected using immunohistochemistry in greater than 95% GISTs, and show a less than perfect concordance with the presence of the mutation [53]. Indeed, *KIT* "wild-type" tumors, i.e., those tumors that lack a detectable mutation in one of the kinase genes, often express the protein.

9.2.2 Mutations in RTKs: PDGFRA

While *KIT* mutations account for the majority of GISTs, mutually exclusive mutations occur in the *PDFGRA* gene, which codes for another type III receptor tyrosine kinase, in 10 – 30% of tumors lacking the *KIT* mutation, i.e., approximately 7% of all GISTs (Fig. 9.1) [54]. The most commonly involved site is exon 18 and rarely mutations occur in exons 12 and 14 (which correspond to the kinase II domain, juxtamembrane domain and the kinase I domain respectively) [1, 8, 11, 12, 16, 43–45, 55]. GISTs with these mutations preferentially localize to the stomach, contain epithelioid cells that tend not to express an immunohistochemically detectable KIT protein and demonstrate more indolent behavior [55]. It should also be noted that mutations producing the D842V alteration (a change from aspartic acid to valine at codon 842) in exon 18, occur most often and characterize GISTs that tend to be non-responsive to imatinib therapy [16, 42].

9.2.3 *BRAF* Mutation

BRAF, which codes for a serine/threonine protein kinase, is mutated in several malignancies particularly melanoma and papillary thyroid carcinoma, with most mutations involving the V600E amino-acid substitution in exon 15 [56]. Approximately 4–13% of *KIT/PDGFRA* mutation negative GISTs demonstrate the *BRAF* V600E mutation [18,

20]. These tumors tend to occur in the small bowel and overexpress the KIT protein (CD117), though they lack the *KIT* mutation.

While the precise molecular pathogenesis of these *BRAF* mutant tumors is unclear, it is interesting that these mutations result in constitutional activation of MAPK and PI3K/AKT pathways, quite like that present in most GISTs. The mutation potentially opens up an additional line of therapy as BRAF targeted agents are currently in trials for other cancers (melanoma, renal cancer).

9.3 Hereditary, Syndromic and Variant GISTs

9.3.1 Hereditary/Familial GISTs

A very small proportion of GISTs occur as a consequence of highly penetrant autosomal dominant activating mutations, predominantly in *KIT* and rarely in *PDGFRA* [4, 13, 57–62]. These tumors are multiple and typically present over a decade earlier than sporadic GISTs (Fig. 9.3). In general, these GISTs appear not to alter the lifespans of the individuals manifesting them. Concomitant skin hyperpigmentation and gastrointestinal symptoms have also been reported [63–65].

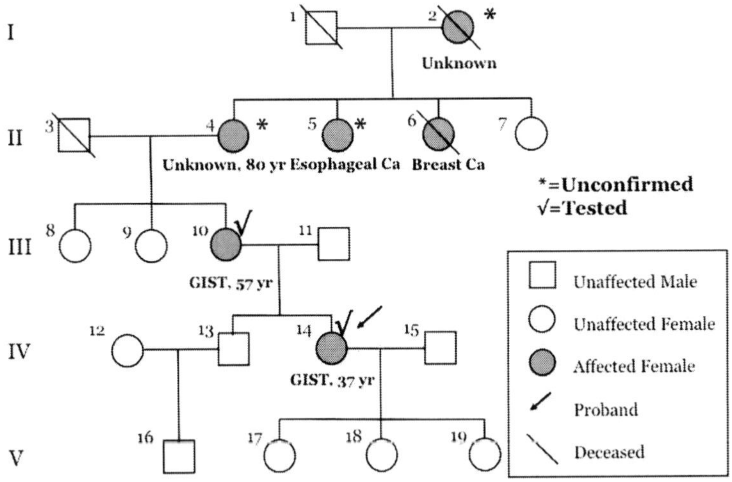

Figure 9.3 Pedigree of a KIT mutant kindred (adapted from [13] and Tarn and Godwin, unpublished data).

9.3.2 Neurofibromatosis I (NF1)-Associated GISTs

Patients with NF1 have a 7% risk of developing GISTs [66]. This autosomal dominant syndrome more frequently manifests with tumors of the nervous system. GISTs related to NF1 are typically multiple, occur in the small intestine and express the KIT protein but lack *KIT/PDGFRA* mutations except for a small minority (approximately 11%) that demonstrate them [67–71]. Instead these tumors are associated with inactivating somatic NF1 mutations resulting in lack of neurofibromin and overactivation of the MAPK pathway [68].

9.3.3 Carney's Triad-associated GISTs

This rare triad comprises gastric GIST, paraganglioma and pulmonary chondroma and has been reported in less than a 100 patients [72]. These GISTs lack mutations in *KIT/PDGFRA/BRAF* and are unusual in that they are multifocal tumors affecting the stomachs of young female patients. Lymph node or liver metastases may occur but do not necessarily portend a rapidly progressive malignant course. These features are similar to pediatric GISTs (described in Section 9.3.5), raising the likelihood that pediatric GISTs may be an attenuated or atypical form of Carney's triad. KIT protein expression has been variably reported in the GISTs of Carney's triad [43, 73–75].

9.3.4 Carney–Stratakis Syndrome (Carney's Dyad)

As opposed to Carney's triad, patients with Carney–Stratakis Syndrome have only two neoplasms: GIST (typically multiple and gastric in location) and paraganglioma. These individuals may show germline mutations in the genes for succinate dehydrogenase, subunits B, C, or D [76] although it is not clear if this is truly distinct syndrome.

9.3.5 Pediatric GISTs

Accounting for only 1–2% of all GISTs, GISTs in the pediatric population represent a group of tumors that differ from their older adult counterparts in both clinical and molecular characteristics. These tumors are preferentially diagnosed in females and occur as multifocal neoplasms with epithelioid or mixed morphology within the stomach. These tumors tend to recur and metastasize, and metastatic sites may include lymph nodes in addition to sites typically seen in older adults (such as liver and peritoneum) [77]. Despite this tendency, pediatric GISTs have favorable long-term prognoses [77, 78]. The risk stratification system (Table 9.1) used for adult GISTs is less predictive when applied to pediatric tumors [78]. Young adult GIST patients are a mixed group, with some more typical of pediatric GISTs and others more characteristic of GISTs in older adults.

While KIT protein expression by immunohistochemistry is the norm, the vast majority of these pediatric tumors are *KIT/PDGFRA* wild-type, with only rare mutant tumors being reported [78, 79]. Though lacking mutations in *KIT/PDGFRA*, these tumors show similar KIT activation (phosphorylation) levels to those found in *KIT* mutant tumors and adult WT GISTs [25, 77]. In the small number of pediatric GISTs tested, mutations of *BRAF* have not been detected.

While late onset mutant GISTs progressing to malignancy demonstrate several chromosomal alterations (such as losses of 9p and 13p, gains in 5p, 8q, 17q and 20q), pediatric wild-type GISTs do not exhibit large-scale chromosomal changes [19, 25] (Fig. 9.4). Two gene expression profiling studies have shown that wild-type pediatric and young adult GISTs cluster together and separate from late onset GISTs (whether mutant or wild-type). Several differentially regulated genes in the pediatric/young adult group were identified in these studies but their only common finding was upregulation of IGF-1R [77, 79]. Indeed IGF-1R protein and mRNA are more frequently overexpressed in wild-type adult and pediatric GISTs in comparison with mutant GISTs, raising the possibility of a therapeutic target in this subset of tumors [75, 80]. Overall, the genetic changes in GISTs, both adult and pediatric, without *KIT/PDGFRA/BRAF* mutations is not known, so we and others are currently using modern day deep sequencing approaches to further explore the GIST genome in these patients to uncover additional clues to help define therapy.

Table 9.1 Guidelines for risk assessment of primary gastrointestinal stromal tumor

Tumor parameters		**Risk of progressive disease# (%)**			
Mitotic Index	*Size*	*Gastric*	*Duodenum*	*Jejunum/Ileum*	*Rectum*
≤5 per 50 high-power fields (HPF)	≤2 cm	None (0%)	None (0%)	None (0%)	None (0%)
	>2 – ≤5 cm	Very low (1.9%)	Low (8.3%)	Low (4.3%)	Low (8.5%)
	>5 – ≤10 cm	Low (3.6%)	(Insufficient data)	Moderate (24%)	(Insufficient data)
	>10 cm	Moderate (10%)	High (34%)	High (52%)	High (57%)
>5 per 50 HPF	≤2 cm	None##	(Insufficient data)	High##	High (54%)
	>2 – ≤5 cm	Moderate (16%)	High (50%)	High (73%)	High (52%)
	>5 – ≤10 cm	High (55%)	(Insufficient data)	High (85%)	(Insufficient data)
	>10 cm	High (86%)	High (86%)	High (90%)	High (71%)

Defined as metastasis or tumor-related death.

Denotes small number of cases.

Note:

1. Data based on long-term follow-up of 1055 gastric, 629 small intestinal, 144 duodenal, and 111 rectal GISTs from the pre-imatinib era [146–149].
2. The required total count of mitoses is per 5 mm^2 on the glass slide section. With the use of older model microscopes, 50 HPF is equivalent to 5 mm^2. Most modern microscopes with wider 40× lenses/fields require only 20 HPF to embrace 5 mm^2.
3. Adapted with permission from Miettinen and Lasota [3]. Copyright 2006 by Elsevier.

Source: http://www.cap.org/apps/docs/committees/cancer/cancer_protocols/2011/GIST_11protocol.pdf

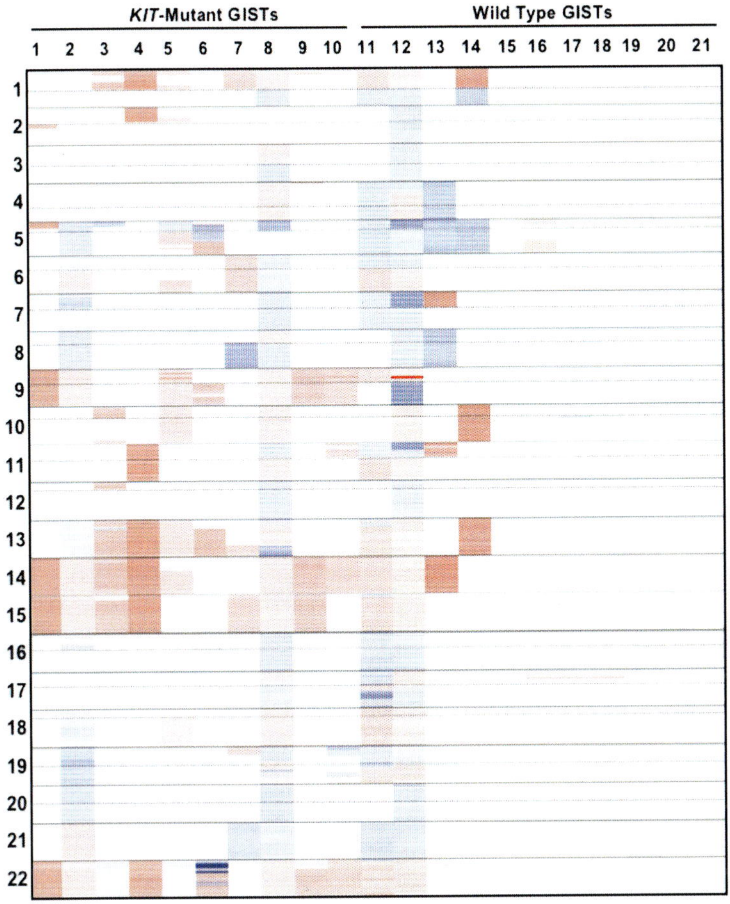

Figure 9.4 Heat map ($\log_2$ ratios of SNP copy number) for GIST samples analyzed by Affymetrix Genome-Wide Human SNP Array 6.0. The color scale ranges from bright red (copy number loss) to white to bright blue (copy number gain). Case numbers and mutation status are indicated at the top. Chromosome numbers are shown to the left of the figure, p and q arms are separated by hash marks. Lane #11 represents a *BRAF* mutation carrier and #15 represents a pediatric GIST samples (adapted from [19] and Belinsky and Godwin, unpublished data).

9.3.6 Risk Assessment

Accurate risk stratification of GISTs has become increasingly important as effective adjuvant therapy emerges. The National

Institutes of Health (NIH) consensus classification system, based on tumor size and mitotic count, is commonly used to assess patient prognosis after surgical resection (Table 9.1). Large retrospective cohort studies from several countries indicate that the NIH classification carries substantial prognostic value. Patients with high-risk GIST (approximately 44% of all) have substantially poorer outcome than those with intermediate-risk (24%) or low/very low-risk GIST (32%), whose survival is not markedly inferior to that of the general population in some studies. Gastric GISTs are associated with better outcome and have a lower risk of recurrence than non-gastric tumors of the same size and mitotic count. Tumor rupture carries an increased risk. These 2 important risk stratification factors are not considered in the NIH classification. Patients with certain non-gastric tumors (2.1–5 cm and > 5 mitoses per 50 high-power fields or 5.1–10 cm and < or = 5 per 50 high-power fields) and those with tumor rupture are proposed to be included in the NIH high-risk category. High-risk patients defined by the proposed modified system have more than 15% to 20% risk of disease recurrence. The use of this system may aid in identifying which patients might potentially benefit from adjuvant therapy.

9.4 Treatment of GIST

Systemic treatment with chemotherapy is generally ineffective in the treatment of GIST [81–84]. In a multicenter, randomized trial evaluating the efficacy of imatinib in advanced GIST, Demetri and colleagues reported that none of the patients previously treated with chemotherapy demonstrated any objective response to any regimen [29]. With the introduction of tyrosine kinase inhibitors, the median survival of patients with advanced GIST has substantially increased [28].

9.4.1 Imatinib Mesylate

Imatinib is a selective small molecule competitive inhibitor of KIT, PDGFRA, and other tyrosine kinases downstream signaling [85–88]. Imatinib was originally approved by the FDA for chronic myeloid leukemia (CML). It inhibited proliferation of leukemic cells expressing BCR-ABL, as well as leukemic cells and GIST cells with activated KIT [89, 90]. In 2001, Joensuu published a case report of a 50-year-old woman with rapidly progressing advanced GIST who demonstrated dramatic and sustained tumor regression with use of imatinib [91].

9.4.1.1 Efficacy of imatinib in patients with advanced GIST

The B2222 trial was a randomized, open-label, multicenter trial to determine efficacy of imatinib in 147 patients with advanced GIST who were randomized to two daily doses of imatinib, either 400 mg or 600 mg. This trial was based on doses felt to be safe from the experience in CML studies. Partial response was seen in 53% of patients and another 28% has stable disease. No significant difference in activity was seen between the two doses of imatinib. It was also observed that in patients who had a response, improvement in positron emission tomography (PET) scan was seen as early as 24 hours after the initiation of imatinib therapy. Edema, diarrhea and fatigue were the most common toxicities, but imatinib was generally well tolerated [29]. The results of this study led to the approval of imatinib in advanced or unresectable GIST.

9.4.1.2 Dose and efficacy

The efficacy of imatinib was examined in a phase I open-label trial in 36 patients with advanced GIST using imatinib 400 to 1000 mg daily. Therapeutic effect was observed in all dose levels. All but four patients (69%) experienced objective response. Dose-limiting toxicities included severe nausea, vomiting, edema, and rash [92].

To further investigate the dose dependency of response and progression-free survival with imatinib in advanced GIST, the European Organization for Research and Treatment of Cancer (EORTC) Soft Tissue and Bone Sarcoma Group performed a multicenter, randomized, phase II trial of imatinib with a maximum tolerated dose of 800 mg daily. Twenty-seven of the 51 patients had GIST and 71% of the patients had received prior chemotherapy. Anemia, periorbital edema, rash, fatigue, nausea, granulocytopenia and diarrhea were the most common side effects but they were generally mild to moderate and no patient was taken off study due to side effects. Objective response was achieved in 71% of patients with additional 18% of patients with stable disease. In GIST patients, the response rates were 4% complete remission, 67% partial remission,

18% stable disease and 11% progression. 73% of GIST patients were free from progression at 1 year [93].

In order to better understand optimal dosing of imatinib, the EORTC performed a phase III clinical trial with 946 patients who were randomly assigned to imatinib 400 mg once a day or twice a day. At a median follow-up of about 27 months, 56% (263/473) of the patients randomized to receive imatinib at 400 mg once a day had progressed compared with 50% (235/473) of those who received imatinib at 400 mg twice daily. There was no difference in the overall response rates observed. The median maximum response was seen at about 3.5 months [94].

The intergroup trial Southwest Oncology Group (SWOG) S0033/Cancer and Leukemia Group B (CALGB) 150105 carried out an open-label, phase III trial of 746 with advanced GIST from 148 US and Canadian centers which evaluated the progression-free survival and overall survival with standard imatinib dose of 400 mg daily versus high-dose of 400 mg twice daily. The patients on the standard-dose arm could re-register and receive the high-dose imatinib regimen at the time of disease progression. With a median follow-up of 4.5 years there were no statistically significant differences in objective response rates, progression-free survival, or overall survival. The median progression-free survival was 18 months for patients on the standard-dose arm, and 20 months for those receiving high-dose imatinib. The median overall survival was 55 months on the standard dose and 51 months for the high dose. Thirty-three percent of patients who crossed over to the high-dose imatinib regimen achieved either an objective response or stable disease. High-dose imatinib was associated with more grade 3–5 toxicities. The results of this trial suggest that a reasonable approach is to start GIST patients with imatinib at 400 mg daily and to consider dose escalation at the time of disease progression [95].

9.4.1.3 Duration of therapy

To address optimal duration of therapy in the treatment of advanced GIST using imatinib, a prospective, randomized, multicenter trial was designed to compare continuous compared with interrupted imatinib beyond 1 year of treatment in patients with advanced GIST with tumor control from imatinib. The primary end point was progression-free survival. The trial was closed early after only 58 patients were randomized because of the observation that the risk of progression was significantly worse in the intermittent arm even in patients who had complete response. Twenty-six of the 32 (81%) patients in the intermittent arm progressed compared with only 8 of 26 (31%) in the continuous arm. The median progression-free survival in randomly assigned patients were 6 months and 18 months, respectively. No difference in survival, incidence of imatinib resistance, or quality of life was observed between the two arms. Tumor control was achieved with imatinib reintroduction in 24 of 26 patients in the intermittent arm [96]. In another report, 50 patients with non-progressive disease after completing 3 years of imatinib therapy were randomized to continue or interrupt imatinib therapy. After a median follow-up of 35 months, the 2-year progress free survival (PFS) were 16% and 80% in those who continued therapy versus those who did not. The incidence rates of grade 3 or higher adverse event were similar in both groups. Tumor control after reintroduction of imatinib was seen in all patients. The development of secondary resistance to imatinib did not appear to be increased with the interruption of therapy [97]. Because of rapid progression associated with imatinib interruption, intermittent therapy is not advised unless the patient is experiencing significant toxicities. Continuous therapy until progression or lifelong therapy is recommended.

9.4.1.4 Management of toxicities

Imatinib is generally well tolerated. Fluid retention, muscle cramps, nausea, diarrhea, fatigue, abdominal pain, and rash are the most common side effects associated with imatinib use [93]. Periorbital edema is generally worse in the morning but improves through the day. Occasionally diuretic may be indicated for significant peripheral edema. Muscle cramps may be the most bothersome side effect of imatinib associated with long-term use. Some patients find some relief with use of tonic water (quinine) as well as intake of supplemental calcium and magnesium. Nausea may improve if imatinib is taken with food. The absorption of imatinib is generally not affected by food. Diarrhea is usually mild although loperamide use may be indicated. Generalized maculopapular rash may develop but this improves with longer exposure to imatinib. Anemia is mild although some patients may require erythropoietin and less frequently, blood transfusion. Severe neutropenia is rare. Serum phosphate levels should be routinely monitored especially with use of imatinib at higher dose. Hypothyroidism has been reported with imatinib use and thyroid functions should also be monitored. Of additional note, imatinib may cause alteration in liver function tests. Severe hepatotoxicity has been observed in a patient with CML in blast crisis that was using large doses of acetaminophen during imatinib therapy and therefore concomitant use of acetaminophen should be avoided [98]. There is a 5% risk of gastrointestinal bleeding in patients with large bulky tumors. Therefore it is recommended that the hemoglobin should be monitored particularly in the first 2 months of imatinib therapy [99].

Karkela published a report describing cardiotoxicity in 10 patients receiving imatinib [100]. In the aforementioned EORTC 62005 study there was no increased cardiac event observed in the patient population treated for a median time of 24 months. In the adjuvant American College of Surgeons Oncology Group (ACOSOG) Z9001 trial, imatinib use was not associated with increased cardiotoxicity [101]. Although cardiotoxicity is uncommon, patients who are being considered for imatinib therapy should be informed of the potential risk of cardiotoxicity. Appropriate diagnostic and therapeutic intervention should be done for patients with symptoms suggestive of left ventricular dysfunction.

9.4.2 Sunitinib

Sunitinib malate is a receptor tyrosine kinase inhibitor of KIT and PDGFR. Sunitinib also acts on vascular endothelial growth factor receptors (VEGFR1–3), Fms-related tyrosine kinase (FLT3), colony-stimulating factor (CSF)-1R, and the rearranged during transfection (RET) protooncogene. Thus, sunitinib possesses potential antiangiogenic activity in addition to antitumor action related to receptor tyrosine kinase inhibition TKI. In the laboratory, sunitinib inhibits some KIT mutant isoforms that are resistant to imatinib.

One of the largest trials on safety and efficacy of sunitinib in GIST was published by Demetri and colleagues [102]. Patients with GIST resistant to imatinib or patients intolerant of imatinib were included in this trial. The primary end-point of this randomized, double-blind, placebo-controlled study was time to tumor progression. Sunitinib was given at 50 mg orally daily in 6-week cycles with 4 weeks on and 2 weeks off treatment. 312 patients were randomized in a 2:1 ratio to receive sunitinib or placebo. The trial was terminated early after planned interim analysis showed significantly longer time to tumor progression with sunitinib. The median time to tumor progression was 27.3 weeks (95% CI 16.0–32.1) in patients receiving sunitinib and 6.4 weeks (4.4–10.0) in those assigned to placebo (hazard ratio 0.33; $p < 0.0001$). The objective response rate was 7% in the sunitinib arm and tolerability to the drug was acceptable. The most common treatment-related adverse events were fatigue, diarrhea, skin discoloration, and nausea [102]. Patients who experienced progression of their disease while on placebo crossed over to sunitinib. Despite this crossover, the survival was found to be significantly better in patients in the sunitinib arm compared with placebo arm.

To assess the anti-tumor activity, safety, pharmacokinetics and pharmacodynamics of continuous daily sunitinib dosing in patients with imatinib-resistant/intolerant GIST and to assess morning dosing versus evening dosing an open-label phase II study was done. This showed that the antitumor activity and safety of sunitinib were generally similar with morning and evening dosing. Continuous daily sunitinib dosing achieved and sustained effective drug concentrations without additional accumulation across cycles. Decreases from baseline in plasma levels of soluble KIT after 20 and 24 weeks of dosing correlated with longer overall survival [103].

The activity of sunitinib is influenced by specific mutations. A phase I/II trial of sunitinib in 97 patients with metastatic, imatinib-resistant/intolerant GIST was undertaken to evaluate the impact of primary and secondary kinase genotype on sunitinib activity. *KIT*/*PDGFRA* mutational status was analyzed for 78 patients by using tumor specimens obtained before and after prior imatinib therapy. Sunitinib was associated with clinical benefit (partial response or stable disease) for the three most common primary GIST genotypes: *KIT* exon 9 (58%), *KIT* exon 11 (34%), and *KIT*/*PDGFRA* wild-type (56%). PFS was significantly longer for patients with primary *KIT* exon 9 mutations ($p = 0.0005$) or with a wild-type genotype ($p = 0.0356$) than for those with *KIT* exon 11 mutations. Similar trend was observed for overall survival. PFS and OS were longer for patients with secondary *KIT* exon 13 or 14 mutations (which involve the KIT-adenosine triphosphate binding pocket) than for those with exon 17 or 18 mutations (which involve the KIT activation loop). Biochemical profiling studies confirmed the clinical results. The authors concluded that the clinical activity of sunitinib after imatinib failure is significantly influenced by both primary and secondary mutations in the predominant pathogenic kinases, which has implications for optimization of the treatment of patients with GIST [104].

Similar to imatinib, PET scan allowed earlier determination of response compared with CT scan [105, 106]. After extended initial response to sunitinib, development of secondary mutations has been reported as a mechanism of resistance in sunitinib therapy similar to mechanisms identified in imatinib resistance [107].

9.4.2.1 Management of sunitinib toxicities

Fatigue, nausea, vomiting, diarrhea, anemia, neutropenia, abdominal pain, mucositis, anorexia, hypothyroidism, and discoloration of hair are the more common side effects of sunitinib. Hypertension, bleeding, hand-foot syndrome, fever, myelosuppression, renal toxicity, change in urine colon, elevation of amylase and lipase, and decrease in left ventricular ejection are less common toxicities associated with sunitinib [102, 108]. A pre-eclampsia like syndrome characterized by reversible hypertension and proteinuria has been described in the use of multitargeted kinase inhibitors including sunitinib and sorafenib [109]. The pathophysiology of this syndrome is not completely understood but vascular endothelial growth factor receptor inhibition is implicated and patients should be closely monitored for hypertension and evidence of cardiac toxicity. Dose reductions may be necessary for recurrent neutropenia. Hypothyroidism is common. The risk is increased with longer duration of therapy usually occurring at 50 weeks of therapy. Inhibition of iodine uptake is suspected as the mechanism of hypothyroidism induced by sunitinib. More patients receiving sunitinib for renal cell cancer appear to have destructive thyroiditis with transient thyrotoxicosis followed by hypothyroidism. Unlike Grave's disease, this transient thyrotoxicosis is associated with low radioiodine uptake [110]. Thyroid functions should be closely monitored at baseline then at monthly intervals [15, 111–113]. GI bleeding, bowel perforation, and wound healing problems have been reported with sunitinib use. It is recommended that initiation of therapy be delayed for four weeks after major surgery and two weeks after minor surgery.

9.4.3 Sorafenib and Other Tyrosine Kinase Inhibitors

Investigators from the University of Chicago performed a multicenter phase II consortium clinical trial of 32 patients with KIT expressing advanced GISTs with disease progression on imatinib. The protocol was amended to require progression on sunitinib after the FDA approval of sunitinib. The primary endpoint was response rate. Thirteen percent of these patients had partial response and 55% had stable disease with disease control rate of 68%. The median PFS was 5.2 months and median overall survival was 11.6 months. Grade 3 hand-foot syndrome and hypertension were observed in 45 and 21 % of patients, respectively. Diarrhea, hypophosphatemia, gastrointestinal bleed, thrombosis, gastrointestinal perforation, and intracranial hemorrhage were the other adverse events observed. Dose reductions were required [114].

Clinical trials are under way to determine the role of other tyrosine kinase inhibitors tested in the treatment of GISTs. These include nilotionib, dasatinib, masatinib, sirolimus, and regofenib. Nilotinib is a new oral aminopyrimidine-derivative second-generation tyrosine kinase inhibitor similar inhibitory activity as imatinib in CML cells through competitive inhibition at the ATP-binding site of BCR-ABL, leading to the inhibition of tyrosine phosphorylation of proteins that are involved in the intracellular signal transduction. Nilotinib is more selective and has higher binding affinity for the ABL kinase than does imatinib with 20 to 50 times the inhibitory activity of imatinib in imatinib-sensitive CML cell lines and 3 to 7 times the activity in imatinib-resistant cell lines [115, 116]. A retrospective analysis on patients who have failed imatinib and sunitinib given access to nilotinib on a compassionate use was reported by Montemurro *et al.* Thirty-six patients from five European countries treated in eight centers received nilotinib at starting dose of 400 mg BID. Partial response with nilotinib treatment was seen in 5% of patients and disease stabilization in 37% with disease control rate of 42%. The median PFS was 12 weeks and median overall survival was 34 weeks. Nilotinib was observed to be generally well tolerated with discontinuation rate of 12% [117]. A number of clinical trials are under way to determine the role of nilotinib I the treatment of GISTs. Dasatinib is a novel, oral tyrosine kinase inhibitor of KIT, PDGFR, ABL, and SRC. Trent and colleagues [118] reported the results of a phase II trial assessing the antitumor activity of dasatinib in imatinib- and sunitinib-refractory advanced GIST. The study showed that dasatinib has did not meet the predefined 6-month PFS rate of 30%. Masatinib is a novel tyrosine kinase inhibitor with activity for both wild-type and mutated KIT. The French Sarcoma Group reported activity of this drug in advanced GIST in a phase II trial [119]. A 4-year follow-up was recently reported showing prolonged PFS [120]. A phase III trial is ongoing. Metabolic response measured by FDG-positron emission tomography (PET)/Computed tomography (CT) was seen in GIST patients treated with regorafenib, an oral tyrosine kinase inhibitor [121]. Further analyses are under way.

9.4.3.1 Assessment of response to therapy

The pattern of radiographic response to tyrosine kinase inhibitor consists of an initial decrease in tumor density followed by a slow tumor regression. The standard RECIST (Response Evaluation Criteria in Solid Tumors), which is based on tumor measurements has limited utility in assessing response to TKI therapy. Tumor progression may show as a hyperdense "nodule-within-a-mass" [122]. It may also present as a new or enlarging tumor mass as filling-in of a previously hypodense lesion. The identification of this entity allows for ablation of these masses, which are reported to harbor resistant clones while continuing imatinib to control remaining imatinib sensitive disease [123]. Since RECIST is insensitive in evaluating imatinib-treated GIST, this has lead to new criteria for defining response. The Choi criteria are defined as a 10% decrease in size or a 15% decrease in density on contrast-enhanced CT. Using the Choi criteria investigators demonstrated in a small training set of patients who showed response as measured by PET, that the modified criteria were more predictive of time to tumor progression (TTP) than response by RECIST [124]. These investigators designed a second trial to validate these observations in an independent data set. Fifty-eight patients with imatinib-treated GISTs were evaluated by RECIST and Choi criteria and TTP was compared with TTP in the training set. Patients were analyzed initially with follow-up for 28 months, extended to 60 months for survival analysis. Patients who met Choi response criteria on CT at 2 months had significantly better TTP than those who did not whereas response group by RECIST was not significantly correlated with TTP. Disease-specific survival was also significantly correlated with response group by Choi criteria but not with response group by RECIST [125].

During the early stages of imatinib therapy, decrease in tumor size may not follow changes in tumor density but patients may experience clinical benefit with improvement of symptoms even in the absence of tumor shrinkage. Tumor size may even increase in size in case of intratumoral bleeding or myxoid degeneration. Peritoneal tumor implants may resolve rapidly whereas tumors in the liver may take longer time to regress. Maximum response may be seen until 6 to 12 months or longer with imatinib. Stable disease in CT has been demonstrated to predict time to treatment failure. The results of a pilot study on the predictive power of change in CT bidimensional measurements, change in PET SUV_{max}, and PET SUV_{max} at 1 month after initiation of treatment was reported by Holdsworth and colleagues [126]. They found that no reduction in the CT bidimensional tumor measurements at 1 month is an effective indicator of prolonged clinical benefit in patients with advanced GIST. PET seems to detect earlier and greater number of responses than CT. Responses can be observed within 24 hours of initiating imatinib.

9.4.3.2 Adjuvant therapy

The primary therapy for localized GIST is surgery with complete resection and negative margins. Although the majority of patients with localized GIST are able to undergo complete surgical resection, only about one-half are free of recurrence at 5 years or longer. In a single institution report of 200 patients with GIST, patient, tumor, and treatment variables were analyzed to identify patterns of tumor recurrence and factors that predict survival. Tumor size was found to be predictive of disease-specific survival in patients with primary disease who undergo complete gross resection [127, 128].

9.4.4 Biomarkers That Predict Benefit, Response, and Resistance to Therapy

9.4.4.1 Response predictors in GIST

As discussed in Section 9.2, the majority of GISTs harbor activating mutations of *KIT, PDGFRA or BRAF*. A number of randomized clinical trials have demonstrated that the mutational status of these oncoproteins, especially *KIT*, is predictive of clinical response to imatinib. The type of mutation in *KIT* and *PDGFRA* also correlate with clinical outcome to imatinib. In the B2222 phase II trial, activating mutations in *KIT* and *PDGFRA* were observed in 88 and 4.7 percent, respectively. All KIT mutant isoforms were associated with positive response to imatinib. However, only a subset of patients with *PDGFRA* mutation responded to imatinib. The patients with exon 11 *KIT* mutations experienced partial response rate of 86%, whereas patients with tumors containing exon 9 *KIT* mutations or no detectable mutation of *KIT* or *PDGFRA* had partial response rates of 48% ($p = .0006$) and 0.0% ($p < .0001$), respectively [28]. Patients whose tumors demonstrated exon 11 *KIT* mutations had a longer event-free and overall survival than those whose tumors expressed either exon 9 *KIT* mutations or had no detectable kinase mutation (Fig. 9.5).

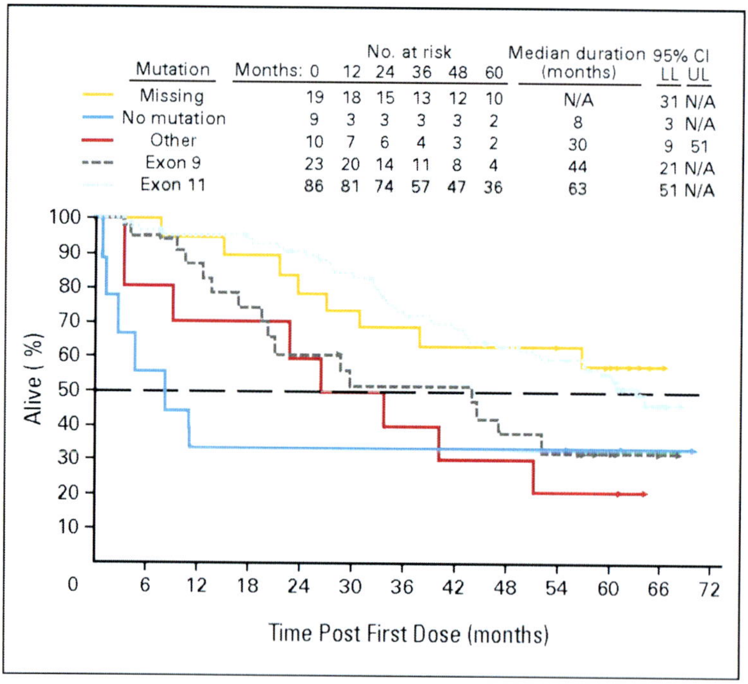

Figure 9.5 Tumor genotype is of major prognostic significance for overall survival in patients treated with imatinib for advanced GISTs. Overall survival according to *KIT* and *PDGFR* mutational status. LL, lower limit; UL, upper limit; N/A, not available. Adapted with permission from Blanke [28].

The investigators of the phase III clinical trial EORTC 62005 analyzed pre-treatment samples of GISTs from 377 patients to determine if tumor mutational status correlates with response to imatinib. The strongest negative predictive factor for response to imatinib was found to be the presence of exon 9 mutations in *KIT*. The relative risk of progression was increased by 171% ($p < 0.0001$) and the relative risk of death by 190% ($p < 0.0001$) when compared with KIT exon 11 mutants. Additionally, treatment with the high-dose regimen resulted in a significantly superior progression-free survival ($p = 0.0013$), with a reduction of the relative risk of 61% in patients with an exon 9 *KIT* mutation. In patients without detectable *KIT* or *PDGFRA* mutations, the relative risk of progression was increased by 108% ($p < 0.0001$) and the relative risk of death by 76% ($p = 0.028$). The authors concluded that tumor genotype carries major prognostic importance for progression-free survival and overall survival in patients with advanced GIST treated with imatinib [129].

The SWOG S0033/CALGB 150105 trial confirmed the findings of the B2222 and EORTC 62005 trials demonstrating that *KIT* exon 11 mutations are associated with more favorable outcome compared with *KIT* exon 9 mutations or wild-type GIST. The objective response rates for *KIT* exon 11 mutations, KIT exon 9 mutations or no detectable kinase mutation were 84%, 48% and 0%, respectively. Longer time to disease progression was also observed in patients with *KIT* exon 11 mutations (25 vs. 17 vs. 13 months, respectively). The median overall survival was 60 months versus 38 months versus 49 months, respectively. Although the response rate was better with high-dose imatinib in patients with *KIT* exon 9 mutations, unlike the EORTC trial, there was no demonstrable improvement in PFS with high dose compared with standard dose imatinib [95].

The Gastrointestinal Stromal Tumor Meta-Analysis Group (MetaGIST) analyzed the data from EORTC 62005 and SWOG S0033/CALGB 150105 comparing the two doses of imatinib (400 mg vs. 800 mg per day) in 1,640 patients. The study end-points were progression-free survival and overall survival. At a median follow-up of 45 months, a small but significant progression-free survival was observed in the 800 mg arm. The overall survival was not different between the two treatment arms. Overall, the presence of *KIT* exon 9 mutations was predictive for PFS benefit to high dose imatinib resulting in significantly longer progression-free survival and higher objective response rate [130].

In the treatment of advanced GIST, mutational status has prognostic importance and may also be predictive of response to TKI therapy. The National Comprehensive Cancer Network (NCCN) currently recommends initial dosing of imatinib and 400 mg once daily in patients with unresectable or advanced GIST. However, for those patients with a known exon 9 mutation, NCCN supports the use of higher imatinib dose at 800 mg daily. In comparison, the European Society of Medical Oncology (ESMO) recommends routine use of mutational testing for all patients with advanced GIST and use of imatinib at an initial dose of 800 mg daily for those patients whose tumors harbor exon 9 mutation [99, 131].

9.4.4.2 Imatinib plasma levels

The occurrence of adverse events correlates with free plasma levels of imatinib. It has been reported that significant variability in plasma imatinib area under the curve (AUC) levels was observed in patients who did not develop significant side effects than those who did [132]. The inter-patient variability in imatinib pharmacokinetic exposure may explain the lack of benefit with higher doses of imatinib. Higher imatinib AUC predicted for better response to imatinib with the strongest association seen in patients whose tumors harbor exon 9 mutation or wild-type KIT [133].

In the previously described B2222 trial, clinical outcome was correlated with plasma imatinib trough levels in 73 patients in whom plasma trough levels were available at the start of therapy and at steady state. The patients were grouped into quartiles of imatinib plasma trough levels. A high interpatient variability in imatinib pharmacokinetic exposure was observed. The median time to progression was 11.4 months for patients in the lowest trough level quartile compared with more than 30 months for those in the other three quartiles. The overall objective benefit rate (complete response, partial response, and stable disease) was also lower in the lowest quartile. A mutation in *KIT* exon 11 was associated with overall objective benefit rate of 67% for the lowest quartile versus 100% for the other quartiles [134]. In a small population-based trial, low imatinib trough levels were also reported in prior major gastrectomy, increased creatinine clearance, and high serum albumin levels [135]. The variability in free plasma levels of imatinib may also contribute to drug resistance. In a small pharmacokinetic study, Judson observed a trend towards increased imatinib clearance after chronic exposure over 12 months with 42 % reduction in systemic exposure compared with the start of therapy. It is uncertain as to whether these decreased levels are truly from increased clearance or a factor of waning patient compliance [136]. Overall, it is not clear if dosing of imatinib based on target plasma levels will result in more favorable outcome in imatinib therapy for GIST and thus monitoring imatinib plasma levels is not currently recommended for routine use. A clinical trial is under way aimed at prospectively addressing this issue.

9.4.4.3 Drug interactions

Imatinib is metabolized by the cytochrome P450 (CYP) enzyme system. The main enzyme responsible for imatinib metabolism is the CYP3A4 in the liver. Plasma levels of imatinib may be affected by concomitant use of drugs that can potentially interact with CYP3A4. Ketoconazole, itraconazole, grapefruit juice, orpomegranade juice are considered CYP3A4 inhibitors and may increase plasma levels of imatinib. CYP3A4 inducers may decrease imatinib concentration in the blood. Rifampin, phenytoin, and St. John's worth are examples of these drugs. Imatinib is a competitive inhibitor of CYP3A4. Warfarin and midazolam are metabolized by CYP3A4 is enzymes. Dose adjustments may be needed when administering imatinib with these CYP3A4 substrates [137].

9.4.4.4 Imatinib resistance and intolerance

The majority of patients with advanced GIST will initially respond to imatinib. However, primary resistance to imatinib may be observed in patients who fail to respond to imatinib with disease progression in the first 6 months of imatinib therapy. This is commonly seen in GIST patients whose tumors carrying a mutations in *KIT* exon 9, *PDGFRA* exon 18, or lack detectable mutations in KIT/PDGFR/BRAF [138]. Secondary resistance to imatinib are thought to develops when new kinase mutations are acquired [139]; however, other mechanisms are likely. Patients on imatinib for more than 6 months with initial response to imatinib and later develop progression of their GIST are considered to have secondary resistance to imatinib. Novel kinase mutations are commonly located in exons encoding tyrosine kinase domain (exon 13, 14, and 17). These mutations are not random and they confer resistance to imatinib by either of two ways. They can directly alter the ATP binding pocket (V654A and T670I mutations). These acquired mutations can cause conformational changes in the activation loop of the kinase domain (D820Y and N822K mutations) thus making the ATP binding pocket less accessible for imatinib. Genomic amplification of the target receptor is another mechanism of secondary resistance to imatinib but this is uncommon [15, 111–113].

In the B2222 trial, 92 of the 147 patients enrolled in the original trial had documented imatinib resistance. Molecular studies were performed using specimens from 10 patients with primary and 33 patients with secondary resistance. Imatinib-resistant tumors had levels of activated KIT that were similar to or greater than those typically found in untreated GISTs. Secondary kinase mutations were rare in GISTs with primary resistance but frequently found in GISTs with secondary resistance (10% vs. 67%; $p = .002$). Secondary kinase mutations were nonrandom and were associated with decreased imatinib sensitivity compared with typical KIT exon 11 mutations. Imatinib-resistant GIST cells remain dependent on KIT kinase activity for activation of critical downstream signaling pathways.

Imatinib resistance can be addressed in a number of ways. One is to increase the dose of imatinib to 800 mg daily. The median time to progression is about 11 months. The other approach is to change therapy to sunitinib. It is not clear as to which of this approach is associated with the better outcome. In patients who progress within 2 months of initiating imatinib, it is unlikely that increasing the dose of imatinib will benefit these patients.

9.4.4.5 Benefit of imatinib as a function of risk stratification

ACOSOG Z9000, a phase II trial, enrolled 107 patients with completely resected GIST with 10 cm or larger tumor in size. The patients with ruptured, bleeding, or multifocal GIST (<5 sites total) received imatinib at 400 mg daily dose for a year. At a median follow-up of 4 years, the 1-, 2-, and 3-year relapse-free survival which is the primary end-point of the study were 94, 73, and 61%, respectively. ACOSOG Z9001 is a phase III, double blind multicenter trial which enrolled 713 adults with completely resected GIST at least 3 cm in size and are KIT positive by IHC. The trial was terminated early during a planned interim analysis. With a median follow-up of 20 months, the 1-year RFS was 98% versus 83% in favor of the imatinib arm. Subgroup analysis showed significantly longer RFS with imatinib in all risk categories based on size, mitotic rate, and location of the tumor. Subsequent analysis showed that the benefit was greatest with the high-risk disease group with relapse rate of 19% in the imatinib arm and 47% in the placebo arm [101].

SSGVIII/AIO trial compared the use of adjuvant imatinib for 36 months versus 12 months in patients with high risk GIST (defined as tumor greater than 5 cm in size or high mitotic rate of more than 5 mitoses/50 HPF) after surgery. There were 200 patients in each arm. With a median follow-up of 54 months, the 5-year RFS (66% vs. 48%, $p < 0.0001$) and OS (92% vs. 82%, respectively; $p = 0.019$) were longer in the 36-month arm compared with the 12-month arm. These data provide strong evidence that at least 3 years of adjuvant imatinib therapy is safe and effective in reducing recurrence and improves overall survival in patients with high-risk GIST [140]. It is not clear, however, whether longer duration of therapy is necessary.

9.4.4.6 Molecular biomarkers of therapeutic response

Frolov and colleagues in 2003 were the first to identify genetic markers that could predict the response of patients with metastatic/recurrent GIST to imatinib using multiple cell lines and clinical trial samples [141]. They hypothesized that by evaluating gene expression profiles in treated GIST cells and then utilizing these data to evaluate specimens from GIST patients taken before and after imatinib therapy (CSTI571-B2222 clinical trial), they would identify novel genetic biomarkers of this therapy and subsequently define additional downstream mediators of response. A total of 148 genes or ESTs were found to be differentially regulated; while 7 genes displayed a durable response following imatinib treatment, i.e., *SPRY4A, FZD8, PDE2A, RTP801, FLJ20898, ARHGEF2* were downregulated, and *MAFbx* was upregulated. Their studies also confirmed that both AKT- and ERK1/2-signaling pathways are rapidly inhibited after exposure to imatinib, but suggested that other signaling pathways may also be affected by imatinib treatment, which were further defined in subsequent studies [142, 143]. Expanding these studies, Rink and colleagues directly assessed pre-treatment biopsy samples from a prospective neoadjuvant phase II trial (RTOG 0132) and identified an expanded 38-gene signature that included 18 *KRAB-ZNF91* subfamily members, 10 of which mapped to a single locus on chromosome 19p

[144]. siRNA synthetic lethal screens demonstrated that members of this gene signature may not only have predictive value, but functional relevance to enhance imatinib activity.

In other studies Belinsky and colleagues have shown that not all GISTs that lack mutations in *KIT* or *PDGFRA*, the so-called "wild-type" tumors, are alike. Using whole-genome SNP-arrays they were able to obtain high-resolution maps of the GIST's genome, both tumors lacking *KIT/PDGFRA* mutations as well as mutant GISTs. Importantly, these studies found that the majority of *KIT/PDGRFA* mutation-negative, both pediatric and adult GISTs which are clinically more resistant to imatinib-based therapies, generally do not possess the large-scale chromosomal aberrations that are commonly found in mutant GISTs [19, 80]. Based on these molecular studies, tumors can be stratified into three groups: *KIT/PDGFRA* mutation negative/ genome stable GISTs, and *KIT/PDGFRA* mutation positive/genome instable GISTs, and *KIT/PDGFRA* mutation negative/genome instable GISTs (Fig. 9.4). Furthermore, these *KIT/PDGFRA* mutation negative/ genome stable GISTs were shown to greatly overexpress the insulin-like growth factor receptor-1 (IGF1R) relative to other GISTs. Tarn *et al.* demonstrated that IGF1R may be an interesting therapeutic target to consider in GIST patients, especially in pediatric and adult *KIT/PDGFRA* mutation negative cases in which treatment options are extremely limited. *KIT/PDGFRA* mutation negative GISTs were shown to greatly overexpress IGF1R (both at the protein and mRNA levels) (Fig. 9.6) [75, 80, 145] and drugs that target IGF1R can inhibit the ability of GIST cells to grow [75]. Clinical trials are being developed to explore IGF1R as a target based on these and other molecular studies. Finally, a proportion of the previous classified "wild-type" GISTs that demonstrate genome instability have been shown to carry mutations in the *BRAF* gene [19]. There are now several additional reports indicating that the *BRAF* gene is mutated in ~4–13% of GISTs without *KIT* or *PDGFRA* mutations (e.g., [18]). However, the genetic changes in GISTs without *KIT/PDGFRA/BRAF* mutations is not known, so we and others are currently using modern day deep sequencing approaches to further explore the GIST genome in these patients to uncover additional clues to help uncover the etiology and define effective therapies. These are a few examples of how work at the bench can be translated into a better understanding of the disease and suggest ways to improve therapeutic modalities influencing how patients will be treated at their bedside.

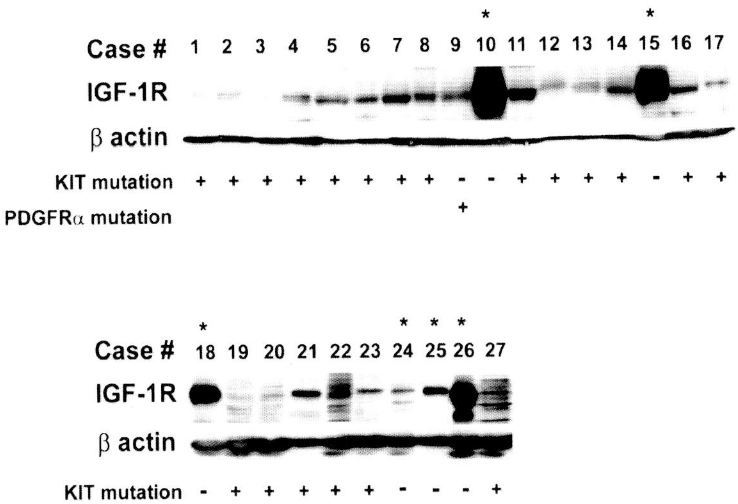

Figure 9.6 IGF1R expression in GIST biopsies. Immunoblot assays of 17 consecutive fresh-frozen GIST biopsies using anti-phospho-IGF1R and anti-IGF1R antibodies. c-*KIT* or *PDGFRA* genotype for each GIST is listed below (+ or -). Asterisk indicate WT GISTs (adapted from [75]).

9.5 Summary

GISTs are the most common mesenchymal tumor of the digestive tract and the majority are associated with gain-of-function mutations in the TKRs, KIT and PDGFR, and to a lesser extent the serine/threonine kinase, BRAF. Tyrosine kinase inhibitors such as imatinib have significantly improved the outcomes for GIST patients, who previously faced very poor prognosis. Clinicians at the bedside of their patients have seen the impact of these drugs and have also had to face the challenges of what to do when a novel, promising therapy fails. Unfortunately, about half of the metastatic GIST patients treated with imatinib will have their tumor start to grow again by 2 years. Furthermore, the driving force in the pathogenesis of most early-onset GISTs is unknown and the treatment options limited. For patients with localized resectable GIST, cytoreductive surgery remains the primary option. This surgery seeks to remove the entire gross tumor, and frequently requires total or subtotal organ resection, depending on tumor location and size. However, surgery has limited success for locally recurrent or metastatic GIST and clinical response of young patients to systemic therapy using novel biologics such as imatinib mesylate and sunitinib is abysmal. Recent advancement in the molecular characterization of GISTs has lead to the discovery of other pathways that may be important in this disease and the ability to molecularly stratify (Fig. 9.7). Clinical trials are currently being designed and implemented to target some of these pathways. Only time will tell if these new therapeutic approaches will have the type of impact seen when first treated introduced to imatinib.

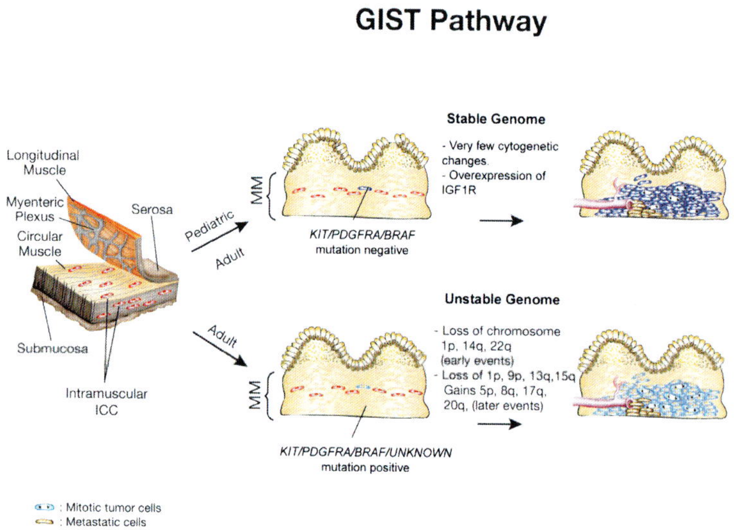

Figure 9.7 Genetic and Cytogenetic Changes Associated with Pathogenesis of Pediatric and Adult GISTs. GISTs are thought to arise from the Interstitial Cells of Cajal (ICC), the pacemaker cells of the gut, or from interstitial mesenchymal precursor stem cells that also have the potential of giving rise to cells in the omentum and peritoneal surfaces. Most pediatric and some adult GISTs are thought to develop independent from acquired mutations in *KIT, PDGFRA, and BRAF*. These tumors are frequently associated with increased expression of IGF1R [19, 75, 80]. Adult and for some young adult GISTs develop as a result of an acquired mutation in *KIT*, *PDGFRA* or *BRAF* and a series of sequential genetic and genomic abnormalities. Loss of chromosome 1p and monosomy 14 and 22 are two of the earliest cytogenetic changes that have been detected cytogenetically. Also depicted in Fig. 9.4. MM – Muscularis Mucosae, CNVs – copy number variants (adapted from [43]).

Acknowledgments

Grant supports: This work was supported in part by NIH grants (CA106588 and a supplement to U10 CA21661) to A.K.G. The authors acknowledge support from The University of Kansas Cancer Center and the Kansas Bioscience Authority Eminent Scholar Program. A.K.G. is the Chancellors Distinguished Chair in Biomedical Sciences endowed Professor.

References

1. Corless CL, Fletcher JA, Heinrich MC. Biology of gastrointestinal stromal tumors. *J Clin Oncol* 2004; 22(18): 3813–3825.
2. Miettinen M, El-Rifai W, L HLS, Lasota J. Evaluation of malignancy and prognosis of gastrointestinal stromal tumors: a review. *Hum Pathol* 2002; 33(5): 478–483.
3. Miettinen M, Lasota J. Gastrointestinal stromal tumors: pathology and prognosis at different sites. *Semin Diagn Pathol* 2006; 23(2): 70–83.
4. Hirota S, Isozaki K, Moriyama Y, Hashimoto K, Nishida T, Ishiguro S, *et al.* Gain-of-function mutations of c-kit in human gastrointestinal stromal tumors. *Science* 1998; 279(5350): 577–580.
5. Sarlomo-Rikala M, Kovatich AJ, Barusevicius A, Miettinen M. CD117: a sensitive marker for gastrointestinal stromal tumors that is more specific than CD34. *Mod Pathol* 1998; 11(8): 728–734.
6. Besmer P, Murphy JE, George PC, Qiu FH, Bergold PJ, Lederman L, *et al.* A new acute transforming feline retrovirus and relationship of its oncogene v-kit with the protein kinase gene family. *Nature* 1986; 320(6061): 415–421.
7. Yarden Y, Kuang WJ, Yang-Feng T, Coussens L, Munemitsu S, Dull TJ, *et al.* Human proto-oncogene c-kit: a new cell surface receptor tyrosine kinase for an unidentified ligand. *EMBO J* 1987; 6(11): 3341–3351. PMCID: 553789.
8. Heinrich MC, Corless CL, Duensing A, McGreevey L, Chen CJ, Joseph N, *et al.* PDGFRA activating mutations in gastrointestinal stromal tumors. *Science* 2003; 299(5607): 708–710.
9. Lux ML, Rubin BP, Biase TL, Chen CJ, Maclure T, Demetri G, *et al.* KIT extracellular and kinase domain mutations in gastrointestinal stromal tumors. *Am J Pathol* 2000; 156(3): 791–795. PMCID: 1876850.
10. Lasota J, Wozniak A, Sarlomo-Rikala M, Rys J, Kordek R, Nassar A, *et al.* Mutations in exons 9 and 13 of KIT gene are rare events in

gastrointestinal stromal tumors. A study of 200 cases. *Am J Pathol* 2000; 157(4): 1091–1095. PMCID: 1850182.

11. Hirota S, Ohashi A, Nishida T, Isozaki K, Kinoshita K, Shinomura Y, *et al.* Gain-of-function mutations of platelet-derived growth factor receptor alpha gene in gastrointestinal stromal tumors. *Gastroenterology* 2003; 125(3): 660–667.
12. Subramanian S, West RB, Corless CL, Ou W, Rubin BP, Chu KM, *et al.* Gastrointestinal stromal tumors (GISTs) with KIT and PDGFRA mutations have distinct gene expression profiles. *Oncogene* 2004; 23(47): 7780–7790.
13. Tarn C, Merkel E, Canutescu AA, Shen W, Skorobogatko Y, Heslin MJ, *et al.* Analysis of KIT mutations in sporadic and familial gastrointestinal stromal tumors: therapeutic implications through protein modeling. *Clin Cancer Res* 2005; 11(10): 3668–3677.
14. Heinrich MC, Corless CL. Gastric GI stromal tumors (GISTs): the role of surgery in the era of targeted therapy. *J Surg Oncol* 2005; 90(3): 195–207; discussion
15. Antonescu CR, Besmer P, Guo T, Arkun K, Hom G, Koryotowski B, *et al.* Acquired resistance to imatinib in gastrointestinal stromal tumor occurs through secondary gene mutation. *Clin Cancer Res* 2005; 11(11): 4182–4190.
16. Corless CL, Schroeder A, Griffith D, Town A, McGreevey L, Harrell P, *et al.* PDGFRA mutations in gastrointestinal stromal tumors: frequency, spectrum and *in vitro* sensitivity to imatinib. *J Clin Oncol* 2005; 23(23): 5357–5364.
17. Agaimy A, Terracciano LM, Dirnhofer S, Tornillo L, Foerster A, Hartmann A, *et al.* V600E BRAF mutations are alternative early molecular events in a subset of KIT/PDGFRA wild-type gastrointestinal stromal tumours. *J Clin Pathol* 2009; 62(7): 613–616.
18. Agaram NP, Wong GC, Guo T, Maki RG, Singer S, Dematteo RP, *et al.* Novel V600E BRAF mutations in imatinib-naive and imatinib-resistant gastrointestinal stromal tumors. *Genes Chromosomes Cancer* 2008; 47(10): 853–859. PMCID: 2902874.
19. Belinsky MG, Skorobogatko YV, Rink L, Pei J, Cai KQ, Vanderveer LA, *et al.* High density DNA array analysis reveals distinct genomic profiles in a subset of gastrointestinal stromal tumors. *Genes Chromosomes Cancer* 2009; 48(10): 886–896. PMCID: 2830286.
20. Hostein I, Faur N, Primois C, Boury F, Denard J, Emile JF, *et al.* BRAF mutation status in gastrointestinal stromal tumors. *Am J Clin Pathol* 2010; 133(1): 141–148.

21. Sircar K, Hewlett BR, Huizinga JD, Chorneyko K, Berezin I, Riddell RH. Interstitial cells of Cajal as precursors of gastrointestinal stromal tumors. *Am J Surg Pathol* 1999; 23(4): 377–389.

22. Kindblom LG, Remotti HE, Aldenborg F, Meis-Kindblom JM. Gastrointestinal pacemaker cell tumor (GIPACT): gastrointestinal stromal tumors show phenotypic characteristics of the interstitial cells of Cajal. *Am J Pathol* 1998; 152(5): 1259–1269. PMCID: 1858579.

23. Huizinga JD, Thuneberg L, Kluppel M, Malysz J, Mikkelsen HB, Bernstein A. W/kit gene required for interstitial cells of Cajal and for intestinal pacemaker activity. *Nature* 1995; 373(6512): 347–349.

24. Robinson TL, Sircar K, Hewlett BR, Chorneyko K, Riddell RH, Huizinga JD. Gastrointestinal stromal tumors may originate from a subset of CD34-positive interstitial cells of Cajal. *Am J Pathol* 2000; 156(4): 1157–1163. PMCID: 1876891.

25. Janeway KA, Liegl B, Harlow A, Le C, Perez-Atayde A, Kozakewich H, *et al.* Pediatric KIT wild-type and platelet-derived growth factor receptor alpha-wild-type gastrointestinal stromal tumors share KIT activation but not mechanisms of genetic progression with adult gastrointestinal stromal tumors. *Cancer Res* 2007; 67(19): 9084–9088.

26. Isozaki K, Hirota S, Nakama A, Miyagawa J, Shinomura Y, Xu Z, *et al.* Disturbed intestinal movement, bile reflux to the stomach, and deficiency of c-kit-expressing cells in Ws/Ws mutant rats. *Gastroenterology* 1995; 109(2): 456–464.

27. Gottlieb JA, Benjamin RS, Baker LH, O'Bryan RM, Sinkovics JG, Hoogstraten B, *et al.* Role of DTIC (NSC-45388) in the chemotherapy of sarcomas. *Cancer Treat Rep* 1976; 60(2): 199–203.

28. Blanke CD, Demetri GD, von Mehren M, Heinrich MC, Eisenberg B, Fletcher JA, *et al.* Long-term results from a randomized phase II trial of standard- versus higher-dose imatinib mesylate for patients with unresectable or metastatic gastrointestinal stromal tumors expressing KIT. *J Clin Oncol* 2008; 26(4): 620–625.

29. Demetri GD, von Mehren M, Blanke CD, Van den Abbeele AD, Eisenberg B, Roberts PJ, *et al.* Efficacy and safety of imatinib mesylate in advanced gastrointestinal stromal tumors. *N Engl J Med* 2002; 347(7): 472–480.

30. Thomas RM, Sobin LH. Gastrointestinal cancer. *Cancer* 1995; 75(1 Suppl): 154–170.

31. Perez EA, Livingstone AS, Franceschi D, Rocha-Lima C, Lee DJ, Hodgson N, *et al.* Current incidence and outcomes of gastrointestinal mesenchymal tumors including gastrointestinal stromal tumors. *J Am Coll Surg* 2006; 202(4): 623–629.

32. Stenman G, Eriksson A, Claesson-Welsh L. Human PDGFA receptor gene maps to the same region on chromosome 4 as the KIT oncogene. *Genes Chromosomes Cancer* 1989; 1(2): 155–158.

33. Blume-Jensen P, Claesson-Welsh L, Siegbahn A, Zsebo KM, Westermark B, Heldin CH. Activation of the human c-kit product by ligand-induced dimerization mediates circular actin reorganization and chemotaxis. *EMBO J* 1991; 10(13): 4121–4128. PMCID: 453162.

34. Casteran N, De Sepulveda P, Beslu N, Aoubala M, Letard S, Lecocq E, *et al.* Signal transduction by several KIT juxtamembrane domain mutations. *Oncogene* 2003; 22(30): 4710–4722.

35. Duensing A, Medeiros F, McConarty B, Joseph NE, Panigrahy D, Singer S, *et al.* Mechanisms of oncogenic KIT signal transduction in primary gastrointestinal stromal tumors (GISTs). *Oncogene* 2004; 23(22): 3999–4006.

36. Kitayama H, Kanakura Y, Furitsu T, Tsujimura T, Oritani K, Ikeda H, *et al.* Constitutively activating mutations of c-kit receptor tyrosine kinase confer factor-independent growth and tumorigenicity of factor-dependent hematopoietic cell lines. *Blood* 1995; 85(3): 790–798.

37. Rossi F, Ehlers I, Agosti V, Socci ND, Viale A, Sommer G, *et al.* Oncogenic Kit signaling and therapeutic intervention in a mouse model of gastrointestinal stromal tumor. *Proc Natl Acad Sci U S Am* 2006; 103(34): 12843–12848. PMCID: 1568935.

38. O'Farrell AM, Abrams TJ, Yuen HA, Ngai TJ, Louie SG, Yee KW, *et al.* SU11248 is a novel FLT3 tyrosine kinase inhibitor with potent activity *in vitro* and *in vivo*. *Blood* 2003; 101(9): 3597–3605.

39. Dolci S, Williams DE, Ernst MK, Resnick JL, Brannan CI, Lock LF, *et al.* Requirement for mast cell growth factor for primordial germ cell survival in culture. *Nature* 1991; 352(6338): 809–811.

40. Tsai M, Takeishi T, Thompson H, Langley KE, Zsebo KM, Metcalfe DD, *et al.* Induction of mast cell proliferation, maturation, and heparin synthesis by the rat c-kit ligand, stem cell factor. *Proc Natl Acad Sci U S A* 1991; 88(14): 6382–6386. PMCID: 52087.

41. Liegl-Atzwanger B, Fletcher JA, Fletcher CD. Gastrointestinal stromal tumors. *Virchows Arch* 2010; 456(2): 111–127.

42. Heinrich MC, Corless CL, Demetri GD, Blanke CD, von Mehren M, Joensuu H, *et al.* Kinase mutations and imatinib response in patients with metastatic gastrointestinal stromal tumor. *J Clin Oncol* 2003; 21(23): 4342–4349.

43. Rink L, Godwin AK. Clinical and molecular characteristics of

gastrointestinal stromal tumors in the pediatric and young adult population. *Curr Oncol Rep* 2009; 11(4): 314–321. PMCID: 2822338.

44. Tarn C, Godwin AK. Molecular research directions in the management of gastrointestinal stromal tumors. *Curr Treat Options Oncol* 2005; 6(6): 473–486.

45. Tarn C, Godwin AK. The molecular pathogenesis of gastrointestinal stromal tumors. *Clin Colorectal Cancer* 2006; 6 Suppl 1: S7–S17.

46. Wardelmann E, Losen I, Hans V, Neidt I, Speidel N, Bierhoff E, *et al.* Deletion of Trp-557 and Lys-558 in the juxtamembrane domain of the c-kit protooncogene is associated with metastatic behavior of gastrointestinal stromal tumors. *Int J cancer J Inte du cancer* 2003; 106(6): 887–895.

47. Antonescu CR, Sommer G, Sarran L, Tschernyavsky SJ, Riedel E, Woodruff JM, *et al.* Association of KIT exon 9 mutations with nongastric primary site and aggressive behavior: KIT mutation analysis and clinical correlates of 120 gastrointestinal stromal tumors. *Clin Cancer Res* 2003; 9(9): 3329–3337.

48. Lasota J, Dansonka-Mieszkowska A, Stachura T, Schneider-Stock R, Kallajoki M, Steigen SE, *et al.* Gastrointestinal stromal tumors with internal tandem duplications in 3' end of KIT juxtamembrane domain occur predominantly in stomach and generally seem to have a favorable course. *Mod Pathol* 2003; 16(12): 1257–1264.

49. Kinoshita K, Isozaki K, Hirota S, Nishida T, Chen H, Nakahara M, *et al.* c-kit gene mutation at exon 17 or 13 is very rare in sporadic gastrointestinal stromal tumors. *J Gastroenterol Hepatol* 2003; 18(2): 147–151.

50. Lasota J, Corless CL, Heinrich MC, Debiec-Rychter M, Sciot R, Wardelmann E, *et al.* Clinicopathologic profile of gastrointestinal stromal tumors (GISTs) with primary KIT exon 13 or exon 17 mutations: a multicenter study on 54 cases. *Mod Pathol* 2008; 21(4): 476–484.

51. McAuliffe JC, Wang WL, Pavan GM, Pricl S, Yang D, Chen SS, *et al.* Unlucky number 13? Differential effects of KIT exon 13 mutation in gastrointestinal stromal tumors. *Mol Oncol* 2008; 2(2): 161–163.

52. O'Riain C, Corless CL, Heinrich MC, Keegan D, Vioreanu M, Maguire D, *et al.* Gastrointestinal stromal tumors: insights from a new familial GIST kindred with unusual genetic and pathologic features. *Am J Surg Pathol* 2005; 29(12): 1680–1683.

53. Medeiros F, Corless CL, Duensing A, Hornick JL, Oliveira AM, Heinrich MC, *et al.* KIT-negative gastrointestinal stromal tumors: proof of

concept and therapeutic implications. *Am J Surg Pathol* 2004; 28(7): 889–894.

54. Corless CL, McGreevey L, Haley A, Town A, Heinrich MC. KIT mutations are common in incidental gastrointestinal stromal tumors one centimeter or less in size. *Am J Pathol* 2002; 160(5): 1567–1572. PMCID: 1850861.

55. Lasota J, Dansonka-Mieszkowska A, Sobin LH, Miettinen M. A great majority of GISTs with PDGFRA mutations represent gastric tumors of low or no malignant potential. *Lab Invest* 2004; 84(7): 874–883.

56. Davies H, Bignell GR, Cox C, Stephens P, Edkins S, Clegg S, *et al.* Mutations of the BRAF gene in human cancer. *Nature* 2002; 417(6892): 949–954.

57. Nishida T, Hirota S, Taniguchi M, Hashimoto K, Isozaki K, Nakamura H, *et al.* Familial gastrointestinal stromal tumours with germline mutation of the KIT gene. *Nat Genet* 1998; 19(4): 323–324.

58. Robson ME, Glogowski E, Sommer G, Antonescu CR, Nafa K, Maki RG, *et al.* Pleomorphic characteristics of a germ-line KIT mutation in a large kindred with gastrointestinal stromal tumors, hyperpigmentation, and dysphagia. *Clin Cancer Res* 2004; 10(4): 1250–1254.

59. Maeyama H, Hidaka E, Ota H, Minami S, Kajiyama M, Kuraishi A, *et al.* Familial gastrointestinal stromal tumor with hyperpigmentation: association with a germline mutation of the c-kit gene. *Gastroenterology* 2001; 120(1): 210–215.

60. Isozaki K, Terris B, Belghiti J, Schiffmann S, Hirota S, Vanderwinden JM. Germline-activating mutation in the kinase domain of KIT gene in familial gastrointestinal stromal tumors. *Am J Pathol* 2000; 157(5): 1581–1585. PMCID: 1885736.

61. Hirota S, Nishida T, Isozaki K, Taniguchi M, Nishikawa K, Ohashi A, *et al.* Familial gastrointestinal stromal tumors associated with dysphagia and novel type germline mutation of KIT gene. *Gastroenterology* 2002; 122(5): 1493–1499.

62. Kim HJ, Lim SJ, Park K, Yuh YJ, Jang SJ, Choi J. Multiple gastrointestinal stromal tumors with a germline c-kit mutation. *Pathol Int* 2005; 55(10): 655–659.

63. Demetri GD, von Mehren M, Antonescu CR, DeMatteo RP, Ganjoo KN, Maki RG, *et al.* NCCN Task Force report: update on the management of patients with gastrointestinal stromal tumors. *J Natl Compr Cancer Netw* 2010;8 Suppl 2: S1–S41; quiz S2–S4.

64. Agarwal R, Robson M. Inherited predisposition to gastrointestinal stromal tumor. *Hematol Oncol Clin North Am* 2009;23(1): 1–13, vii.

65. Kleinbaum EP, Lazar AJ, Tamborini E, McAuliffe JC, Sylvestre PB, Sunnenberg TD, *et al.* Clinical, histopathologic, molecular and therapeutic findings in a large kindred with gastrointestinal stromal tumor. *Int J Cancer* 2008; 122(3): 711–718.

66. Zoller ME, Rembeck B, Oden A, Samuelsson M, Angervall L. Malignant and benign tumors in patients with neurofibromatosis type 1 in a defined Swedish population. *Cancer* 1997; 79(11): 2125–2131.

67. Andersson J, Sihto H, Meis-Kindblom JM, Joensuu H, Nupponen N, Kindblom LG. NF1-associated gastrointestinal stromal tumors have unique clinical, phenotypic, and genotypic characteristics. *Am J Surg Pathol* 2005; 29(9): 1170–1176.

68. Maertens O, Prenen H, Debiec-Rychter M, Wozniak A, Sciot R, Pauwels P, *et al.* Molecular pathogenesis of multiple gastrointestinal stromal tumors in NF1 patients. *Hum Mol Genet* 2006; 15(6): 1015–1023.

69. Miettinen M, Fetsch JF, Sobin LH, Lasota J. Gastrointestinal stromal tumors in patients with neurofibromatosis 1: a clinicopathologic and molecular genetic study of 45 cases. *Am J Surg Pathol* 2006; 30(1): 90–96.

70. Mussi C, Schildhaus HU, Gronchi A, Wardelmann E, Hohenberger P. Therapeutic consequences from molecular biology for gastrointestinal stromal tumor patients affected by neurofibromatosis type 1. *Clin Cancer Res* 2008; 14(14): 4550–4555.

71. Shinomura Y, Kinoshita K, Tsutsui S, Hirota S. Pathophysiology, diagnosis, and treatment of gastrointestinal stromal tumors. *J Gastroenterol* 2005; 40(8): 775–780.

72. Stratakis CA, Carney JA. The triad of paragangliomas, gastric stromal tumours and pulmonary chondromas (Carney triad), and the dyad of paragangliomas and gastric stromal sarcomas (Carney–Stratakis syndrome): molecular genetics and clinical implications. *J Intern Med* 2009; 266(1): 43–52.

73. Sauseng W, Benesch M, Lackner H, Urban C, Kronberger M, Gadner H, *et al.* Clinical, radiological, and pathological findings in four children with gastrointestinal stromal tumors of the stomach. *Pediatr Hematol Oncol* 2007; 24(3): 209–219.

74. Horenstein MG, Hitchcock TA, Tucker JA. Dual CD117 expression in gastrointestinal stromal tumor (GIST) and paraganglioma of Carney triad: a case report. *Int J Surg Pathol* 2005; 13(1): 87–92.

75. Tarn C, Rink L, Merkel E, Flieder D, Pathak H, Koumbi D, *et al.* Insulin-like growth factor 1 receptor is a potential therapeutic target for

gastrointestinal stromal tumors. *Proc Natl Acad Sci U S A.* 2008; 105(24): 8387–8392. PMCID: 2448846.

76. Pasini B, McWhinney SR, Bei T, Matyakhina L, Stergiopoulos S, Muchow M, *et al.* Clinical and molecular genetics of patients with the Carney-Stratakis syndrome and germline mutations of the genes coding for the succinate dehydrogenase subunits SDHB, SDHC, and SDHD. *Eur J Hum Genet* 2008; 16(1): 79–88.

77. Agaram NP, Laquaglia MP, Ustun B, Guo T, Wong GC, Socci ND, *et al.* Molecular characterization of pediatric gastrointestinal stromal tumors. *Clin Cancer Res* 2008; 14(10): 3204–3215.

78. Miettinen M, Lasota J, Sobin LH. Gastrointestinal stromal tumors of the stomach in children and young adults: a clinicopathologic, immunohistochemical, and molecular genetic study of 44 cases with long-term follow-up and review of the literature. *Am J Surg Pathol* 2005; 29(10): 1373–1381.

79. Prakash S, Sarran L, Socci N, DeMatteo RP, Eisenstat J, Greco AM, *et al.* Gastrointestinal stromal tumors in children and young adults: a clinicopathologic, molecular, and genomic study of 15 cases and review of the literature. *J Pediatr Hematol/Oncol* 2005; 27(4): 179–187.

80. Belinsky MG, Rink L, Cai KQ, Ochs MF, Eisenberg B, Huang M, *et al.* The insulin-like growth factor system as a potential therapeutic target in gastrointestinal stromal tumors. *Cell Cycle* 2008; 7(19): 2949–2955. PMCID: 2626174.

81. Blanke CD, Corless CL. State-of-the art therapy for gastrointestinal stromal tumors. *Cancer Invest* 2005; 23(3): 274–280.

82. Edmonson JH, Marks RS, Buckner JC, Mahoney MR. Contrast of response to dacarbazine, mitomycin, doxorubicin, and cisplatin (DMAP) plus GM-CSF between patients with advanced malignant gastrointestinal stromal tumors and patients with other advanced leiomyosarcomas. *Cancer Invest* 2002; 20(5–6): 605–612.

83. Ryan DP, Puchalski T, Supko JG, Harmon D, Maki R, Garcia-Carbonero R, *et al.* A phase II and pharmacokinetic study of ecteinascidin 743 in patients with gastrointestinal stromal tumors. *Oncologist* 2002; 7(6): 531–538.

84. Trent JC, Beach J, Burgess MA, Papadopolous N, Chen LL, Benjamin RS, *et al.* A two-arm phase II study of temozolomide in patients with advanced gastrointestinal stromal tumors and other soft tissue sarcomas. *Cancer* 2003; 98(12): 2693–2699.

85. Buchdunger E, Zimmermann J, Mett H, Meyer T, Muller M, Druker BJ, *et al.* Inhibition of the Abl protein-tyrosine kinase *in vitro* and *in vivo*

by a 2-phenylaminopyrimidine derivative. *Cancer Res* 1996; 56(1): 100–104.

86. Dewar AL, Cambareri AC, Zannettino AC, Miller BL, Doherty KV, Hughes TP, *et al.* Macrophage colony-stimulating factor receptor c-fms is a novel target of imatinib. *Blood* 2005; 105(8): 3127–3132.

87. Heinrich MC, Griffith DJ, Druker BJ, Wait CL, Ott KA, Zigler AJ. Inhibition of c-kit receptor tyrosine kinase activity by STI 571, a selective tyrosine kinase inhibitor. *Blood* 2000; 96(3): 925–932.

88. Okuda K, Weisberg E, Gilliland DG, Griffin JD. ARG tyrosine kinase activity is inhibited by STI571. *Blood* 2001; 97(8): 2440–2448.

89. Druker BJ, Tamura S, Buchdunger E, Ohno S, Segal GM, Fanning S, *et al.* Effects of a selective inhibitor of the Abl tyrosine kinase on the growth of Bcr-Abl positive cells. *Nat Med* 1996; 2(5): 561–566.

90. Tuveson DA, Willis NA, Jacks T, Griffin JD, Singer S, Fletcher CD, *et al.* STI571 inactivation of the gastrointestinal stromal tumor c-KIT oncoprotein: biological and clinical implications. *Oncogene* 2001; 20(36): 5054–5058.

91. Joensuu H, Roberts PJ, Sarlomo-Rikala M, Andersson LC, Tervahartiala P, Tuveson D, *et al.* Effect of the tyrosine kinase inhibitor STI571 in a patient with a metastatic gastrointestinal stromal tumor. *N Engl J Med* 2001; 344(14): 1052–1056.

92. van Oosterom AT, Judson I, Verweij J, Stroobants S, Donato di Paola E, Dimitrijevic S, *et al.* Safety and efficacy of imatinib (STI571) in metastatic gastrointestinal stromal tumours: a phase I study. *Lancet* 2001; 358(9291): 1421–1423.

93. Verweij J, van Oosterom A, Blay JY, Judson I, Rodenhuis S, van der Graaf W, *et al.* Imatinib mesylate (STI-571 Glivec, Gleevec) is an active agent for gastrointestinal stromal tumours, but does not yield responses in other soft-tissue sarcomas that are unselected for a molecular target. Results from an EORTC Soft Tissue and Bone Sarcoma Group phase II study. *Eur J Cancer* 2003; 39(14): 2006–2011.

94. Verweij J, Casali PG, Zalcberg J, LeCesne A, Reichardt P, Blay JY, *et al.* Progression-free survival in gastrointestinal stromal tumours with high-dose imatinib: randomised trial. *Lancet* 2004; 364(9440): 1127–1134.

95. Heinrich MC, Owzar K, Corless CL, Hollis D, Borden EC, Fletcher CD, *et al.* Correlation of kinase genotype and clinical outcome in the North American Intergroup Phase III Trial of imatinib mesylate for treatment of advanced gastrointestinal stromal tumor: CALGB 150105 Study by

Cancer and Leukemia Group B and Southwest Oncology Group. *J Clin Oncol* 2008; 26(33): 5360–5367. PMCID: 2651078.

96. Blay JY, Le Cesne A, Ray-Coquard I, Bui B, Duffaud F, Delbaldo C, *et al.* Prospective multicentric randomized phase III study of imatinib in patients with advanced gastrointestinal stromal tumors comparing interruption versus continuation of treatment beyond 1 year: the French Sarcoma Group. *J Clin Oncol* 2007; 25(9): 1107–1113.

97. Le Cesne A, Ray-Coquard I, Bui BN, Adenis A, Rios M, Bertucci F, *et al.* Discontinuation of imatinib in patients with advanced gastrointestinal stromal tumours after 3 years of treatment: an open-label multicentre randomised phase 3 trial. *Lancet Oncol* 2010; 11(10): 942–949.

98. Ferrero D, Pogliani EM, Rege-Cambrin G, Fava C, Mattioli G, Dellacasa C, *et al.* Corticosteroids can reverse severe imatinib-induced hepatotoxicity. *Haematologica* 2006; 91(6 Suppl): ECR27.

99. Demetri GD, von Mehren M, Antonescu CR, DeMatteo RP, Ganjoo KN, Maki RG, *et al.* NCCN Task Force report: update on the management of patients with gastrointestinal stromal tumors. *J Natl Compr Cancer Netw* 2010;8 Suppl 2: S1–S41; quiz S2–S4.

100. Kerkela R, Grazette L, Yacobi R, Iliescu C, Patten R, Beahm C, *et al.* Cardiotoxicity of the cancer therapeutic agent imatinib mesylate. *Nat Med* 2006; 12(8): 908–916.

101. Blackstein ME. Risk assessment for tumor recurrence after surgical resection of localized primary gastrointestinal stromal tumor (GIST): North America Intergroup phase III trial ACOSOG Z9001 (abstract). 2010 ASCO Gastrointestinal Cancers Symposium. 2010.

102. Demetri GD, van Oosterom AT, Garrett CR, Blackstein ME, Shah MH, Verweij J, *et al.* Efficacy and safety of sunitinib in patients with advanced gastrointestinal stromal tumour after failure of imatinib: a randomised controlled trial. *Lancet* 2006; 368(9544): 1329–1338.

103. George S, Blay JY, Casali PG, Le Cesne A, Stephenson P, Deprimo SE, *et al.* Clinical evaluation of continuous daily dosing of sunitinib malate in patients with advanced gastrointestinal stromal tumour after imatinib failure. *Eur J Cancer* 2009; 45(11): 1959–1968.

104. Heinrich MC, Maki RG, Corless CL, Antonescu CR, Harlow A, Griffith D, *et al.* Primary and secondary kinase genotypes correlate with the biological and clinical activity of sunitinib in imatinib-resistant gastrointestinal stromal tumor. *J Clin Oncol* 2008; 26(33): 5352–5359. PMCID: 2651076.

105. Demetri GD, Heinrich MC, Fletcher JA, Fletcher CD, Van den Abbeele AD, Corless CL, *et al.* Molecular target modulation, imaging, and clinical

evaluation of gastrointestinal stromal tumor patients treated with sunitinib malate after imatinib failure. *Clin Cancer Res* 2009; 15(18): 5902–5909.

106. Prior JO, Montemurro M, Orcurto MV, Michielin O, Luthi F, Benhattar J, *et al.* Early prediction of response to sunitinib after imatinib failure by 18F-fluorodeoxyglucose positron emission tomography in patients with gastrointestinal stromal tumor. *J Clin Oncol* 2009; 27(3): 439–445.

107. Guo T, Hajdu M, Agaram NP, Shinoda H, Veach D, Clarkson BD, *et al.* Mechanisms of sunitinib resistance in gastrointestinal stromal tumors harboring KITAY502-3ins mutation: an in vitro mutagenesis screen for drug resistance. *Clin Cancer Res* 2009; 15(22): 6862–6870. PMCID: 2783687.

108. Zhu X, Stergiopoulos K, Wu S. Risk of hypertension and renal dysfunction with an angiogenesis inhibitor sunitinib: systematic review and meta-analysis. *Acta Oncol* 2009; 48(1): 9–17.

109. Patel TV, Morgan JA, Demetri GD, George S, Maki RG, Quigley M, *et al.* A preeclampsia-like syndrome characterized by reversible hypertension and proteinuria induced by the multitargeted kinase inhibitors sunitinib and sorafenib. *J Natl Cancer Inst* 2008; 100(4): 282–284.

110. Grossman M, Premaratne E, Desai J, Davis I. Thyrotoxicosis during sunitinib treatment for renal cell cancer. *Clin Endocrinol* 2008; 69(4): 669–672.

111. Chen LL, Trent JC, Wu EF, Fuller GN, Ramdas L, Zhang W, *et al.* A missense mutation in KIT kinase domain 1 correlates with imatinib resistance in gastrointestinal stromal tumors. *Cancer Res* 2004; 64(17): 5913–5919.

112. Debiec-Rychter M, Cools J, Dumez H, Sciot R, Stul M, Mentens N, *et al.* Mechanisms of resistance to imatinib mesylate in gastrointestinal stromal tumors and activity of the PKC412 inhibitor against imatinib-resistant mutants. *Gastroenterology* 2005; 128(2): 270–279.

113. Heinrich MC, Corless CL, Blanke CD, Demetri GD, Joensuu H, Roberts PJ, *et al.* Molecular correlates of imatinib resistance in gastrointestinal stromal tumors. *J Clin Oncol* 2006; 24(29): 4764–4774.

114. Campbell NP, Wroblewski, K., Maki, RG, D'Adamo DR, Chow WA, Gandara DR, Antonescu C, Stadler WM, Vokes EE, Kindler HL. Final results of the University of Chicago phase II consortium trial of sorafenib (SOR) in patients (pts) with imatinib (IM)- and sunitinib (SU)-resistant (RES) gastrointestinal stromal tumor (GIST) (abstract). ASCO GI Cancers Symposium. 2011.

115. Kantarjian H, Giles F, Wunderle L, Bhalla K, O'Brien S, Wassmann B, *et al.* Nilotinib in imatinib-resistant CML and Philadelphia chromosome-positive ALL. *N Engl J Med* 2006; 354(24): 2542–2551.

116. Verstovsek S, Golemovic M, Kantarjian H, Manshouri T, Estrov Z, Manley P, *et al.* AMN107, a novel aminopyrimidine inhibitor of p190 Bcr-Abl activation and of in vitro proliferation of Philadelphia-positive acute lymphoblastic leukemia cells. *Cancer* 2005; 104(6): 1230–1236.

117. Montemurro M, Schoffski P, Reichardt P, Gelderblom H, Schutte J, Hartmann JT, *et al.* Nilotinib in the treatment of advanced gastrointestinal stromal tumours resistant to both imatinib and sunitinib. *Eur J Cancer* 2009; 45(13): 2293–2297.

118. Trent JC, Wathen K, von Mehren M, Samuels BL, Staddon AP, Choy E, Butrynski JE, Chugh R, Chow WA, Rushing DA, Forscher CA, Baker LH, Schuetz S , editor. A phase II study of dasatinib for patients with imatinib-resistant gastrointestinal stromal tumor (GIST). 2011 ASCO Annual Meeting; 2011; Chicago, IL.

119. Le Cesne A, Blay J, Bui NB, Bouché O, Adenis A, Domont J, Cioffi A, Moussy A, Hermine O, editor. Masatinib mesylate in imatinib-naive locally advanced or metastatic gastrointestinal stromal tumor (GIST): Results of the French Sarcoma Group phase II trial. 2009 ASCO Annual Meeting; 2009; Chicago, IL.

120. Blay J, Le Cesne A, Bin Bui N, Bouche O, Adenis A, Julien D, Cioffi A, Moussy A, Hermine O, editor. Overall survival benefit with masitinib mesylate in imatinib-naive, locally advanced, or metastatic gastrointestinal stromal tumor (GIST): 4-years follow-up of the French Sarcoma Group phase II trial. 2011 Gastrointestinal Cancers Symposium 2011; San Francisco, CA.

121. Van Den Abbeele AD, Tanaka Y, Locascio T, Sakellis C, Heinrich MC,von Mehren M, Choy E, Tap WD, Manola J, Demetri GD, George S, Yap JT , editor. Assessment of regorafenib activity with FDG-PET/CT in a multicenter phase II study in patients (pts) with advanced gastrointestinal stromal tumor (GIST) following failure of standard therapy (Rx). 2011 ASCO Annual Meeting; 2011; Chicago, IL.

122. Shankar S, vanSonnenberg E, Desai J, Dipiro PJ, Van Den Abbeele A, Demetri GD. Gastrointestinal stromal tumor: new nodule-within-a-mass pattern of recurrence after partial response to imatinib mesylate. *Radiology* 2005; 235(3): 892–898.

123. Avritscher R, Gupta S. Gastrointestinal stromal tumor: role of interventional radiology in diagnosis and treatment. *Hematol Oncol Clin North Am* 2009; 23(1): 129–137.

124. Choi H. Critical issues in response evaluation on computed tomography: lessons from the gastrointestinal stromal tumor model. *Curr Oncol Rep* 2005; 7(4): 307–311.

125. Choi H, Charnsangavej C, Faria SC, Macapinlac HA, Burgess MA, Patel SR, *et al.* Correlation of computed tomography and positron emission tomography in patients with metastatic gastrointestinal stromal tumor treated at a single institution with imatinib mesylate: proposal of new computed tomography response criteria. *J Clin Oncol* 2007; 25(13): 1753–1759.

126. Holdsworth CH, Badawi RD, Manola JB, Kijewski MF, Israel DA, Demetri GD, *et al.* CT and PET: early prognostic indicators of response to imatinib mesylate in patients with gastrointestinal stromal tumor. *AJR Am J Roentgenol* 2007; 189(6): W324–W330.

127. DeMatteo RP, Lewis JJ, Leung D, Mudan SS, Woodruff JM, Brennan MF. Two hundred gastrointestinal stromal tumors: recurrence patterns and prognostic factors for survival. *Ann Surg* 2000; 231(1): 51–58. PMCID: 1420965.

128. Joensuu H. Risk stratification of patients diagnosed with gastrointestinal stromal tumor. *Hum Pathol* 2008; 39(10): 1411–1419.

129. Debiec-Rychter M, Sciot R, Le Cesne A, Schlemmer M, Hohenberger P, van Oosterom AT, *et al.* KIT mutations and dose selection for imatinib in patients with advanced gastrointestinal stromal tumours. *Eur J Cancer* 2006; 42(8): 1093–1103.

130. (MetaGIST) GSTM-AG. omparison of two doses of imatinib for the treatment of unresectable or metastatic gastrointestinal stromal tumors: a meta-analysis of 1,640 patients. *J Clin Oncol* 2010; 28(1247–1253).

131. Casali PG, Jost L, Reichardt P, Schlemmer M, Blay JY. Gastrointestinal stromal tumours: ESMO clinical recommendations for diagnosis, treatment and follow-up. *Ann Oncol* 2009; 20 Suppl 4: 64–67.

132. Delbaldo C, Chatelut E, Re M, Deroussent A, Seronie-Vivien S, Jambu A, *et al.* Pharmacokinetic-pharmacodynamic relationships of imatinib and its main metabolite in patients with advanced gastrointestinal stromal tumors. *Clin Cancer Res* 2006; 12(20 Pt 1): 6073–6078.

133. Widmer N, Decosterd LA, Leyvraz S, Duchosal MA, Rosselet A, Debiec-Rychter M, *et al.* Relationship of imatinib-free plasma levels and target genotype with efficacy and tolerability. *Br J Cancer* 2008; 98(10): 1633–1640. PMCID: 2391118.

134. Demetri GD, Wang Y, Wehrle E, Racine A, Nikolova Z, Blanke CD, *et al.* Imatinib plasma levels are correlated with clinical benefit in patients

with unresectable/metastatic gastrointestinal stromal tumors. *J Clin Oncol* 2009; 27(19): 3141–3147.

135. Yoo C, Ryu MH, Kang BW, Yoon SK, Ryoo BY, Chang HM, *et al.* Cross-sectional study of imatinib plasma trough levels in patients with advanced gastrointestinal stromal tumors: impact of gastrointestinal resection on exposure to imatinib. *J Clin Oncol* 2010; 28(9): 1554–1559.

136. Judson I, Ma P, Peng B, Verweij J, Racine A, di Paola ED, *et al.* Imatinib pharmacokinetics in patients with gastrointestinal stromal tumour: a retrospective population pharmacokinetic study over time. EORTC Soft Tissue and Bone Sarcoma Group. *Cancer Chemother Pharmacol* 2005; 55(4): 379–386.

137. Peng B, Lloyd P, Schran H. Clinical pharmacokinetics of imatinib. *Clin Pharmacokinet* 2005; 44(9): 879–894.

138. Benjamin RS, Debiec-Rychter M, Le Cesne A, Sleijfer S, Demetri GD, Joensuu H, *et al.* Gastrointestinal stromal tumors II: medical oncology and tumor response assessment. *Semin Oncol* 2009; 36(4): 302–311.

139. Maleddu A, Pantaleo MA, Nannini M, Di Battista M, Saponara M, Lolli C, *et al.* Mechanisms of secondary resistance to tyrosine kinase inhibitors in gastrointestinal stromal tumours (Review). *Oncol Rep* 2009; 21(6): 1359–1366.

140. Joensuu H, Eriksson M, Hatrmann J, Sundby Hall K, Schutte J, Reichardt A, Schlemmer M, Wardelmann E, Ramadori G, Al-Batran S, NIlsson BE, Monge O, Kallio R, Sarlomo-Rikala M, Bono P, Leinonen M, Hohenberger P, Alvegard T, Reichardt P , editor. Twelve versus 36 months of adjuvant imatinib (IM) as treatment of operable GIST with a high risk of recurrence: Final results of a randomized trial (SSGXVIII/AIO). 2011 ASCO Annual Meeting; 2011; Chicago, IL.

141. Frolov A, Chahwan S, Ochs M, Arnoletti JP, Pan ZZ, Favorova O, *et al.* Response markers and the molecular mechanisms of action of Gleevec in gastrointestinal stromal tumors. *Mol Cancer Ther* 2003; 2(8): 699–709.

142. Ochs MF, Rink L, Tarn C, Mburu S, Taguchi T, Eisenberg B, *et al.* Detection of treatment-induced changes in signaling pathways in gastrointestinal stromal tumors using transcriptomic data. *Cancer Res* 2009; 69(23): 9125–9132. PMCID: 2789202.

143. Tarn C, Skorobogatko YV, Taguchi T, Eisenberg B, von Mehren M, Godwin AK. Therapeutic effect of imatinib in gastrointestinal stromal tumors: AKT signaling dependent and independent mechanisms. *Cancer Res* 2006; 66(10): 5477–5486.

144. Rink L, Skorobogatko Y, Kossenkov AV, Belinsky MG, Pajak T, Heinrich MC, *et al.* Gene expression signatures and response to imatinib mesylate in gastrointestinal stromal tumor. *Mol Cancer Ther* 2009; 8(8): 2172–2182. PMCID: 2822341.

145. Janeway KA, Zhu MJ, Barretina J, Perez-Atayde A, Demetri GD, Fletcher JA. Strong expression of IGF1R in pediatric gastrointestinal stromal tumors without IGF1R genomic amplification. *Int J cancer* 2010; 127(11): 2718–2722. PMCID: 2947557.

146. Miettinen M, Sobin LH, Lasota J. Gastrointestinal stromal tumors of the stomach: a clinicopathologic, immunohistochemical, and molecular genetic study of 1765 cases with long-term follow-up. *Am J Surg Pathol* 2005; 29(1): 52–68.

147. Miettinen M, Furlong M, Sarlomo-Rikala M, Burke A, Sobin LH, Lasota J. Gastrointestinal stromal tumors, intramural leiomyomas, and leiomyosarcomas in the rectum and anus: a clinicopathologic, immunohistochemical, and molecular genetic study of 144 cases. *Am J Surg Pathol* 2001; 25(9): 1121–1133.

148. Miettinen M, Kopczynski J, Makhlouf HR, Sarlomo-Rikala M, Gyorffy H, Burke A, *et al.* Gastrointestinal stromal tumors, intramural leiomyomas, and leiomyosarcomas in the duodenum: a clinicopathologic, immunohistochemical, and molecular genetic study of 167 cases. *Am J Surg Pathol* 2003; 27(5): 625–641.

149. Miettinen M, Makhlouf H, Sobin LH, Lasota J. Gastrointestinal stromal tumors of the jejunum and ileum: a clinicopathologic, immunohistochemical, and molecular genetic study of 906 cases before imatinib with long-term follow-up. *Am J Surg Pathol* 2006; 30(4): 477–489.

Chapter 10

PML/RARα Fusion Gene and Response to Retinoic Acid and Arsenic Trioxide Treatment

Alicja M. Gruszka[a] **and Myriam Alcalay**[a,b]
[a]*Department of Experimental Oncology, European Institute of Oncology, Via Adamello 16, 20 139 Milan, Italy*
[b]*Dipartimento di Medicina, Chirurgia e Odontoiatria, University of Milan, Via A. di Rudini 8, 20142, Milan, Italy*
myriam.alcalay@ifom-ieo-campus.it

The leukaemia-associated fusion gene, promyelocytic leukaemia (PML)-retinoic acid receptor-α (RARα), causes acute promyelocytic leukaemia (APL) by blocking differentiation and increasing self-renewal of leukaemic progenitor cells. The current standard of APL's induction therapy comprises all-trans retinoic acid (ATRA) and chemotherapy, but arsenic trioxide also cures many patients and is used routinely in the treatment of relapsed patients. When combined with ATRA and chemotherapy, a 5-year overall survival of 90% is achieved. Molecularly, ATRA at pharmacological doses unblocks the transcriptional repression brought upon by the PML/RARα chimeric protein. Arsenic, on the other hand, binds thiol residues and induces

Handbook of Therapeutic Biomarkers in Cancer
Edited by Sherry X. Yang and Janet E. Dancey
Copyright © 2013 Pan Stanford Publishing Pte. Ltd.
ISBN 978-981-4364-65-2 (Hardcover), 978-981-4364-66-9 (eBook)
www.panstanford.com

the formation of reactive oxygen species, consequently affecting numerous signalling pathways. Both agents target the fusion protein for degradation, albeit via diverse mechanisms. Moreover, ATRA synergises with arsenic in eliminating leukaemia-initiating cells (LICs) also referred to as leukaemic stem cells (LSC). The progress made in the treatment of patients with APL constitutes an example of a success story of translational haematology towards which both basic scientists and clinicians have contributed over the years.

10.1 Introduction

Acute promyelocytic leukaemia, commonly called APL, is a malignancy of the bone marrow in which there is a deficiency of mature blood cells of the myeloid lineage and an excess of immature cells blocked at the promyelocyte stage of the differentiation. APL accounts for 5–10% of cases of acute myeloid leukaemia. The peak incidence of APL is ~40 years of age. It was classified as the M3 variant of AML of the French-American-British (FAB) Classification [5].

APL was first recognised as a distinct disease entity in 1957 [27]. The original description underlined a very rapid downhill course, with survival rates of only few weeks from diagnosis, mainly due to the severe bleeding tendency caused by fibrinolysis and thrombocytopaenia. Indeed, from 1950s until the introduction of anthracyclins into the chemotherapeutic regimens in the 1970s [6], APL had a 100% mortality rate, as there was no effective treatment.

10.1.1 Description of Acute Promyelocytic Leukaemia

The signs and symptoms of APL are non-specific and include fatigue, minor infections and a haemorrhagic diathesis manifesting itself as petechiae, small ecchymosis, epistaxis, or in more extreme cases, bleeding from the mouth or venipunctures and haematuria. The haemorrhagic diathesis may precede the diagnosis of leukaemia by 2–8 weeks [17].

Microscopically, typical promyelocytic malignant cells are found in bone marrow aspirates, and less frequently, in the peripheral blood while the white blood cell count (WBC) remains low. The predominant malignant cells are abnormal promyelocytes, with an immature nucleus and copious cytoplasm filled with azurophilic granules. An abundance of large granules sometimes covers or masks the nucleus. Auer rods are seen in some cells, grouped into what are called faggots. There are two further rare morphological variants of APL: hypogranular and basophilic microgranular [4].

In addition to the specific blast morphology, their immunophenotypic profile is characteristic and comprises CD33, CD9, CD13, CD117, rarely CD34 or the aberrant presence of the T-cell–associated antigen [40].

10.1.2 Modern Therapeutic Approaches

From the historical perspective, the evolution of treatment of APL can be divided into two main eras: the pre- all-trans-retinoic acid (ATRA), chemotherapy alone period and then, following the discovery of the efficacy of vitamin A derivatives (that include ATRA), the era of ATRA, alone or in combination with chemotherapy, and/or more recently, arsenic trioxide [17]. The use of ATRA in APL constitutes a paradigm of "differentiation therapy". Its introduction into the treatment regimens from 1985 onwards has changed the fate of APL patients.

Upon the suspicion of a diagnosis of APL, according to modern standard operating procedures, three actions must be undertaken: immediate ATRA administration, supportive care composed of transfusions of platelets and plasma and verification of the molecular diagnosis [54]. In principle, ATRA can induce a complete remission in most patients with APL. However there is a caveat to the use of this agent, as ATRA cannot eliminate the leukaemic clone, and moreover, its prolonged administration can lead to disease resistance to treatment [59]. ATRA is therefore used in combination with chemotherapy including an anthracyclin as the standard induction treatment for newly diagnosed patients. There is less standardisation concerning the composition and duration of the consolidation therapy; however, most protocols include ATRA and anthracyclin and/or mitoxantrone or Ara-C. Maintenance treatment with ATRA, and possibly with low-dose chemotherapy, further reduces the incidence of relapse. In case of rare relapses, ATRA and arsenic trioxide are used together (all therapeutic options are reviewed in [54]).

10.1.3 Treatment Complications and Prognosis

The differentiation syndrome (previously called "ATRA syndrome") is a potentially deadly complication of the induction therapy in cases with APL. It occurs in approximately 25% of patients treated with ATRA or arsenic trioxide. Early recognition and aggressive management are fundamental [51].

The prognosis for APL depends on a number of factors including the WBC count at diagnosis, elderly age, expression of CD56 surface marker and severity of bleeding complications [37]. Overall, at present more than 90% of patients with newly diagnosed APL can achieve complete remission, and nearly 80% can be cured by the combination of ATRA and chemotherapy [54].

10.2 Molecular Pathogenesis

10.2.1 Translocation (15;17) and Cloning of the Fusion Gene

In over 95% of cases, APL occurs due to the t(15;17)(q22;q11–q12) translocation, reported by Rowley *et al.* as a consistent feature of this leukaemia subtype in 1977 [52]. This translocation is not a mere marker of APL, but a causative event in the development of the disease.

At a molecular level, as determined by Borrow *et al.* [7], Alcalay *et al.* [3] and de The *et al.* [16] in 1991, t(15;17) disrupts and fuses the promyelocytic leukaemia (PML) gene located on chromosome 15 to the retinoic acid receptor α (RARα) gene on chromosome 17, leading to the formation of the PML/RARα fusion gene, and consequently, of a chimeric protein. In view of the reciprocity of the translocation also a RARα/PML fusion forms; however, its presence is not reliably detected in all of the patients suggesting that PML/RARα, and not its reciprocal translocation product, plays a more important role in the leukaemogenesis [2].

Importantly, all diagnostic tests take advantage of the presence of this unique chromosomal translocation in APL and its product – PML/RARα fusion gene. The methods include cytogenetic analysis, fluorescence *in situ* hybridisation (FISH), PML immunofluorescence (IF) staining and reverse-transcriptase polymerase chain reaction

(RT-PCR) [40, 49]. IF using anti-PML antibodies is the most rapid and economical test for the confirmation of APL through the detection of the numerous nuclear "microspeckles" that differ from the normal pattern of few large nuclear dots [21]. In contrast to conventional karyotyping, which fails to detect t(15;17) in up to 10% of presumed APLs, the use of FISH and RT-PCR allows for the analysis of interphase cells, samples with no or poor-quality metaphases as well as for the identification of cryptic or complex rearrangements leading to the formation of PML/RARα in the absence of the classical t(15;17) [49]. Currently, RT-PCR for the PML/RARα fusion constitutes a mandatory element of the laboratory work-up in APL as a fast confirmatory test and as a means for defining precise genetic breakpoints locations needed for subsequent treatment outcome monitoring and minimal residual disease assessment [49].

In addition, in a small percentage of patients (<5%) rare translocations occur. The common denominator of these fusions is the presence of RARα and partner genes include: PLZF, NPM, NuMA and Stat5a; however, for the purpose of this chapter, they will not be discussed any further [42].

10.2.2 Partner Genes and Their Physiological Function

The physiological roles of the two partners are multiple and complex. The PML protein is a member of the tripartite motif (TRIM) family [8]. The TRIM motif includes three zinc-binding domains, a RING, two B-boxes and a coiled-coil region (called jointly RBCC domain). Both the RING and coiled-coil moieties of PML are required for its localisation in complex subnuclear structures called nuclear bodies (NB), and for its interaction with other proteins within NBs such as p53, CBP, BLM, eIF-4, DAXX, Ubc9, pRb and sp100 [8]. PML protein, the exact functions of which are still elusive, has been implicated in a variety of processes, such as tumour suppression, senescence, apoptosis and telomere maintenance [35]. RARα, on the other hand, is a nuclear hormone receptor that confers the transcriptional response to retinoic acid (RA) [11]. RARα binds to specific DNA sequences, referred to as RA responsive elements (RARE), consisting of a PuGGTCA sequence spaced by two or five nucleotides (DR2, DR5), in a heterodimeric complex with retinoid-X receptor (RXRα). RARs have a common domain structure consisting of an N-terminal DNA-binding, a linker and a C-terminal ligand (RA)-binding domains.

In the absence of RA, RARα/RXRα complexes repress transcription through interactions with NCoR and SMRT co-repressors that do not possess an enzymatic activity themselves, but recruit complexes with histone deacetylase activity (HDAC) generating a condensed chromatin structure. Physiological concentrations of RA induce a conformational change in the RARs causing release of these co-repressors and recruitment of transcriptional co-activators from the p160 family, ultimately leading to activation of gene transcription through the association with histone acetyltransferases (HATs) and histone methyltransferases (HMT) [50]. Retinoid signalling has been implicated in the regulation of development, differentiation, apoptosis, granulopoeisis, stem cell differentiation and transcription of clock genes [26].

The most frequent PML/RARα chimeric fusion (70% of cases) contains most of the RARα protein (with the exception of the ligand-independent activation domain) fused to all but the Serine-Proline-rich domain of the PML protein, although there is a degree of heterogeneity within the PML breakpoints giving rise to three distinctive variants [42].

10.2.3 Mechanisms of Action of PML/RARα

The expression of PML/RARα is critical for the transformation of haematopoietic stem cells and/or progenitors *in vitro* and causes leukaemia in transgenic murine models that faithfully recapitulates the features of the human malignancy. In general, PML/RARα, through its RARα portion, binds RA, RXRα and DNA with the same affinity as RARα, whilst the PML portion confers the ability to form heterodimers with PML and homodimers with PML/RARα itself, both mediated by the PML homodimerisation domain [55]. At present, the classical belief that PML/RARα transformed haematopoietic progenitors due to transcriptional repression of RARα target genes [42] in a dominant negative fashion has evolved into a much more complex model [15]. The novel functions of PML/RARα include: transcriptional regulation of a variety of new targets due to degenerate DNA binding [31, 41], sequestration of various proteins [8, 58], increased self-renewal of haematopoietic progenitors [61] and protein misfolding [32], although the division of these effects seems quite artificial and reflects simply the focus of the different studies dedicated to the dissection of PML/RARα function.

Accordingly, the revised model of APL pathogenesis postulates that PML/RARα is a potent gain-of-function mutant that performs many tasks that neither translocation partner does alone [55].

10.2.3.1 Global transcriptional repression

PML/RARα behaves as a potent repressor of the RA signalling pathway (Fig. 10.1). The traditional model postulated that PML/RARα acted as a constitutive transcriptional repressor that altered the normal RARα signalling in APL cells, as the chimeric protein is unable to respond to physiological fluctuations of RA [42]. The transcriptional repression was shown to be the consequence of greatly enhanced binding to the SMRT/NCoR co-repressors and HDACs [23]. PML/RARα homodimerises and binds to DNA at the RARE sites even in the absence of its normal heterodimeric partner RXR. The homodimerisation is thought to enhance the binding of the physiological RARα interactors. Simplistically, enhanced co-repressors binding depends upon the fact that the homodimer harbours two co-repressor docking sites and not just one as in RXR/RARα heterodimer, leading to a change in stoichiometry of association of PML/RARα with co-repressors and chromatin modifiers [39]. However, in addition, the formation of homodimers leads to the creation of novel binding interfaces. Histone methyltransferase SUV39H1, responsible for trimethylation of lysine 9 of histone H3 is one of the chimera-specific partners responsible for imposing a heterochromatin-like structure on target genes, thereby establishing permanent transcriptional silencing [9]. Similarly, polycomb repressive complex 2 (PRC2) represents another example of a new PML/RARα interactor. It has been found that PRC2 is recruited to tumour suppressor genes causing and maintaining their silencing during the initial steps of PML/RARα driven leukaemogenesis [57].

In the current model, the transcriptional repression is not restricted to the RARα target genes but the epigenetic regulation is extended to areas of the genome that do not possess the classical DR2 and DR5 repeats in their promoter. Recent genome-wide studies have shown that PML/RARα binds to regions containing atypical DR1, DR3, DR4 motifs, as well as, sites with other spacing and orientation [31]. It is thought that this massively enhanced DNA binding is underscored by the formation of heterotetramers (PML/RARα-RXRα)x2 giving rise to a complex with four DNA binding

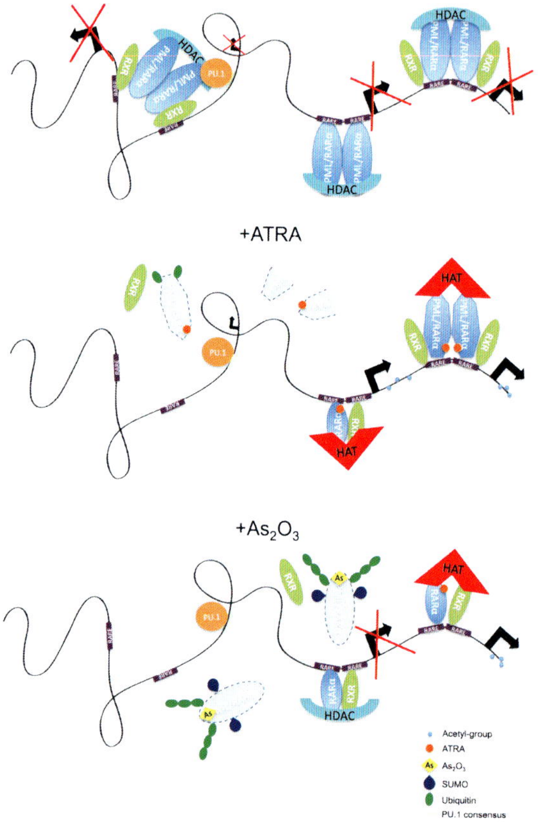

Figure 10.1 Molecular pathogenesis of APL and mechanisms of action of ATRA and arsenic trioxide. Top panel: PML/RARα binds to DNA as a homodimer or as a heterotetradimer with RXR to canonical or atypical RARE or indirectly via protein–protein interaction with other transcription factors, e.g. PU.1, recruits repressive chromatin-modifiers and blocks transcription of RA-target genes and other sequences. Middle panel: ATRA causes conformational change of the PML/RARα molecule allowing for the recruitment of HATs that mediate acetylation of histones and open the chromatin structure leading to de-repression of blocked transcription. In addition, the presence of ATRA causes the degradation of PML/RARα via cleavage and proteasomal destruction detaching PML/RARα from the atypical RAREs. Bottom panel: Arsenic binds to the PML portion of PML/RARα and causes its SUMOylation followed by poly-ubiquitination and proteasomal destruction. Some transcriptional re-activation is present.

domains that allows for the binding to a vast variety of *de novo* DNA binding sites. The novel binding sites include recognition sites for other nuclear receptors, such as vitamin D and thyroid hormones receptors. The property of non-RARE binding constitutes a powerful gain-of-function for the PML/RARα fusion [31, 41]. It has been shown, that binding of RXR to the PML/RARα homodimer is necessary for the full leukaemogenic potential of the fusion protein in murine models and for the execution of the transcriptional program on its target genes validating this aspect of the model of pathogenesis [44].

Finally, yet another mechanism of transcriptional repression mediated by posttranslational modification, namely SUMOylation, of the PML/RARα protein has been proposed. Death-domain-associated protein (DAXX), a potent transcriptional repressor, is recruited to the SUMOylated oncogene and contributes to the formation of repressive chromatin status [15].

10.2.3.2 Transcriptional activation

PML/RARα function as a transcriptional repressor has been known and studied extensively. However, a possibility exists that it can also act as an activator of the transcription of certain genes. It must be pointed out that this function has been much less explored than the transcriptional repression counterpart. Comparison of the gene expression patterns set by PML-RARα in the absence of ATRA revealed a transcriptional activation, specifically, of the Wnt/β-Catenin and Jagged/Notch pathways, which promote self-renewal of the leukaemic cells [1, 45]. However, gene activation seems to be indirect, perhaps due to the aforementioned sequestration of co-repressors and of proteins essential for the normal RA signalling, such as RXR, other RARα factors by the fusion protein complexes [38].

An important example of a gene upregulated by PML/RARα expression is p21, as described by Viale *et al.* [56]. Haematopoietic cells are subjected to DNA damage upon PML/RARα expression that, if too extensive, could potentially become incompatible with the survival of the cell. Hence, the oncogene induces expression of p21, and therefore, a checkpoint, which leads to reversible cell-cycle arrest allowing for DNA repair. In this way, excessive DNA damage, incompatible with survival, and stem cell exhaustion are prevented.

10.2.3.3 Deregulation of other haematopoietic transcription factors

PML/RARα is thought to interact with many proteins forming high-molecular weight complexes [24]. Amongst the various interactors are the transcription factors regulating the process of haematopoiesis, such as GATA-2, several AP-1 factors and PU.1. By binding to these factors, PML/RARα alters their function. For example, PML/RARα chimera cooperates with c-Jun and c-Fos to activate the transcription of both synthetic and natural reporter genes containing an AP-1 site. In the absence of RA, a condition in which RARα has no effect on AP-1 activity, PML/RARα is an inhibitor [19].

The interaction between PML/RARα and transcription factors is not limited to inter-protein binding, but is detected also at the level of DNA binding. A very recent genome-wide study of PML/RARα binding sites revealed that the fusion protein binds to additional regulators of haematopoiesis, including RUNX1, RUNX3 and GFI1. Significantly, these are not classical targets of RA signalling [53].

A curious, protein-protein and protein-DNA interaction has been put forward as an explanation of the binding between PML/RARα and PU.1. The latter is a transcription factor essential to the process of differentiation and lineage commitment. Bioinformatics analysis performed by Wang *et al.* revealed that the majority of the PML-RARα binding sites contained at least one RARE half site in the vicinity of the PU.1 motif. RARE bound PML/RARα binds to PU.1 and gets tethered to DNA at the PU.1 binding sites. The binding of PML/RARα to regions containing both PU.1 and RARE consensus sequences leads to the repression of the PU.1 target genes. The same study identified PML/RARα binding sites proximal to consensus sequences specific for other transcription factors including ETS, and AP-1, both of which had previously been found to interact with PML/RARα at the protein level, suggesting that PML/RARα collaborates with other factors for DNA binding in the manner suggested for PU.1 [58]. Nonetheless, a functional confirmation of the importance of this mechanism is still needed.

Taken together, there is evidence that PML/RARα interacts with a number of transcription factors involved in the regulation of haematopoiesis, thus extending the repertoire of regulated genes to many non RA-regulated sites (see the section on transcriptional activation).

10.2.3.4 Protein misfolding

A fairly novel mechanism of action of PML/RARα (reviewed in [32]) comprises protein misfolding that has conventionally been linked to the pathogenesis of diverse neurodegenerative disorders. However, emerging evidence suggests that protein misfolding may also play an essential role in neoplasms caused by fusion proteins. PML/RARα, a misfolded protein itself, promotes misfolding of the NCoR protein, a co-repressor essential for the function of several tumour-suppressor proteins, by inducing aberrant post-translational modifications. Misfolded NCoR subsequently contributes to differentiation arrest and enhanced survival of APL [32].

10.2.4 Consequences of the Expression of PML/RARα

The classical phenotypes associated with PML/RARα expression include the block of differentiation and the promotion of survival or self-renewal [25]. More recently, a role in the maintenance of leukaemia stem cells has been ascribed to the fusion.

10.2.4.1 Differentiation block

APL blasts resemble promyelocytes, their normal cellular counterparts, which are a distinct stage of the myeloid/granulocytic differentiation process occurring between the myeloblast and the myelocyte thereby suggesting that PML/RARα blocks the normal maturation programme. Indeed, *in vitro* differentiation assays performed on cells ectopically expressing PML/RARα in various model systems reproduced this phenotype [15]. Clearly, all of the mechanisms described above can contribute to a differentiation block.

10.2.4.2 Enhanced self-renewal

As mentioned above, gene expression microarray analysis of APL showed upregulation of genes involved in self-renewal. Murine haematopoietic progenitor cells transduced with PML/RARα show an increase in serial replating efficiency pointing to an enhanced proliferative potential or self-renewal [43, 48]. Evaluation of their proliferative potential by serial replating of methylcellulose colonies revealed the ability of PML/RARα–expressing cells to re-form colonies, in contrast to control cells, which lost the capacity to grow in methylcellulose after a few platings due to terminal differentiation [43].

Recently, Welch *et al.* [61] have generated a new murine model whereby PML/RARα expression is driven by the endogenous *Pml* promoter in the context of *Pml* haploinsufficiency. This system testifies to the role of the chimera in the enhancement of self-renewal without myeloproliferation showing that, at physiological doses, PML/RARα primarily acts by increasing haematopoietic progenitor self-renewal, expanding a cell population that may acquire secondary mutations necessary for the progression to leukaemia.

Furthermore, the expression of PML/RARα confers increased self-renewal to committed progenitors [62].

10.2.4.3 Apoptosis resistance

In agreement with the crucial role of PML in apoptosis, PML/RARα has a pronounced anti-apoptotic activity that depends, at least partially, on its ability to deregulate physiological PML functions [22]. In the presence of PML/RARα, the nuclear architecture is disrupted and the PML protein delocalised from the NBs to micro-speckles with the consequent disappearance of NBs. Their disruption leads to haploinsufficiency of PML and inhibits the normal function of PML as a tumour suppressor gene. NBs host a large number of proteins that are recruited there, where they undergo posttranslational modifications with PML acting as scaffold, keeping the partners together. The most studied posttranslational modifications are those of p53. The net effect of their action is the enhancement of the function of this tumour suppressor gene. PML/RARα, not only disrupts the NBs, interfering with the stabilisation of p53, but also inhibits p53 by inducing posttranslational modifications with an opposite effect on the stability (reviewed in [35]). Specifically, the oncogene associates with p53 and causes its HDACs driven deacetylation and successive degradation [28].

Furthermore, the expression of PML/RARα results in the delocalisation of the aforementioned DAXX from the NBs. Physiologically, the activity of DAXX is regulated by its sequestration in the NBs. However, since both pro- and anti-apoptotic properties have been implicated for this protein, the exact effect of its delocalisation, other than its contribution to the repressive chromatin status, is not clear [35]. A possibility exists that DAXX ultimately modulates the function of another tumour suppressor gene deregulated in the presence of PML/RARα, namely, PTEN, an inhibitor of the PI3K/Akt oncogenic pathway [10, 35].

Interestingly, the anti-apoptotic effect of PML/RARα action is cell-type specific as the non-haematopoietic cell lines tested turned out to be non-permissive for PML/RARα expression due to a direct growth inhibitory effect of the fusion protein and apoptosis. On the contrary, oncogene expression permissive haematopoietic cells display a decreased rate of apoptosis [28].

10.2.5 The Role of Cooperating Mutations

A "two hit" model hypothesises that PML/RARα has leukaemogenic potential, but requires one or more further mutations to trigger overt malignancy. Consequently, the expression of the chimera is not sufficient to transform cells. In all model systems, additional genetic/epigenetic events are required for the completion of the transformation process. The long latencies and incomplete penetrance rates observed in the published murine models of APL testify to this hypothesis [33]. Initially, the fusion protein provokes a so-called "preleukaemic" state, characterised, *in vivo*, by a mild impairment in myeloid differentiation. Following the accumulation of cooperating mutations, the occurrence of which is favoured by PML/RARα expression, the animals succumb to a fatal leukaemia. To date little is known about the nature and identity of these "second hits". Several studies have performed an analysis of leukaemic karyotypes deriving from the different murine models [33]. The frequency and spectrum of the abnormalities differed from one model to the next, suggesting that there is more than one way of evolving from preleukaemic to overt leukaemic state. For example, the gain of chromosome 15 was one of the recurrent abnormalities identified in the hMRP8-PML/RARα mice and in mice co-expressing PML/RARα and BCL2 [36]. Murine chromosome 15 harbours the *Myc* protooncogene leading to the speculation of its involvement in the leukaemogenesis. Indeed, it has been shown that Myc cooperates with PML/RARα to generate a very rapid myeloid leukaemia in 100% of mice [30]. Wartman *et al.*, on the other hand, have sequenced a murine APL genome and identified a somatic non-synonymous mutation in the *Jak1* gene as well as a deletion in the Lysine-specific demethylase 6A (*KDM6A*). Both anomalies seem pertinent to the human pathology, as corresponding mutations have been found in patients' samples [60].

10.3 Mechanisms of Action of ATRA and Arsenic Trioxide

10.3.1 Mechanisms of ATRA Action

In 1985, a 5-year-old girl with APL in Shanghai became the first patient to be treated with ATRA, a natural ligand of the retinoic acid receptors, and obtained a durable complete remission [59]. The presence of PML/RARα in the leukaemic cells is responsible for both the disease pathogenesis and the remarkable sensitivity of the blast cells to the therapy with ATRA. At the level of cellular morphology, ATRA confers a striking change in the phenotype of APL cells that shift from being immature promyelocytes to becoming short-lived granulocytes. The differentiation effect has been described both *in vivo* and *in vitro* [15].

PML/RARα, in the context of being a mutated RARα, requires high (10^{-7}–10^{-6}M) doses of ATRA in order to be transactivated [47, 59]. This pharmacological concentration of ATRA causes a conformational change of the chimera, just as the physiological concentration does to the normal RARα. ATRA binds the fusion protein through the ligand-binding domain of the RARα portion. As a result, the co-repressor complex dissociates from the chimera and co-activators, HATs and HMTs are recruited leading to the de-repression of the RA-signalling target genes through the opening of the chromatin structure. Accordingly, genome-wide epigenetic studies revealed that treatment with pharmacological doses of ATRA induces increase in H3 acetylation at the PML/RARα-RXR binding sites or at nearby target genes, although, little or no changes to the levels of H3K27me3, H3K9me3, or DNA methylation [41].

It was first thought that the profound curative effect originated from ATRA's ability to reverse transcriptional repression of genes responsible for myeloid differentiation. However, ATRA induces a proteasome- and caspase-mediated degradation of the oncogene as an additional effect. A kinetics analysis shows that ATRA causes a biphasic PML/RARα degradation. A rapid decrease in PML/RARα expression is observed already within 1 h, whereas a second degradative step occurs after 12 h and is characterised by the appearance of an 85-kDa PML/RARα cleavage product (ΔPML/RARα). Not surprisingly, the ligand-binding domain required for transcriptional activation in also required for the receptor catabolism, as a way of turning off transcription factors following activation [65]. Interestingly, UBE2D3 ubiquitin-conjugated and UBE1L ubiquitin-activating enzymes are both ATRA-induced [14]. Specifically, proteasome-mediated degradation depends on the binding of SUG1 proteasome subunit to the AF-2 transactivation domain of RARα [59]. Clearly, the degradation of the oncogene leads to the de-repression of the non-RARE sites as well as to the liberation of the proteins sequestered within complexes with PML/RARα. It has to be pointed out that the relative contribution of the two mechanisms, transcriptional re-activation and protein destruction to the therapeutic effect of ATRA remains controversial [34] (Fig. 10.1).

Unfortunately, ATRA does not eliminate leukaemia-initiating cells (LICs) as cells treated with this drug still transplant leukaemia in mouse model systems. While promyelocytes differentiate to mature granulocytes, the LICs persist, albeit at a reduced number [46]. This may explain a relatively high rate of relapse in patients treated with ATRA alone. The combined therapy of ATRA with arsenic trioxide constitutes a more effective cure for APL [47]. Table 10.1 compares the effects exerted by ATRA and arsenic.

10.3.2 Mechanisms of Arsenic Trioxide Action

Arsenic trioxide, a naturally occurring chemical element, found in both organic and inorganic compounds, is a poison-become-medicine used for more than 2000 years [59]. Arsenic affects many cellular pathways through different targets. The interaction of trivalent arsenic with the thiol (-SH) groups of cysteine-rich proteins constitutes the basic reaction that underlies the its biological actions [20]. Arsenic promotes degradation of normal PML, as well as the destruction of the PML/RARα fusion protein through its PML moiety. It has been recently shown that arsenic binds directly to -SH groups of PML and generates reactive oxygen species (ROS). The RBCC domain of the PML protein contains cysteine-rich zinc fingers (ZFs) that in normal conditions bind atoms of zinc. The interaction of the ZFs with arsenic leads to slight changes in their structure ultimately causing oligomerisation of PML proteins. Arsenic binds to the cysteine residues of ZFs within the same molecule or by forming ROS-induced disulphide-links with the RBCC domains in the homodimer, facilitating further oligomerisation of PML and PML/RARα. The resultant multimers form matrix-associated NBs that are subsequently SUMOylated by UBC9 enzyme followed by RNF4 ubiquitin E3 ligase-mediated ubiquitination and proteasomal degradation [29, 63] (Fig. 10.1).

Table 10.1 The comparison of effects of treatment with ATRA and arsenic trioxide

	ATRA	Arsenic trioxide
Effect on blast cells	Differentiation	Partial differentiation and apoptosis
Effect on the PML/ RARα protein	Degradation of PML/ RARα by caspase-driven cleavage and ubiquitin/proteasome system Autophagy	SUMOylation of PML/ RARα that targets it for ubiquitin-mediated degradation Autophagy
Target moiety	Acts on the RARα moiety	Acts on the PML moiety
Effect on LICs	Decreases the number of LICs, but rarely eradicates them	Eradicates LICs, especially when used in combination with ATRA
Effect on NBs	Reformation of NBs	Reformation of NBs
Overall outcome of therapy	Frequent relapses and rarely curative when used on its own	Cases of cure described in literature
Effect on transcriptome	Release of the transcriptional block–potent transcriptional regulation	Less genes regulated, many of the genes regulated by arsenic are also regulated by ATRA and to a greater extend
Effect on proteome	Little effect	Profound change in proteome pattern

Low doses of arsenic ($<0.5 \times 10^{-6}$ M) tend to promote differentiation of APL blasts, whereas high doses (0.5×10^{-6} to 2×10^{-6} M) induce apoptosis. The duration of exposure also plays a role in the outcome of treatment. The exact molecular mechanism by which arsenic causes terminal differentiation remains obscure. It is possible that the process relies on the de-repression of a critical set of differentiation specific genes owed to the rapid degradation of the oncogene. The induction of apoptosis, instead, has been linked to multiple pathways, including the activation of the caspase cascade, the decrease of the mitochondrial membrane potential and the production of ROS. ROS are damaging to DNA, RNA, proteins and lipids and arsenic causes both their production and prevents their detoxification by inhibiting anti-oxidant enzymes. Moreover, arsenic seems to promote apoptosis by downregulating BCL2 expression and to inhibit the survival promoting NFkB pathway [12, 20].

There is no certain explanation for the selective cytotoxicity of arsenic against leukaemic promyelocytes. One possible reason is that drug specificity depends, at least in part, on impairment of osmoregulation leading to an increased drug uptake as compared to non-APL cells [18]. Furthermore, the sensitivity of APL to arsenic has been linked to its poor anti-oxidant capacity [20].

There is evidence that arsenic targets LICs to a greater extent than ATRA. Arsenic-treated cells fail to form colonies in methylcellulose and do not form spleen colonies in the CFU-spleen assay [46, 64]. Table 10.1 compares the effects exerted by ATRA and by arsenic.

10.3.3 Synergy between ATRA and Arsenic Trioxide

Since both drugs cause the catabolism of the PML/RARα fusion protein, therapeutic strategies combining the two agents have been devised. In animal APL models and in the NB4 cell line xenograft model the combination of ATRA and arsenic drastically prolongs the survival of the animals and in some cases, eradicates the disease. Although, ATRA and arsenic induce the degradation of PML/RARα through distinct pathways, a possibility exists that there may be a degree of cross talk between the two pathways through the cyclic AMP and cAMP-dependent protein kinase signalling as arsenic synergises with cAMP in inducing differentiation of ATRA-sensitive and ATRA-resistant cells and ATRA triggers an increase in the intracellular levels of cAMP [13].

The efficacy of this combined treatment was also seen in the clinical setting [59].

10.4 Conclusions

The progress achieved in the treatment of APL, a disease described by its discoverer, Dr Leif K. Hallestad, as "the most malignant form of acute leukaemia", is a success story in the field of haematology. Today, more than 50 years since the original description, over 90% of APL patients are cured, some of them without the use of highly toxic chemotherapeutic agents. The key to this success stems from the observation of a recurrent chromosomal translocation and the subsequent cloning of the PML/RARα leukaemogenic fusion. PML/RARα was soon considered not only a molecular hallmark of the disease, suitable for diagnosis and treatment evaluation, but also the genetic abnormality responsible, although not alone, for the disease development, and moreover, for its unique sensitivity to ATRA and more recently to arsenic trioxide.

Strikingly, a more rapid clearance of PML/RARα transcripts and higher rates of remission and survival in newly diagnosed APL cases are achieved when the two agents are combined, as compared to either monotherapy, making APL a curable disease. Thus, the story of APL can serve as a model for the development of novel approaches towards the treatment of leukaemia and cancer in general. It suggests that molecularly synergistic therapies are potent tools against neoplasms. Twenty-years on from the original cloning of PML/RARα, it is clear that careful studies of disease pathogenesis are fundamental for devising efficient molecular target-based therapies.

References

1. Alcalay, M., Meani, N., Gelmetti, V., Fantozzi, A., Fagioli, M., Orleth, A., Riganelli, D., Sebastiani, C., Cappelli, E., Casciari, C., Sciurpi, M. T., Mariano, A. R., Minardi, S. P., Luzi, L., Muller, H., Di Fiore, P. P., Frosina, G., and Pelicci, P. G. (2003). Acute myeloid leukemia fusion proteins deregulate genes involved in stem cell maintenance and DNA repair, *J Clin Invest*, **112**, pp. 1751–1761.
2. Alcalay, M., Zangrilli, D., Fagioli, M., Pandolfi, P. P., Mencarelli, A., Lo Coco, F., Biondi, A., Grignani, F., and Pelicci, P. G. (1992). Expression pattern of the RAR alpha-PML fusion gene in acute promyelocytic leukemia, *Proc Natl Acad Sci U S A*, **89**, pp. 4840–4844.

3. Alcalay, M., Zangrilli, D., Pandolfi, P. P., Longo, L., Mencarelli, A., Giacomucci, A., Rocchi, M., Biondi, A., Rambaldi, A., Lo Coco, F., *et al.* (1991). Translocation breakpoint of acute promyelocytic leukemia lies within the retinoic acid receptor alpha locus, *Proc Natl Acad Sci U S A*, **88**, pp. 1977–1981.

4. Avvisati, G., Lo Coco, F., and Mandelli, F. (2001). Acute promyelocytic leukemia: clinical and morphologic features and prognostic factors, *Semin Hematol*, **38**, pp. 4–12.

5. Bennett, J. M., Catovsky, D., Daniel, M. T., Flandrin, G., Galton, D. A., Gralnick, H. R., and Sultan, C. (1976). Proposals for the classification of the acute leukaemias. French-American-British (FAB) co-operative group, *Br J Haematol*, **33**, pp. 451–458.

6. Bernard, J., Weil, M., Boiron, M., Jacquillat, C., Flandrin, G., and Gemon, M. F. (1973). Acute promyelocytic leukemia: results of treatment by daunorubicin, *Blood*, **41**, pp. 489–496.

7. Borrow, J., Goddard, A. D., Sheer, D., and Solomon, E. (1990). Molecular analysis of acute promyelocytic leukemia breakpoint cluster region on chromosome 17, *Science*, **249**, pp. 1577–1580.

8. Brown, N. J., Ramalho, M., Pedersen, E. W., Moravcsik, E., Solomon, E., and Grimwade, D. (2009). PML nuclear bodies in the pathogenesis of acute promyelocytic leukemia: active players or innocent bystanders?, *Front Biosci*, **14**, pp. 1684–1707.

9. Carbone, R., Botrugno, O. A., Ronzoni, S., Insinga, A., Di Croce, L., Pelicci, P. G., and Minucci, S. (2006). Recruitment of the histone methyltransferase SUV39H1 and its role in the oncogenic properties of the leukemia-associated PML-retinoic acid receptor fusion protein, *Mol Cell Biol*, **26**, pp. 1288–1296.

10. Carracedo, A., Ito, K., and Pandolfi, P. P. (2011). The nuclear bodies inside out: PML conquers the cytoplasm, *Curr Opin Cell Biol*, **23**, pp. 360–366.

11. Chambon, P. (1996). A decade of molecular biology of retinoic acid receptors, *FASEB J*, **10**, pp. 940–954.

12. Chen, S. J., Zhou, G. B., Zhang, X. W., Mao, J. H., de The, H., and Chen, Z. (2011). From an old remedy to a magic bullet: molecular mechanisms underlying the therapeutic effects of arsenic in fighting leukemia, *Blood*, **117**, pp. 6425–6437.

13. Chen, Z., Zhao, W. L., Shen, Z. X., Li, J. M., Chen, S. J., Zhu, J., Lallemand-Breittenbach, V., Zhou, J., Guillemin, M. C., Vitoux, D., and de The, H. (2007). Arsenic trioxide and acute promyelocytic leukemia: clinical and biological, *Curr Top Microbiol Immunol*, **313**, pp. 129–144.

14. Collins, S. J. (2008). Retinoic acid receptors, hematopoiesis and leukemogenesis, *Curr Opin Hematol*, **15**, pp. 346–351.

15. de The, H., and Chen, Z. (2010). Acute promyelocytic leukaemia: novel insights into the mechanisms of cure, *Nat Rev Cancer*, **10**, pp. 775–783.

16. de The, H., Lavau, C., Marchio, A., Chomienne, C., Degos, L., and Dejean, A. (1991). The PML-RAR alpha fusion mRNA generated by the t(15;17) translocation in acute promyelocytic leukemia encodes a functionally altered RAR, *Cell*, **66**, pp. 675–684.

17. Degos, L. (2003). The history of acute promyelocytic leukaemia, *Br J Haematol*, **122**, pp. 539–553.

18. Dilda, P. J., Perrone, G. G., Philp, A., Lock, R. B., Dawes, I. W., and Hogg, P. J. (2008). Insight into the selectivity of arsenic trioxide for acute promyelocytic leukemia cells by characterizing Saccharomyces cerevisiae deletion strains that are sensitive or resistant to the metalloid, *Int J Biochem Cell Biol*, **40**, pp. 1016–1029.

19. Doucas, V., Brockes, J. P., Yaniv, M., de The, H., and Dejean, A. (1993). The PML-retinoic acid receptor alpha translocation converts the receptor from an inhibitor to a retinoic acid-dependent activator of transcription factor AP-1, *Proc Natl Acad Sci U S A*, **90**, pp. 9345–9349.

20. Emadi, A., and Gore, S. D. (2010). Arsenic trioxide — An old drug rediscovered, *Blood Rev*, **24**, pp. 191–199.

21. Falini, B., Flenghi, L., Fagioli, M., Lo Coco, F., Cordone, I., Diverio, D., Pasqualucci, L., Biondi, A., Riganelli, D., Orleth, A., Liso, A., Martelli, M. F., Pelicci, P. G., and Pileri, S. (1997). Immunocytochemical diagnosis of acute promyelocytic leukemia (M3) with the monoclonal antibody PG-M3 (anti-PML), *Blood*, **90**, pp. 4046–4053.

22. Ferrucci, P. F., Grignani, F., Pearson, M., Fagioli, M., Nicoletti, I., and Pelicci, P. G. (1997). Cell death induction by the acute promyelocytic leukemia-specific PML/RARalpha fusion protein, *Proc Natl Acad Sci U S A*, **94**, pp. 10901–10906.

23. Grignani, F., De Matteis, S., Nervi, C., Tomassoni, L., Gelmetti, V., Cioce, M., Fanelli, M., Ruthardt, M., Ferrara, F. F., Zamir, I., Seiser, C., Lazar, M. A., Minucci, S., and Pelicci, P. G. (1998). Fusion proteins of the retinoic acid receptor-alpha recruit histone deacetylase in promyelocytic leukaemia, *Nature*, **391**, pp. 815–818.

24. Grignani, F., Ferrucci, P. F., Testa, U., Talamo, G., Fagioli, M., Alcalay, M., Mencarelli, A., Peschle, C., Nicoletti, I., *et al.* (1993). The acute promyelocytic leukemia-specific PML-RAR alpha fusion protein

inhibits differentiation and promotes survival of myeloid precursor cells, *Cell*, **74**, pp. 423–431.

25. Grignani, F., Gelmetti, V., Fanelli, M., Rogaia, D., De Matteis, S., Ferrara, F. F., Bonci, D., Nervi, C., and Pelicci, P. G. (1999). Formation of PML/RAR alpha high molecular weight nuclear complexes through the PML coiled-coil region is essential for the PML/RAR alpha-mediated retinoic acid response, *Oncogene*, **18**, pp. 6313–6321.
26. Gudas, L. J., and Wagner, J. A. (2010). Retinoids regulate stem cell differentiation, *J Cell Physiol*, **226**, pp. 322–330.
27. Hillestad, L. K. (1957). Acute promyelocytic leukemia, *Acta Med Scand*, **159**, pp. 189–194.
28. Insinga, A., Pelicci, P. G., and Inucci, S. (2005). Leukemia-associated fusion proteins. Multiple mechanisms of action to drive cell transformation, *Cell Cycle*, **4**, pp. 67–69.
29. Jeanne, M., Lallemand-Breitenbach, V., Ferhi, O., Koken, M., Le Bras, M., Duffort, S., Peres, L., Berthier, C., Soilihi, H., Raught, B., and de The, H. (2010). PML/RARA oxidation and arsenic binding initiate the antileukemia response of As_2O_3, *Cancer Cell*, **18**, pp. 88–98.
30. Jones, L., Wei, G., Sevcikova, S., Phan, V., Jain, S., Shieh, A., Wong, J. C., Li, M., Dubansky, J., Maunakea, M. L., Ochoa, R., Zhu, G., Tennant, T. R., Shannon, K. M., Lowe, S. W., Le Beau, M. M., and Kogan, S. C. (2010). Gain of MYC underlies recurrent trisomy of the MYC chromosome in acute promyelocytic leukemia, *J Exp Med*, **207**, pp. 2581–2594.
31. Kamashev, D., Vitoux, D., and De The, H. (2004). PML-RARA-RXR oligomers mediate retinoid and rexinoid/cAMP cross-talk in acute promyelocytic leukemia cell differentiation, *J Exp Med*, **199**, pp. 1163–1174.
32. Khan, M. (2010). Interplay of protein misfolding pathway and unfolded-protein response in acute promyelocytic leukemia, *Expert Rev Proteomics*, **7**, pp. 591–600.
33. Kogan, S. C. (2007). Mouse models of acute promyelocytic leukemia, *Curr Top Microbiol Immunol*, **313**, pp. 3–29.
34. Kogan, S. C. (2009). Curing APL: differentiation or destruction?, *Cancer Cell*, **15**, pp. 7–8.
35. Lallemand-Breitenbach, V., and de The, H. (2010). PML nuclear bodies, *Cold Spring Harb Perspect Biol*, **2**, pp. a000661.
36. Le Beau, M. M., Bitts, S., Davis, E. M., and Kogan, S. C. (2002). Recurring chromosomal abnormalities in leukemia in PML-RARA transgenic

mice parallel human acute promyelocytic leukemia, *Blood*, **99**, pp. 2985–2991.

37. Lengfelder, E., Saussele, S., Weisser, A., Buchner, T., and Hehlmann, R. (2005). Treatment concepts of acute promyelocytic leukemia, *Crit Rev Oncol Hematol*, **56**, pp. 261–274.
38. Licht, J. D. (2006). Reconstructing a disease: What essential features of the retinoic acid receptor fusion oncoproteins generate acute promyelocytic leukemia?, *Cancer Cell*, **9**, pp. 73–74.
39. Lin, R. J., and Evans, R. M. (2000). Acquisition of oncogenic potential by RAR chimeras in acute promyelocytic leukemia through formation of homodimers, *Mol Cell*, **5**, pp. 821–830.
40. Lo-Coco, F., Ammatuna, E., Montesinos, P., and Sanz, M. A. (2008). Acute promyelocytic leukemia: recent advances in diagnosis and management, *Semin Oncol*, **35**, pp. 401–409.
41. Martens, J. H., Brinkman, A. B., Simmer, F., Francoijs, K. J., Nebbioso, A., Ferrara, F., Altucci, L., and Stunnenberg, H. G. (2010). PML-RARalpha/RXR Alters the Epigenetic Landscape in Acute Promyelocytic Leukemia, *Cancer Cell*, **17**, pp. 173–185.
42. Melnick, A., and Licht, J. D. (1999). Deconstructing a disease: RARalpha, its fusion partners, and their roles in the pathogenesis of acute promyelocytic leukemia, *Blood*, **93**, pp. 3167–3215.
43. Minucci, S., Monestiroli, S., Giavara, S., Ronzoni, S., Marchesi, F., Insinga, A., Diverio, D., Gasparini, P., Capillo, M., Colombo, E., Matteucci, C., Contegno, F., Lo-Coco, F., Scanziani, E., Gobbi, A., and Pelicci, P. G. (2002). PML-RAR induces promyelocytic leukemias with high efficiency following retroviral gene transfer into purified murine hematopoietic progenitors, *Blood*, **100**, pp. 2989–2995.
44. Minucci, S., and Pelicci, P. G. (2007). Determinants of oncogenic transformation in acute promyelocytic leukemia: the hetero-union makes the force, *Cancer Cell*, **12**, pp. 1–3.
45. Muller-Tidow, C., Steffen, B., Cauvet, T., Tickenbrock, L., Ji, P., Diederichs, S., Sargin, B., Kohler, G., Stelljes, M., Puccetti, E., Ruthardt, M., deVos, S., Hiebert, S. W., Koeffler, H. P., Berdel, W. E., and Serve, H. (2004). Translocation products in acute myeloid leukemia activate the Wnt signaling pathway in hematopoietic cells, *Mol Cell Biol*, **24**, pp. 2890–2904.
46. Nasr, R., and de The, H. (2010). Eradication of acute promyelocytic leukemia-initiating cells by PML/RARA-targeting, *Int J Hematol*, **91**, pp. 742–747.

47. Nasr, R., Lallemand-Breitenbach, V., Zhu, J., Guillemin, M. C., and de The, H. (2009). Therapy-induced PML/RARA proteolysis and acute promyelocytic leukemia cure, *Clin Cancer Res*, **15**, pp. 6321–6326.

48. Occhionorelli, M., Santoro, F., Pallavicini, I., Gruszka, A., Moretti, S., Bossi, D., Viale, A., Shing, D., Ronzoni, S., Muradore, I., Soncini, M., Pruneri, G., Rafaniello, P., Viale, G., Pelicci, P. G., and Minucci, S. (2011). The self-association coiled-coil domain of PML is sufficient for the oncogenic conversion of the retinoic acid receptor (RAR) alpha, *Leukemia*, **25**, pp. 814–820.

49. Reiter, A., Lengfelder, E., and Grimwade, D. (2004). Pathogenesis, diagnosis and monitoring of residual disease in acute promyelocytic leukaemia, *Acta Haematol*, **112**, pp. 55–67.

50. Rochette-Egly, C., and Germain, P. (2009). Dynamic and combinatorial control of gene expression by nuclear retinoic acid receptors (RARs), *Nucl Recept Signal*, **7**, pp. e005.

51. Rogers, J. E., and Yang, D. (2011). Differentiation syndrome in patients with acute promyelocytic leukemia, *J Oncol Pharm Pract*, **18**, pp. 109-114

52. Rowley, J. D., Golomb, H. M., and Dougherty, C. (1977). 15/17 translocation, a consistent chromosomal change in acute promyelocytic leukaemia, *Lancet*, **1**, pp. 549–550.

53. Saeed, S., Logie, C., Stunnenberg, H. G., and Martens, J. H. (2011). Genome-wide functions of PML-RARalpha in acute promyelocytic leukaemia, *Br J Cancer*, **104**, pp. 554–558.

54. Sanz, M. A., and Lo-Coco, F. (2011). Modern approaches to treating acute promyelocytic leukemia, *J Clin Oncol*, **29**, pp. 495–503.

55. Scaglioni, P. P., and Pandolfi, P. P. (2007). The theory of APL revisited, *Curr Top Microbiol Immunol*, **313**, pp. 85–100.

56. Viale, A., De Franco, F., Orleth, A., Cambiaghi, V., Giuliani, V., Bossi, D., Ronchini, C., Ronzoni, S., Muradore, I., Monestiroli, S., Gobbi, A., Alcalay, M., Minucci, S., and Pelicci, P. G. (2009). Cell-cycle restriction limits DNA damage and maintains self-renewal of leukaemia stem cells, *Nature*, **457**, pp. 51–56.

57. Villa, R., Pasini, D., Gutierrez, A., Morey, L., Occhionorelli, M., Vire, E., Nomdedeu, J. F., Jenuwein, T., Pelicci, P. G., Minucci, S., Fuks, F., Helin, K., and Di Croce, L. (2007). Role of the polycomb repressive complex 2 in acute promyelocytic leukemia, *Cancer Cell*, **11**, pp. 513–525.

58. Wang, K., Wang, P., Shi, J., Zhu, X., He, M., Jia, X., Yang, X., Qiu, F., Jin, W., Qian, M., Fang, H., Mi, J., Xiao, H., Minden, M., Du, Y., Chen, Z., and

Zhang, J. (2010). PML/RARalpha targets promoter regions containing PU.1 consensus and RARE half sites in acute promyelocytic leukemia, *Cancer Cell*, **17**, pp. 186–197.

59. Wang, Z. Y., and Chen, Z. (2008). Acute promyelocytic leukemia: from highly fatal to highly curable, *Blood*, **111**, pp. 2505–2515.

60. Wartman, L. D., Larson, D. E., Xiang, Z., Ding, L., Chen, K., Lin, L., Cahan, P., Klco, J. M., Welch, J. S., Li, C., Payton, J. E., Uy, G. L., Varghese, N., Ries, R. E., Hoock, M., Koboldt, D. C., McLellan, M. D., Schmidt, H., Fulton, R. S., Abbott, R. M., Cook, L., McGrath, S. D., Fan, X., Dukes, A. F., Vickery, T., Kalicki, J., Lamprecht, T. L., Graubert, T. A., Tomasson, M. H., Mardis, E. R., Wilson, R. K., and Ley, T. J. (2011). Sequencing a mouse acute promyelocytic leukemia genome reveals genetic events relevant for disease progression, *J Clin Invest*, **121**, pp. 1445–1455.

61. Welch, J. S., Yuan, W., and Ley, T. J. (2011). PML-RARA can increase hematopoietic self-renewal without causing a myeloproliferative disease in mice, *J Clin Invest*, **121**, pp. 1636–1645.

62. Wojiski, S., Guibal, F. C., Kindler, T., Lee, B. H., Jesneck, J. L., Fabian, A., Tenen, D. G., and Gilliland, D. G. (2009). PML-RARalpha initiates leukemia by conferring properties of self-renewal to committed promyelocytic progenitors, *Leukemia*, **23**, pp. 1462–1471.

63. Zhang, X. W., Yan, X. J., Zhou, Z. R., Yang, F. F., Wu, Z. Y., Sun, H. B., Liang, W. X., Song, A. X., Lallemand-Breitenbach, V., Jeanne, M., Zhang, Q. Y., Yang, H. Y., Huang, Q. H., Zhou, G. B., Tong, J. H., Zhang, Y., Wu, J. H., Hu, H. Y., de The, H., Chen, S. J., and Chen, Z. (2010). Arsenic trioxide controls the fate of the PML-RARalpha oncoprotein by directly binding PML, *Science*, **328**, pp. 240–243.

64. Zheng, X., Seshire, A., Ruster, B., Bug, G., Beissert, T., Puccetti, E., Hoelzer, D., Henschler, R., and Ruthardt, M. (2007). Arsenic but not all-trans retinoic acid overcomes the aberrant stem cell capacity of PML/RARalpha-positive leukemic stem cells, *Haematologica*, **92**, pp. 323–331.

65. Zhu, J., Gianni, M., Kopf, E., Honore, N., Chelbi-Alix, M., Koken, M., Quignon, F., Rochette-Egly, C., and de The, H. (1999). Retinoic acid induces proteasome-dependent degradation of retinoic acid receptor alpha (RARalpha) and oncogenic RARalpha fusion proteins, *Proc Natl Acad Sci U S A*, **96**, pp. 14807–14812.

Chapter 11

Dihydropyrimidine Dehydrogenase Deficiency and 5-Fluorouracil Toxicity

Eva Gross[a] **and André B. P. van Kuilenburg**[b]
[a]*Technische Universität München, Clinic for Gynecology and Obstetrics, München, Germany*
[b]*Academic Medical Center, University of Amsterdam, Emma Children's Hospital, Department of Clinical Chemistry, Amsterdam, The Netherlands*
a.b.vankuilenburg@amc.uva.nl

11.1 Introduction

After five decades since its approval, the anticancer drug 5-fluorouracil (5FU) still remains one of the most prescribed drugs in the chemotherapy of common cancers such as breast, colorectal, esophagogastric and head and neck cancers. Today, combination therapies of 5FU with for example anthracyclines or platinum drugs are predominantly being used, leading to improved therapeutic response rates. In addition, increased usage of oral prodrugs of 5-FU such as capecitabine is expected. However, therapeutic success is often limited by acute drug-adverse events, observed in up to

Handbook of Therapeutic Biomarkers in Cancer
Edited by Sherry X. Yang and Janet E. Dancey
Copyright © 2013 Pan Stanford Publishing Pte. Ltd.
ISBN 978-981-4364-65-2 (Hardcover), 978-981-4364-66-9 (eBook)
www.panstanford.com

15–20% of the patients treated with 5FU. Frequently, these severe toxicities result in interruption of the chemotherapy and imply the risk of disease progression.

Drug disposition and the efficacy/toxicity balance of 5FU are highly influenced by the action of dihydropyrimidine dehydrogenase (DPD), which constitutes the initial and rate-limiting step in the degradation of pyrimidines and fluoropyrimidines [10]. As such, high intratumoral DPD concentrations have been associated with low efficacy of a 5FU therapy. Furthermore, low tumoral DPD levels together with low thymidine phosphorylase (TP) and thymidylate synthase (TS) expression have been associated with a favorable treatment response [12,37]. Since more than 80% of administered 5FU is rapidly degraded by liver DPD, a general deficiency of this enzyme results in markedly prolonged exposure of the drug [10,28,45]. The main toxic reactions caused by reduced clearance of 5FU are myelosuppression, mucositis, diarrhea and neuropathy including also life-threatening events. Recent studies investigating the DPD status suggest that reduced DPD function may account for around 50% of the toxicity cases observed under 5FU therapy [8,49]. Besides DPD, variations in downstream acting enzymes of the pyrimidine catabolism such as dihydropyrimidase (DHP) and beta-ureidopropionase (BUP-1) may also contribute to some extent to an altered capacity to catabolize 5FU [41,42,55].

Regarding the high number of patients receiving 5FU-based therapies per year and the deleterious effects that are exerted by severe toxicities on their quality of life and disease cure, it is of major clinical interest to reduce the incidence of 5FU-related adverse events. To establish strategies for identifying high-risk patients, pharmacogenetic studies have mainly investigated the role of the gene encoding DPD (*DPYD*) and its mutational inactivation.

11.2 Variability of the *DPYD* Gene

DPD deficiency has been described as an autosomal recessive disorder with about 3% carriers in the Caucasian population and 0.02% homozygous deficient patients [13]. A higher incidence of DPD deficiency has been observed in African-Americans [30]. The human *DPYD* gene has been mapped to chromosome 1p21.3 and contains 23 exons. It encompasses at least 950 kb in length with 3 kb of coding sequence [57,60]). To date, more than 50 different sequence variants have been identified within the *DPYD* coding regions and exon/intron borders. Among these, the splice site mutation IVS14+1G>A (*DPYD*2A)* affecting the highly conserved splice-donor consensus sequence of the exon 14/intron 14 boundary was found to be the most prevalent mutation in DPD deficient patients. Numerous studies described the deleterious effect of this aberration, which causes exon 14 skipping, on the DPD protein and its activity [36,53,57]. Moreover, a high prevalence (50%) of the *DPYD*2A* allele has been reported in patients showing grade 4 neutropenia after 5FU administration [49]. Its prominent role in the occurrence of severe toxicities is also reflected by a relatively high frequency of 1.8–3.5% in the normal European population [35,50]. Nevertheless, inconsistent results about the impact of IVS14+1G>A on drug-related adverse events are present in the literature implicating also IVS14+1G>A-carriers with 5FU tolerance [9,27,33,39]. This phenomenon most likely reflects the fact that some heterozygous carriers of the IVS14+1G>A mutation possessed low normal DPD activity [49,54].

Besides direct sequencing, high-throughput technologies such as DHPLC (14,40) and pyrosequencing methods [33] have been established today to enable rapid screening of the entire *DPYD* coding region. It turned out that deleterious mutations are spread all over the coding sequence of the gene. In addition to IVS14+1G>A, a further splice site mutation, IVS11+1G>T [48], and various frameshift mutations, i.e., c.295–298del TCAT [52], c.812delT [58], c.1039–1042delTG [44], c.1109delTA [19] and c.1897delC [56], were reported. While most of these truncating disease-causing mutations are rarely found, the majority of sequence aberrations observed in *DPYD* are missense mutations, many of them of unknown relevance. Predicting the pathogenicity of non-synonymous mutations is therefore a challenge and has been facilitated by the determination of the crystal structure of pig DPD, homology modeling tools and site-directed mutagenesis strategies [11,20,25,44]. In this way, evidence for a deleterious role of the missense mutations c.1679T>G (I560S) and c.2846A>T (D949V) could be obtained by analysis of the three-dimensional protein indicating that I560S is likely to destabilize the protein structure and D949V directly interferes with cofactor binding and electron transport [44].

Morel *et al.* (2006) proposed testing of the three most common deleterious mutations mentioned above (IVS14+1G>A, c.2846A>T, and c.1679T>G) to predict 5FU related side effects in clinical practice. In this study of 487 patients, they could identify 27% of the high-risk patients with a positive predictive value of 0.62 and a negative predictive value of 0.98. In line with these results, a British study detected 19% of the high-grade toxicities associated with this same set of sequence aberrations [24]. Genetic association studies and haplotype-based strategies were introduced as a further approach to establish distinct *DPYD* variants as putative predictive markers for severe 5FU-toxicity [19,23,26]. In a recent study [19], an attributable risk of 56.9% was found for a haplotype containing the non-synonymous polymorphism c.496A>G (M166V). Amstutz *et al.* identified another risk-haplotype predominantly associated with severe toxicities that contained the silent mutation c.1236G>A [2].

The observation that a *DPYD* haplotype, not containing any non-synonymous or splice-site mutations was associated with 5FU toxicity, suggested the presence of additional genetic variations in the noncoding region of *DPYD*. Indeed, an interesting finding with respect to this recently reported haplotype [2] was a deep intronic mutation, c.1129–5923C>G, which affects DPD pre-mRNA splicing [47]. The variation localized in intron 10 created a cryptic splice donor site and as a consequence, a 44 bp fragment of intron 10 was inserted in the mature DPD mRNA thereby disrupting the reading frame of the mRNA sequence. Present data suggest that the ratio of alternatively spliced transcripts to wild-type transcripts is about 140% in homozygous and 30% in heterozygous carriers, respectively. Increased susceptibility for 5FU toxicity has been confirmed in a retrospective analysis of 203 cancer patients showing that the c.1129–5923C>G mutation was significantly enriched in patients with severe side effects. The apparently high prevalence of the intronic variant in the normal Dutch (2.6%) and German (3.3%) population speaks in favor of including this mutation in a set of predictive markers for 5FU toxicity.

Previously, it has been suggested that *DPYD* might be prone to high recombination rates due to the common fragile site FRA1E, which extends over 370 kb within the *DPYD* locus [21]. Common fragile sites represent chromosome structures that are particularly prone to breakage under replication stress and the genomic instability can give rise to deletions, translocations and amplifications. Following the detection of such *DPYD* rearrangements in pediatric patients with a profound DPD deficiency [46], evidence for a large genomic deletion within *DPYD* was further obtained in a breast cancer patient who presented with grade 4 5FU toxicity and reduced DPD activity [47]). The deleted region, encompassing exons 21–23, was located in proximity to the common fragile site FRA1E. The patient suffered from severe leukopenia and febrile neutropenia, which led to the interruption of the chemotherapy and in the course to the development of brain metastases (unpublished results). So far, this was the only reported case of a large genomic *DPYD* rearrangement associated with drug-adverse events. The fact that several studies did not observe large intragenic aberrations within *DPYD* in patients suffering from high-grade 5FU toxicity ([34,38,43] suggests a minor role of this mechanism in the susceptibility for 5FU toxicity. However, it should be considered in severe cases of drug intolerance when conventional PCR-based methods do not reveal any mutations.

11.3 Epigenetic and Non-Genetic Effects on *DPYD* Dysregulation

While *DPYD* mutations have been unambiguously shown to provoke potentially life-threatening toxicities after 5FU administration, they do not sufficiently explain the total number of adverse events associated with reduced DPD function. Currently, there is a debate whether epigenetic mechanisms may play a role in DPD deficiency [1,15,64]. However, most of the studies dealing with this issue did not provide evidence for DNA hypermethylation within the *DPYD* promoter as a cause of 5FU-related side effects (1,38). Scarce knowledge is available for a possible role of alterations at the transcriptional level of the *DPYD* gene. In this context, Zhang *et al.* [63] identified the Sp1 and Sp3 transcription factors as activators of the *DPYD* promoter. An interesting finding in hepatocellular carcinoma (which might be exploited therapeutically) was related to a novel gene, astrocyte elevated gene-1 (AEG-1), which was shown to enhance *DPYD* expression. Knockdown of AEG-1 led to increased sensitivity towards 5FU *in vitro* [61].

Additionally, drug to drug interactions and circadian variations are non-genetic mechanisms, which might contribute to the large interpatient variability observed in DPD activity [31,62]. Altogether, these effects could lead to an underestimation of DPD deficient patients if screening for DPD deficiency is only based on the genotype [7].

11.4 Functional Testing of the DPD Status

Due to the limits of genotype-based strategies to assess high-risk patients for 5FU toxicity, there is a growing interest in technologies for analyzing the functionality of the DPD protein. Initially, direct measurement of the DPD enzyme activity has been established in peripheral blood mononuclear cells (PBMC) as a surrogate tissue for the liver enzyme. While this technique usually involves time-consuming separation of radiolabeled metabolites using reversed-phase HPLC [22,51], several methodologies have been introduced today, which might be better suited for clinical routines. The determination of the dihydrouracil/uracil (UH_2/U) ratio in plasma has been proposed as a surrogate marker for the DPD activity [3,8,31]. This type of analysis is simple and inexpensive with acceptable run times to allow pretherapeutically analysis of the DPD status in patients scheduled for a 5FU-based therapy. Recent clinical studies [6,8,59] suggested a clinical benefit for DPD deficient patients when the DPD phenotypic status is determined prior to treatment and subsequent dose-tailoring of 5FU is achieved. Marked reduction in the occurrence of grade 4 toxicities including also lethal outcome was observed [8]. However, clear threshold levels for patients at risk are not available yet for the UH_2/U test.

Since there is a close relationship between 5FU plasma concentrations and high-grade toxicities [17,18,32], pharmacokinetically guided, FU dose adjustment has been previously introduced as another potent tool to avoid severe drug adverse events. Gamelin *et al.*, 2008, demonstrated a significant decrease of grade 3/4 toxicities and improved response rates in colorectal cancer patients, when individual dose adjustment based on a weekly single point measurement of 5FU plasma levels at steady-state was performed. This method has the advantage of directly controlling drug exposure during treatment, thereby ensuring that optimal 5FU area under the curve (AUC) values for treatment efficacy are reached. Further approaches to indirectly measure the functional status of DPD—including also alterations in downstream acting enzymes of the pyrimidine catabolism such as dihydropyrimidase and beta-ureidopropionase—are the uracil-breath test method [29] and the 5FU test dose method [4,5,45]. The latter uses a reduced 5FU test injection before starting the therapy to determine alterations of 5FU pharmacokinetics.

11.5 Conclusion

Regarding severe drug-adverse effects, the function of the 5-FU-degrading enzyme DPD is of clinical relevance for a large number of cancer patients. A wide range of methods has been established to assess the genetic and functional status of DPD. As specific sequence variations in the *DPYD* gene have been clearly associated with impaired breakdown of 5FU followed by severe toxicities [45], genotyping of *DPYD* remains an important tool to identify DPD deficient patients. However, its suitability for routine pretreatment testing is controversially discussed, not least due to conflicting genotype-phenotype relations in mutation carriers and relatively low positive predictive values [33]. For that reason, genotyping of *DPYD* is not yet recommended by health organizations. In addition to genetic screening, a number of phenotype-based methods have now been introduced, which appear to be well suited for clinical laboratories and which might be an attractive option for (pretreatment) monitoring of the DPD status. As DPD-based dose tailoring gave promising results towards improved clinical outcome of 5FU-treated patients [5,18,59] systematic testing for DPD deficiency may come more and more into the focus as part of the clinical management of cancer patients scheduled for 5FU.

References

1. Amstutz, U., Farese, S., Aebi, S., and Largiader, C.R. (2008): Hypermethylation of the DPYD promoter region is not a major predictor of severe toxicity in 5-fluorouracil based chemotherapy. *J. Exp. Clin. Cancer Res.*, 27: 54.
2. Amstutz, U., Farese, S., Aebi, S., and Largiader, C.R. (2009): Dihydropyrimidine dehydrogenase gene variation and severe 5-fluorouracil toxicity: a haplotype assessment. *Pharmacogenomics.*, 10: 931–944.

3. Ben, F.R., Gross, E., Ben, A.S., Hassine, H., and Saguem, S. (2009): The dihydrouracil/uracil ratio in plasma, clinical and genetic analysis for screening of dihydropyrimidine dehydrogenase deficiency in colorectal cancer patients treated with 5-fluorouracil. *Pathol. Biol.(Paris)*, 57: 470–476.
4. Bocci, G., Barbara, C., Vannozzi, F., Di, P.A., Melosi, A., Barsanti, G., Allegrini, G., Falcone, A., Del, T.M., and Danesi, R. (2006): A pharmacokinetic-based test to prevent severe 5-fluorouracil toxicity. *Clin. Pharmacol. Ther.*, 80: 384–395.
5. Bocci, G., Di, P.A., Barbara, C., Masi, G., Fornaro, L., Loupakis, F., Allegrini, G., Falcone, A., Del, T.M., and Danesi, R. (2009): Pharmacokinetics, a main actor in a many-sided approach to severe 5-FU toxicity prediction. *Br. J. Clin. Pharmacol.*, 67: 132–134.
6. Capitain, O., Boisdron-Celle, M., Poirier, A.L., Abadie-Lacourtoisie, S., Morel, A., and Gamelin, E. (2008): The influence of fluorouracil outcome parameters on tolerance and efficacy in patients with advanced colorectal cancer. *Pharmacogenomics. J.*, 8: 256–267.
7. Ciccolini, J., Gross, E., Dahan, L., Lacarelle, B., and Mercier, C. (2010): Routine dihydropyrimidine dehydrogenase testing for anticipating 5-fluorouracil-related severe toxicities: hype or hope? *Clin. Colorectal Cancer*, 9: 224–228.
8. Ciccolini, J., Mercier, C., Evrard, A., Dahan, L., Boyer, J.C., Duffaud, F., Richard, K., Blanquicett, C., Milano, G., Blesius, A., Durand, A., Seitz, J.F., Favre, R., and Lacarelle, B. (2006): A rapid and inexpensive method for anticipating severe toxicity to fluorouracil and fluorouracil-based chemotherapy. *Ther. Drug Monit.*, 28: 678–685.
9. Collie-Duguid, E.S., Etienne, M.C., Milano, G., and McLeod, H.L. (2000): Known variant DPYD alleles do not explain DPD deficiency in cancer patients. *Pharmacogenetics*, 10: 217–223.
10. Diasio, R.B., Beavers, T.L., and Carpenter, J.T. (1988): Familial deficiency of dihydropyrimidine dehydrogenase. Biochemical basis for familial pyrimidinemia and severe 5-fluorouracil-induced toxicity. *J. Clin. Invest*, 81: 47–51.
11. Dobritzsch, D., Schneider, G., Schnackerz, K.D., and Lindqvist, Y. (2001): Crystal structure of dihydropyrimidine dehydrogenase, a major determinant of the pharmacokinetics of the anti-cancer drug 5-fluorouracil. *EMBO J.*, 20: 650–660.
12. Etienne, M.C., Cheradame, S., Fischel, J.L., Formento, P., Dassonville, O., Renee, N., Schneider, M., Thyss, A., Demard, F., and Milano, G. (1995): Response to fluorouracil therapy in cancer patients: the role of tumoral

dihydropyrimidine dehydrogenase activity. *J. Clin. Oncol.*, 13: 1663–1670.

13. Etienne, M.C., Lagrange, J.L., Dassonville, O., Fleming, R., Thyss, A., Renee, N., Schneider, M., Demard, F., and Milano, G. (1994): Population study of dihydropyrimidine dehydrogenase in cancer patients. *J. Clin. Oncol.*, 12: 2248–2253.
14. Ezzeldin, H., Johnson, M.R., Okamoto, Y., and Diasio, R. (2003): Denaturing high performance liquid chromatography analysis of the DPYD gene in patients with lethal 5-fluorouracil toxicity. *Clin. Cancer Res.*, 9: 3021–3028.
15. Ezzeldin, H.H., Lee, A.M., Mattison, L.K., and Diasio, R.B. (2005): Methylation of the DPYD promoter: an alternative mechanism for dihydropyrimidine dehydrogenase deficiency in cancer patients. *Clin. Cancer Res.*, 11: 8699–8705.
16. Fischer, J., Schwab, M., Eichelbaum, M., and Zanger, U.M. (2003): Mutational analysis of the human dihydropyrimidine dehydrogenase gene by denaturing high-performance liquid chromatography. *Genet. Test.*, 7: 97–105.
17. Gamelin, E., Boisdron-Celle, M., Delva, R., Regimbeau, C., Cailleux, P.E., Alleaume, C., Maillet, M.L., Goudier, M.J., Sire, M., Person-Joly, M.C., Maigre, M., Maillart, P., Fety, R., Burtin, P., Lortholary, A., Dumesnil, Y., Picon, L., Geslin, J., Gesta, P., Danquechin-Dorval, E., Larra, F., and Robert, J. (1998): Long-term weekly treatment of colorectal metastatic cancer with fluorouracil and leucovorin: results of a multicentric prospective trial of fluorouracil dosage optimization by pharmacokinetic monitoring in 152 patients. *J. Clin. Oncol.*, 16: 1470–1478.
18. Gamelin, E., Delva, R., Jacob, J., Merrouche, Y., Raoul, J.L., Pezet, D., Dorval, E., Piot, G., Morel, A., and Boisdron-Celle, M. (2008): Individual fluorouracil dose adjustment based on pharmacokinetic follow-up compared with conventional dosage: results of a multicenter randomized trial of patients with metastatic colorectal cancer. *J. Clin. Oncol.*, 26: 2099–2105.
19. Gross, E., Busse, B., Riemenschneider, M., Neubauer, S., Seck, K., Klein, H.G., Kiechle, M., Lordick, F., and Meindl, A. (2008): Strong association of a common dihydropyrimidine dehydrogenase gene polymorphism with fluoropyrimidine-related toxicity in cancer patients. *PLoS. One.*, 3: e4003.
20. Gross, E., Ullrich, T., Seck, K., Mueller, V., de, W.M., von, S.C., Meindl, A., Schmitt, M., and Kiechle, M. (2003): Detailed analysis of five mutations

in dihydropyrimidine dehydrogenase detected in cancer patients with 5-fluorouracil-related side effects. *Hum. Mutat.*, 22: 498.

21. Hormozian, F., Schmitt, J.G., Sagulenko, E., Schwab, M., and Savelyeva, L. (2007): FRA1E common fragile site breaks map within a 370kilobase pair region and disrupt the dihydropyrimidine dehydrogenase gene (DPYD). *Cancer Lett.*, 246: 82–91.

22. Johnson, M.R., Yan, J., Shao, L., Albin, N., and Diasio, R.B. (1997): Semi-automated radioassay for determination of dihydropyrimidine dehydrogenase (DPD) activity. Screening cancer patients for DPD deficiency, a condition associated with 5-fluorouracil toxicity. *J. Chromatogr. B Biomed. Sci. Appl.*, 696: 183–191.

23. Kleibl, Z., Fidlerova, J., Kleiblova, P., Kormunda, S., Bilek, M., Bouskova, K., Sevcik, J., and Novotny, J. (2009): Influence of dihydropyrimidine dehydrogenase gene (DPYD) coding sequence variants on the development of fluoropyrimidine-related toxicity in patients with high-grade toxicity and patients with excellent tolerance of fluoropyrimidine-based chemotherapy. *Neoplasma*, 56: 303–316.

24. Loganayagam, A., Arenas-Hernandez, M., Fairbanks, L., Ross, P., Sanderson, J.D., and Marinaki, A.M. (2010): The contribution of deleterious DPYD gene sequence variants to fluoropyrimidine toxicity in British cancer patients. *Cancer Chemother. Pharmacol.*, 65: 403–406.

25. Lohkamp, B., Voevodskaya, N., Lindqvist, Y., and Dobritzsch, D. (2010): Insights into the mechanism of dihydropyrimidine dehydrogenase from site-directed mutagenesis targeting the active site loop and redox cofactor coordination. *Biochim. Biophys. Acta*, 1804: 2198–2206.

26. Maekawa, K., Saeki, M., Saito, Y., Ozawa, S., Kurose, K., Kaniwa, N., Kawamoto, M., Kamatani, N., Kato, K., Hamaguchi, T., Yamada, Y., Shirao, K., Shimada, Y., Muto, M., Doi, T., Ohtsu, A., Yoshida, T., Matsumura, Y., Saijo, N., and Sawada, J. (2007): Genetic variations and haplotype structures of the DPYD gene encoding dihydropyrimidine dehydrogenase in Japanese and their ethnic differences. *J. Hum. Genet.*, 52: 804–819.

27. Magne, N., Etienne-Grimaldi, M.C., Cals, L., Renee, N., Formento, J.L., Francoual, M., and Milano, G. (2007): Dihydropyrimidine dehydrogenase activity and the IVS14+1G>A mutation in patients developing 5FU-related toxicity. *Br. J. Clin. Pharmacol.*, 64: 237–240.

28. Maring, J.G., van Kuilenburg, A.B.P., Haasjes, J., Piersma, H., Groen, H.J.M., Uges, D.R.A., van Gennip, A.H., and de Vries, E.G.E. (2002): Reduced 5-FU

clearance in a patient with low DPD activity due to heterozygosity for a mutant allele of the DPYD gene. *Br. J. Cancer*, 86: 1028–1033.

29. Mattison, L.K., Ezzeldin, H., Carpenter, M., Modak, A., Johnson, M.R., and Diasio, R.B. (2004): Rapid identification of dihydropyrimidine dehydrogenase deficiency by using a novel 2-13C-uracil breath test. *Clin. Cancer Res.*, 10: 2652–2658.

30. Mattison, L.K., Fourie, J., Desmond, R.A., Modak, A., Saif, M.W., and Diasio, R.B. (2006): Increased prevalence of dihydropyrimidine dehydrogenase deficiency in African-Americans compared with Caucasians. *Clin. Cancer Res.*, 12: 5491–5495.

31. Mercier, C., and Ciccolini, J. (2006): Profiling dihydropyrimidine dehydrogenase deficiency in patients with cancer undergoing 5-fluorouracil/capecitabine therapy. *Clin. Colorectal Cancer*, 6: 288–296.

32. Milano, G., Roman, P., Khater, R., Frenay, M., Renee, N., and Namer, M. (1988): Dose versus pharmacokinetics for predicting tolerance to 5-day continuous infusion of 5-FU. *Int. J. Cancer*, 41: 537–541.

33. Morel, A., Boisdron-Celle, M., Fey, L., Soulie, P., Craipeau, M.C., Traore, S., and Gamelin, E. (2006): Clinical relevance of different dihydropyrimidine dehydrogenase gene single nucleotide polymorphisms on 5-fluorouracil tolerance. *Mol. Cancer Ther.*, 5: 2895–2904.

34. Pare, L., Paez, D., Salazar, J., Del, R.E., Tizzano, E., Marcuello, E., and Baiget, M. (2010): Absence of large intragenic rearrangements in the DPYD gene in a large cohort of colorectal cancer patients treated with 5-FU-based chemotherapy. *Br. J. Clin. Pharmacol.*, 70: 268–272.

35. Raida, M., Schwabe, W., Hausler, P., van Kuilenburg, A.B.P., van Gennip, A.H., Behnke, D., and Hoffken, K. (2001): Prevalence of a common point mutation in the dihydropyrimidine dehydrogenase (DPD) gene within the 5'-splice donor site of intron 14 in patients with severe 5-fluorouracil (5-FU)- related toxicity compared with controls. *Clin. Cancer Res.*, 7: 2832–2839.

36. Saif, M.W., Ezzeldin, H., Vance, K., Sellers, S., and Diasio, R.B. (2007): DPYD*2A mutation: the most common mutation associated with DPD deficiency. *Cancer Chemother. Pharmacol.*, 60: 503–507.

37. Salonga, D., Danenberg, K.D., Johnson, M., Metzger, R., Groshen, S., Tsao-Wei, D.D., Lenz, H.J., Leichman, C.G., Leichman, L., Diasio, R.B., and Danenberg, P.V. (2000): Colorectal tumors responding to 5-fluorouracil have low gene expression levels of dihydropyrimidine dehydrogenase, thymidylate synthase, and thymidine phosphorylase. *Clin. Cancer Res.*, 6: 1322–1327.

38. Savva-Bordalo, J., Ramalho-Carvalho, J., Pinheiro, M., Costa, V.L., Rodrigues, A., Dias, P.C., Veiga, I., Machado, M., Teixeira, M.R., Henrique, R., and Jeronimo, C. (2010): Promoter methylation and large intragenic rearrangements of DPYD are not implicated in severe toxicity to 5-fluorouracil-based chemotherapy in gastrointestinal cancer patients. *BMC. Cancer*, 10: 470 doi:10.1186/1471-2407-10-470.

39. Schwab, M., Zanger, U.M., Marx, C., Schaeffeler, E., Klein, K., Dippon, J., Kerb, R., Blievernicht, J., Fischer, J., Hofmann, U., Bokemeyer, C., and Eichelbaum, M. (2008): Role of genetic and nongenetic factors for fluorouracil treatment-related severe toxicity: a prospective clinical trial by the German 5-FU Toxicity Study Group. *J. Clin. Oncol.*, 26: 2131–2138.

40. Seck, K., Riemer, S., Kates, R., Ullrich, T., Lutz, V., Harbeck, N., Schmitt, M., Kiechle, M., Diasio, R., and Gross, E. (2005): Analysis of the DPYD gene implicated in 5-fluorouracil catabolism in a cohort of Caucasian individuals. *Clin. Cancer Res.*, 11: 5886–5892.

41. Thomas, H.R., Ezzeldin, H.H., Guarcello, V., Mattison, L.K., Fridley, B.L., and Diasio, R.B. (2007): Genetic regulation of dihydropyrimidinase and its possible implication in altered uracil catabolism. *Pharmacogenet. Genomics*, 17: 973–987.

42. Thomas, H.R., Ezzeldin, H.H., Guarcello, V., Mattison, L.K., Fridley, B.L., and Diasio, R.B. (2008): Genetic regulation of beta-ureidopropionase and its possible implication in altered uracil catabolism. *Pharmacogenet. Genomics*, 18: 25–35.

43. Ticha, I., Kleiblova, P., Fidlerova, J., Novotny, J., Pohlreich, P., and Kleibl, Z. (2009): Lack of large intragenic rearrangements in dihydropyrimidine dehydrogenase (DPYD) gene in fluoropyrimidine-treated patients with high-grade toxicity. *Cancer Chemother. Pharmacol.*, 64: 615–618.

44. van Kuilenburg, A.B.P., Dobritzsch, D., Meinsma, R., Haasjes, J., Waterham, H.R., Nowaczyk, M.J., Maropoulos, G.D., Hein, G., Kalhoff, H., Kirk, J.M., Baaske, H., Aukett, A., Duley, J.A., Ward, K.P., Lindqvist, Y., and van Gennip, A.H. (2002): Novel disease-causing mutations in the dihydropyrimidine dehydrogenase gene interpreted by analysis of the three-dimensional protein structure. *Biochem. J.*, 364: 157–163.

45. van Kuilenburg, A.B.P., Maring, J.G., Schalhorn, A., Terborg, C., Schmalenberg, H., Behnke, D., Schwabe, W., Jabschinsky, K., and Hausler, P. (2008): Pharmacokinetics of 5-fluorouracil in patients heterozygous for the IVS14+1G > A mutation in the dihydropyrimidine dehydrogenase gene. *Nucleosides Nucleotides Nucleic Acids*, 27: 692–698.

46. van Kuilenburg, A.B.P., Meijer, J., Mul, A.N., Hennekam, R.C., Hoovers, J.M., de Die-Smulders, C.E., Weber, P., Mori, A.C., Bierau, J., Fowler, B., Macke, K., Sass, J.O., Meinsma, R., Hennermann, J.B., Miny, P., Zoetekouw, L., Vijzelaar, R., Nicolai, J., Ylstra, B., and Rubio-Gozalbo, M.E. (2009): Analysis of severely affected patients with dihydropyrimidine dehydrogenase deficiency reveals large intragenic rearrangements of DPYD and a de novo interstitial deletion del(1)(p13.3, p21.3). *Hum. Genet.*, 125: 581–590.

47. van Kuilenburg, A.B.P., Meijer, J., Mul, A.N., Meinsma, R., Schmid, V., Dobritzsch, D., Hennekam, R.C., Mannens, M.M., Kiechle, M., Etienne-Grimaldi, M.C., Klumpen, H.J., Maring, J.G., Derleyn, V.A., Maartense, E., Milano, G., Vijzelaar, R., and Gross, E. (2010): Intragenic deletions and a deep intronic mutation affecting pre-mRNA splicing in the dihydropyrimidine dehydrogenase gene as novel mechanisms causing 5-fluorouracil toxicity. *Hum. Genet.*, 128: 529–538.

48. van Kuilenburg, A.B.P., Meinsma, R., Beke, E., Bobba, B., Boffi, P., Enns, G.M., Witt, D.R., and Dobritzsch, D. (2005): Identification of three novel mutations in the dihydropyrimidine dehydrogenase gene associated with altered pre-mRNA splicing or protein function. *Biol. Chem.*, 386: 319–324.

49. van Kuilenburg, A.B.P., Meinsma, R., Zoetekouw, L., and van Gennip, A.H. (2002): Increased risk of grade IV neutropenia after administration of 5-fluorouracil due to a dihydropyrimidine dehydrogenase deficiency: high prevalence of the IVS14+1g>a mutation. *Int. J. Cancer*, 101: 253–258.

50. van Kuilenburg, A.B.P., Muller, E.W., Haasjes, J., Meinsma, R., Zoetekouw, L., Waterham, H.R., Baas, F., Richel, D.J., and van Gennip, A.H. (2001): Lethal outcome of a patient with a complete dihydropyrimidine dehydrogenase (DPD) deficiency after administration of 5-fluorouracil: frequency of the common IVS14+1G>A mutation causing DPD deficiency. *Clin. Cancer Res.*, 7: 1149–1153.

51. van Kuilenburg, A.B.P., van, L.H., Tromp, A., Veltman, P.C., and van Gennip, A.H. (2000): Pitfalls in the diagnosis of patients with a partial dihydropyrimidine dehydrogenase deficiency. *Clin. Chem.*, 46: 9–17.

52. van Kuilenburg, A.B.P., Vreken, P., Abeling, N.G., Bakker, H.D., Meinsma, R., van, L.H., De Abreu, R.A., Smeitink, J.A., Kayserili, H., Apak, M.Y., Christensen, E., Holopainen, I., Pulkki, K., Riva, D., Botteon, G., Holme, E., Tulinius, M., Kleijer, W.J., Beemer, F.A., Duran, M., Niezen-Koning, K.E., Smit, G.P., Jakobs, C., Smit, L.M., and van Gennip, A.H. (1999): Genotype and phenotype in patients with dihydropyrimidine dehydrogenase deficiency. *Hum. Genet.*, 104: 1–9.

53. van Kuilenburg, A.B.P., Vreken, P., Beex, L.V., Meinsma, R., van, L.H., De Abreu, R.A., and van Gennip, A.H. (1997): Heterozygosity for a point mutation in an invariant splice donor site of dihydropyrimidine dehydrogenase and severe 5-fluorouracil related toxicity. *Eur. J. Cancer*, 33: 2258–2264.

54. van Kuilenburg, A.B.P., Meinsma, J.R., Zoetekouw, L., and van Gennip, A.H. (2002): High prevalence of the IVS14 + 1G>A mutation in the dihydropyrimidine dehydrogenase gene of patients with severe 5-fluorouracil-associated toxicity. *Pharmacogenetics*, 12: 555–558.

55. van Kuilenburg, A.B.P., Meinsma, J.R., Zonnenberg, B.A., Zoetekouw, L., Baas, F., Matsuda, K., Tamaki, N., and van Gennip, A.H. (2003): Dihydropyrimidinase deficiency and severe 5-Fluorouracil toxicity. *Clin. Cancer Res.*, 9: 4363–4367.

56. Vreken, P., van Kuilenburg, A.B.P., Meinsma, R., and van Gennip, A.H. (1997): Dihydropyrimidine dehydrogenase (DPD) deficiency: identification and expression of missense mutations C29R, R886H and R235W. *Hum. Genet.*, 101: 333–338.

57. Wei, X., McLeod, H.L., McMurrough, J., Gonzalez, F.J., and Fernandez-Salguero, P. (1996): Molecular basis of the human dihydropyrimidine dehydrogenase deficiency and 5-fluorouracil toxicity. *J. Clin. Invest*, 98: 610–615.

58. Yamaguchi, K., Arai, Y., Kanda, Y., and Akagi, K. (2001): Germline mutation of dihydropyrimidine dehydrogenese gene among a Japanese population in relation to toxicity to 5-Fluorouracil. *Jpn. J. Cancer Res.*, 92: 337–342.

59. Yang, C.G., Ciccolini, J., Blesius, A., Dahan, L., Bagarry-Liegey, D., Brunet, C., Varoquaux, A., Frances, N., Marouani, H., Giovanni, A., Ferri-Dessens, R.M., Chefrour, M., Favre, R., Duffaud, F., Seitz, J.F., Zanaret, M., Lacarelle, B., and Mercier, C. (2011): DPD-based adaptive dosing of 5-FU in patients with head and neck cancer: impact on treatment efficacy and toxicity. *Cancer Chemother. Pharmacol.*, 67: 49–56.

60. Yokota, H., Fernandez-Salguero, P., Furuya, H., Lin, K., McBride, O.W., Podschun, B., Schnackerz, K.D., and Gonzalez, F.J. (1994): cDNA cloning and chromosome mapping of human dihydropyrimidine dehydrogenase, an enzyme associated with 5-fluorouracil toxicity and congenital thymine uraciluria. *J. Biol. Chem.*, 269: 23192–23196.

61. Yoo, B.K., Gredler, R., Vozhilla, N., Su, Z.Z., Chen, D., Forcier, T., Shah, K., Saxena, U., Hansen, U., Fisher, P.B., and Sarkar, D. (2009): Identification of genes conferring resistance to 5-fluorouracil. *Proc. Natl. Acad. Sci. U. S. A*, 106: 12938–12943.

62. Zhang, R., Lu, Z., Liu, T., Soong, S.J., and Diasio, R.B. (1993): Relationship between circadian-dependent toxicity of 5-fluorodeoxyuridine and circadian rhythms of pyrimidine enzymes: possible relevance to fluoropyrimidine chemotherapy. *Cancer Res.*, 53: 2816–2822.

63. Zhang, X., Li, L., Fourie, J., Davie, J.R., Guarcello, V., and Diasio, R.B. (2006): The role of Sp1 and Sp3 in the constitutive DPYD gene expression. *Biochim. Biophys. Acta*, 1759: 247–256.

64. Zhang, X., Soong, R., Wang, K., Li, L., Davie, J.R., Guarcello, V., and Diasio, R.B. (2007): Suppression of DPYD expression in RKO cells via DNA methylation in the regulatory region of the DPYD promoter: a potentially important epigenetic mechanism regulating DPYD expression. *Biochem. Cell Biol.*, 85: 337–346.

Chapter 12

UGT1A1 Polymorphisms and Mutations Lead to Irinotecan-Induced Toxicity

K. M. Reece[a] and W. D. Figg[b]
[a]*Molecular Pharmacology Section and* [b]*Clinical Pharmacology Core, Medical Oncology Branch, Center for Cancer Research, National Cancer Institute, National Institutes of Health, 9000 Rockville Pike, Bethesda, MD 20892/USA*
kelie.reece@nih.gov, wdfigg@helix.nih.gov

12.1 Irinotecan

Irinotecan is a topoisomerase I inhibitor that is widely used for the treatment of metastatic and recurrent colorectal cancer [1–9]. It is a semisynthetic, water-soluble derivative of camptothecin, an alkaloid which was isolated from the bark and wood of the Chinese tree *Camptotheca acuminata* [2–4, 6,7]. Camptothecin advanced to clinical trials in the 1970s, but despite some antitumor activity, its poor water solubility and unpredictable toxicity profile led to the termination of its further evaluation [7]. Renewed interest in irinotecan was prompted by the identification of topoisomerase I as its principal target in the 1980s [6,7]. The discovery of this "novel mechanism of action" fueled efforts directed at the synthesis of more

Handbook of Therapeutic Biomarkers in Cancer
Edited by Sherry X. Yang and Janet E. Dancey
Copyright © 2013 Pan Stanford Publishing Pte. Ltd.
ISBN 978-981-4364-65-2 (Hardcover), 978-981-4364-66-9 (eBook)
www.panstanford.com

water-soluble analogues that had a favorable toxicity profile as well as consistent antitumor activity [7]. The survival advantages associated with irinotecan was the basis for full FDA approval of irinotecan as a second-line therapy for patients with metastatic colorectal cancer in September of 1998 although it has more recently been approved for use as first-line treatment as well, and is often used in combination with 5-fluorouracil, leucovorin, bevacizumab, and/or cetuximab [2,6,8–11,19].

The metabolism of irinotecan is summarized in Fig. 12.1. Irinotecan is hydrolyzed in the liver and intestine by the carboxy-

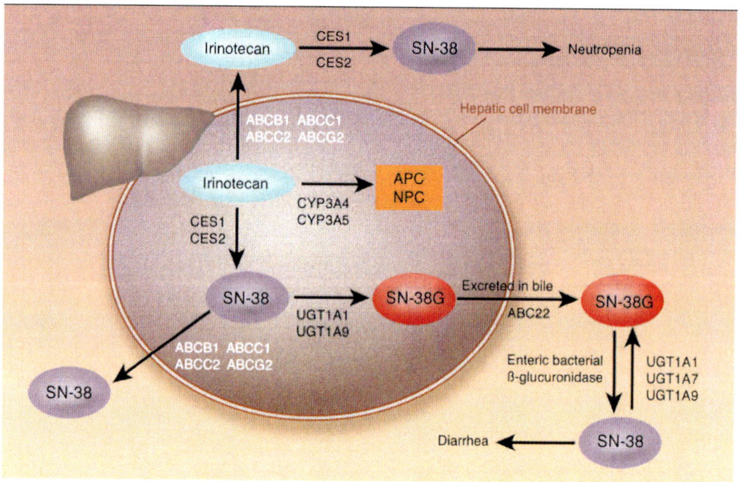

Figure 12.1 The conversion of the irinotecan to the active metabolite, 7-ethyl-10-hydroxycamptothecin (SN-38) is catalyzed by carboxyesterase (CES) 1 and 2. Prolonged levels of SN-38 in the plasma leads to neutropenia. The inactivation of SN-38 to the glucuronide form (SN-38G) is catalyzed by UGT1A1 and UGT1A9. SN-38G is excreted in the bile, where its conversion back to SN-38 is catalyzed by enteric bacterial ß-glucuronidase. Increased levels of SN-38 in the intestine can lead to severe diarrhea. Extrahepatic SN-38 is inactivated to SN-38G through a reaction catalyzed by UGT1A1, UGT1A7, and UGT1A9. Cytochrome P-450 (CYP) 3A4 and 3A5 catalyze the conversion of irinotecan to the inactive metabolites 7-ethyl-10-[4-*N*-(5-aminopentanoic acid)-1-piperidino] carbonyloxycamptothecin (APC) and 7-ethyl-10-[4-amino-1-piperidino] carbonyloxycamptothecin (NPC). Adenosine triphosphate-binding cassette (ABC) B1, C1, C2, and G2 function catalyze the transport of irinotecan out of the liver. Modified from [2,9].

lesterases CES1 and CES2, into the active metabolite hydroxycamptothecin or SN-38, which is approximately 1,000 times more potent than the parent drug [2]. Irinotecan is also oxidized by the cytochrome P450 (CYP) 3A4 to two metabolites – APC (7-ethyl-10-[4-N-(5-aminopentanoic acid)-1-piperidino]-carbonyloxycamptothecin) and NPC (7-ethyl-10-[4-(1-piperidino)-1-amino]-carbonyloxycamptothecin). Of these, APC is a major oxidative metabolite and is inactive, however NPC can be hydrolyzed by CES to SN-38. SN-38 itself is further glucuronidated in the liver and intestine to form the inactive SN-38 glucuronide (SN-38G). SN-38G is excreted in the small intestine via bile where it can be deconjugated back to SN-38 in the gut and intestine via the action of bacterial β-glucuronidase enzymes. SN-38 may be conjugated to SN-38G again after reabsorption into the intestinal tissues. The fate of irinotecan is also affected by the cytochrome P450 enzyme system, primarily CYP3A4, which produces an inactive form of irinotecan, the organic anion transporting polypeptide c (OATP1B1) which transports SN-38 from the plasma into the liver, as well as the adenosine triphosphate-binding cassette (ABC) transporter system that mediates drug efflux [2–7,9,10,12–14].

Irinotecan exerts its chemotherapeutic effects by interacting with the enzyme topoisomerase I and inhibiting its function [2–7,10,15]. Topoisomerase I is an enzyme in the cell nucleus that regulates DNA topology and facilitates nuclear processes such as DNA replication, recombination, and repair [6,10,15]. During these processes, topisomerase I relieves torsional strain in DNA by inducing reversible single-strand breaks, allowing single DNA strands to pass through the break. The 3′-DNA terminus of the broken DNA strands binds covalently with the topoisomerase enzyme to form a catalytic intermediate called a cleavable complex. After the DNA is sufficiently relaxed and the strand passage reaction is complete, DNA topoisomerase I reattaches the broken DNA strands to form the chemically unaltered topoisomers that allows transcription to proceed. Irinotecan and its active metabolite SN-38 binds to the topoisomerase I-DNA complex and prevents religation of these single-strand breaks, thus stabilizing the DNA-topoisomerase I complex in the cleaved DNA form, leading to breaks in both DNA chains. As a result, DNA damage is not efficiently repaired, ultimately leading to DNA fragmentation and cell death [2–7,10,15]. The main toxicities associated with irinotecan therapy are myleosuppression and diarrhea [2–10,12,16–19]. Myelosuppression is linked to elevated and prolonged levels of SN-38 in the plasma, and manifests itself as grade 3 to 4 neutropenia, leukopenia, severe anemia, or severe thrombocytopenia [4]. Irinotecan can also cause early or delayed diarrhea [4,6,9]. Early-onset diarrhea occurs during the drug's infusion and is not associated with metabolism, but is caused by a cholinergic-mediated event and can thus be prevented or ameliorated with atropine [9]. In contrast, late-onset diarrhea which occurs more than 24 hours after irinotecan administration, can be severe and life-threatening [4,6,9,13]. The mechanism behind this is not clearly understood but is thought to arise when biliary-excreted SN-38G is converted back to the active metabolite, SN-38, in the gut and intestine by bacterial β-glucuronidase, thus increasing SN-38 levels in the intestine which can then cause enteric injury [3,4,9,12,13].

12.2 The UGT1A Gene Complex

The major route of SN-38 elimination and the final step of irinotecan metabolism is the inactivation of SN-38 by glucuronidation to SN-38 glucuronide (SN-38G) by hepatic uridine diphosphate glucuronyl transferase (UGT) enzymes [3,13,14,19]. UGTs are membrane-bound enzymes of the glucuronidation pathway that aid in the elimination of small lipophilic molecules such as steroids, bilirubin, hormones, and drugs. During the glucuronidation reaction, UGTs catalyze the transfer of a glycosyl group from a UTP-sugar to small hydrophobic molecules that contain oxygen, nitrogen, sulfur or carboxyl functional groups. The resulting glucuronide is more hydrophilic and thus more easily excreted than the substrate molecule [3,4,9,14,18]. The UGT enzymes have been classified into two families based on their amino acid sequences, UGT1 and UGT2, which are further subdivided into the three subfamilies, UGT1A, UGT2A, and UGT2B [3,4,8,9,12,14,18]. The *UGT1A* gene is located on the long (q) arm of chromosome 2 at position 37, and codes for nine functional proteins (UGT1A1, UGT1A3, UGT1A4, UGT1A5, UGT1A6, UGT1A7, UGT1A8, UGT1A9, and UGT1A10), as depicted in Fig. 12.2. The enzyme isoforms coded within the *UGT1A* locus consists of four shared exons (exons 2-5) and one unique exon (exon1) that contains the substrate-binding domain [3,4,8,12,14,18,20]. Specifically, the 5′ region of the *UGT1A* complex contains thirteen unique first exons, followed by four common exons

[3,4,8,14,18,20]. Four of the first exons are considered pseudogenes and have been designated UGT1A2p, UGT1A11p, UGT1A12p, and UGT1A13p [3,4,8,12].

Each of the remaining nine 5′ exons are independently spliced to common exons 2 through 5, resulting in nine proteins with different N-termini and identical C-termini [8,12,18,20]. The N-terminal region encoded by each unique first exon is regulated by its own promoter and determines acceptor substrate specificity; the preferred substrate of this enzyme is bilirubin, although it also has moderate activity with simple phenols, flavones, and C18 steroids [12,18,20]. The 246-amino acid C-terminal region encoded by the 4 common exons specifies interactions with the common donor substrate, UDP-glucuronic acid [20]. Figure 12.2 shows the relative positions of the exons in the *UGT1A* locus. Given this gene structure, single nucleotide polymorphisms (SNPs) within the four shared exons are common to all the UGT1A enzymes, whereas those found in the variable exon1 and the upstream promoter regions are unique to each UGT1A isoform [8]. Of all the UGTs, the 1A1 variant has been the most comprehensively studied [3,9,10], and is the predominant enzyme responsible for the metabolism of irinotecan. It is expressed highly in the liver, moderately in regions of the small intestine and rectum, and weakly in the thyroid, spinal cord, trachea, uterus, and esophagus [4,21].

12.3 Pharmacogenetics of UGT1A1

UGT1A1 is the enzyme primarily responsible for irinotecan glucuronidation [2–6,15,16]. Its levels are regulated primarily through transcriptional control, and variation in promoter structure influences its rate of transcription [2,8,16]. In particular, a series of TA repeats in the proximal promoter which vary in length, are important regulators of enzyme activity. The wild-type allele of *UGT1A1* (*UGT1A1*1*) has six TA repeats, [$(TA)_6$TAA], but three variant alleles with five, seven, and eight TA repeats in the TATA sequence-$(TA)_5$, $(TA)_7$, and $(TA)_8$-respectively, have also been identified [2–4,8–10,16,24]. With the exception of the $(TA)_5$ polymorphism which leads to normal to high enzyme function, these polymorphisms are associated with reduced UGT1A1 expression and decreased glucuronidation of SN-38, ultimately resulting in increased toxicity due to increased blood levels of the active metabolite [4,8,10,24].

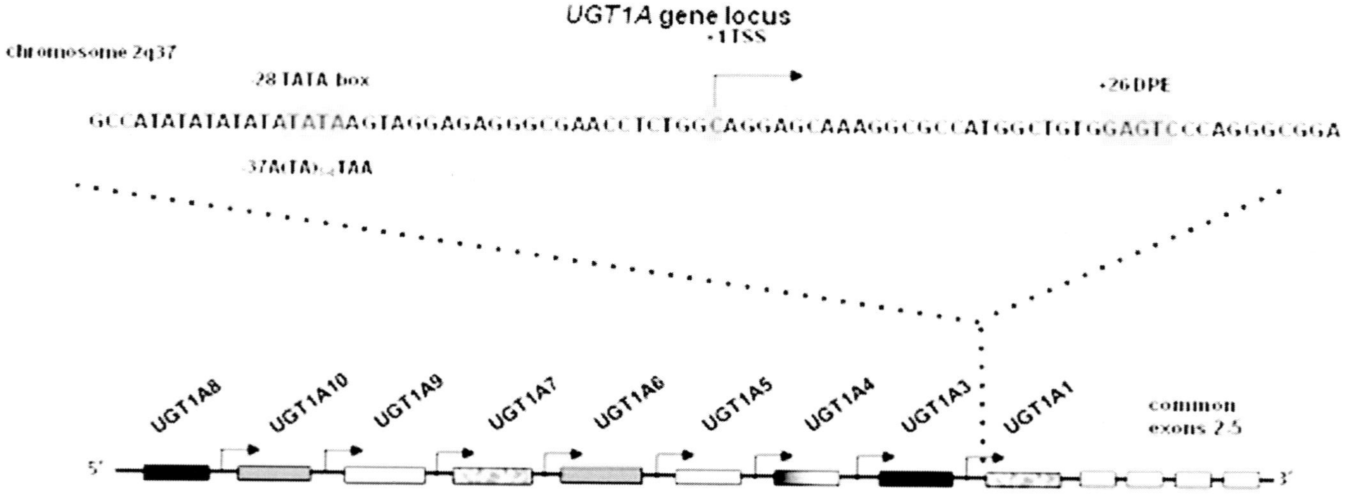

Figure 12.2 The organization of the UGT1A locus is depicted (not drawn to scale). Individual first exons are positioned at the 5′ end of the chromosome and common exons 2–5 are positioned at the 3′ end. The DNA sequence of the *UGT1A1* RNA polymerase II core promoter is shown. A TATA box (TATAA) is located at −28 from the transcription start site (TSS), and a downstream core promoter element (DPE) with an AGTCC is found at +26 from the TSS. A dinucleotide polymorphism of $A(TA)_{5-8}TAA$ in the TATA box is common in all ethnic populations, with $A(TA)_6TAA$ as the major allele.

Although more than 60 genetic variations in the promoter and coding regions of the UGT1A1 gene are currently known [4,8], the *UGT1A1*28* polymorphism has been extensively evaluated in pharmacogenetic studies as it has a significant impact on irinotecan metabolism and toxicity [4,8]. It is characterized by an extra TA repeat in the TATA sequence of the *UGT1A1* promoter, [$(TA)_7$TAA] instead of $(TA)_6$TAA], and the phenotypic alteration that results from this allelic variation is reduced UGT1A1 expression and glucuronidation of SN-38 in human liver microsomes. This results in increased toxicity because of increased blood levels of the active metabolite [3,6,15,18,46]. The presence of the 28 allele on only one chromosome results in a 25% decrease in enzyme activity, and when present in a homozygous fashion, UGT1A1 transcription is reduced by 70%, and is associated with an increased frequency of irinotecan-associated neutropenia [8,24]. Thus the *UGT1A1*28* polymorphism is a compelling pharmacogenetic biomarker for the prediction of irinotecan toxicity, and the association between this genotype and irinotecan-related toxicity is now referenced in the Camptosar label [8–10,12,14–16,24]. Specifically, in July 2005, The US Food and Drug Administration (FDA) approved irinotecan labeling changes to include consideration of a reduction by one dose level in patients homozygous for *UGT1A1*28* and subsequently approved a genetic test to help clinicians to identify these patients [9,10,12,14,15,16,24].

12.4 Ethnic Differences in UGT1A1 Variants

Studies of the distribution of *UGT1A1* polymorphisms by race have shown that the *UGT1A1* genotype varies according to ethnicity [2,3,14]. *UGT1A1*28* is quite common, occurring at a frequency in Caucasian and African populations of 26–31% and 42–56% respectively, and at lower but appreciable frequencies in Asian populations (9–16%) [22,23,27]. Two additional alleles at this location, $(TA)_5$ or *UGT1A1*36* and $(TA)_8$ or *UGT1A1*37*, have been identified in persons of African ancestry [14,24,27]. Given that UGT1A1 activity increases with shorter (TA) repeats in its promoter, the *UGT1A1*36* polymorphism leads to an increase in UGT1A1 activity and SN-38 metabolism; the *UGT1A1*37* variant leads to lower enzymatic and decreased UGT1A1 transcriptional activity [14,24]. *UGT1A1*6* is a single nucleotide substitution of G by A, resulting in the amino acid change of glycine to arginine (G71R) [3,14]. This nonsynomous variant, is the most common *UGT1A1* coding variant among Asian populations, with an allele frequency of 13% in Japanese populations and a frequency of 23% in both Korean and Chinese populations [14,25]. This variant is virtually absent in Caucasian populations and populations of African origin [3]. The *UGT1A1*6* polymorphism results in a 47% decrease in glucuronidation efficiency compared to wild-type, thus individuals with this variant have a reduced capacity to metabolize SN-38 and suffer from irinotecan toxicity [26]. Another nonsynonymous variant observed in Asian populations, *UGT1A1*27* P229Q occurs at even lower frequencies (<3%) but almost completely abolishes *UGT1A1* activity [19,25]. Two other common promoter variants, *UGT1A1*60* and *UGT1A1*93*, are associated with lower UGT1A1 levels, and are also suspected to influence irinotecan toxicity [27,28]. They are both located in the Phenobarbital Response Enhancer Module (PBREM), a regulatory element positioned 3kb upstream of the *UGT1A1* gene [27]. *UGT1A1*60* also referred to as -3263T>G, has an allele frequency of about 33% in African Americans and a much lower frequency in the Caucasian population (9%) [30]. Interestingly, it occurs at an estimated frequency of 83% in Japanese populations [27,28]. *UGT1A1*93*, also referred to as -3156G>A has an allele frequency of 30% in African American and Caucasian populations and a frequency of 10–12% in Asian populations [28,29]. With the exception of the *UGT1A1*28* allele, irinotecan labeling does not address dosing in patients with these other polymorphisms or mutations in the *UGT1A1* gene that may impair irinotecan detoxification.

12.5 Crigler–Najjar Syndrome

UGT1A1 polymorphisms are associated with Crigler–Najjar syndrome, an inherited state of hyperbilirubinemia [12]. Crigler–Najjar Syndrome is a rare disorder affecting the metabolism of bilirubin, and is divided into two types: type I and type II, with the latter sometimes called Arias syndrome [32,38]. More than 40 different deleterious mutations distributed both in the unique and common exons of the UGT1A1 gene have been found in Crigler–Najjar syndrome types I and II [31]. Crigler–Najjar syndrome type I is a very rare disease with less than 1 case per million live births, and an autosomal recessive pattern of inheritance [32,33]. In Crigler-Najjar syndrome type I, most patients have a mutation in one of the common exons (2 to 5) of UGT1A1 and have complete absence of enzyme activity. Thus patients have difficulties conjugating several substrates (i.e., drugs and xenobiotics) [6,7,32–34]. A smaller percentage of patients have mutations limited to the bilirubin-specific A1 exon; their conjugation defect is mostly restricted to bilirubin itself [34,35]. This results in severe hyperbilirubinemia, which is characterized by serum bilirubin levels that usually range from 20–50 mg/dL (the normal range for total bilirubin is 0.1–1.0 mg/dL) [38]. As a result, intense jaundice appears in the first days of life and persists thereafter. Some affected infants die within the first year of life of kernicterus (bilirubin accumulation in the basal ganglia and brainstem nuclei), while others survive with little or no neurologic defect [32–38]. Patients do not respond to phenobarbital treatment, which induces glucuronyl transferase induction, and only traces of unconjugated bilirubin can be found in their bile which is often colorless [32,38]. Phototherapy is the preferred treatment for Crigler–Najjar type I disease. However, since kernicterus can develop at a later age, phototherapy sometimes becomes inadequate at keeping serum bilirubin levels below 15 mg/dL, at which point liver transplantation may be required [32,38–40].

A milder variant of the classical Crigler–Najjar syndrome has been termed Crigler–Najjar syndrome type II, and is characterized by a partial deficiency of the glucoronyl transferase enzyme [32,38]. As in type I, genetic lesions are found in the coding region of UGT1A1, however these are always single base pair mutations which significantly reduces UGT1A1 activity without abolishing it [32,38]. Thus, kernicterus is rare due to lower total serum bilirubin levels which generally range from 6 to 20 mg/dL, and in some cases are only detected later in life. Patients with Crigler–Najjar syndrome type II have pigmented bile which contains significant fractions of bilirubin glucoronides [32,28]. Treatment with phenobarbital is effective because UGT1A1 is present at reduced but detectable levels, generally with a decrease of at least 25% in serum bilirubin [32]. The inheritance pattern of Crigler–Najjar syndrome type II has been difficult to determine and both dominant and recessive inheritance patterns have been described [38]. Due to the fact that type II patients have some bilirubin transferase activity, they have less severe hyperbilirubinemia, are subsequently less severely

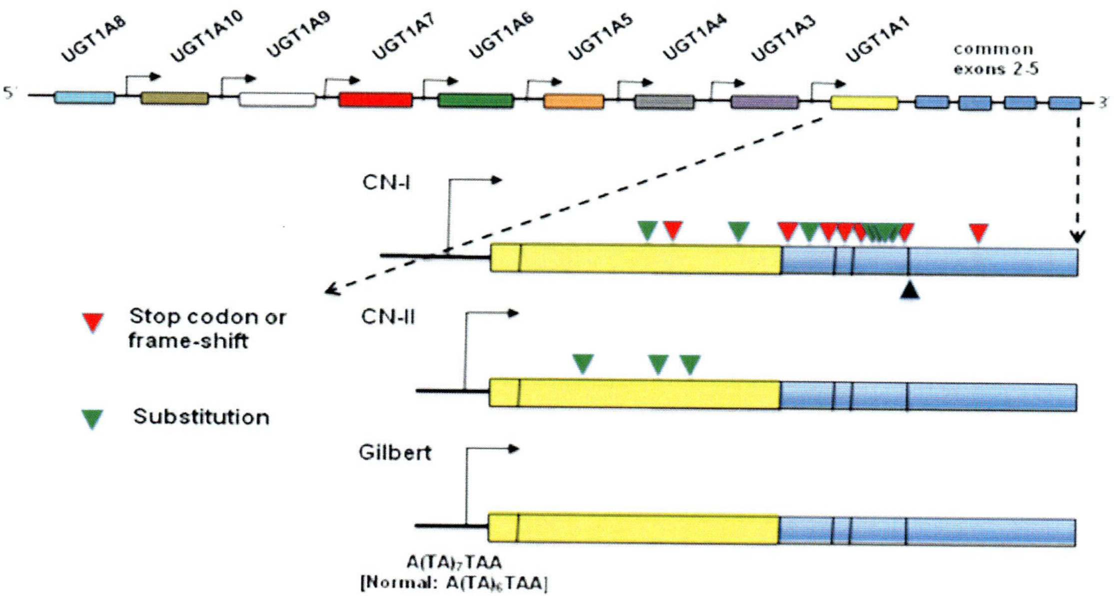

Figure 12.3 UGT1A1 polymorphisms are associated with Crigler–Najjar syndrome and Gilbert's syndrome. In Crigler–Najjar syndrome type I, most patients have a mutation in one of the common exons (2 to 5) of UGT1A1 which result in a premature stop codon or a single amino acid substitution, ensuing in a complete absence of enzyme activity. In Crigler–Najjar type II, genetic lesions are found in the coding region of UGT1A1; these are always single base pair mutations that result in the substitution of a single amino acid, significantly reducing UGT1A1 activity without abolishing it. In contrast to Crigler–Najjar syndrome types I or II, Gilbert's syndrome is caused by a promoter abnormality; alleles containing the Gilbert-type promoter are termed UGT1A1*28.

jaundiced, and generally survive into adulthood without neurologic or intellectual impairment, although bilirubin encephalopathy may develop later in life [32,38].

12.6 Gilbert's Syndrome

Mutations in the UGT1A1 gene are also responsible for the more common and mild hyperbilirubinemia known as Gilbert's syndrome that is found in 3-7% of the U.S. population [41]. In contrast to a normal UGT1A1 gene which has a promoter region TATA box containing the genetic subsequence $(TA)_6TAA$, Gilbert's syndrome is most commonly associated with homozygous $(TA)_7TAA$ alleles, referred to as UGT1A1*28 as previously described [6,7,12,42,45]. This leads to a 70% reduction in UGT1A1 activity [43,44]. This reduction in enzyme activity leads to an elevated level of unconjugated bilirubin in the bloodstream, however serum bilirubin concentrations rarely exceed 5 mg/dL and usually fluctuate between 1.3 and 3 mg/dL [38]. There are normally no serious consequences associated with this disease; mild jaundice may appear under conditions of exertion, stress, fasting, and infections, but the condition is otherwise usually asymptomatic. Thus Gilbert's syndrome is considered an entirely benign and clinically inconsequential entity, requiring neither treatment nor long-term medical attention, although the mild hyperbilirubinemia that occurs is sometimes mistaken for a sign of occult, chronic, or progressive liver disease [45]. Gilbert's syndrome is generally considered to be an autosomal recessive disorder. However, some cases of heterozygosity and compound heterozygosity have been reported in patients with Gilbert syndrome, predominantly among the Asian population [33].

12.7 Conclusion

Irinotecan is now regarded as the most active drug for the treatment of colorectal cancer [3]. However, one of the most difficult issues that face oncologists is deciding the optimal dosage for therapy, as each individual shows different outcomes even at the same dose in regard to treatment-related adverse events, which range from no toxicity to a lethal event. Thus the identification of predictive markers in irinotecan therapy has been a significant goal of pharmacogenetic research. Polymorphisms within the drug-metabolizing enzyme UGT1A1, namely the *UGT1A1*28* variant, is now recognized as a known marker of irinotecan toxicity 3,46]. Specifically, patients with this allele have a 70% reduction in enzyme activity and thus clear irinotecan less quickly from their body than the rest of the population [8,24]. They effectively receive a greater exposure to the drug from the same dose and as a consequence are at higher risk of considerable side effects such as neutropenia and diarrhea [3–6,9,12,13]. In light of this information, the labeling of irinotecan was recently changed and now includes a warning of greater irinotecan toxicity in patients with the *UGT1A1*28* genotype [8–10,12,14–16, 24]. Irinotecan is clearly an example of personalized medicine in oncology. This is an area that continues to grow as the understanding of the mechanisms of action of anti cancer agents, and the significance of genetic mutations in drug metabolism/exposure becomes more established.

References

1. Toffoli, G., Cecchin, E., Coron, G., Russo, A., Buonadonna, A., D'Andrea, M., Pasetto, L.M., Pessa, S., Errante, D., De Pangher, V., Giusto, M., Medici, M., Gaion, F., Sandri, P, Galligioni, E., Bonura, S., Boccalon, M., Biaon, P., Frustaci, S. (2006). The role of UGT1A1*28 polymorphism in the Pharmacodynamics and pharmacokinetics of irinotecan in patients with metastatic colorectal cancer, *J. Clin. Oncol.* 24(19): 3061–3068.
2. Pangilinan, J.M., Khan, G.N., Zalupski, M.M. (2008). Irinotecan pharmacogenetics: an overview for the community oncologist, *Commun. Oncol.* 5(2): 99–114.
3. Shimoyama, S. (2010). Pharmacogenetics of irinotecan: an ethnicity-based prediction of irinotecan adverse events, *World J. Gastrointest. Surg.* 2(1): 14–21.
4. Nagar, S., Blanchard, R.L. (2006). Pharmacogenetics of uridine diphosphoglucuronosyltransferase *UGT) 1A family members and its role in patient response to irinotecan, *Drug Metab. Rev.* 38: 393–409.
5. Iyer, L, Das, S., Janisch, L., Wen, M., Ramirez, J., Karrison, T., Gleming, G.F., Vokes, E.E., Schilsky, R.L., Ratain, M.J. (2002). UGT1A1*28 polymorphism as a determinant of irinotean disposition and toxicity, *Pharmacogenomics J.* 2: 43–47.
6. Smith, N.F., Figg, W.D., Sparreboom, A. (2006). Pharmacogenetics of irinotecan metabolism and transport: an update, *Toxicol. in Vitro* 20: 163–167.

7. Mathijssen, R., Van Alphen, R.J., Verweij, J., Loos, W.J., Nooter, K., Stoter, G., Sparreboom, A. (2001). Clinical pharmacokinetics and metabolism of irinotecan (CPT-11). *Clin. Cancer Res.* 7: 2182–2194.

8. Perera, M.A., Innocenti, F., Ratain, M.J. (2008). Pharmacogenetic testing for uridine diphosphate glucuronosyltransferase 1A1 polymorphisms: are we there yet? *Pharmacotherapy,* 28(6): 755–768.

9. Hahn, K.K., Wolff, J.J., Kolesar, J.M. (2007). Pharmacogenetics and irinotecan therapy. *Am. J., Health-Syst. Ph.* 63(22): 2211–2217.

10. Marsh, S., Hoskins, J.M. (2010). Irinotecan pharmacogenomics. *Pharmacogenomics,* 11(7): 1003–1010.

11. Saltz, L.B. (1998). Irinotecan in the first-line treatment of colorectal cancer. *Oncology,* 12: 54–58.

12. Hasegawa, Y., Ando, Y., Ando, M., Hashimoto, N., Imaizumi, K., Shimokata, K. (2006). Pharmacogenetic approach for cancer treatment-tailored medicine in practice. *Ann. N.Y. Acad. Sci.* 1086: 223–232.

13. Innocenti, F., Ratain, M.J. (2006). Pharmacogenetics of irinotecan: clinical perspectives on the utility of genotyping. *Pharmacogenomics,* 7(8): 1211–1221.

14. Marques, S.C., Ikediobi, O.N. (2010). The clinical application of UGT1A1 pharmacogenetic testing: gene-environment interactions. *Hum. Genom.* 4(4): 238–249.

15. Bhusha, S., McLeod, H., Walko, C.M. (2009). Role of pharmacogenetics as predictive biomarkers of response and/or toxicity in the treatment of colorectal cancer. *Clin. Colorectal Cancer,* 8(1): 15–21.

16. O'Dwyer, P.J. (2006). Uridine diphosphate glucuronosyltransferase (UGT) 1A1 and irinotecan: practical pharmacogenomics arrives in cancer therapy. *J. Clin. Oncol.* 24(28): 45344538.

17. Weekes, J., Lam, A.K., Sebesan, S., Ho, Y. (2009). Irinotecan therapy and molecular targets in colorectal cancer: a systemic review. *World. J. Gastorenterol.* 15(29): 3597–3602.

18. McLeod, H.L., Watters, J.W. (2004). Irinotecan pharmacogenetics: is it time to intervene: *J. Clin. Oncol.* 22(8): 1356–1359.

19. Gagne, J. Montminy, V. Belanger, P. Journault, K., Gaucher, G., Guillemett, C. (2002). Common human UGT1A1 polymorphisms and the altered metabolism of irinotecan active metabolite 7-ethyl-10-hydroxycamptothecin (SN-38). *Mol. Pharm.* 62(3): 608–617.

20. Gong, Q.H., Cho, J.W., Huang, T., Potter, C., Gholami, N., Basu, N.K., Kubota, S., Carvalho, S., Pennington, M.W., Owens, I.S., Popescu, N.C. (2001). Thirteen UDP-glucuronosyltransferase genes are encoded at

the human UGT1 gene complex locus. *Pharmacogenetics,* 11(4): 357–368.

21. Ritter, J.K., Crawford, J.M., Owens, I.S. (1991). Cloning of two human liver bilirubin UDP-glucuronosyltransferase cNAs with expression in COS-1 cells. *J. Biol. Chem.* 226(2): 1043–1047.

22. Hall, D., Ybazeta, G., Destro-Bisol, G., Petzl-Erler, M.L., Di Rienzo, A. (1999). Variability at the uridine diphosphateglucuronosyltransferase 1A1 promoter in human populations and primates. *Pharmacogenetics,* 9(5): 591–599.

23. Beutler, E., Gelbart, T., Demina, A. (1998). Racial variability in the UDP-glucuronosyltransferase 1 (UGT1A1) promoter: a balanced polymorphism for regulation of bilirubin. *PNAS.* 95(14): 8170–8174.

24. Ames, M.M., Gibbs, C.D., Goetz, M.P., Kim, W.R., Majewski, D.W., Mitchell, P.S., O'Kane, D.J., Yao, J.D. (2006). UGT1A1 genotyping: an important tool for identifying patients at risk for irinotecan toxicity. *Communique,* 31(7): 1–11.

25. Akaba, K., Kimura, T., Sasaki, A., Tanabe, S., Ikegami, T., Hashimoto, M., Umeda, H., Yoshida, H., Umetsu, K., Chiba, H., Yuasa, I., Hayasaka, K. (1998). Neonatal hyperbilirubinemia and mutation of the bilirubin uridine diphosphate-glucuronosyltransferase gene: a common missense mutation among Japanese, Koreans and Chinese. *Biochem. Mol. Biol. Int.* 46(1): 21–26.

26. Hinno, J., Tanaka-Kagawa, T., Hanioka, N., Saeki, M., Ishida, S., Nishimura, T., Ando, M., Saito, Y., Ozawa, S., Sawada, J. (2003). Glucuronidation of 7-ethyl-10-hydroxycamptothecin (SN-38), an active metabolite of irinotecan (CPT-11), by human UGT1A1 variants, G71R, P22Q, and Y486D. *Drug. Metab. Dispos.* 31(1): 108–113.

27. Sugatani, J., Yamakawa, K., Yoshinari, K., Machida, T., Takagi, H., Mori, M., Kakizaki, S., Sueyoshi, T., Negishi, M., Miwa, M. (2002). Identification of a defect in the UGT1A1 gene promoter and its association with hyperbilirubinemia. *Biochem. Biophys. Res. Commun.* 292(2): 492–497.

28. Innocenti, F., Grimsley, C., Das, S., Ramírez, J., Cheng, C., Kuttab-Boulos, H., Ratain, M.J., Di Rienzo, A. Haplotype structure of the UDP-glucuronosyltransferase 1A1 promoter in different ethnic groups. *Pharmacogenetics,* 2(9): 725–733.

29. Innocenti, F., Liu, W., Chen, Pl, Desai, A.A., Das, S., Ratain, M.J. (2005). Haplotypes of variants in the UDP-glucuronosyltransferase1A9 and 1A1 genes. *Pharmacogenet. Genomics,* 15(5): 295–301.

30. Kaniwa, N., Kurose, K., Jinno, H., Tanaka-Kagawa, T., Saito, Y., Saeki, M., Sawada, J., Tohkin, M., Hasegawa, R. (2005). Racial variability in haplotype frequencies of UGT1A1 and glucuronidation activity of a novel single nucleotide polymorphism 686C>T (P229L) found in an African-American. *Drug. Metab. Dispos.* 33: 458–465.

31. Mackenzie, P.I., Owens, I.S., Burchell, B., Bock, K.W., Bairoch, A., Belanger, A., Fournel-Gigleux, S., Green, M., Hum, D.W., Iyanagi, T., Lancet, D., Louisot, P., Magalou, J., Roy Chowdhury, J., Ritter, J.K., Schachter, H., Tephly, T.R., Tipton, K.E., Nebert, D.W. (2007). The UDP glycosyltransferase gene superfamily: recommended nomenclature update based on evolutionary divergence. *Pharmacogenetics* 7: 255–269.

32. Agarwal, R., Silva, W.D. (2008). Crigler Najjar syndrome type 2. *Asian Biomed.* 2(5): 289–391.

33. Chowdhury, J.R., Wolkoff, A.W., Chowdhury, N.R., Arias, I.M. (2001). *Hereditary jaundice and disorders of bilirubin metabolism.* 2: 8th eds. Scriver, C.R., Beaudet, A.L., Sly, W.S., Valle, D. "The Metabolic and Molecular Bases of Inherited Disease" (McGraw-Hill, NY) pp: 3063–3101.

34. Jansen, P.L. (1999). Diagnosis and management of Crigler–Najjar syndrome. *Eur. J. Pediatr.* 158 suppl 2: S89–S94.

35. Owens, I.D., Basu, N.K., Banerjee, R. (2005). UDP-glucuronosyltransferases: gene structures of UGT1 and UGT2 families. *Methods Enzymol.,* 400: 1–22.

36. Toietta, G., Mane, V.P., Norona, W.S., Finegold, M.J., Ng, P., McDonagh, A.F., Beaudet, A.L., Lee, B. (2005). Lifelong elimination of hyperbilirubinemia in the Gunn rat with a single injection of helper-dependent adenoviral vector. *Proc. Natl. Acad. Sci. U.S.A.* 102(11): 3930–3935.

37. Crigler, J.F., Najjar, V. (1952). Congenital familial nonhemolytic jaundice with kernicterus. *Pediatrics,* 10(2): 169–180.

38. Bhagavan, N.V. (2002). *Metabolism of Iron and Heme.* Hayhurst, J. "Medical Biochemistry." (Harcourt Academic Press, CA) pp: 671–697.

39. Shevell, M.I., Bernard, B., Adelson, J.W., Doody, D.P., Laberge, J.-M., Guttman, F.M. (1987). Crigler-Najjar syndrome type I: treatment by home phototherapy followed by orthotopic hepatic transplantation. *J. Pediat.* 110: 429–431.

40. Jansen, P.L.M., Mulder, G.J., Burchell, B., Bock, K.W. New developments in glucuronidation research: report of a workshop on 'glucuronidation, its role in health and disease.' (1992). *Hepatology* 15: 532–544.

41. Manandhar, S.R., Gurubacharya, R.L., Baral, M.R., Manandhar, D.S. (2003). A case report of Gilbert's syndrome. *Kathmandu Univ. Med. J.*, 1(3): 187–189.

42. Lankisch, T.O., Schulz, C., Zwingers, T., Erichsen, T.J., Manns, M.P., Heinemann, V., Strassburg, C.P. (2008). Gilbert's syndrome and irinotecan toxicity: combination with UDT-glucuronosyltransferase 1A7 variants increases risk. *Cancer. Epidemiol. Biomarkers Prev.* 17(3): 695–701.

43. Raijmakers, M.T., Jansen, P.L., Steegers, E.A., Peters, W.H. (2000). Association of human liver bilirubin UDP-glucuronyltransferase activity with a polymorphism in the promoter region of the UGT1A1 gene. *J. Hepatol.* 33(3): 348–351.

44. Bosma, P.J., Chowdhury, J.R., Bakker, C., Gantla, S., De Boer, A., Oostra, B.A., Lindhout, D., Tytgat, G.N., Jansen, P.L., Oude Elferink, R.P., Chowdhury, N.R. (1995). The genetic basis of the reduced expression of bilirubin UDP-glucuronosyltransferase 1 in Gilbert's syndrome. *N. Engl. J. Med.* 333(18): 1171–1175.

45. Schmid, R. (1995). Gilbert's syndrome—a legitimate genetic anomaly? *N. Engl. J. Med.* 333: 1217–1218.

46. Biason, P., Masier, S., Toffoli, G., (2008). UGT1A1*28 and other UGT1A polymorphisms as determinants of irinotecan toxicity. *J. Chemother.* 20(2): 158–165.

Chapter 13

The 21-Gene Recurrence Score and Benefit of Chemotherapy in Estrogen Receptor-Positive Breast Cancer

Petra Rietschel and Joseph A. Sparano
Albert Einstein College of Medicine, Montefiore Medical Center-Weiler Division, 1825 Eastchester Road, Bronx, NY 10461, USA
preitsch@montefiore.org, jsparano@montefiore.org

13.1 Introduction

Breast cancer is a heterogeneous disease associated with variable clinical outcomes and response to therapy that are not accurately predicted by classical clinicopathologic factors. Gene expression profiling has facilitated the development of "multiparameter" assays that may identify breast cancer subtypes associated with distinct clinical outcomes that were not previously recognized, or provide more accurate information about prognosis or response to specific therapies. Breast cancer is the most common cancer in women in the United States and worldwide and is the leading cause of cancer death in women worldwide, with nearly 460,000 deaths estimated in

Handbook of Therapeutic Biomarkers in Cancer
Edited by Sherry X. Yang and Janet E. Dancey
Copyright © 2013 Pan Stanford Publishing Pte. Ltd.
ISBN 978-981-4364-65-2 (Hardcover), 978-981-4364-66-9 (eBook)
www.panstanford.com

2008,[1] including about 40,000 deaths in the United States.[2] Although breast cancer mortality rates have declined in the United States due to screening and wider use of more effectives systemic adjuvant therapies,[3] a higher proportion of women are receiving adjuvant therapy who might have been cured without it and are thus being overtreated. Diagnostic tools that could identify which patients derived greatest benefit from specific therapies are needed.

13.2 Genomics

Genomics is defined as the study of all of the nucleotide sequences in an organism. "Genomic profiling" in breast cancer and other cancers has focused on the evaluation of gene expression, or the translation of the information encoded in genomic DNA into an RNA transcript. Although RNA transcripts include messenger RNAs which are translated into proteins and various other RNAs (e.g., transfer RNA, ribosomal RNA, micro RNA, and noncoding RNA) that have important biologic functions, most studies of gene expression profiling in breast cancer have focused on mRNA expression. The same principles may be applied to the study of the epigenome,[4,5] microRNAs,[6] proteins,[7] or integrative approaches that evaluate combinations of profiling methods.[8] In addition, high-throughput massively parallel sequencing is now feasible, allowing direct sequencing of cDNAs which may provide not only absolute gene expression levels, but also information on alternatively spliced isoforms, mutations and novel transcripts arising from fusion genes.

13.3 Development and Validation of Multiparameter Assays

The promise and pitfalls in developing multiparameter assays has been reviewed elsewhere,[9–12] specific criteria have been proposed for the level of evidence required to define and support their clinical utility.[13] There are several steps in the development of a marker, which may broadly classified as (1) conceptualization, (2) clinical development, (3) technical development, (4) validation, and (5) application. Development of an accurate assay is largely a function of the interplay between sample size and classification difficulty.[14] Criteria have been developed for assessing and reporting multiparameter assays (and other tumor markers) called the REMARK guidelines (REporting recommendations for tumor MARKer prognostic studies).[15] Most peer-reviewed journals require that reports describing tumor markers, including multiparameter assays, follow the REMARK guidelines in order to be considered for publication. In addition, standards termed "minimal information about a microarray experiment" (MIAME) for reporting the data have been established by the Microarray Gene Expression Data Society,[16] and most journal require that the gene expression data described in the publication be deposited in a publicly available data base (e.g., Gene Expression Omnibus (http://www.ncbi.nlm.nih.gov/geo). Approval of multiparameter breast cancer assays is regulated under the under the provisions of the Clinical Laboratory Improvement Act of 1988 (CLIA), which applies to laboratories that examine human specimens for the diagnosis, prevention, or treatment of any disease or assessment of health. The US Food and Drug Administration issued guidance document in 2007 indicating that multiparameter assays also fall under regulatory jurisdiction of the agency under regulations governing approval of medical devices (http://www.fda.gov/MedicalDevices/DeviceRegulationandGuidance/GuidanceDocuments/ucm079163.htm#1). A gene expression profiling test system for breast cancer prognosis was defined as a device that measures the RNA expression level of multiple genes and combines this information to yield a signature (pattern or classifier or index) to aid in prognosis. Approval provided by this mechanism is commonly known as "510(k)" clearance. Several multiparameter assays have now been approved for breast cancer by CLIA or FDA in the United States and Europe, and their characteristics are summarized in Table 13.1, including the 21-gene Recurrence Score.

13.4 Development of the 21-Gene Recurrence Score

The genes included in the Oncotype DX assay and algorithm used to calculate the Recurrence Score (RS) are summarized in Table 13.2. There were several critical steps in the development of this assay, including: (1) development of methodologies to extract RNA from formalin fixed paraffin embedded tissue (FFPE), to perform quantitative reverse-transcriptase-polymerase-chain-reaction

Table 13.1 Multiparameter assays for breast cancer approved by regulatory agencies

Assay (company)	Method	Tissue type	Approval	Patient population	Prognosis/prediction
Mammaprint 70-Gene Assay (Agendia)	DNA Microarray	Fresh or Frozen	Europe and US (FDA)	ER-Pos/Neg stage I-II breast cancer	Prognostic for distant recurrence
Oncotype DX 21-Gene Assay (Genomic Health)	qRT-PCR	FFPE	Europe and US (CLIA)	ER-Pos stage I-II breast cancer	Prognostic for distant recurrence Predictive of chemotherapy benefit if RS high
Theros H/I ^{SM} 2-Gene Ratio (Biotheranostics)	qRT-PCR	FFPE	US (CLIA)	ER-Pos, Lymph node Negative Breast Cancer	Prognostic for distant recurrence
Theros MGI ^{SM} 5-Gene Molecular Grade	qRT-PCR	FFPE	US (CLIA)	ER-pos, Grade 2 tumors	Prognosis – reclassification of tumors from grade 2 to grade 1 or 3

Table 13.2 Recurrence score algorithm

- The score of each group is calculated as follows using quantitative RNA expression levels (ranging from 0–15) for each gene:
  - o *GRB7* group score = 0.9 × *GRB7*+0.1 × *HER2*
    - ▪ if the result is less than 8, then the group score is considered 8
  - o ER group score = (0.8 × ER) + (1.2 × PGR) + BCL2 + SCUBE2
  - o Proliferation group score = Ki-67 + STK15 + Survivin + Cyclin B1 + MYBL2
    - ▪ if the result is less than 6.5, then the proliferation group score is considered 6.5
  - o Invasion group score = CTSL2 + MMP11
  - o Other genes: GSTM1 (drug metabolism), BAG1 (apoptosis), and CD68 (immune function)
- The unscaled RS is calculated with the use of coefficients that are predefined on the basis of regression analysis of gene expression and recurrence in the three training studies:
  - o +0.47 × GRB7group score
  - o −0.34 × ER group score
  - o +1.04 × proliferation group score
  - o +0.10 × invasion group score
  - o +0.05 × CD68
  - o −0.08 × GSTM1
  - o −0.07 × BAG1
- RS is rescaled from the unscaled RS as follows:
  - o RS = 0 if RSU < 0
  - o RS = 20 × (unscaled RS-6.7) if 0 ≤ unscaled RS ≤ 100
  - o RS = 100 if unscaled RS > 100.

Note: For the coefficients used to calculate the RS, "+" sign indicates that increased expression is associated with an increased risk of recurrence, and a "−" sign indicates that increased expression is associated with a decreased risk of recurrence.

(qRT-PCR) using RNA derived from FFPE, and to reduce sources of preanalytic and analytic variability, (2) identification of genes associated with recurrence and response to therapy in training set populations, (3) selection of the genes and development of algorithm for the RS and (4) prospective validation of the algorithm. Since the requirement for fresh or snap-frozen tissue had limited the clinical application of other arrays, methods were developed to extract RNA from FFPE and perform qRT-PCR using this material.[17]

Then, by searching the published literature, genomic databases, pathway analysis, and microarray-based gene expression profiling experiments, a panel of 250 candidate genes likely to be associated with outcome was selected from three independent data sets including a total of 447 patients.[18] The training studies described above identified a panel of 16-cancer related genes selected on the strength of their performance in all three studies and the consistency of primer of probe performance in the assay. These include in ER signaling, proliferation, HER2 signaling, invasion, and genes involved in immune function, drug metabolism, and apoptosis (Table 13.2). The expression of these genes is normalized to the expression of 5 reference genes not included in the calculation of the RS. The RS algorithm reports results on a scale from 0 to 100 (higher scores indicate a higher likelihood of recurrence). A score is generated for each gene ranging from 0 to 15, with each integer of 1 corresponding to a twofold increase in RNA expression level. Tumors were classified in to the following categories: low risk (RS less than 18), intermediate risk (RS 18–30) and high risk (RS 31 or higher).

13.5 Recurrence Score and Prognosis

The assay has been prospectively validated in several studies, as summarized in Table 13.3. The first prospective validation study was performed using samples from patients enroll in the NSABP B-14 trial, which assessed tamoxifen use in patients with ER-positive breast cancer and negative axillary nodes.[19] Between 1982 and 1988, 2892 patients enrolled on the B14 trial were randomly assigned to receive placebo or tamoxifen, followed by additional 1235 patients were assigned to receive tamoxifen.[20] Of the 2617 patients treated with tamoxifen, paraffin blocks containing sufficient tissue were available in 675 patients, and qRT-PCR was successful in 668 of the 675 blocks. Fifty-one percent of the patients had a low RS of less than 18, 22% had an intermediate RS of 18–31, and 27% had a high RS of higher than 31. RS was associated with a significantly higher risk of distant recurrence whether evaluated as a continuous variable (i.e., risk of recurrence for a specific RS) or as a categorical variable (i.e., low vs. intermediate vs. high). The rate of distant recurrence was 6.8% in the low RS group, 14.3% in the intermediate RS group, and 30.5% in the high RS group. In a multivariate Cox model, the RS score was significantly associated with distant recurrence independent of age and tumor size.

Table 13.3 Prospective validation studies of oncotype DX recurrence score in ER-positive breast cancer

Trial/ author	Population	No.	Adjuvant treatment	Summary of key findings
Prognosis				
B14 Paik *et al.*	Node negative	668	Tamoxifen	RS significantly associated with distant recurrence
Kaiser Habel *et al.*	Node-negative	790	Tamoxifen or no adjuvant therapy	RS significantly associated with breast cancer death
TransATAC Dowsett *et al.*	Node-negative or positive	1231	Tamoxifen or Anastraszole	RS significantly associated with distant recurrence in AI or tamoxifen-treated patients irrespective of nodal status
E2197 Goldstein *et al.*	Node-negative or 1–3 positive nodes	465	Chemotherapy plus endocrine therapy	RS significantly associated with recurrence (distant and local)
Prediction of chemotherapy benefit				
B20 Paik *et al.*	Node-negative	651	Tamoxifen +/− CMF/MF	Chemotherapy benefit associated with high RS
S8814 Albain *et al.*	Postmenopausal, node-positive	367	Tamoxifen +/− CAF	Chemotherapy benefit associated with high RS

A second prospective validation study including a population-based study involving a case:cohort sample chosen from 4964 patients with node negative breast cancer diagnosed between 1985 and 1994 who did not receive adjuvant chemotherapy.[21] Cases included 220 patients who died from breast cancer, and were individually matched to 570 controls with respect to age, race, adjuvant tamoxifen, medical facility and diagnosis year and were alive at the date of death of their matched case. About 50% of patients had a low RS irrespective of tamoxifen used. After adjusting for tumor size and grade, RS was significantly associated with risk of breast cancer death in ER-positive, tamoxifen-treated and -untreated patients ($P = 0.003$ and $P = 0.03$, respectively). At 10 years, the risks for breast cancer death in ER-positive, tamoxifen-treated patients were 2.8% (95% confidence interval [CI] 1.7–3.9%), 10.7% (95% CI 6.3–14.9%), and 15.5% (95% CI 7.6–22.8%) for those in the low, intermediate and high-risk RS groups, respectively. For those with ER-positive tumors not treated with tamoxifen, the 10 year risk of death was 6.2% (95% CI 4.5–7.9%), 17.8% (95% CI 11.8–23.3%), and 19.9% (95% CI 14.2–25.2%), respectively. This study therefore provided additional validation of the RS in a population-based study of lymph node-negative, ER-positive breast cancer patients not treated with chemotherapy, showing that RS was strongly associated breast cancer death among ER-positive, tamoxifen-treated and -untreated patients.

A third study provided information about the prognostic utility of RS in patients treated with aromatase inhibitors (AIs), which have been shown to be more effective than tamoxifen in preventing recurrence in post-menopausal women.[22] One of the pivotal trials showing benefit of AIs was the Arimidex, Tamoxifen, Alone or in Combination (ATAC) trial, which compared anastrozole (1 mg) with tamoxifen (20 mg), or their combination, given orally every day for 5 years as adjuvant treatment for postmenopausal women with early-stage breast cancer. After a median of 120 months, there were significant improvements in the anastrozole group (but not the combination group) compared with the tamoxifen group for disease-free survival (hazard ratio [HR] 0.91, 95% CI 0.83–0.99; $p = 0.04$), time to recurrence (0.84, 0.75–0.93; $p = 0.001$), and time to distant recurrence (0.87, 0.77–0.99; $p = 0.03$).[23] The Trans-ATAC trial evaluated the prognostic utility of RS in the 1231 treated on the two monotherapy arms.[24] RS was significantly associated with distant recurrence when adjusted for tumor size, grade, age and treatment in the 872 patients with negative axillary nodes (HR 5.25, $p < 0.001$) and the 359 patients with positive axillary nodes (HR 3.47, $p < 0.002$). The rates of disease distant recurrence at 9 years for patients with a low, intermediate, and high RS were 4%, 12% and 25%, respectively for patients with negative nodes, and 17%, 28%, and 49%, respectively, for those with positive axillary nodes. Overall survival rates at 9 years were 88%, 84% and 73% for patients with negative nodes and 74%, 69% and 54% for patients with positive nodes. The prognostic value of RS was similar in anastrozole- and tamoxifen-treated patients, thereby providing information about the performance of RS in an AI-treated population. In addition, the prognostic information from RS was independent of the prognostic effect of Adjuvant! and vice versa; Adjuvant! is Web-based tool that provided accurate prognostic information by integrating clinical and treatment information using a validated algorithm.[25] In a model adjusted for treatment, RS and Adjuvant!, each provided a comparable degree of mutually independent prognostic information, indicating that RS and Adjuvant! provide complementary information.

A fourth study evaluated prognostic utility of RS in the context of adjuvant chemotherapy and endocrine therapy. The study included patients enrolled in ECOG trial E2197, which included 2885 patients with stage I-III operable breast cancer (and up to 3 positive axillary nodes) who were randomly assigned to receive four 3 week cycles of doxorubicin and cyclophosphamide or docetaxel plus endocrine therapy.[26] Tamoxifen was initially recommended to all patients after completion of chemotherapy, but 40% of patients eventually took and AI after the protocol when amended to allow AI therapy in postmenopausal women based upon emerging data. Of the 465 patients with hormone receptor positive disease included in the analysis, continuous RS was a highly significant predictor of recurrence (including distant and local) in both the node-negative and the node-positive patients ($p < 0.001$).[27] Forty-six percent of patients had low RS tumors, indicating that node-positive disease was associated with similar rates of low RS as node negative disease. RS also predicted recurrence more accurately than clinical variables when integrated by an algorithm modeled after Adjuvant! that was adjusted to 5-year outcomes.

13.6 Recurrence Score and Prediction of Chemotherapy Benefit

Two prospective validation studies have shown that patients with high RS benefit from chemotherapy, patients with a low RS are unlikely to benefit from chemotherapy, and the role of chemotherapy in patients with a mid-range RS is uncertain.

The first trial was NSABP trial B20, which included 2363 patients with ER-positive, lymph node negative disease who were randomized to receive tamoxifen alone or tamoxifen plus six cycles of CMF (cyclophosphamide, methotrexate and fluorouracil) or MF (methotrexate and fluorouracil). This study showed that the addition of chemotherapy reduced the risk of recurrence by about 30%.[20] Of the 2299 eligible patients, blocks containing sufficient invasive breast cancer were available for 670 patients. Gene expression results were obtained in 651 patients, including 227 of 770 tamoxifen-treated and 424 of 1529 chemotherapy-treated patients. Among the 651 assessable patients included in this analysis, the proportion without distant recurrence at 10 years was 92.2% for those treated with tamoxifen plus chemotherapy and 87.8% for those treated with tamoxifen alone. Of the 651 patients, 353 (54.2%) had a tumor with low RS, 134 (20.6%) had a tumor with an intermediate RS, and 164 (25.2%) had a tumor with a high RS. The test for interaction between chemotherapy treatment and RS was statistically significant ($P = 0.038$). Patients with high RS had a large benefit from chemotherapy (HR 0.26; 95% CI, 0.13 to 0.53; absolute decrease in 10-year distant recurrence rate: mean, 27.6%; SE, 8.0%). Patients with low-RS (<18) tumors derived minimal, if any, benefit from chemotherapy treatment (HR 1.31; 95% CI, 0.46 to 3.78; absolute decrease in distant recurrence rate at 10 years: mean, −1.1%; SE, 2.2%). Patients with intermediate-RS tumors did not appear to have a large benefit, but the uncertainty in the estimate could not exclude a clinically important benefit. The magnitude of chemotherapy benefit appeared to increase continuously as the RS increased, but there were an insufficient number of patients treated to identify with precision a threshold above which chemotherapy benefit could be identified.

Although the B20 trial provided convincing evidence that high RS was associated with chemotherapy benefit, the trial did include patients who had previously been used in the training set for development of the RS, raising the concern about overfitting. However, a second trial which did not share this deficiency found similar results. The analysis included specimens derived from 1477 postmenopausal women with lymph node positive, ER-positive operable breast cancer enrolled on SWOG trial S8814 who were randomized to receive tamoxifen alone, or tamoxifen beginning concurrently with or sequentially following adjuvant chemotherapy including cyclophosphamide, doxorubicin, and 5-flourourail (CAF). In a subset of 367 patients treated either with tamoxifen alone ($N = 148$) or sequential CAF followed by tamoxifen ($N = 219$), there was a significant improvement in disease free survival (DFS) from CAF for patients with a high RS (HR = 0.59; $p = 0.033$), but not a low (HR = 1.02, $p = 0.97$) or intermediate RS (HR = 0.72, $p = 0.48$).[24] For patients with tumors associated with a low RS, the 10-year DFS estimates 64% for the CAF-tamoxifen group compared with 60% in the tamoxifen alone group, indicating that although patients were at high risk for an event, there did not appear to be any benefit from chemotherapy.

13.7 Impact of RS on Clinical Decision Making

Several reports have indicated that application of 21-gene assay results in a change in therapeutic recommendations approximately up to 50% of patients, usually in the direction of sparing of adjuvant chemotherapy, as summarized in Table 13.4.

The first reported study was retrospective analysis of 68 patients with node negative, ER-positive breast cancer treated by four oncologists for whom the RS was ordered and treatment recommendations had been documented before knowledge of the RS.[28] The oncologists changed their treatment recommendation in 14 patients (21%) after the RS became available, including seven patients in whom the recommendation was changed from endocrine therapy alone to chemoendocrine therapy, and seven patients in whom the recommendation was changed from chemoendocrine therapy to endocrine therapy alone. The actual treatment administered changed in 25% of patients, of whom 82% changed from chemoendocrine therapy to endocrine therapy alone. This discrepancy was due to the patient's involvement in the shared decision making process;

a physician's recommendation to change spare chemotherapy was accepted more often than a recommendation to add chemotherapy.

Table 13.4 Studies evaluating effect of 21-gene assay on clinical decision making

Author	Patient population	No. of patients	No. of physicians	Summary of key findings
Oratz *et al.*[28]	Node-negative	68	4	Treatment administered changed for 25% of patients after knowledge of the RS, of whom 21% changed from chemoendocrine therapy to endocrine therapy alone
Lo *et al.*[29]	Node-negative	89	17	Physician recommendation changed in 32% of patients after knowledge of RS, of whom 23% changed from chemoendocrine therapy to endocrine therapy alone
Oratz *et al.*[30]	Node-positive	160	160	Physician recommendation changed in 51% of patients after knowledge of RS, of whom 33% changed from chemoendocrine therapy to endocrine therapy alone, and 9% changed from endocrine therapy alone to chemoendocrine therapy

A second report involved a prospective trial which included 89 assessable patients who were treated by 17 medical oncologists at one community and three academic practices.[29] The oncologists stated their adjuvant treatment recommendations before and after obtaining a RS, and patients likewise indicated their treatment choice before and after being informed of the RS result. The treatment recommendations of the physicians changed in 32% of the cases with the knowledge of the RS, usually (23%) from chemotherapy plus endocrine therapy to endocrine therapy alone. Likewise, 27% of patients changed their treatment decision with knowledge of the RS, including 10% who chose endocrine therapy alone instead of chemotherapy followed by endocrine therapy. Physician's confidence in their treatment recommendation increased with the RS in 68% of cases. More importantly patient anxiety and decisional conflicts were significantly lower with the knowledge of their recurrence score. This is consistent with past research that demonstrated that women with breast cancer want to participate in treatment decision making.

A third report described a survey of 160 medical oncologists who had ordered an Oncotype DX assay for patients with node-positive disease, were invited to participate in a survey, and agreed to participate.[30] The majority of the respondents (71%) were in the community and had practiced for a median of 11 years. The proportion of patients who had 1, 2, 3, or 4 or more positive lymph nodes was 69%, 18%, 6%, and 3% of patients, respectively. Eighty-six percent made treatment recommendations before obtaining the RS, of whom 51% changed their recommendation after receiving the RS. The treatment recommendation was changed from chemotherapy plus endocrine therapy to endocrine therapy alone in 33%, and from endocrine therapy alone to chemoendocrine therapy in 9%.

13.8 Gene Expression Profiles and Expert Panels

Two expert reviews have indicated that evidence supporting the clinical utility of gene expression profiles for breast cancer is insufficient.[31,32] On the other hand, expert panels convened by the American Society of Clinical Oncology (ASCO) and National Comprehensive Cancer Network (NCCN) concluded that certain multiparameter assays such as the 21-gene assay demonstrate clinical utility and recommend their use in specific clinical scenarios.[33,34] These recommendations are based largely upon validation studies demonstrating a benefit from chemotherapy in high risk subjects as identified by multiparameter gene expression assay.[35] It is biologically plausible that high risk populations identified by other assays might also benefit from chemotherapy, since poor risk signatures are driven by high expression of proliferation associated genes that predict response to chemotherapy (particularly in ER-positive, HER2-negative disease).[36]

13.9 Prospective Clinical Trials Evaluating Multiparameter Assays

Two clinical trials are prospectively evaluating multiparameter assays in clinical practice.[37] In the TAILORx trial (Trial Assigning Individualized Options for Treatment), patients 75 years of age or younger with ER-positive, HER2-negative, axillary node negative breast cancer who meet established NCCN guidelines for adjuvant chemotherapy have the treatment assigned or randomized on the basis of the 21-gene assay results (NCT00310180). Patients with a low RS are directed to endocrine therapy and with high RS are directed to chemoendocrine therapy, whereas those with an indeterminate mid-range RS (11–25) are randomized to receive chemoendocrine therapy (the standard treatment arm) or endocrine therapy alone (the experimental arm).[38] In the MINDACT trial (Microarray in Node-Negative Disease May Avoid Chemotherapy) trial (NCT00433589), patients with stage I-II breast cancer undergo risk assessment by the 70-gene assay and Adjuvant online, are assigned to endocrine therapy alone if both genomic and clinical criteria are concordant for low risk disease and chemoendocrine if concordant for high risk disease; patients who risk is discordant by genomic and clinical features are randomized to treatment assigned by clinical or genomic criteria.

13.10 Conclusion

The 21-gene RS assay brings a high degree of precision and standardization to the biologic characteristics of ER-positive breast cancer. Although the test identifies patients most likely to benefit from chemotherapy, there is clearly bias in use of the assay. For example, clinicians are less likely to order the test in patients who have very favorable or very unfavorable characteristics, and more likely to order it in situations where there is therapeutic equipoise (e.g., node negative disease, 1–2 cm tumor, intermediate histologic grade). For those selected to undergo testing, about 20% or more may be spared chemotherapy. Some analyses have suggested that use of the test is cost effective.[39–41] Uncertainty remains regarding the optimal management of patients with a mid-range RS, and a large clinical trial has been completed that was designed to address this issue.

References

1. Jemal A, Bray F, Center MM, *et al.* Global cancer statistics. *CA Cancer J Clin* 61: 69–90, 2011.
2. Jemal A, Siegel R, Ward E, *et al.* Cancer statistics, 2008. *CA Cancer J Clin* 58: 71–96, 2008.
3. Berry DA, Cronin KA, Plevritis SK, *et al.* Effect of screening and adjuvant therapy on mortality from breast cancer. *N Engl J Med* 353: 1784–1792, 2005.
4. Hatchwell E, Greally JM: The potential role of epigenomic dysregulation in complex human disease. *Trends Genet* 23: 588–595, 2007.
5. Thompson RF, Suzuki M, Lau KW, *et al.* A pipeline for the quantitative analysis of CG dinucleotide methylation using mass spectrometry. *Bioinformatics* 25: 2164–2170, 2009.
6. Iorio MV, Casalini P, Tagliabue E, *et al.* MicroRNA profiling as a tool to understand prognosis, therapy response and resistance in breast cancer. *Eur J Cancer* 44: 2753–2759, 2008.
7. Gast MC, Schellens JH, Beijnen JH: Clinical proteomics in breast cancer: a review. *Breast Cancer Res Treat* 116: 17–29, 2009.
8. Figueroa ME, Reimers M, Thompson RF, *et al.* An integrative genomic and epigenomic approach for the study of transcriptional regulation. *PLoS ONE* 3: e1882, 2008.
9. Ransohoff DF: Rules of evidence for cancer molecular-marker discovery and validation. *Nat Rev Cancer* 4: 309–314, 2004.
10. Dobbin K, Simon R: Sample size determination in microarray experiments for class comparison and prognostic classification. *Biostatistics* 6: 27–38, 2005.

11. Dobbin KK, Simon RM: Sample size planning for developing classifiers using high dimensional DNA microarray data. *Biostatistics* 8: 101–117, 2006.

12. Simon R, Radmacher MD, Dobbin K, *et al.* Pitfalls in the use of DNA microarray data for diagnostic and prognostic classification. *J Natl Cancer Inst* 95: 14–18, 2003.

13. Hayes DF, Bast RC, Desch CE, *et al.* Tumor marker utility grading system: a framework to evaluate clinical utility of tumor markers. *J Natl Cancer Inst* 88: 1456–1466, 1996.

14. Popovici V, Chen W, Gallas BG, *et al.* Effect of training sample size and classification difficulty on the accuracy of genomic predictors. *Breast Cancer Res* 12: R5 doi: 10.1186/bcr2468, 2010.

15. McShane LM, Altman DG, Sauerbrei W, *et al.* Reporting recommendations for tumor marker prognostic studies (REMARK). *J Natl Cancer Inst* 97: 1180–1184, 2005.

16. Ball CA, Sherlock G, Parkinson H, *et al.* Standards for microarray data. *Science* 298: 539, 2002.

17. Cronin M, Pho M, Dutta D, *et al.* Measurement of gene expression in archival paraffin-embedded tissues: development and performance of a 92-gene reverse transcriptase-polymerase chain reaction assay. *Am J Pathol* 164: 35–42, 2004.

18. Cobleigh MA, Tabesh B, Bitterman P, *et al.* Tumor gene expression and prognosis in breast cancer patients with 10 or more positive lymph nodes. *Clin Cancer Res* 11: 8623–8631, 2005.

19. Paik S, Shak S, Tang G, *et al.* A multigene assay to predict recurrence of tamoxifen-treated, node-negative breast cancer. *N Engl J Med* 351: 2817–2826, 2004.

20. Fisher B, Jeong JH, Dignam J, *et al.* Findings from recent National Surgical Adjuvant Breast and Bowel Project adjuvant studies in stage I breast cancer. *J Natl Cancer Inst Monogr* (30): 62–66, 2001.

21. Habel LA, Shak S, Jacobs MK, *et al.* A population-based study of tumor gene expression and risk of breast cancer death among lymph node-negative patients. *Breast Cancer Res* 8: R25 doi:10.1186/bcr1412, 2006.

22. Winer EP, Hudis C, Burstein HJ, *et al.* American Society of Clinical Oncology technology assessment on the use of aromatase inhibitors as adjuvant therapy for postmenopausal women with hormone receptor-positive breast cancer: status report 2004. *J Clin Oncol* 23: 619–629, 2005.

23. Cuzick J, Sestak I, Baum M, *et al.* Effect of anastrozole and tamoxifen as adjuvant treatment for early-stage breast cancer: 10-year analysis of the ATAC trial. *Lancet Oncol* 11: 1135–1141, 2010.

24. Dowsett M, Cuzick J, Wale C, *et al.* Prediction of risk of distant recurrence using the 21-gene recurrence score in node-negative and node-positive postmenopausal patients with breast cancer treated with anastrozole or tamoxifen: a TransATAC study. *J Clin Oncol* 28: 1829–1834.

25. Olivotto IA, Bajdik CD, Ravdin PM, *et al.* Population-based validation of the prognostic model ADJUVANT! for early breast cancer. *J Clin Oncol* 23: 2716–2725, 2005.

26. Goldstein LJ, O'Neill A, Sparano JA, *et al.* Concurrent doxorubicin plus docetaxel is not more effective than concurrent doxorubicin plus cyclophosphamide in operable breast cancer with 0 to 3 positive axillary nodes: North American Breast Cancer Intergroup Trial E 2197. *J Clin Oncol* 26: 4092–4099, 2008.

27. Goldstein LJ, Gray R, Badve S, *et al.* Prognostic utility of the 21-gene assay in hormone receptor-positive operable breast cancer compared with classical clinicopathologic features. *J Clin Oncol* 26: 4063–4071, 2008.

28. Oratz R, Paul D, Cohn AL, *et al.* Impact of a commercial reference laboratory test recurrence score on decision making in early-stage breast cancer. *J Oncol Pract* 3: 182–186, 2007.

29. Lo SS, Mumby PB, Norton J, *et al.* Prospective multicenter study of the impact of the 21-gene recurrence score assay on medical oncologist and patient adjuvant breast cancer treatment selection. *J Clin Oncol* 28: 1671–1676, 2010.

30. Oratz R, Kim B, Chao C, *et al.* Physician survey of the effect of the 21-gene recurrence score assay results on treatment recommendations for patients with lymph node-positive, estrogen receptor-positive breast cancer. *J Oncol Pract* 7: 94–99, 2011.

31. Marchionni L, Wilson RF, Wolff AC, *et al.* Systematic review: gene expression profiling assays in early-stage breast cancer. *Ann Intern Med* 148: 358–369, 2008.

32. Evaluation of Genomic Applications in Practice and Prevention (EGAPP) Working Group. Recommendations from the EGAPP Working Group: can tumor gene expression profiling improve outcomes in patients with breast cancer? *Genet Med* 11: 66–73, 2009.

33. Harris L, Fritsche H, Mennel R, *et al.* American Society of Clinical Oncology 2007 update of recommendations for the use of tumor markers in breast cancer. *J Clin Oncol* 25: 5287–5312, 2007.

34. Carlson RW, Allred DC, Anderson BO, *et al.* Breast cancer. Clinical practice guidelines in oncology. *J Natl Compr Cancer Netw* 7: 122–192, 2009.

35. Paik S, Tang G, Shak S, *et al.* Gene expression and benefit of chemotherapy in women with node-negative, estrogen receptor-positive breast cancer. *J Clin Oncol* 24: 3726–3734, 2006.

36. Desmedt C, Haibe-Kains B, Wirapati P, *et al.* Biological processes associated with breast cancer clinical outcome depend on the molecular subtypes. *Clin Cancer Res* 14: 5158–5165, 2008.

37. Sotiriou C, Pusztai L: Gene-expression signatures in breast cancer. *N Engl J Med* 360: 790–800, 2009.

38. Sparano JA: TAILORx: trial assigning individualized options for treatment (Rx). *Clin Breast Cancer* 7: 347–350, 2006.

39. Klang SH, Hammerman A, Liebermann N, *et al.* Economic implications of 21-gene breast cancer risk assay from the perspective of an Israeli-managed health-care organization. *Value Health* 13: 381–387, 2010.

40. Kondo M, Hoshi SL, Yamanaka T, *et al.* Economic evaluation of the 21-gene signature (Oncotype DX) in lymph node-negative/positive, hormone receptor-positive early-stage breast cancer based on Japanese validation study (JBCRG-TR03). *Breast Cancer Res Treat* 127: 739–749, 2011.

41. Vanderlaan BF, Broder MS, Chang EY, *et al.* Cost-Effectiveness of 21-gene assay in node-positive, early-stage breast cancer. *Am J Manag Care* 17: 455–464, 2011.

Chapter 14

MammaPrint for Individualized Recurrence Risk Assessment and Treatment Recommendations for Early-Stage Breast Cancer Patients

Sonal J. Desai and Tianhong Li
University of California Davis Comprehensive Cancer Center, Division of Hematology & Oncology, Sacramento, CA 95817, USA
sonal.desai@ucdmc.ucdavis.edu, tianhong.li@ucdmc.ucdavis.edu

14.1 Introduction

Breast cancer is the second most common cancer and the second leading cause of cancer-related deaths in women worldwide. In 2013 an estimated 232,340 women are expected to be diagnosed with new cases of invasive breast cancers in the United States, and 39,620 women are expected to die from the disease [1, 2]. The majority of these deaths are due to disease recurrence or distant metastasis after initial treatment. Adjuvant systemic therapy with either endocrine therapy and/or chemotherapy has been shown to reduce the risk of distant recurrence and death from invasive breast cancer after local treatment with surgery with or without radiation therapy. To save lives, existing guidelines, aimed at avoiding under use of adjuvant

Handbook of Therapeutic Biomarkers in Cancer
Edited by Sherry X. Yang and Janet E. Dancey
Copyright © 2013 Pan Stanford Publishing Pte. Ltd.
ISBN 978-981-4364-65-2 (Hardcover), 978-981-4364-66-9 (eBook)
www.panstanford.com

systemic therapy in early-stage breast cancer patients, recommend considering adjuvant therapy in all age of patients whose tumors are at least 1.0 cm [3, 4]. The decision to administer adjuvant therapy is based on several prognostic factors, including patient age, co-morbidities, tumor size, tumor grade, number of involved axillary lymph nodes, and possibly HER2 tumor status. These clinical and pathological features of breast cancer can be calculated using several Web-based tools such as Adjuvant! Online (www.adjuvantonline.com). Under this guideline, only 15–20% of node-negative breast cancer patients whose tumors are 1.0 cm or less are considered "low-risk," given that the 10-year recurrence risk is 10% and less. As only 30% of node-negative and 70% of node-positive breast cancer patients have recurrent disease, this recommendation has led to overtreatment with chemotherapy in many breast cancer patients.

The current clinical and pathological risk assessment tool also does not take into account the influence of tumor genetics or biology on disease recurrence for individual patients. With the advances in molecular technologies, molecular profiling has been increasingly used to subtype breast cancers and stratify patients into different prognostic and/or predictive subsets for risk assessment and individualized cancer therapy for both standard and novel targets [5]. While many of these molecular profiling assays are still in preclinical development, two diagnostic tests, Oncotype Dx and MammaPrint®, which use multi-gene expression signatures of breast tumors to assess the genetic risk of breast cancer recurrence in individual patients, have reached clinical testing. Like other new diagnostic biomarker assays, the development of these tests occurs in several phases before widespread clinical use, i.e., discovery, retrospective clinical validation, analytic validation, and prospective clinical validation. The utility of these tests in guiding clinical treatment decision are being evaluated in two large-scale, prospective randomized trials, the TAILORx for Oncotype Dx and the MINDACT for MammaPrint. In this chapter, we will summarize the development and clinical use of the MammaPrint assay. The Oncotype Dx test is reviewed in another chapter of this book.

14.2 Discovery of MammaPrint

The MammaPrint assay uses microarray technology to assess a 70-gene expression profile to assess a breast cancer patient's risk of developing recurrence or distant metastases or death [4, 6–8]. The 70-gene expression profile was discovered by researchers at the Netherlands Cancer Institute, and is now performed and marketed by Agendia (www.agendia.com). This test was developed using an unbiased 3-step supervised biostatistical method to identify common gene expression patterns in 78 primary breast tumors from lymph-node negative patients [7]. First, gene expression microarray analysis using an Agilent Hu25K array, which contains approximately 25,000 human genes, identified 5,000 genes that were commonly regulated genes in at least 80% of samples. Next, 231 genes were selected based on significant association with disease outcome from patient data using the magnitude of the correlation coefficient. Finally, a leave-one-out method was used to optimize this gene list, resulting in a final list of 70 genes. The expression level of these 70 genes was used to calculate a recurrence score that is either low risk or high risk. "Low Risk" MammaPrint result means that a patient has a 10% chance that her cancer will recur within 10 years without any additional adjuvant treatment, either hormonal therapy or chemotherapy. A "High Risk" MammaPrint result means that a patient has a 29% chance that her cancer will recur within 10 years without any additional adjuvant treatment, either hormonal therapy or chemotherapy [9]. Patients at low risk might be safely spared adjuvant systemic chemotherapy, while patients at high risk might benefit from adjuvant systemic chemotherapy. The positive predictive value of tumor recurrence in high-risk patients at 5 and 10 years is 23% and 29%, respectively. The negative predictive value of tumor recurrence in low-risk patients at 5 and 10 years is 95% and 90%, respectively. Thus, compared to the current clinical-pathological risk assessment tools, the MammaPrint test more accurately identifies breast cancer patients with low-risk for recurrence than patients with high risk for recurrence.

14.3 Retrospective Clinical Validation

The MammaPrint test was first validated in a study of lymph-node negative breast cancer patients that were less than 56 years old with tumors less than 5 cm [7]. To date, the MammaPrint assay has been validated in various other retrospective studies in more than 5,600 patients [8–27]. These studies support that MammaPrint could accurately identify those patients with low risk for distant recurrence and thus could avoid unnecessary adjuvant systemic chemotherapy. Table 14.1 summarizes the published validation tests of the MammaPrint assay in different populations of breast cancer patients, including various European (Dutch, German, Italian, Spanish), American, and Japanese cohorts.

Small tumors (T1 or ≤2 cm) are generally considered to have a low risk of recurrence after surgical resection. However, some stage 1 breast cancers do metastasize, and it has been a challenge to try and identify those small tumors with metastatic potential. MammaPrint evaluation of 964 patients with T1 tumors showed that 46% had a poor prognosis signature, suggesting that MammaPrint is a better tool for selecting early-stage breast cancer patients with smaller tumors for adjuvant therapy compared to current clinical-pathological risk assessment tools [15, 23]. The MammaPrint test was developed using fresh tumor specimens obtained at the time of surgery. However, more recently, Mayordomo and colleagues showed that sufficient RNA could be isolated from 14-gauge core biopsies to perform array analysis [22, 28]. This broadens the clinical applicability of MammaPrint as it eliminates the reliance on surgical specimens, which are often fixed immediately. Further optimization is required for smaller tumors since one in seven tumor specimens taken from tumors smaller than 2 cm had insufficient tumor material for analysis. Recently, studies were conducted to compare gene signatures obtained from MammaPrint using matched samples from fresh frozen tissue and samples from formalin-fixed, paraffin embedded tissue. Results showed that the gene signatures from both samples were highly similar, increasing the potential resources available for the MammaPrint test [29]. While most of the validation studies have retrospective analyses, one study by Bueno-de-Mesquita *et al.* showed that MammaPrint could be used prospectively for risk assessment and that MammaPrint is a feasible option for use in community hospitals [12].

MammaPrint has also been validated in many different clinical scenarios, such as in patients ranging from 35 to 70 years of age, in pre- or postmenopausal women, and in Stage T0–T4 breast cancers that are either node positive or node negative (Table 14.1). Multivariate analyses show that MammaPrint risk assessment is independent of ER, PR and HER2 status [10].

Table 14.1 Summary of published validation tests of MammaPrint assay

First author (year)	Patient population	Period of specimen collection	No. of patients	Age of patient	Histopathology	Median follow-up (years)	Distant metastasis-free survival (DMFS)			Overall survival (OS)		
							Good prognostic signature	Poor prognostic signature	Hazard ratio (HR)	Good prognostic signature	Poor prognostic signature	Hazard ratio (HR)
Van de Vijver (2002)	Dutch	1984–1995	295	<53	pTI-pT2, N0 or N+	6.7	$n = 115$	$n = 180$	N.A.			
							(5y) 94.7% ± 2.1%	(5y) 60.5% ± 3.8	<5y: HR 8.8; 95% CI 3.8–20; $P < 0.001$	(5y) 97.4 ± 1.5%	(5y) 74.1 ± 3.3%	HR 8.6; 95% CI 4–19; $P < 0.001$
							(10y) 85.2% ± 4.3%	(10y) 50.6 % ± 4.5	>5y: HR 1.8; 95% CI 0.69–4.5 $P = 0.24$	(10y) 94.5 ± 2.6	(10y) 54.6 ± 4.4%	N.A
Buyse (2006)	European	1980–1998	307	<61	T1-T2, N0, ER-	13.6	$n = 113$	$n = 194$	N.A.			
							(5y) 96%	(5y) 83%	N.A.			
							(10y) 88%	(10y) 71%	(10y) HR 1.5 95% CI 1.04–2.16 $P < 0.001$	(10y) 88%	(10y) 69%	(10y) HR 2.79: 95% CI 1.6–4.87; $P = 0.032$

(Continued)

Table 14.1 (*Continued*)

First author (year)	Patient population	Period of specimen collection	No. of patients	Age of patient	Histopathology	Median follow-up (years)	Distant metastasis-free survival (DMFS)			Overall survival (OS)		
							Good prognostic signature	Poor prognostic signature	Hazard ratio (HR)	Good prognostic signature	Poor prognostic signature	Hazard ratio (HR)
Bueno-de-Mesquita (2007)	Dutch	2004–2006	427	<61	T1-4, N0, M0	1.16 (14 months)	*n* = 219	*n* = 208	N.A.	N.A.	N.A.	N.A
Bueno-de-Mesquita (2008) [Update report of Van de Vijver (2002)]	Dutch	1984–1995	151	<53	N0	10.2	*n* = 60	*n* = 91	N.A.			
							(10y) 86 ± 5%	(10y) 50 ± 6%	HR 5.5; 95% CI 2.5–12 *P* < 0.01	(10y) 94 ± 3%	(10y) 51 ± 5%	HR 10.7; 95% CI 3.9–30; *P* < 0.01
Bueno-de-Mesquita (2008)	Dutch	1996–1999	123	<56 years	pT1-T2, N0	5.8	*n* = 64	*n* = 59				
							(5y) 98 ± 2%	(5y) 78 ± 6%	HR 5.7; 95% CI 1.6–20; *P* = 0.007	(5y) 97 ± 2%	(5y) 82 ± 5%	HR 3.4; 95% CI 1.2–9.6; *P* = 0.021

First author (year)	Patient population	Period of specimen collection	No. of patients	Age of patient	Histopathology	Median follow-up (years)	Distant metastasis-free survival (DMFS)			Overall survival (OS)		
							Good prognostic signature	Poor prognostic signature	Hazard ratio (HR)	Good prognostic signature	Poor prognostic signature	Hazard ratio (HR)
Mayordomo (2008)	Spain	2006–2008	35	all ages	T2-T3 (Core needle biopsy)	N.A.	$n = 4$	$n = 31$				
Wittner (2008)	US	1985–1997	100	median age 63	pT1-T3, pN0	11.3	$n = 27$	$n = 73$				
							(5y) 93%	(5y) 51%	N.A.	85%	52%	N.A
							(10y) 75%	(10y) 72%	N.A.	N.A.	N.A.	N.A
Mook (2008)	European	1994–2001	241	<71	T1-T3, N1-N3	7.8	$n = 99$	$n = 142$				
							(5y) 98 ± 2%	(5y) 80 ± 4%	N.A.	(5y) BCSS 96 ± 2%	(5y) BCSS 75 ± 4%	BCSS HR 5.7; 95% CI 2.01–16.23; $P = 0.001$
							(10y) 91 ± 4%	(10y) 76 ± 4%	HR 4.13; 95% CI 1.72–9.96; $P = 0.002$	(10y) BCSS 99 ± 1%	(10y) BCSS 88 ± 3%	(Overall) HR 5.4; 95% CI 2.11–13.8; $P < 0.001$
Mook (2008)	Dutch	1984–1996	148	<71	T1-T2, N0	12.5	$n = 91$	$n = 57$				

(*Continued*)

Table 14.1 *(Continued)*

First author (year)	Patient population	Period of specimen collection	No. of patients	Age of patient	Histopathology	Median follow-up (years)	Distant metastasis-free survival (DMFS)			Overall survival (OS)		
							Good prognostic signature	Poor prognostic signature	Hazard ratio (HR)	Good prognostic signature	Poor prognostic signature	Hazard ratio (HR)
							(5y) 93 ± 3%	(5y) 72 ± 6%	(5y) HR 4.6; 95% CI 1.8–12; *P* = 0.001	(5y) BCSS 99 ± 1%	(5y) BCSS 80 ± 5%	HR 19.1; 95% CI 2.5–148; *P* = 0.005
							(overall DMFS) 81%	(overall DMFS) 68%	(Overall) HR 1.8; 95% CI 0.9–3.5; *P* = 0.07	N.A.	N.A.	(Overall) HR 2.0; 95% CI 1.0–4.0; *P* = 0.04
Generali (2008)	Italy	N.A	13	N.A.	T2-T4, N0-N1, ER+, Postmenapausal	N.A.	*n* = 4	*n* = 9	N.A.	N.A.	N.A.	N.A
Kunz (2009)	German	2005–2008	140	N.A.	T1-T2	N.A.	*n* = 77	*n* = 59	N.A.	N.A.	N.A.	N.A
Saghatchian (2009)	European	N.A	153	N.A.	T1-T3, *N* = 4–6	N.A.	*n* = 65	*n* = 88				
							N.A.	N.A.	N.A.	(5y) 97 ± 2%	(5y) 79 ± 5%	HR 2.4; 95% CI 1.0–5.5; *P* = 0.04

First author (year)	Patient population	Period of specimen collection	No. of patients	Age of patient	Histopathology	Median follow-up (years)	Distant metastasis-free survival (DMFS)			Overall survival (OS)		
							Good prognostic signature	Poor prognostic signature	Hazard ratio (HR)	Good prognostic signature	Poor prognostic signature	Hazard ratio (HR)
							N.A.	N.A.	N.A.	(10y) 91 ± 4%	(10y) 74 ± 5%	N.A
Bender (2009)	European	N.A	1637	N.A.	T1-T2, N0 or N+	7.1	$n = 772$	$n = 865$				
							(10y) 90%	(10y) 30%	HR 3.88; 95% CI 1.99–7.58; $P < 0.01$	N.A.	N.A.	N.A
Knauer (2009) [Update report of Bender (2009)]	European	1984–2009	541	N.A.	pT1-T3, N0-N1, M0	7.1	$n = 252$	$n = 289$	N.A.	N.A.	N.A.	N.A
							95%	82%	HR 3.88; 95% CI 1.99–7.58; $P < 0.01$	(5y) BCSS 97%	(5y) BCSS 87%	HR 4.01; 95% CI 1.98–11.67; $P = 0.05$
Ishitobi (2010)	Japanese	1998–2001	102	<70	N0	7.1	$n = 20$	$n = 82$				

(Continued)

Table 14.1 *(Continued)*

First author (year)	Patient population	Period of specimen collection	No. of patients	Age of patient	Histopathology	Median follow-up (years)	Distant metastasis-free survival (DMFS)			Overall survival (OS)		
							Good prognostic signature	Poor prognostic signature	Hazard ratio (HR)	Good prognostic signature	Poor prognostic signature	Hazard ratio (HR)
							(5y) 100%	(5y) 94%	N.A.	N.A.	N.A.	N.A
Straver (2010)	Dutch	2000–2008	162	<69	T1-T2	1.04 (25 months)	*n* = 23	*n* = 144	N.A.	N.A.	N.A.	N.A
Kunz (2010)	German	2004–2008	44	<56	T1-3, N0-3	N.A.	*n* = 29	*n* = 15	N.A.	N.A.	N.A.	N.A
Somlo (2010)	US	2006–2009	68	<70	T2-T3 (Core needle biopsy)	N.A.	*n* = 15	*n* = 53	N.A.	N.A.	N.A.	N.A
Mook (2010)	Dutch	N.A	964	N.A.	N.A.	7.1	*n* = 525	*n* = 439				
							(5y) 95 ± 1%	(5y) 80 ± 2%	N.A.	(5y) BCSS 99 ± 1%	(5y) BCSS 88 ± 2%	N.A
							(10y) 87 ± 2%	(10y) 72 ± 3%	(10y) HR 2.7; 95% CI 1.88–3.88; *P* < 0.001	(10y) BCSS 91 ± 2%	(10y) BCSS 72 ± 2%	(10y) HR 4.22: 95% CI 2.7–6.6; *P* < 0.001

Abbreviations: BCSS, breast cancer specific survival; N.A., not available.

14.4 Analytic Development for MammaPrint as a Diagnostic Test

All MammaPrint tests are performed using a customized microarray in the Agendia's Clinical Laboratory Improvement Amendments (CLIA)-accredited service laboratories at the company's headquarters in Amsterdam, the Netherlands since 2004 and in Irvine, California since 2008. In February 2007, it became the first *in vitro* diagnostic multivariate index assay cleared by the US Food and Drug Administration (FDA) to assess the risk of tumor recurrence in lymph node negative breast cancer patients under 61 years of age with tumors or less than 5 cm, who were within 10 years of diagnosis. FDA clearance under the *in vitro* diagnostic multivariate index assay (IVDMIA) guidelines requires clinical and analytical validation and reporting systems to ensure patient safety issues are addressed. The FDA label indicates that as a diagnostic tool, MammaPrint has a 98.9% degree of accuracy in classifying patients as Low Risk or High Risk and technical reproducibility of 98.5%. It is recommended that MammaPrint® results are used by physicians as a prognostic marker only, along with other clinicopathological factors in planning treatment with adjuvant chemotherapy.

Since its initial FDA clearance, several modifications of the cleared device have been made such as changing the specimen type from fresh frozen tissue to fresh tissue stored in a specific RNA preservative solution (i.e., RNARetain room temperature tissue fixative) and XPrint software v1.33 to v1.40. Recently, the FDA cleared the MammaPrint test for breast cancer recurrence in all ages and all stage I and II breast cancer patients, including patients with negative and up to three positive lymph nodes. Agendia received its fifth FDA clearance for MammaPrint in early 2011 to allow the test to be performed using two additional Agilent Microarray scanners and two Agilent Bioanalyzers in CLIA- and College of American Pathologists (CAP)-accredited central laboratories in Amsterdam, the Netherlands. With this approval, Agendia could provide a consistent testing service in case one machine is down and could accommodate the increasing need for clinical testing.

It is currently covered by Centers for Medicare and Medicaid Services and most major insurance carriers in the US. In Europe, MammaPrint has been approved for all ages since 2004. The cost of the assay in the US is $4,200. In Europe, the test costs EUR 2675. Some insurance companies will pay for the total cost of the MammaPrint test, while others may pay a portion of the cost. Agendia created the Agendia Cares Program to help with insurance and payment issues.

Table 14.2 summarizes the current approved indication for MammaPrint in the United States and Europe.

Table 14.2 Current approved indication for MammaPrint in the United States and Europe

Eligibility	United States	Europe and others
Staging and histology of breast cancer	stage I or stage II invasive hormone-receptor-positive AND hormone-receptor-negative smaller than 5 centimeters lymph node: 0–3	stage I or stage II invasive hormone-receptor-positive AND hormone-receptor-negative smaller than 5 centimeters lymph node: 0–3
Age	Women diagnosed with cancer must be 61 or younger All (since 2011)	All (since 2004)
Date of Approval	February 2007	2004
Cost	$4,200	EUR 2675
Coverage	Most of the insurance carriers	Depending on the country

14.5 Prospective Clinical Validation of MammaPrint

MINDACT (Microarray In Node-negative and 1 to 3 positive lymph node Disease may Avoid ChemoTherapy), as illustrated in Fig. 14.1, is a prospective randomized study comparing the 70-gene signature MammaPrint assay with the common clinical-pathological criteria in selecting patients for adjuvant chemotherapy in breast cancer with 0–3 positive nodes (EORTC Protocol 10041 -BIG 3–04; http://www.eortc.be/services/unit/mindact/MINDACT_websiteii.asp; ClinicalTrials.gov Identifier: NCT00433589) [30, 31]. The objective of MINDACT is to test whether patients with a low risk signature may

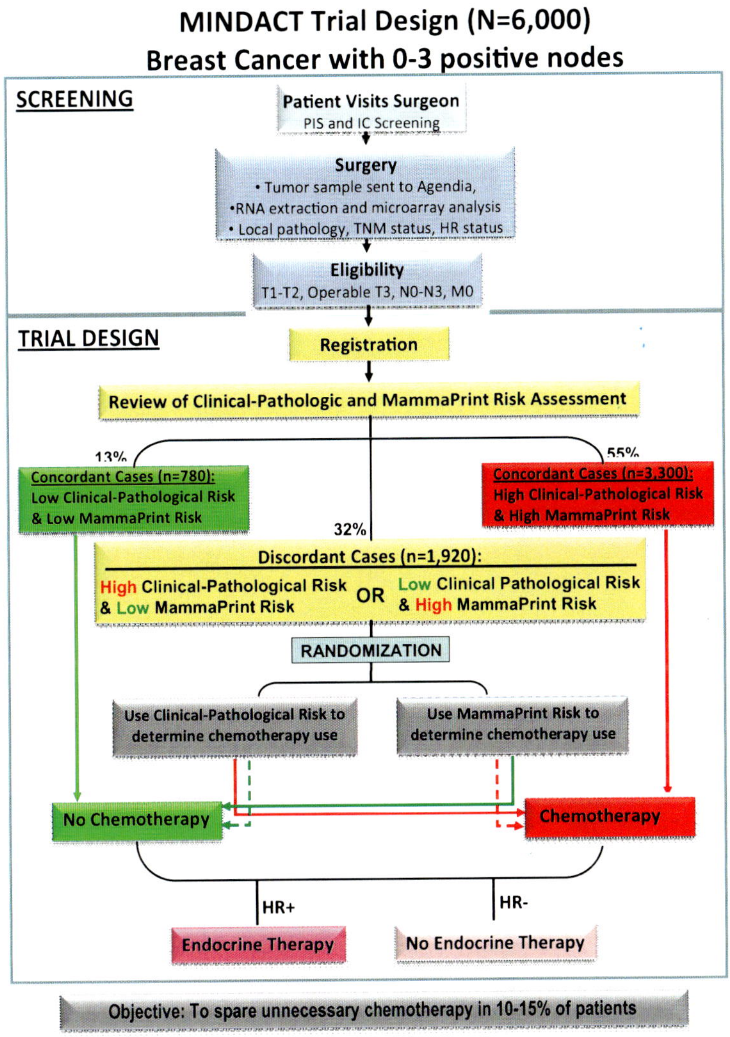

Figure 14.1 Schema of MINDACT study. *Abbreviations:* PIS, patient information sheet; IC, informed consent; T: tumor stage; N: nodes; M: metastasis; HR: hormone receptor.

safely be spared the toxicities of chemotherapy without affecting their survival. The study is complex in that it further tests which of two chemotherapy treatments offers better survival and which of two endocrine regimes is more effective. Each enrolled breast cancer patient will be assessed using both MammaPrint and conventional prognostic tools. For most women, the results of the two tests are expected to be the same. Chemotherapy, either anthracycline- or docetaxel-based, will be offered to those who are categorized as high risk by both assays. It will not be offered if they are shown to be at low risk by both assays. Women for whom the conventional and MammaPrint tests do not agree will be randomized according to conventional or MammaPrint test results and allocated to either adjuvant chemotherapy or no adjuvant chemotherapy. Additionally, all women who are estrogen receptor positive will be offered one of two endocrine treatment regimens: 7 years of single agent letrozole or the sequential strategy of 2 years of tamoxifen followed by 5 years of letrozole. It is hoped that not only will 10% to 20% of patients be spared treatment but that there will be considerable savings for the national health services. The MINDACT was first activated in the Netherlands on March 22, 2007, and was amended to increase the sample size to 6600. By July 2011, the study reached the accrual goal with 6700 breast cancer patients have been enrolled from 119 participating institutions in 9 European Countries. From 119 participating institutions in 9 European Countries. Similarly to the TAILORx trial in the United States, the MINDACT study has progressed well in accruing patients, suggesting the general acceptance of using new molecular risk assessment tools in guiding clinical decisions for treatment planning. The results from these two studies are highly anticipated.

14.6 Biologic Implication of MammaPrint Results

14.6.1 Understanding of Tumor Biology

It is interesting that a prognostic signature could be developed using the transcriptional profile of primary breast tumors. This suggests that the metastatic potential is inherently expressed in the initial tumor rather than being acquired at a later stage. A recent report provides functional annotation of the 70 genes in the MammaPrint with the hallmarks of cancer [32]. Figure 14.2 illustrates the genes that are known to be functionally involved in recurrence, such as signal transduction and cell cycle, invasion, metastasis and angiogenesis genes that are found to be significantly upregulated in the poor prognosis signature. Many of these candidate genes should be investigated to test their feasibility as new therapeutic targets. It is interesting to note that individual genes that have been previously correlated with disease outcome, such as cyclin D1, ERS1, HER2, c-myc, UPA and PAI-1 are not present in the 70-gene MammaPrint profile. This is most likely due to the overlap and redundancy inherent in biological processes as well as our own lack of knowledge. Indeed, genes regulated by ERα and HER2 are represented in the 70-gene profile. A network association map of the 70 genes that comprise MammaPrint shows that key players in cancer, such as p53, Rb, c-myc, Jun and CDKN2, are central regulators although their change in expression is not integral to the profile [32]. It is known that clinical factors, such as co-morbidities (e.g., diabetes, chronic inflammation, etc.) affect patient's risk of tumor recurrence and survival. Currently, it is not known how these clinical factors affect the results of 70-gene expression in MammaPrint test.

14.6.2 Revealing New Therapeutic Targets

Most of the genes in the MammaPrint profile are not targets of current drug discovery efforts; however, many can be categorized into one of the six "Hallmarks of Cancer," as defined by Hanahan and Weinberg [32, 33]. Eleven genes fall into the sustained proliferative signaling hallmark, 5 into the evading growth suppressors hallmark, 15 into the enabling replicative immortality hallmark, 5 into the resisting cell death hallmark, 8 into the activating invasion and metastasis hallmark and 12 into the inducing angiogenesis hallmark (Fig. 14.2). Often, the most druggable targets are the genes involved in signal transduction pathways, and several were found to be included in the MammaPrint profile, including FLT1, HRASLS, STK32B, TGFB3, RASSF7, MELK, IGFBP5, FGF18, and DCK. Four of these genes, each of which represents a different signaling pathway, stand out as pathway-specific drug targets for therapy. FLT1, a member of the VEGF pathway, is a receptor tyrosine kinase that binds to VEGFR-A, VEGFR-B and placental growth factor and plays an important role in angiogenesis and vasculogenesis [34]. TGFβ3 is part of the TGFβ pathway and is involved in cell differentiation [35]. IGFBP5 is a member of the insulin pathway and is linked to senescence and autophagy, which are often associated with apoptosis [36, 37]. FGF18 is a growth factor in the FGF signaling pathway and is thought to promote cell survival under stress, through autocrine and paracrine signaling, and angiogenesis [38]. While these represent mechanisms to target individual pathways, targeted therapy could be successful since they may be involved in multiple hallmark functions. While these are individual pathways, targeted therapy could be successful since they can overlap in their hallmark function. For instance, targeting FLT1 could prevent tumor cells from evading apoptosis and initiating/maintaining angiogenesis, and antibodies against TGFβ3 could resensitize cells to growth inhibitory therapies.

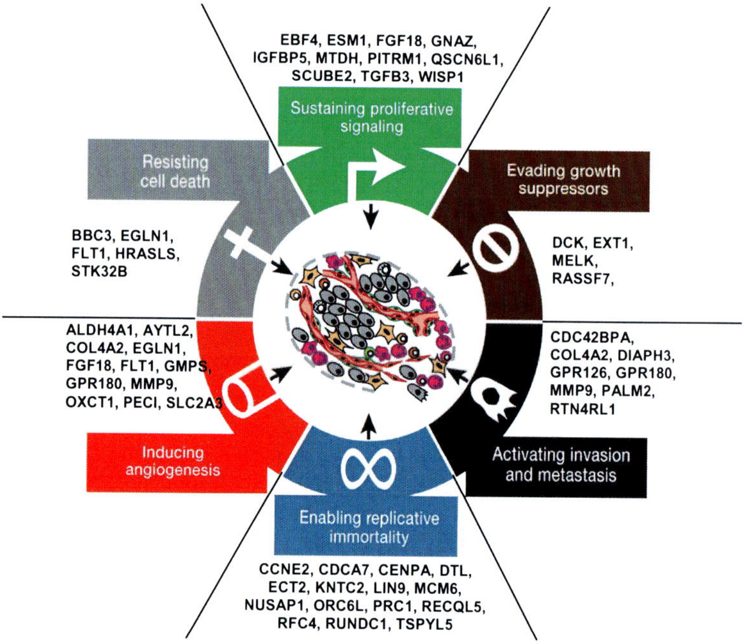

Figure 14.2 Biological function of MammaPrint® 70 genes and the hallmarks of cancer. The 70 genes are involved in at least eight well-defined hallmarks of cancer, in tumor progression and metastasis related biological processes, as well as epithelial-mesenchymal transition. Adapted from Biomark Insights. 2010, Tian *et al.*, Biological Functions of the Genes in the MammaPrint Breast Cancer Profile Reflect the Hallmarks of Cancer, 5: 129–138. Copyright © 2010 the author(s), publisher and licensee Libertas Academica Ltd. Modified with permission.

In addition to genes in specific pathways, there were several genes that may play a role in multiple signaling pathways. HRASLS is a RAS tumor suppressor [39], STK32B has been linked to the NFκB pathway [40], RASSF7 is a JNK inhibitor and involved in mitosis [41, 42], MELK inhibits apoptosis and has been associated with poor prognosis in breast cancer [43, 44], and DCK is involved in metabolism of nucleosides and is clinically important because of its relationship to drug resistance and sensitivity [45]. The advantage of targeting these genes is that signaling pathways often converge on downstream signaling molecules thereby targeting multiple pathways in a single agent.

Finally, several other genes are worth mentioning due to their importance in breast tumor biology. For instance, CCNE2 is a member of the cyclin family, is tightly controlled and mediates the G1/S transition during the cell cycle. It binds to CDK2, is regulated by estrogen, and has been shown to be upregulated in tumor cells [46, 47]. CENPA is a key player in mitosis and is required for recruitment of most proteins to the centromere [48]. COL4A2 is a type IV collagen and is a basement membrane protein that contributes to tumor structure, is degraded during tumor invasion and metastasis, and is overexpressed on ER-negative breast cancer patients [49]. MMP9 is a metalloproteinase extracellular matrix protein that is very important in metastasis. It is involved in the degradation of type IV and V collagens [50]. These are all potential drug targets. A recent update to the "Hallmarks of Cancer" added four new categories: inflammation, evading the immune system, unstable DNA, and deregulated metabolism [51], of which MammaPrint genes fit into the two latter categories (deregulated metabolism: ALDH4A1, AYTL2, DCK, GMPS, GSTM3, OXCT1, PECI, PITRM1, SLC2A3 and unstable DNA: CCNE2, CENPA, MCM6, NUSAP1, ORC6L, PRC1, RASSF7 and RFC4) as illustrated in Fig. 14.3. It is unclear yet how these new categories will aid in the development of more potent and beneficial therapy.

14.6.3 Prediction for Response or Resistance to Chemotherapy

MammaPrint has been established and validated as an accurate prognostic tool that categorizes patients according to their recurrence risk (low or high) for breast cancer and thus identifies patients for whom adjuvant therapy would be most beneficial. The next clinical challenge now becomes how to manage the high-risk

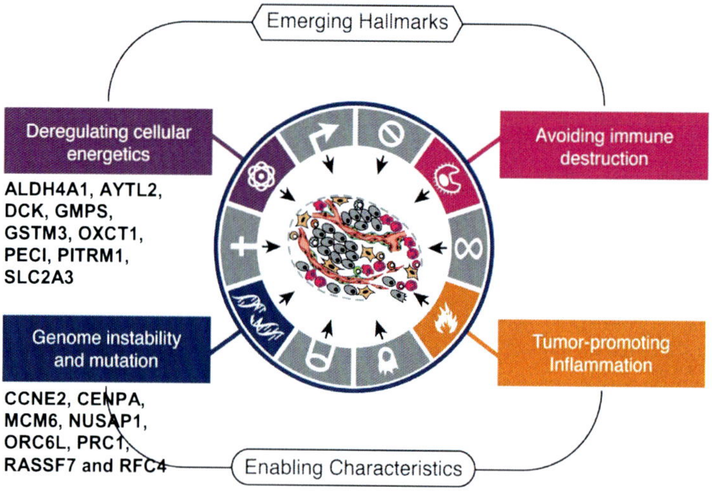

Figure 14.3 Biological function of MammaPrint® 70 genes and new hallmarks of cancer.

patients characterized by MammaPrint test. Breast cancer patients who were at high risk for recurrence had a 32% pathological complete response to chemotherapy in a recent neoadjuvant trial. In contrast, those patients who were low-risk had a 9% pathological complete response rate [26]. To determine the predictive value of MammaPrint for neoadjuvant chemotherapy, Knauer *et al.* reported on a meta-analysis of data gathered from seven large studies, comparing 541 patients that were given endocrine therapy alone or chemotherapy plus endocrine therapy [10, 17, 19]. MammaPrint analysis identified 252 of the 541 patients as low risk, with a 5-year distant disease-free survival (DDFS) of 95%, and 289 patients as high risk, with a 5-year DDFS of 82%. In the low-risk group, those who received chemotherapy plus endocrine therapy had only a slight increase in their 5-year DDFS (99% vs. 93%). In contrast, high-risk patients who received the combination therapy demonstrated a significantly higher 5-year DDFS compared to those who received endocrine therapy alone (88% vs. 76%). Multivariate analysis of clinical pathologic criteria that may have affected the results showed that there was a statistically significant correlation between high-risk signatures and tumor size, number of positive lymph nodes and positive PR status. Similar studies conducted with smaller numbers of patients also showed that patients with a low-risk gene signature are less sensitive to chemotherapy compared to patients with a high-risk gene signature, further confirming that neoadjuvant therapy is the most beneficial to patients with poor prognosis profiles, and that MammaPrint may be useful for predicting patient response to chemotherapy. MammaPrint assays have been incorporated in the two ongoing, large-scale neoadjuvant studies I-SPY1 Trial (Investigation of Serial Studies to Predict Your Therapeutic Response with Imaging And moLecular Analysis or The I-SPY Trial) and I-SPY2 Trial (Neoadjuvant and Personalized Adaptive Novel Agents to Treat Breast Cancer; ClinicalTrials.gov Identifier: No. NCT01042379). While I-SPY I aims to evaluate biomarkers and imaging for predicting response to standard neoadjuvant chemotherapy, I-SPY II will use biomarkers to stratify patients based on their predicted likelihood of response to treatment with molecularly targeted drugs in combination with standard chemotherapy in the neoadjuvant setting. More insights will be highly anticipated from these studies.

14.6.4 Elucidation of Resistant Mechanisms to Chemotherapy

MammaPrint is a good prognostic tool for predicting the risk of recurrence and metastasis in early-stage breast cancer patients. However the question remains: How do tumors become resistant to chemotherapy? Since the probability of recurrence and metastasis can be determined from the primary tumor, the possible mechanisms of resistance may also be part of the MammaPrint gene signature. The first steps of metastasis require cells to undergo a series of biochemical changes including cytoskeletal rearrangement and loss of cell polarity, breaking down the extracellular matrix and invasion into surrounding tissue. During these processes the cells go from an epithelial phenotype to a mesenchymal phenotype, called the epithelial to mesenchymal transition (EMT), also known to be part of the developmental process. Genes that mediate EMT may not be targeted by current chemotherapy regiments. There are several EMT related genes that are expressed in the MammaPrint gene signature. TGFβ signaling is known to induce EMT, and TGFβ1 and TGFβ3 have been shown to play complementary roles in the process [35]. IGFBP-5 is a member of the insulin pathway, but has also been linked to tissue remodeling by inducing the TGFβ pathway [52]. In addition, IGFBP-5 has been found to be involved in different mechanisms mediating resistances to tamoxifen in cell lines and mouse models [53]. MMP9, as mentioned above, is a matrix metalloproteinase that degrades extracellular matrix (ECM) proteins and may be upregulated by TGFβ signaling [54]. COL42A, another ECM protein mentioned in previous section, may also be involved in the metastasis process. While EMT is thought to mitigate the first steps of metastasis, the reverse process, mesenchymal to epithelial transition (MET) is required to establish the tumor cells in their distal locations. Re-depositing basement membrane protein at that time is consistent with re-establishing cell polarity, and the surrounding stromal environment. The WNT signaling pathway has also been implicated in EMT. WISP1, a downstream signaling molecule of WNT pathway, is expressed in fibroblasts and is thought to promote tumor cell migration as well as prevent apoptosis [55]. The role of EBF4 gene in EMT is unknown. However, it is a transcription factor involved in neuronal development, and therefore may modulate at least some part of the EMT process [56].

After tumor cells undergo EMT, they intravasate into blood vessels, extravasate to the metastatic site, initiate and sustain angiogenesis and through the process continue to survive through anti-apoptotic signals. Several proteins related to angiogenesis have already been described (see Fig. 14.2) including FLT1, FGF18, COL4A2, and MMP9. Other relevant proteins include: GPR180, a G protein-coupled receptor, which regulates vascular remodeling [57]; GPR126, a G protein-coupled receptor that is increased in human umbilical vein endothelial cells; ESM1, a protein secreted in endothelial cells [58], and SCUBE2, a developmental protein expressed in the vascular endothelium [59]. The role of these proteins in metastasis has yet to be elucidated.

14.7 Potential Advantages of MammaPrint as a Prognostic Test

Approximately 30% of node-negative patients will need chemotherapy to reduce the risk of recurrence. Currently, 47% of breast cancer patients in the US are identified as node-negative at the time of diagnosis; this could reach to 60–70% worldwide in regions with widespread breast cancer screening and disease awareness [60]. As the tumor biology is inherited in individual tumors, the expected increase in the prevalence of breast cancer will need a reliable risk assessment tool for clinical management of these patients. A substantial decrease, from 63% in 2009 to 49% in 2010, in the use of chemotherapy for node-negative breast cancer patients, has seemingly been driven by clinician's use of the two prognostic genomic assays Oncotype Dx and MammaPrint. This corresponds to the determination that, at 10-year follow-up, low-risk patients with node positivity gained absolutely no benefit from chemotherapy with 5-fluorouracil, doxorubicin (Adriamycin), and cyclophosphamide (FAC) chemotherapy with or without tamoxifen [61].

Compared to conventional risk assessment tools, microarray data are much more quantitative and reproducible as they are subjected to less human variability. In support of this, Paik *et al.* compared the performance of the 21-gene recurrence score with the histologic grading performance of three pathologists [62]. They concluded that the 21-gene recurrence score was more robust in the multivariate analyses as substantial inter-observer variation (discordance 57%, kappa 0.23–0.36) existed between the three pathologists in that study. As long as inter-observer variation between pathologists is large [62, 63] there is a need for molecular prognostic tests such as the 70-gene prognosis signature. In patients who have intermediate-risk features (e.g., grade 2) in which inter-observer variation is most distinct, the prognosis signature could potentially add valuable prognostic information.

14.8 Challenges in Clinical Application of MammaPrint

The most limiting factor for widespread clinical use of MammaPrint is that the tumor sample needs to be fresh or fresh frozen tissue. Unlike the Oncotype DX assay, MammaPrint analyses cannot be obtained from tumor tissues that have been preserved in a formaldehyde solution and embedded in wax, which is a common way to preserve tissue samples in routine pathology labs. Therefore, the decision to have a MammaPrint test must be made before surgery. It is recommended that a tumor sample be taken from an unfixed tumor specimen using a 3-mm punch within an hour of surgery and immediately shipped to Agendia. This process, which allows a fresh sample to be obtained, is not the normal process whereby samples are surgically removed and immediately fixed. Furthermore, the immediate degradation of the fresh samples has the potential to negatively affect downstream processes such and IHC, DNA and RNA isolation. There is a pressing need to establish a quality control measurement to ensure preanalytic variables for sample collection and processing could be tracked and controlled. The other issue is the time for initiating the test. Although Agendia has made every effort to expedite the sample collection and turnaround time of the test, the need to perform the test in a single company in the Netherlands has the potential to impede the ability to remain competitive, with regards to extra time and high cost, with other emerging tests that could be performed locally at CLIA-certified laboratories.

14.9 Summary and Perspectives

Tailoring therapy to an individual breast cancer patient by state-of-art technologies is a promising approach for selecting the most appropriate therapeutic regimen to maximize efficacy and minimize unwanted toxicity. The MammaPrint assay is the first FDA-approved, multi-gene molecular profiling test using fresh breast cancer tissue samples to qualitatively assess a patient's 5- to 10-year risk for distant metastasis. Over the past few years, it has been validated in several cohorts of early-stage breast cancer patients. Early-stage breast cancer patients with a low MammaPrint risk score are at low risk (≤10%) of developing distant metastases or death, and therefore might safely be spared chemotherapy after surgical treatment alone. Currently, the MammaPrint assay is performed at two Agendia laboratories in the Netherlands and approved by global regulatory agencies for all age of breast cancer patients with 0–3 lymph node metastasis. MammaPrint is currently being tested prospectively for the ability to predict benefit of chemotherapy in the MINDACT trial for its feasibility as a prognostic (versus predictive) test. As the prevalence of early-stage breast cancer with node-negative or 1–3 positive lymph nodes is expected to increase with screening and education, the MammaPrint assay is potentially an important tool to avoid over treatment with cytotoxic chemotherapy in breast cancer patients with good prognostic features. Further studies are needed to assess whether systemic therapy targeting the key genes in the 70-gene signature will reduce the risk of breast cancer recurrence and how co-morbidities might affect the expression of 70-gene signature.

References

1. Seigel, R., Naishadham, D., and Jemal, E. (2013) Cancer statistics, 2013. *CA Cancer J Clin* 63, 11–30.
2. Cancer Facts and Figures 2013. *Atlanta* American Cancer Society, page 4.
3. Gnant, M., Harbeck, N., and Thomssen, C. (2011) St. Gallen 2011: Summary of the Consensus Discussion. *Breast Care (Basel)* 6, 136–141
4. (2011) NCCN Clinical Practise Guidelines in Oncology, Breast Cancer. National Comprehensive Cancer Network.
5. de Snoo, F., Bender, R., Glas, A., and Rutgers, E. (2009) Gene expression profiling: decoding breast cancer. *Surg Oncol* 18, 366–378.
6. Glas, A. M., Floore, A., Delahaye, L. J., Witteveen, A. T., Pover, R. C., Bakx, N., Lahti-Domenici, J. S., Bruinsma, T. J., Warmoes, M. O., Bernards, R., Wessels, L. F., and Van't Veer, L. J. (2006) Converting a breast cancer microarray signature into a high-throughput diagnostic test. *BMC Genomics* 7, 278 doi:10.1186/1471-2164-7-278.
7. van 't Veer, L. J., Dai, H., van de Vijver, M. J., He, Y. D., Hart, A. A., Mao, M., Peterse, H. L., van der Kooy, K., Marton, M. J., Witteveen, A. T., Schreiber, G. J., Kerkhoven, R. M., Roberts, C., Linsley, P. S., Bernards, R., and Friend, S. H. (2002) Gene expression profiling predicts clinical outcome of breast cancer. *Nature* 415, 530–536.
8. van de Vijver, M. J., He, Y. D., van't Veer, L. J., Dai, H., Hart, A. A., Voskuil, D. W., Schreiber, G. J., Peterse, J. L., Roberts, C., Marton, M. J., Parrish, M., Atsma, D., Witteveen, A., Glas, A., Delahaye, L., van der Velde, T., Bartelink, H., Rodenhuis, S., Rutgers, E. T., Friend, S. H., and Bernards, R. (2002) A gene-expression signature as a predictor of survival in breast cancer. *N Engl J Med* 347, 1999–2009.
9. Buyse, M., Loi, S., van't Veer, L., Viale, G., Delorenzi, M., Glas, A. M., d'Assignies, M. S., Bergh, J., Lidereau, R., Ellis, P., Harris, A., Bogaerts, J., Therasse, P., Floore, A., Amakrane, M., Piette, F., Rutgers, E., Sotiriou, C., Cardoso, F., and Piccart, M. J. (2006) Validation and clinical utility of a 70-gene prognostic signature for women with node-negative breast cancer. *J Natl Cancer Inst* 98, 1183–1192.
10. Bender, R. A., Knauer, M., Rutgers, E. J., Glas, A. M., de Snoo, F. A., Linn, S. C., and Van 't Veer, L. J. (2009) The 70-gene profile and chemotherapy

benefit in 1,600 breast cancer patients. *ASCO Meeting Abstracts* 27, 512.

11. Bueno-de-Mesquita, J. M., Linn, S. C., Keijzer, R., Wesseling, J., Nuyten, D. S., van Krimpen, C., Meijers, C., de Graaf, P. W., Bos, M. M., Hart, A. A., Rutgers, E. J., Peterse, J. L., Halfwerk, H., de Groot, R., Pronk, A., Floore, A. N., Glas, A. M., Van't Veer, L. J., and van de Vijver, M. J. (2009) Validation of 70-gene prognosis signature in node-negative breast cancer. *Breast Cancer Res Treat* 117, 483–495.
12. Bueno-de-Mesquita, J. M., van Harten, W. H., Retel, V. P., van't Veer, L. J., van Dam, F. S., Karsenberg, K., Douma, K. F., van Tinteren, H., Peterse, J. L., Wesseling, J., Wu, T. S., Atsma, D., Rutgers, E. J., Brink, G., Floore, A. N., Glas, A. M., Roumen, R. M., Bellot, F. E., van Krimpen, C., Rodenhuis, S., van de Vijver, M. J., and Linn, S. C. (2007) Use of 70-gene signature to predict prognosis of patients with node-negative breast cancer: a prospective community-based feasibility study (RASTER). *Lancet Oncol* 8, 1079–1087.
13. de Snoo, F. A., Knauer, M., Bender, R. A., Stork-Sloots, L., Rutgers, E. J., Glas, A. M., Linn, S. C., and Van 't Veer, L. J. (2009) Outcome prediction by the 70-gene profile in the context of the National Comprehensive Cancer Network (NCCN) guidelines. *ASCO Meeting Abstracts* 27, 535.
14. Generali, D. G., Bazzola, L., Foroni, C., Berruti, A., Dogliotti, L., Fox, S. B., and Bottini, A. (2008) MammaPrint: Predictive tool of response and resistance to primary chemo-endocrine treatment in elderly breast cancer (BC) patients. In *33rd ESMO Congress* Vol. 19 pp. viii77-viii88, Abstract 203P, Early Breast Cancer Session, Abstract Book in Annals of Oncology, Stockholm, Sweden.
15. Glas, A. M., Roepman, P., De Snoo, F., and van 't Veer, L. J. (2008) MammaPrint predicts survival in samll T1 breast cancer tumors. In *33rd ESMO Congress* Vol. 19 pp. viii77-viii88, Abstract 189PD, Early Breast Cancer Session, Abstract Book in Annals of Oncology, Stockholm, Sweden.
16. Ishitobi, M., Goranova, T. E., Komoike, Y., Motomura, K., Koyama, H., Glas, A. M., van Lienen, E., Inaji, H., Van't Veer, L. J., and Kato, K. (2010) Clinical utility of the 70-gene MammaPrint profile in a Japanese population. *Jpn J Clin Oncol* 40, 508–512.
17. Knauer, M., Mook, S., Rutgers, E. J., Bender, R. A., Hauptmann, M., van de Vijver, M. J., Koornstra, R. H., Bueno-de-Mesquita, J. M., Linn, S. C., and van 't Veer, L. J. (2010) The predictive value of the 70-gene signature for adjuvant chemotherapy in early breast cancer. *Breast Cancer Res Treat* 120, 655–661.

18. Knauer, M., Rutgers, E., Mook, S., Cardoso, F., Van de Vijver, M., Viale, G., Glas, A., Saghatchian, M., Kok, M., Bueno-de-Mesquita, J. M., Linn, S. C., and van 't Veer, L. J. (2009) The 70-gene MammaPrint signature for optimal risk stratification in endocrine responsive breast cancer. In *11th International Conference Primary Therapy of Early Breast Cancer, St Gallen Oncology Conferece* Vol. 18, Supplement 1 pp. S36, Abstract 0072, The Breast, St. Gallen, Switzerland.

19. Knauer, M., Straver, M. E., Rutgers, E., Bender, R., Cardoso, F., Mook, S., Van de Vijver, M., Saghatchian, M., Koornstra, R. H., Bueno-de-Mesquita, J. M., Rodenhuis, S., Linn, S. C., and van 't Veer, L. J. (2009) The 70-gene MammaPrint signature is predictive for chemotherapy benefit in early breast cancer. In *11th International Conference Primary Therapy of Early Breast Cancer, St Gallen Oncology Conference* Vol. 18, Supplement 1 pp. S36, Abstract 0073, The Breast, St. Gallen, Switzerland.

20. Kunz, G., Glas, A., De Snoo, F., and Simon, I. (2009) Use of the genomic test MammaPrint in daily clinical practice to assist in risk stratification of young breast cancer patients. In *11th International Conference Primary Therapy of Early Breast Cancer, St Gallen Oncology Conference* Vol. 18, Supplement 1 pp. S37, Abstract 0075, The Breast, St. Gallen, Switzerland.

21. M. Saghatchian, M., Mook, S., Pruneri, G., Viale, G., Glas, A., Eekhout, I., Knauer, M., and van 't Veer, L. J. (2009) The 70-gene MammaPrint profile allows to identify a subgroup of very good-prognostic patients with primary breast cancer and 4–9 positive lymph nodes. In *11th International Conference Primary Therapy of Early Breast Cancer, St Gallen Oncology Conference* Vol. 18, Supplement 1 pp. S36, Abstract 0071, The Breast, St. Gallen, Switzerland.

22. Mayordomo, J. I., Roig, A. M., Rolfo, C. D., Morales, S., Glas, A., Floore, A., De Snoo, F., and Stork-Sloots, L. (2008) Feasibility of using core needle biopsies for the 70-gene prognosis signature. In *33rd ESMO Congress* Vol. 19 pp. viii77-viii88, Abstract 202P, Early Breast Cancer Session, Abstract Book in Annals of Oncology, Stockholm, Sweden.

23. Mook, S., Knauer, M., Bueno-de-Mesquita, J. M., Retel, V. P., Wesseling, J., Linn, S. C., Van't Veer, L. J., and Rutgers, E. J. (2010) Metastatic potential of T1 breast cancer can be predicted by the 70-gene MammaPrint signature. *Ann Surg Oncol* 17, 1406–1413.

24. Mook, S., Schmidt, M. K., Viale, G., Pruneri, G., Eekhout, I., Floore, A., Glas, A. M., Bogaerts, J., Cardoso, F., Piccart-Gebhart, M. J., Rutgers, E. T., and Van't Veer, L. J. (2009) The 70-gene prognosis-signature predicts disease outcome in breast cancer patients with 1–3 positive lymph

nodes in an independent validation study. *Breast Cancer Res Treat* 116, 295–302.

25. Mook, S., Schmidt, M. K., Weigelt, B., Kreike, B., Eekhout, I., van de Vijver, M. J., Glas, A. M., Floore, A., Rutgers, E. J., and van 't Veer, L. J. (2010) The 70-gene prognosis signature predicts early metastasis in breast cancer patients between 55 and 70 years of age. *Ann Oncol* 21, 717–722.
26. Straver, M. E., Glas, A. M., Hannemann, J., Wesseling, J., van de Vijver, M. J., Rutgers, E. J., Vrancken Peeters, M. J., van Tinteren, H., Van't Veer, L. J., and Rodenhuis, S. (2010) The 70-gene signature as a response predictor for neoadjuvant chemotherapy in breast cancer. *Breast Cancer Res Treat* 119, 551–558.
27. Wittner, B. S., Sgroi, D. C., Ryan, P. D., Bruinsma, T. J., Glas, A. M., Male, A., Dahiya, S., Habin, K., Bernards, R., Haber, D. A., Van't Veer, L. J., and Ramaswamy, S. (2008) Analysis of the MammaPrint breast cancer assay in a predominantly postmenopausal cohort. *Clin Cancer Res* 14, 2988–2993.
28. De Snoo, F., Glas, A., Floore, A., Mayordomo, J. I., Modollel, A., Rolfo, C. D., van 't Veer, L. J., Rutgers, E., Rodenhuis, S., and Bender, R. (2009) Early prognosis prediction: MammaPrint on core-needle biopsies. In *11th International Conference Primary Therapy of Early Breast Cancer, St Gallen Oncology Conference* Vol. 18, Supplement 1 pp. S37, Abstract 0074, The Breast, St. Gallen, Switzerland.
29. Mittempergher, L., de Ronde, J. J., Nieuwland, M., Kerkhoven, R. M., Simon, I., Rutgers, E. J., Wessels, L. F., and Van't Veer, L. J. (2011) Gene expression profiles from formalin fixed paraffin embedded breast cancer tissue are largely comparable to fresh frozen matched tissue. *PLoS One* 6, e17163.
30. Cardoso, F., Van't Veer, L., Rutgers, E., Loi, S., Mook, S., and Piccart-Gebhart, M. J. (2008) Clinical application of the 70-gene profile: the MINDACT trial. *J Clin Oncol* 26, 729–735.
31. Cardoso, F., Piccart-Gebhart, M., Van't Veer, L., and Rutgers, E. (2007) The MINDACT trial: the first prospective clinical validation of a genomic tool. *Mol Oncol* 1, 246–251.
32. Tian, S., Roepman, P., Van't Veer, L. J., Bernards, R., de Snoo, F., and Glas, A. M. (2010) Biological functions of the genes in the mammaprint breast cancer profile reflect the hallmarks of cancer. *Biomark Insights* 5, 129–138.
33. Hanahan, D., and Weinberg, R. A. (2000) The hallmarks of cancer. *Cell* 100, 57–70.

34. Hiratsuka, S. (2011) Vasculogenensis, angiogenesis and special features of tumor blood vessels. *Front Biosci* 16, 1413–1427.

35. Iordanskaia, T., and Nawshad, A. (2011) Mechanisms of transforming growth factor beta induced cell cycle arrest in palate development. *J Cell Physiol* 226, 1415–1424.

36. Beattie, J., Allan, G. J., Lochrie, J. D., and Flint, D. J. (2006) Insulin-like growth factor-binding protein-5 (IGFBP-5): a critical member of the IGF axis. *Biochem J* 395, 1–19.

37. Kim, K. S., Seu, Y. B., Baek, S. H., Kim, M. J., Kim, K. J., Kim, J. H., and Kim, J. R. (2007) Induction of cellular senescence by insulin-like growth factor binding protein-5 through a p53-dependent mechanism. *Mol Biol Cell* 18, 4543–4552.

38. Gauglhofer, C., Sagmeister, S., Schrottmaier, W., Fischer, C., Rodgarkia-Dara, C., Mohr, T., Stattner, S., Bichler, C., Kandioler, D., Wrba, F., Schulte-Hermann, R., Holzmann, K., Grusch, M., Marian, B., Berger, W., and Grasl-Kraupp, B. (2011) Up-regulation of the fibroblast growth factor 8 subfamily in human hepatocellular carcinoma for cell survival and neoangiogenesis. *Hepatology* 53, 854–864.

39. Ito, H., Akiyama, H., Shigeno, C., and Nakamura, T. (2001) Isolation, characterization, and chromosome mapping of a human A-C1 Ha-Ras suppressor gene (HRASLS). *Cytogenet Cell Genet* 93, 36–39.

40. Trynka, G., Zhernakova, A., Romanos, J., Franke, L., Hunt, K. A., Turner, G., Bruinenberg, M., Heap, G. A., Platteel, M., Ryan, A. W., de Kovel, C., Holmes, G. K., Howdle, P. D., Walters, J. R., Sanders, D. S., Mulder, C. J., Mearin, M. L., Verbeek, W. H., Trimble, V., Stevens, F. M., Kelleher, D., Barisani, D., Bardella, M. T., McManus, R., van Heel, D. A., and Wijmenga, C. (2009) Coeliac disease-associated risk variants in TNFAIP3 and REL implicate altered NF-kappaB signalling. *Gut* 58, 1078–1083.

41. Takahashi, S., Ebihara, A., Kajiho, H., Kontani, K., Nishina, H., and Katada, T. (2011) RASSF7 negatively regulates pro-apoptotic JNK signaling by inhibiting the activity of phosphorylated-MKK7. *Cell Death Differ* 18, 645–655.

42. Sherwood, V., Manbodh, R., Sheppard, C., and Chalmers, A. D. (2008) RASSF7 is a member of a new family of RAS association domain-containing proteins and is required for completing mitosis. *Mol Biol Cell* 19, 1772–1782.

43. Lin, M. L., Park, J. H., Nishidate, T., Nakamura, Y., and Katagiri, T. (2007) Involvement of maternal embryonic leucine zipper kinase (MELK) in mammary carcinogenesis through interaction with Bcl-G, a pro-apoptotic member of the Bcl-2 family. *Breast Cancer Res* 9, R17.

44. Pickard, M. R., Green, A. R., Ellis, I. O., Caldas, C., Hedge, V. L., Mourtada-Maarabouni, M., and Williams, G. T. (2009) Dysregulated expression of Fau and MELK is associated with poor prognosis in breast cancer. *Breast Cancer Res* 11, R60.

45. Tang, K., Zhang, Z., Bai, Z., Ma, X., Guo, W., and Wang, Y. (2011) Enhancement of gemcitabine sensitivity in pancreatic cancer by co-regulation of dCK and p8 expression. *Oncol Rep* 25, 963–970.

46. Payton, M., Scully, S., Chung, G., and Coats, S. (2002) Deregulation of cyclin E2 expression and associated kinase activity in primary breast tumors. *Oncogene* 21, 8529–8534.

47. Caldon, C. E., Sergio, C. M., Schutte, J., Boersma, M. N., Sutherland, R. L., Carroll, J. S., and Musgrove, E. A. (2009) Estrogen regulation of cyclin E2 requires cyclin D1 but not c-Myc. *Mol Cell Biol* 29, 4623–4639.

48. Kops, G. J. (2008) The kinetochore and spindle checkpoint in mammals. *Front Biosci* 13, 3606–3620.

49. Bianchini, G., Qi, Y., Alvarez, R. H., Iwamoto, T., Coutant, C., Ibrahim, N. K., Valero, V., Cristofanilli, M., Green, M. C., Radvanyi, L., Hatzis, C., Hortobagyi, G. N., Andre, F., Gianni, L., Symmans, W. F., and Pusztai, L. (2010) Molecular anatomy of breast cancer stroma and its prognostic value in estrogen receptor-positive and -negative cancers. *J Clin Oncol* 28, 4316–4323.

50. Wang, X., Lu, H., Urvalek, A. M., Li, T., Yu, L., Lamar, J., DiPersio, C. M., Feustel, P. J., and Zhao, J. (2011) KLF8 promotes human breast cancer cell invasion and metastasis by transcriptional activation of MMP9. *Oncogene* 30, 1901–1911.

51. Hanahan, D., and Weinberg, R. A. (2011) Hallmarks of cancer: the next generation. *Cell* 144, 646–674.

52. Sureshbabu, A., Okajima, H., Yamanaka, D., Shastri, S., Tonner, E., Rae, C., Szymanowska, M., Shand, J. H., Takahashi, S., Beattie, J., Allan, G. J., and Flint, D. J. (2009) IGFBP-5 induces epithelial and fibroblast responses consistent with the fibrotic response. *Biochem Soc Trans* 37, 882–885.

53. Ahn, B. Y., Elwi, A. N., Lee, B., Trinh, D. L., Klimowicz, A. C., Yau, A., Chan, J. A., Magliocco, A., and Kim, S. W. (2010) Genetic screen identifies insulin-like growth factor binding protein 5 as a modulator of tamoxifen resistance in breast cancer. *Cancer Res* 70, 3013–3019.

54. Wiercinska, E., Naber, H. P., Pardali, E., van der Pluijm, G., van Dam, H., and Ten Dijke, P. (2011) The TGF-beta/Smad pathway induces breast cancer cell invasion through the up-regulation of matrix metalloproteinase 2 and 9 in a spheroid invasion model system. *Breast Cancer Res Treat* 128, 657–666.

55. Hou, C. H., Chiang, Y. C., Fong, Y. C., and Tang, C. H. (2011) WISP-1 increases MMP-2 expression and cell motility in human chondrosarcoma cells. *Biochem Pharmacol* 81, 1286–1295.

56. Wang, S. S., Betz, A. G., and Reed, R. R. (2002) Cloning of a novel Olf-1/EBF-like gene, O/E-4, by degenerate oligo-based direct selection. *Mol Cell Neurosci* 20, 404–414.

57. Tsukada, S., Iwai, M., Nishiu, J., Itoh, M., Tomoike, H., Horiuchi, M., Nakamura, Y., and Tanaka, T. (2003) Inhibition of experimental intimal thickening in mice lacking a novel G-protein-coupled receptor. *Circulation* 107, 313–319.

58. Chen, L. Y., Liu, X., Wang, S. L., and Qin, C. Y. (2010) Over-expression of the Endocan gene in endothelial cells from hepatocellular carcinoma is associated with angiogenesis and tumour invasion. *J Int Med Res* 38, 498–510.

59. Yang, R. B., Ng, C. K., Wasserman, S. M., Colman, S. D., Shenoy, S., Mehraban, F., Komuves, L. G., Tomlinson, J. E., and Topper, J. N. (2002) Identification of a novel family of cell-surface proteins expressed in human vascular endothelium. *J Biol Chem* 277, 46364–46373.

60. Harbeck, N., and Thomssen, C. (2010) A new look at node-negative breast cancer. *Oncologist* 15 Suppl 5, 29–38.

61. Albain, K. S., Paik, S., and van't Veer, L. (2009) Prediction of adjuvant chemotherapy benefit in endocrine responsive, early breast cancer using multigene assays. *Breast* 18 Suppl 3, S141–S145.

62. Paik, S., Shak, S., Tang, G., Kim, C., Baker, J., Cronin, M., Baehner, F. L., Walker, M. G., Watson, D., Park, T., Hiller, W., Fisher, E. R., Wickerham, D. L., Bryant, J., and Wolmark, N. (2004) A multigene assay to predict recurrence of tamoxifen-treated, node-negative breast cancer. *N Engl J Med* 351, 2817–2826.

63. Bueno-de-Mesquita, J. M., Nuyten, D. S., Wesseling, J., van Tinteren, H., Linn, S. C., and van de Vijver, M. J. (2010) The impact of inter-observer variation in pathological assessment of node-negative breast cancer on clinical risk assessment and patient selection for adjuvant systemic treatment. *Ann Oncol* 21, 40–47.

Chapter 15

BRCA Mutation and PARP Inhibitors

Marcie K. Weil, Shivaani Kummar, James H. Doroshow, and Alice Chen
Division of Cancer Treatment and Diagnosis,
National Cancer Institute, Bethesda, Maryland, USA
chenali@mail.nih.gov.

15.1 Introduction

Cells undergo DNA damage during replication and from environmental exposures. Multiple cellular response mechanisms are involved in the repair of DNA damage, including base excision repair (BER), mismatch repair (MMR), nucleotide excision repair (NER), single strand annealing (SSA), "error free" homologous recombination (HR), and "error prone" nonhomologous end joining (NHEJ). Increasing evidence implicates defects in the DNA repair pathway in carcinogenesis, cancer progression, and response to therapy [10]. *BRCA1* and *BRCA2* mutations are among the most well studied genetic defects involving DNA repair. *BRCA* stands for breast cancer. susceptibility. gene. though mutations in the gene are associated with other cancers as well. These genes are *BRCA1 and BRCA2* involved in the HR pathway for repairing double strand DNA breaks (DSBs) and are

Handbook of Therapeutic Biomarkers in Cancer
Edited by Sherry X. Yang and Janet E. Dancey
Copyright © 2013 Pan Stanford Publishing Pte. Ltd.
ISBN 978-981-4364-65-2 (Hardcover), 978-981-4364-66-9 (eBook)
www.panstanford.com

critical in maintaining genomic stability. Defects in the HR pathway are associated with an increased incidence of cancer, felt to be secondary to the cells being forced to repair DSBs using the error-prone NHEJ pathway resulting in accumulation of genetic abnormalities, leading to either apoptosis or abnormal growth and survival resulting in carcinogenesis [11].

Although mutations in the *BRCA* genes lead to the development of cancer, the presence of such mutations makes these cells more susceptible to cytotoxic injury from chemotherapy, particularly cisplatin [12,13], and more vulnerable to defects or inhibition of other DNA repair pathways via synthetic lethality [14,15]. Poly-ADP-ribose polymerase (PARP) is a family of enzymes involved in single strand DNA break (SSB) repair via the (BER) pathway [16–18], and, to a lesser extent, also involved in DSB repair via NHEJ [19]. Recently PARP inhibitors have been evaluated as anticancer agents in patients with *BRCA* mutations, either germline or somatic. This chapter discusses the biology of *BRCA* gene as it pertains to its role in cancer, the effects of mutations in the *BRCA* genes, the role of PARP in DNA repair, PARP inhibitors and their clinical development as anticancer agents.

15.2 *BRCA*

BRCA1 and *2* are tumor suppressor genes that are located on chromosome 17 and 13, respectively. Wild-type *BRCA* genes code for proteins that are integral for HR. The BRCA1 protein migrates to the DSB site to recruit proteins that aid in DNA repair and is involved in cell cycle replication arrest. *BRCA2* is directly involved in repairing the DNA, at least in part by chaperoning RAD51, the recombination enzyme [7, 11].

Genetic mutations of *BRCA1* or *2* genes predispose to an increased risk of developing cancers, mainly breast, ovarian, prostate, and pancreatic cancer, and at relatively younger ages [21]. Having the *BRCA* mutation increases the risk of contracting ovarian cancer up to 63% and breast cancer up to 87%. *BRCA1* is associated more with ovarian, cervical, uterine, pancreatic, and colon cancer in women, and possibly testicular cancer whereas, *BRCA2* has a higher association with male breast cancer, prostate cancer, pancreatic, stomach, bile duct cancers, and melanoma. *BRCA* mutations are most prevalent in people of Ashkenazi Jewish (Jews from Eastern European) descent. In that population, the predominant mutations are 185delAG and 5382insC in *BRCA1* and 6174delT in *BRCA2*. Although these mutations occur only in 2.5% of the Ashkenazi population, they account for 70–85% of germline mutations in patients with hereditary breast cancer and ovarian cancer. Although the incidence appears high, when the overall population is considered, only 5–10% of breast cancers and 13% of high-grade serous ovarian cancers are associated with germline mutations in the *BRCA* gene [20–24]. Interestingly, 50% of high-grade serous ovarian carcinoma is associated with HR mutations, including *BRCA* mutations and others. In addition, high-grade serous ovarian carcinoma was found to have 22% incidence of *BRCA* mutations, when sporadic mutations were included [13]. It has been reported also that 20% of triple negative breast cancer (TNBC) has *BRCA* mutations [25].

Usually the germline is heterozygous, not homozygous, for the *BRCA* mutation; however, double and even triple heterozygous combinations have been identified in rare instances. These genotypes are not associated with worse prognostic features, like age of onset, lifetime risks, and numbers of tumors, compared with the single heterozygous state. However, for the purposes of assessing risks and providing genetic counseling, it is important to evaluate for a multiple heterozygous state [26]. Patients who have inherited the heterozygous pattern are thought to undergo a second hit within the tumor cells to account for tumorgenesis.

The testing for *BRCA* mutation is a genetic test that analyze for known deleterious polymorphisms. The test is available commercially [27]. There are 500 known mutations in *BRCA 1*. Some *BRCA* variants, such as S1613G, impart increased risk of cancer, whereas others, such as K1183R, have been found to be associated with decreased risk [28]. The reason for testing, of course, is to intervene by increased screening, chemoprevention or prophylactic surgeries to remove the organs at highest risk in patients who carry the abnormal gene. The 2007 American Cancer Society guidelines suggest using MRI in addition to mammography for screening high-risk women. High risk includes patients with *BRCA* mutations, first-degree relatives of *BRCA* carriers who have not been tested for the mutation, and patients with lifetime risk 20% or greater based on clinical history models [29]. Scales have been developed to provide guidance on when to test for mutations in *BRCA1 or 2*, including

FHAT, Manchester Score, Frank, Couch and Bayesian Probabilistic Model (BRCAPRO) The most accurate is the BRCAPRO [30,31]. These scales are based on family history of breast and ovarian cancers, early age of onset of breast and ovarian cancer in the patient and/or family members, multiple tumors in a single patient, breast cancer in men, cancer-free survival in first- or second-degree relatives, and ethnicity. In 2002 a study using a computer model showed that chemoprevention and/or surgical prophylaxis can increase the survival of a 30-year-old woman with *BRCA* mutation by up to 5 years over surveillance alone [32]. Another strategy for determining risk developed by the International Consensus Panel on Breast Cancer Risk includes patients with medullary type breast cancer and triple negative and basal-like breast cancers, especially in patients younger than 50 years old [33].

The incidence of the *BRCA* germline mutations is low in the general population; however, some tumors have somatic mutations or epigenetic changes to the *BRCA* gene or other mutations that impair HR that lead to similar behavior to that of tumors with germline *BRCA* mutations (Fig. 15.1). The similar characteristics of these tumors consist of defective HR and improved response to platinum agents. These tumors are said to express "BRCAness" [2]. Some breast and ovarian cancers without the documented mutations in the *BRCA* gene have decrease in the expression of *BRCA1*. This decreased expression may be due to allelic loss [34,35] or hypermethylation of *BRCA1*, which inhibits transcription of the gene. Hypermethylation has been discovered in sporadic breast and ovarian tumors [3,36–38]. Sporadic breast cancers have been found to have two times less messenger RNA expression and nine times higher level of the *BRCA1* negative regulator, ID4 [39] suggesting that BRCA function is important in preventing the development of cancer. Sixty-three percent of metaplastic breast cancer, a rare form of basal-like breast cancer, had *BRCA1* promoter methylation compared with 12% of noncancerous controls [40]. Microarrays also show similar gene expression patterns between familial *BRCA1* breast cancers and basal-like sporadic breast cancers [40,42]. TNBC, basal-like breast cancers, and *BRCA1* germline breast cancers might all harbor the same tumorgenesis mechanism through *BRCA1* dysfunction. For that reason, these subtypes of breast cancer are being explored as possible histologies that might benefit from PARP inhibitors. Some sporadic ovarian tumors also showed gene expression profiles that were either *BRCA1*-like or *BRCA2* like [43]. Twenty-five percent of *BRCA* mutations in the ovarian tumors and 9–20% of the mutations in high-grade serous ovarian tumors are somatic [44,13, 45], another population that may benefit from the use of PARP inhibitors. Defects in BRCA could be due to genomic, somatic or epigenetic alterations. Currently only certain genetic *BRCA* mutations can be commercially tested to identify patients for appropriate treatment.

Postulated Mechanisms for Imparting BRCAness

1. BRCA mutation, inherent or sporadic
2. Increased expression of proteins that inactivate BRCA, including EMSY [1]
3. BRCA promoter methylation [2–4]
4. FANCF promoter methylation [5–6]
5. Defects in other Fanconi anemia complex proteins [7]
6. Decreased expression of other proteins involved in HR - RAD51, ATR, ATM, CHK2. [8]
7. PTEN deficiency [9]

Figure 15.1 Possible mechanisms for imparting BRCAness.

Other genetic mutations, such as the Fanconi anemia genes [46,47,45] and epigenetic effects, aside from the *BRCA* gene itself, have been found to impart "BRCAness." These genetic mutations and proteins that cause "BRCAness" include the following:

1. Loss of function mutations in genes that code for proteins that affect the HR pathway, such as Fanconi anemia proteins [7], RAD51, ATR, ATM, CHK2,[8] and PTEN [9].

2. Upregulation of *EMSY*, that inactivates BRCA proteins [1].

A heat map showing the gene profile of tumors with and without the *BRCA* mutation differentiated specific patterns for BRCA-like (BL) and non-BRCA-like (NBL) tumors with 94% accuracy in 61 patients. In this study, BRCAness was determined by sensitivity to platinum and the presence of RAD51 foci, as a marker of HR. The BRCAness profile also correlated with PARP inhibitor sensitivity. Interestingly, in this study, two of the three tumors that reverted from nonfunctional to functional *BRCA* status after treatment with cisplatin, coinciding with cisplatin resistance, also showed the change in the genetic expression profile. The BL gene expression profile correlated with improved DFS and OS and was an independent prognostic indicator in multivariate analyses [12].

15.3 Poly (ADP-Ribose) Polymerase (PARP)

PARPs are a family of proteins with enzymatic properties, scaffolding properties, and recruiting abilities for other necessary DNA repair proteins. PARP affects repair for SSBs, primarily as part of BER [19], PARP 1 and PARP 2 are the best studied to date of these proteins, and the targets of most of the PARP inhibitors presently under development for cancer therapy [18]. PARP proteins were initially identified in 1963 [48,16] and PARP inhibitors were proposed to enhance alkylator activity in 1980 [17]. In response, it produces PAR and transfers it to acceptor proteins on the PARP itself and other proteins involved in DNA repair. The negative charge of PAR causes PARP 1 to lose its affinity for DNA. PAR then recruits other repair proteins to the damaged DNA site. PARP 1 retains the largest amount of PAR. PAR also binds XRCC1, the scaffolding protein. PAR regulates histone H1 binding to chromatin, allowing the chromatin to relax. PARP is involved in methylation and transcription of genes coding for cell cycle and stress response, including *p53*. PAR attaches DNA polymerase β to the DNA break site to replace the missing bases. Finally, PAR combines with DNA ligase III to seal the DNA [48–50]. The six proteins fall into three groups:

Group 1: PARP 1, PARP 2, and PARP 3;
Group 2: PARP 4, also called vault PARP;
Group 3: tankyrase (TNKS) and TNKS2, also known as PARP 5 and PARP 6, respectively.

Structurally, the proteins consist of two ribose moieties and two phosphates. PARP 1 is the best understood, existing in a homodimeric form and consisting of three domains, DNA-binding domain (DBD), the automodification domain (AD), and the catalytic domain. The amino-terminal DBD contains three zinc fingers, two that bind PARP 1 to DNA breaks, and a third that couples DNA damage-induced changes in the DBD to catalytic activity. The AD portion contains glutamate and lysine amino acids that accept ADP-ribose units, resulting in self poly(ADP-ribosyl)ation. The AD unit houses the *BRCA1* carboxyl-terminal repeat motif, similar to DNA sequences in other proteins involved in DNA repair. The C-terminal catalytic domain holds the signature sequence containing the most conserved sequence of the PARP family. In this domain ADP-ribose transferase facilitates the transfer of ADP-ribose from nicotinamide adenine dinucleotide (NAD+) to protein acceptors within this domain.

Functionally, DNA strand breaks activate PARP protein. Some PARP proteins are vital for survival of cells, as evidenced by mice dying in the simultaneous absence of PARP 1 and PARP 2 or the simultaneous absence of TNKS or TNKS2. By definition, PARP proteins must be able to transfer ADP-ribose from NAD+ to an acceptor protein and add multiple subunits of Poly(ADP) ribose (pADPr, PAR) [51]. It has not been proven that PARP 3 and PARP 4 form multiple subunits of PAR so they may not be PARP proteins [48]. In the homodimeric state, the DBD recognizes and binds to sites of single strand DNA damage [48–50]. In addition to SSB repair, PARP is also involved in DSB repair. It attaches to DNA-protein kinase catalytic subunit (DNA-PK), Ku 70, and Ku80, to allow DSB repair through NHEJ DNA ligase [19]. PAR recruits ATM, MRE11, and topoisomerase 1, all involved in DSB repair [52,53].

PAR is degraded by Poly(ADP-ribose) glycohydrolase and possibly ADP-ribose hydrolase 3, into ADP-ribose molecules, which are metabolized further to AMP. The increased AMP: ATP ratio catalyzes the metabolic sensor AMP-activated protein kinase (AMPKs). Mammalian target of rapamycin complex 1 (MTORC1) is thereby inhibited, inducing autophagy [54]. Thus, cellular energy homeostasis is regulated. In the process of making PAR, NAD+ is converted to nicotinamide. To replenish the NAD+ from nicotinamide, phosphoribosylpyrophosphate and ATP are converted to AMP and pyrophosphate (Fig. 15.2). In the case of extreme DNA damage, as with ischemia, PARP 1 hyperactivation results in depletion of NAD+ and ATP, resulting in cell death by necrosis or apoptosis [48].

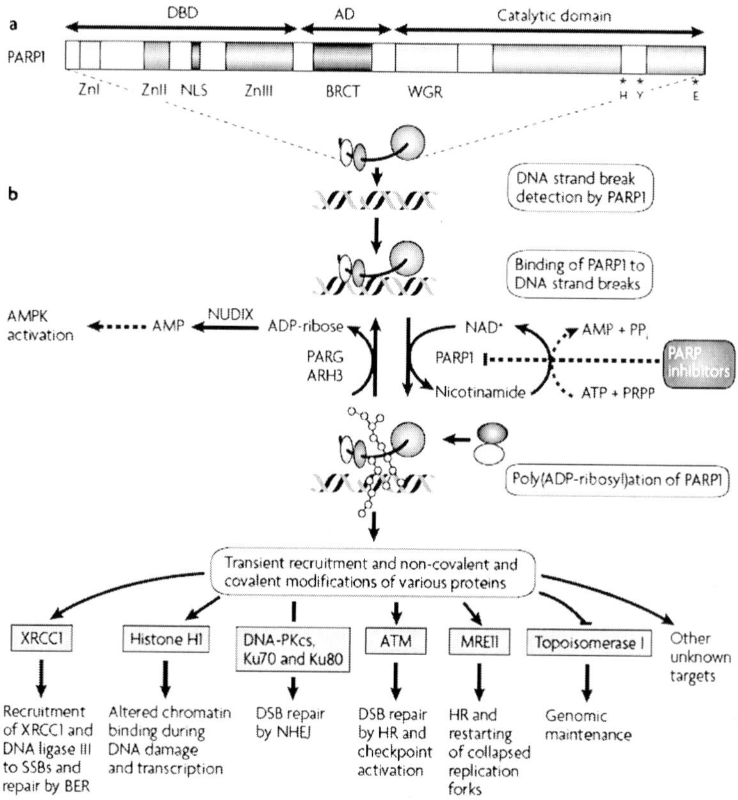

Figure 15.2 (a) Structure of PARP. (b) Activity of PARP. Reprinted by permission from Macmillan Publishers Ltd: *Nature Reviews Cancer*, Rouleau M, PARP Inhibition: PARP1 and Beyond, copyright 2010.

PARP 1 also activates the stress-inducible transcription complex NF-κB, which inhibits apoptosis and promotes proliferation. Cancers demonstrate increased expression of NF-κB. The complex is constitutively activated in some breast cancers, particularly in those that are hormone refractory and those with worse prognoses. NF-κB correlates with progression of disease in breast cancer. NF-κB is also activated by XRT and chemotherapy. Inhibition of NF-κB sensitizes cells to XRT and chemotherapy [55]. In addition, PARP 1 has some effect on the activation of HIF-1α [56]. Because of the interference in DNA repair, the upregulation of autophagy, and the downregulation of NF-κB and HIF-1α, PARP inhibitors are potentially strategic modalities for potentiating the tumor cell death from administration of cytotoxic therapy. Recent data suggest PARP inhibitors may trap PARP 1 and 2 leading to cytotoxicity that is independent to the inactivation of PARP [57a].

PARP inhibitors currently under development include AstraZeneca's olaparib (AZD2281, KU-0059436), Abbott Laboratories' veliparib (ABT 888), Clovis's rucaparib (AG014699, PF 1367338), Tesaro's niraparib (MK 1827) Cephalon's CEP 9722, and Biomarin's BMN673 (LT-673). Sanofi/BiPar's iniparib (BSI 201), which had shown initial activity in TNBC [76], was recently reported not to inhibit PARP [78]. Its mechanism of action is under investigation. Unfortunately, like with other forms of cancer therapy, there appears to be development of resistance to PARP inhibitors.

15.4 PARP Inhibitors as Single Agents to Induce Synthetic Lethality in *BRCA* Tumor Cells

The combination of agents that bring about cell death when neither alone would do so is termed synthetic lethality [8,9]. This technique has been exploited by treating tumors expressing *BRCA* mutations with PARP inhibitors [57]. Cells with mutated *BRCA*, resulting in defective HR repair, are more dependent on other pathways for DNA repair, utilizing BER. If BER is inhibited by a PARP inhibitor, DNA repair cannot occur by either BER or HR. If SSBs cannot be repaired because of defective BER, such as with the use of PARP inhibitors, SSBs result in collapsed replication forks that ultimately result in the formation of DSBs. Typically, DSBs would be repaired via the error free HR pathway; however, in the presence of homozygous mutations in the *BRCA* genes in the tumor, HR is ineffective. Cell death, or apoptosis, and error-prone recombination through the NHEJ pathway result. This phenomenon is thought to occur preferentially in the tumor cells as the normal cells are heterozygous for the BRCA gene and thus retain HR function, while tumor cells are homozygous for the mutation making them HR defective [8, 14,15, 26, 58].

PARP inhibitors, as single agents, kill BRCA2 deficient cells at doses that are not toxic to normal cells both *in vitro* and in xenograft models [14,15]. PARP inhibition in BRCA defective cells leads to NHEJ or SSA, which cause cell arrest in G2 or M phases of the cell cycle followed by apoptosis. Importantly, alterations in *p53* (17p13.1), the tumor suppressor gene that works downstream of PARP 1 after DNA damage, do not interfere with the effect of PARP inhibitors [15, 59].

15.5 PARP Inhibitors in Combination with Cytotoxic Therapy

Cytotoxic chemotherapeutic agents that cause SSBs result in the activation of PARP 1 and thereby the initiation of BER. By repairing the damage, PARP 1 elicits resistance to the chemotherapy [60]. However, in the presence of PARP inhibitors, the SSBs that cannot be repaired lead to DSBs [15]. HR is then overwhelmed by the addition of the converted DSBs. The alternative DNA repair mechanism for handling DSBs, NHEJ, results in error-prone recombination leading to cells unstable for maintaining viability or the normal cellular control [61] In addition, as stated above, PARP inhibitors also inhibit NHEJ. With a weakened DNA repair system, exposure to cytotoxic agents results in cell death.

PARP inhibitors also sensitize MMR-deficient cells, which are resistant to the methylating agent temozolomide [62,63]. Wild-type MMR acts in three ways. First, MMR corrects errors of replication. Second, the wild type causes replication arrest. Third, MMR can result in cell death. If MMR function is abnormal, cells are more likely to survive with abnormal DNA after exposure to the cytotoxic agent [64]. Only tumor cells are known to be deficient in MMR, providing some selectivity in the targeting of the tumor over normal tissues [60, 65]. After exposure to temozolomide, the 0^6-methylguanine incorrectly pairs with thymine. The MMR complex, MutSα (MSH2/MSH6 heterodimer) and the subsequent pathway, recognizes the mismatch pair, and ultimately leads to apoptosis. Deficiency of MMR can result in resistance to temozolomide [66].

PARP inhibitors have a unique interaction with temozolomide. Temozolomide forms methyl adducts in DNA at several sites, including N^7-guanine, O^6-guanine, and N^3-adenine. PARP 1 and 2 can repair the damage at *N*-methylpurines (N^7-MeG and N^3-MeA) so most of the activity of temozolomide results from the damage to the O^6-guanine site. With PARP inhibited, damage to the DNA is increased by preventing repair to the other sites and enhancing cell death [67]. Tumor cell death is increased when temozolomide is administered to patients who have epigenetic silencing of the promoter region on O^6-methylguanine-DNA methyltransferase (MGMT) on chromosome 10q26 (also called O^6-alkylguanine alkyltransferase, OGAT). This gene encodes a DNA repair protein that removes alkyl groups from the O^6 position of guanine. In cancer cells, when MGMT is over-expressed, cells are resistant to temozolomide because of the repair of the O^6-guanine position, but in cells where MGMT has undergone promoter silencing the cells are more sensitive to temozolomide [68]. Researchers are evaluating combining temozolomide and PARP inhibitors in MGMT silenced tumors to maximize the damage to the three main DNA sites affected by temozolomide.

The cytotoxicity of camptothecins, topoisomerase I inhibitors, which cause DNA strand breaks, is also enhanced by PARP inhibitors. The cytotoxic agent is more effective when BER is dysfunctional because of the inability to repair DNA damage. It has been established that the inhibition of PARP's activity with tyrosyl–DNA phosphodiesterase 1 (TDP1) to remove Top1 from the DNA, which is damaging if not removed expeditiously, is one mechanism by which PARP inhibitors potentiate camptothecins. New research has shown that when the endonuclease XPF (xeroderma pigmentosa group F)-ERCC1 (excision repair cross-complementing group 1) is knocked out with siRNA, there is inhibition of the usual formation of γ2HAX foci, a marker of DSBs, in response to a PARP inhibitor given with camptothecin. This finding reveals the effect of PARP inhibitors on this XPF-ERCC1 as an alternate mechanism of action for PARP inhibitors in the setting of camptothecins. Interestingly, the XPF-ERCC1 knockout that was exposed to the PARP inhibitor in combination with camptothecin, although it had reduced γ2HAX, had an enhanced antiproliferation effect, raising the possibility that the maximal antitumor effect of the PARP inhibitor-camptothecin combination might be in XPF-ERCC1 deficient cells [69]. Resistance to camptothecins due to overexpression of XRCC1 (x-ray repair cross-complementing group 1), a polymorphism in the same class as ERCC1, is reversed by PARP inhibitors since the recruitment of XRCC1 to the break site is obstructed. *In vivo*, PARP inhibitors impair the survival of human colon mouse xenografts after exposure to irinotecan [70].

Alkylating agents, including platinum, cyclophosphamide, and temozolomide are also potentiated by PARP inhibitors, again by inhibiting DNA repair necessary for cell survival after the effects of the cytotoxic agent. Various PARP inhibitors have shown synergistic activity with alkylating agents in xenograft models of breast, ovarian, colon, NSCLC cancers and gliomas [71,72,73,60].

15.6 PARP Inhibitors in Combination with Ionizing Radiation (XRT)

Similar to potentiating cytotoxic chemotherapy, PARP inhibitors potentiate ionizing radiation, via the inhibition of BER and NHEJ, as well as possibly through inhibition of NFκB and other inflammatory proteins and altered regulation of cellular metabolism. In cells where PARP is inhibited, exposure to radiation results in accumulation of DSBs from the conversion of SSBs to DSBs via collapse of replication forks. PARP inhibitors preferentially sensitize cells in S-phase, but also increase the sensitivity of cells in growth arrest that are otherwise resistant to radiation. One experiment showed that latent cells subjected to XRT had 73% greater growth inhibition when exposed to the PARP inhibitor, AG 14361, compared to latent cells not exposed to the PARP inhibitor [74] In a mouse colon cancer xenograft, the addition of veliparib to irradiation extended survival from 23 days with radiation alone to 36 days with the combination. One mouse had a complete remission (CR) [71].

15.7 Clinical Development of PARP Inhibitors

15.7.1 Olaparib (AZD 2281, KU-0059436)

Olaparib, an oral PARP inhibitor, is the first PARP inhibitor to show activity in *BRCA*-mutation positive ovarian and breast cancers. Olaparib has been tested in combination with DNA damaging agents, such as topotecan, doxorubicin, carboplatin, carboplatin and paclitaxel, irinotecan, dacarbazine, and gemcitabine and cisplatin, as well as with antiangiogenesis agents and as a single agent. One problem with combining this agent with myelosuppressive chemotherapy has been the enhancement of the myelosuppression, preventing potentially effective dosing of both the chemotherapy agent(s) as well as olaparib. For example, the addition of olaparib to gemcitabine and cisplatin caused significant thrombocytopenia. The MTD was determined to be 100 mg BID of olaparib on day 1, cisplatin 60 mg/m^2 on day 1 and gemcitabine of 500 mg/m^2 on day 1 and 8 on a 21-day cycle [75]. The question is whether the addition of PARP inhibition will overcome the efficacy lost by decreasing the chemotherapy dose.

15.7.1.1 *BRCA*-mutation ovarian cancer and olaparib

A phase I study of olaparib as a single agent enrolled 60 solid tumor patients, including 22 patients with *BRCA* mutations (and another patient who refused testing but had a strong family history). The initial cohort was not restricted to BRCA deficient tumors, but the expansion cohort enriched the study with *BRCA* mutated tumors. The expansion cohort received 200 mg BID continuously. All of the DLTs were reversible, including mood alterations, fatigue, and thrombocytopenia. Other adverse events were nausea, vomiting, diarrhea, dyspepsia, dysgeusia, stomatitis, altered taste, anorexia, dizziness, and anemia. There was no increase in adverse events in *BRCA* mutation carriers compared to non-*BRCA* carriers. The MTD was determined to be 400 mg BID. There was one DLT at this dose with grade 3 mood disorder and fatigue. The 600 mg dose exceeded MTD with grade 4 thrombocytopenia that resolved after 2 weeks and grade 3 somnolence. Two patients died from infections thought not likely to be related to the study drug. Eight of the 15 patients with *BRCA* mutation positive advanced ovarian cancer achieved PRs. There were no responses seen in ovarian cancer patients who did not have *BRCA* mutations. One of the three patients with *BRCA2* breast cancer, who had progressed on anthracyclines, had a CR for over a year. A breast cancer patient had shrinkage of multiple previously untreated subcentimeter brain lesions after olaparib therapy. This is the first clinical study to support the concept of synthetic lethality in the treatment of cancer patients.

The pharmacokinetics and pharmacodynamics of olaparib were measured. The peak concentration was reached in 1–3 hours. The half-life was 5–7 hours. In PBMCs, 90% of PARP was inhibited in patients treated with ≥60 mg BID. Immunoblots of tumor showed PARP inhibition also. Hair follicles had an increase in γH2AX, measuring DSBs, at all dose levels six hours after dosing that remained elevated throughout the entire cycle at doses ≥200 mg BID. The highest increase in γH2AX was in the 100 mg BID dose [76].

The relationship between platinum sensitivity and response to PARP inhibitors was evaluated in these patients. Of the 50 ovarian cancer patients with *BRCA* mutations treated with olaparib, 20 patients (40%) had CR or PR by RECIST and an additional 3 patients had SD > 4 months. The median duration of response was 28 months. Response to olaparib was associated with sensitivity to platinum. There were no responses seen by RECIST in the patients who had been refractory to cisplatin vs. 46% and 33% in the platinum-sensitive and-resistant populations, respectively. No differences were noted in the duration of response or the time to progression between the platinum response groups. These findings suggest that resistance to platinum may be associated with a decrease in sensitivity to olaparib and possibly other PARP inhibitors. One possible explanation may be a restoration of the HR through a second mutation in the *BRCA* gene, restoring its functionality, and thereby, negating synthetic lethality [77].

In ICEBERG 2, a phase II international trial in women with confirmed *BRCA1* or *BRCA2* mutations with recurrent and incurable ovarian cancer, olaparib was given to patients at 400 mg BID on one arm and at 100 mg BID, a dose that had been shown previously to inhibit PARP, on the other arm. The RECIST overall response rate (ORR) for the patients receiving the 400 mg dose was 33%, and for the patients receiving the 100 mg dose was 15.5%. Two patients in the 400 mg cohort had complete responses and none in the lower dose group. The clinical benefit rate (ORR and/or decrease in CA125 by ≥ 50% and SD > 4 months) was 57.6% for the 400 mg patients and 16.7% for the 100 mg patients. This study suggests that there might be off-target mechanisms enhancing tumor response to olaparib since no clear evidence of a difference in the degree of PARP inhibition was demonstrated between the two arms [78].

Another randomized phase II trial compared two doses of olaparib (200 mg and 400 mg BID) vs. pegylated liposomal doxorubicin (PLD) in patients with *BRCA* mutation positive ovarian cancer that had recurred within 12 months of platinum-based chemotherapy. The PFS were 6.5 vs. 8.8 vs. 7.1 months for 200 mg, 400 mg and PDL arms, respectively, which was not clinically significant [79]. Although some have concluded this to be a negative study, one should balance the risk and benefits of oral single agent olaparib vs. cytotoxic PDL.

A trial treating women with high-grade serous ovarian cancer with single agent olaparib noted 11 of 46 (24%) patients without *BRCA* germline mutations had a response, compared to three of 7 (43%) patients with the germline mutations responding [80]. This study suggests that percent of patients with high-grade serous ovarian cancer without germline *BRCA* mutation may have "BRCAness" and may respond to PARP inhibitors. The question is how to screen for these patients who may also benefit from PARP inhibitor therapy.

15.7.1.2 *BRCA*-mutation associated breast cancer and olaprib

A proof of concept study with single agent olaparib in 54 patients with *BRCA1* or *BRCA2* mutation associated breast cancer patients who had at least one prior line of treatment for their breast cancer showed an ORR of 22% vs. 41% and PFS of 3.8 vs. 5.7 months in the 100 mg vs. 400 mg BID cohorts, respectively. The study was designed as a sequential cohort design accruing to the 400 mg first and 100 mg cohort subsequently. There were more grade 3+ nausea, vomiting and fatigue associated with the higher dose cohort. This study again confirms the activity of olaparib as a single agent in the treatment of *BRCA* mutation positive tumors and the concept of synthetic lethality [81]. This study showed the higher dose of 400 mg to be more efficacious than the 100 mg dose suggesting a possible role of off-target effects. However, this was not a randomized trial, and conclusion should be drawn cautiously.

15.7.2 Veliparib (ABT888)

Preclinically, veliparib was shown to be a potent inhibitor of PARP. It has good bioavailability and is formulated for oral administration as a capsule. It crosses the blood brain barrier as seen with the PKs done in rat brains [71].

The first phase 0 study under FDA's new Exploratory IND was performed with veliparib. Phase 0 studies represent a new strategy for expediting drug development. Veliparib was chosen because it has a wide therapeutic index and a validated pharmacodynamic assay was developed. The pharmacokinetics and pharmacodynamics were evaluated over a short time period after a single non-toxic dose of veliparib. Statistically significant decreases in PAR levels in tumor biopsies were noted at 3–6 hours after dosing, demonstrating inhibition of PARP in tumor samples at clinically achievable concentrations [82].

Veliparib is being widely tested in phase I trials with multiple chemotherapy regimens, radiation and chemoradiation. Phase II trials testing the combination with temozolomide and cyclophosphamide are underway.

15.7.2.1 *BRCA* breast and ovarian cancers and TNBC with veliparib

In a phase I study of veliparib in combination with metronomic cyclophosphamide involving 18 heavily pretreated patients with refractory solid tumors and lymphomas, there were 4 PRs, 2 in ovarian cancer patients with *BRCA* mutation tumors, 1 in a breast cancer patient with *BRCA* mutation, and 1 in a patient with TNBC. Activity was seen mostly in *BRCA* mutation associated tumors. There was also a 30% reduction in tumor of a patient with low-grade lymphoma for approximately 2 years. This patient had progressed on eight prior therapies. Adverse events were mild, mainly grade 1 or 2 lymphopenia and neutropenia [83]. A phase II randomized multi-national trial evaluating the role of veliparib in combination with metronomic dosing of cyclophosphamide in patients with *BRCA* mutation positive ovarian cancer, high-grade serous ovarian cancer, fallopian tube cancer, primary peritoneal cancer, TNBC, and low-grade lymphoma, had completed accrual [84].

15.7.3 Rucaparib (AG014699, PF01367338)

AG 014699 (the phosphate salt of AG 14361) was the first PARP inhibitor studied clinically, in 2003. Preclinical data of AG 014699 in ovarian cancer cell lines were obtained as a single agent and in combination with carboplatin, doxorubicin, gemcitabine, paclitaxel or topotecan. The *BRCA* status of these cell lines was not reported; however, it was suggested that activity of AG 014699 in ovarian cancer was not confined to *BRCA* mutated tumors or high-grade serous carcinoma [85].

A phase I study of AG 014699, which is administered intravenously, with oral temozolomide, evaluated dose, toxicity, PK and pharmacodynamic endpoints for the combination. Pharmacokinetics was linear with a mean terminal half-life of 7.4–11.7 hours. The established PARP inhibitory dose (PID) was 12 mg/m^2, which was based on at least 50% inhibition of peripheral blood lymphocyte PARP activity determined by a validated quantitative immunoassay 24 h after AG014669 administration. At this dose, there was 74–97% inhibition of PARP in PBMC and there was no further reduction in PARP activity in PBLs at 18 mg/m^2. There was, however, increased inhibition of PARP at the higher dose in the tumor tissue. The drug remained in all tumor biopsies at 5 hours after administration, with mean level of PARP inhibition in the tumor at 5 hours post administration of 92% (range 46–97%). PARP inhibition was reversible in the PBMCs and PARP function recovered by at least 50% 72 hours after dosing. Significant toxicities included fatigue, infection, hypophosphatemia, and myelosuppression, which was the DLT at the dose of 18 mg/m^2 in combination with TMZ 200 mg/m^2/day both given for 5 days of a 28-day cycle. There was one CR and one PR in melanoma patients who had not received prior therapy and a PR in a patient with desmoid tumor. There were seven patients with stable disease for at least 6 months, including four patients with melanoma, one patient with prostate cancer, and one patient with pancreatic cancer. Four patients were homozygous for CYP2D6 G186A allele, known as CYP2D6 *4, which was expected to decrease the metabolism of AG 014699. Although the AUC was similar for patients with the CYP mutation compared with patients with wild type, 3 of the 4 patients appeared to benefit. The two melanoma patients who had objective responses also carried this mutation. The third patient had prolonged stable disease. The fourth patient with the mutation died of complications of his disease after the test dose making him inevaluable for this purpose [86].

In a phase II trial, 40 patients with chemotherapy-naïve advanced melanoma were treated with AG014699 at 12 mg/m^2 plus TMZ at 200 $mg/m^2 \times$ 5 days in 28-day cycles. Myelosuppression was greater than expected, with five cases (12.5%) of grade 4 thrombocytopenia, six cases (15.0%) of grade 4 neutropenia, and one case (2.5%) of death from febrile neutropenia at the time of the report. Fatigue and mild nausea also occurred. Twelve patients required dose reduction of TMZ to 150 mg, and one patient required a reduction to 100 mg.

There were four partial responses and four prolonged stable diseases [87].

There is a phase I trial testing rucaparib (AG-014699) in combination with carboplatin in solid tumors [88], as well as a phase II trial of this agent in combination with cisplatin in patients with TNBC or *BRCA* mutation associated breast cancer (HER 2 negative, hormone receptors positive) that will begin accrual soon. The study will have a safety run-in testing rucaparib at 16 and 24 mg. The study will randomize between single agent cisplatin and the combination of cisplatin with rucaparib with the objective of evaluating the 2-year disease-free survival [89]. An oral formulation of rucaparib is under evaluation in *BRCA* patients.

15.7.4 Iniparib (BSI 201, NSC-746045; IND-71677)

Iniparib, BSI 201, initially thought to be a PARP inhibitor, has recently been reported not to inhibit PARP 1, *in vitro*. This agent ignited significant interest in PARP inhibitors in 2009 when it was reported to improve PFS and OS in TNBC when given with gemcitabine and carboplatin in a phase II randomized trial; however, the phase III trial did not confirm a statistical advantage to adding iniparib to the cytotoxic combination [90,91].

At AACR 2011, data were presented on veliparib, iniparib, olaparib and MK 4827 demonstrating that iniparib is not a PARP inhibitor. Unlike the other three agents, iniparib did not inhibit PARP 1 as measured by a decrease in PAR levels post treatment. In cell lines with *BRCA* mutations, iniparib and veliparib both caused an increase in γ2HAX, signifying double strand breaks, at different sites of the DNA. Iniparib formed γ2HAX primarily at the telomeres, though to a lesser degree at non-telomere sites compared with the telomeric affect and compared with veliparib, whereas veliparib caused γ2HAX foci at non-telomere sites only. This effect at different sites of the DNA would suggest that the agents recognize different sequences of DNA [92] Clinically, iniparib did not significantly enhance myelosuppression when given with myelosuppressive chemotherapy, as is seen with olaparib and veliparib [75, 87, 93,94].

15.7.5 Niraparib (MK 4827)

Niraparib is an oral PARP 1 and 2 inhibitor with preclinical data showing single agent anti-tumor activity against BRCA mutant cell lines in culture and xenograft models. It is currently being tested in phase 1 clinical trials as a single agent, in advanced solid tumors, ovarian tumors and prostate tumors, and as combination therapy in patients with advanced solid tumors in combination with carboplatin, carboplatin with paclitaxel, and carboplatin with liposomal doxorubicin [95].

A phase I trial of niraparib as a single agent had shown a half-life of approximately 40 hours. PARP inhibition was shown in PBMC of patients treated with doses higher than 110 mg. The MTD was determined to be 300 mg daily. The DLT was thrombocytopenia. In the expansion cohort comprising patients with high-grade serous ovarian cancer treated at the MTD, six ovarian cancer patients achieved PR; five of the six patients had *BRCA* germline mutations [96]. This is the second PARP inhibitor to show activity as a single agent in patients carrying the *BRCA* mutation.

15.8 Acquired Resistance to PARP Inhibitors

As stated above, one defining characteristic of *BRCA1* and *2* mutated tumor cells is that they are sensitive to platinum agents [97]. However, acquired resistance to cisplatin was shown by secondary intragenic mutations in BRCA2 that restore the wild-type BRCA2 reading frame [98]. Secondary mutations that restore BRCA1/2 were associated with resistance to platinum chemotherapy and may predict resistance to PARP inhibitors in hereditary ovarian carcinomas with germline BRCA1/2 mutations [99]. Other mechanisms of platinum resistance aside from restoration of HR function may not result in PAPR inhibitor resistance. For cells that have been converted to wild-type based on a second mutation, it is possible that sensitization to PARP inhibitors may be maintained if another part of the HR pathway was blocked. Because proteasome inhibitors inhibit FANCD2, phospho-ATM, NBS1, BRCA1, FANCD2, and RAD51, which are all necessary in HR, proteasome inhibitors might reinstitute "BRCAness" [98, 100].

Another mechanism of resistance to PARP inhibitors has been reported in a genetically engineered mouse model for *BRCA1* associated breast cancer exposed to AZ2281. Resistance developed secondary to up-regulation of the *Abcb1a/b* gene, encoding for P-glycoprotein efflux pump. With the coadministration of tariquidar, a P-glycoprotein inhibitor, the resistance was reversed [101].

In addition, 53BP1 regulates the repair mechanism in *BRCA* deficient cells and, therefore, could play a role in resistance to PARP inhibitor. In a cell with normal *BRCA1* function, after a DSB, *BRCA1* displaces 53BP1 and HR repair is completed. In the case of *BRCA1* mutation and absence of 53BP1, downstream proteins of HR are still initiated and HR is still activated. It is only when *BRCA1* is mutated and 53BP1 is present that 53BP1 sticks to the site of the DSB preventing HR, but allowing for error-prone NHEJ to takeover. In BRCA-deficient mice bearing mammary tumor, loss of 53BP1 reduced tumor genesis. The presence of 53BP1 and BRCA1 changes the balance between HR and NHEJ. This would suggest PARP inhibitors would be expected to have activity primarily in cells retaining normal 53BP1, where HR is also impeded. Absence of 53BP1 would confer PARP inhibitor resistance since HR is functional could repair DSBs. In addition, an inhibitor of 53BP1 might decrease the tumor risk in carriers of BRCA mutations. However, reducing 53BP1 might be deleterious since it is also necessary for B cell immunoglobulin switching [102].

Recently 6-thioguanine (6-TG) was reported to have activity in cells that are resistant to PARP inhibitors. In a screen for drugs that selectively kill *BRCA2* deficient cells, 6-TG stood out. 6-TG induces DSBs that cannot be repaired by the *BRCA* mutated cells. 6-TG was as efficient as a PARP inhibitor in selectively killing *BRCA2* mutated tumor cells in a xenograft model. It was effective against *BRCA1* defective, PARP inhibitor resistant (PIR) tumors when the resistance was mediated through increased p-glycoprotein since 6-TG is not a substrate for p-glycoprotein. In *BRCA2* deficient tumor cells with PIR through reactivation of HR, the reactivated HR is not sufficient to repair the 6-TG induced lesions to allow tumor cell survival. 6-TG forms both mismatch repair-dependent and independent HR lesions. 6-TG appears to be active in *BRCA* mutated cells that have acquired platinum and/or PARP inhibitor resistance [103].

15.9 Future Directions

PARP inhibitors are an exciting new class of agents that have shown efficacy, especially in *BRCA*-mutation positive ovarian and breast cancers. In addition, activity has been seen in high-grade serous ovarian cancer, presumably due to other defects in the DNA repair pathway and the induction of BRCAness in germline *BRCA* wild-type tumors. There are many PARP inhibitors in development. Research with these agents has varied in their routes and schedules of administration. The similarity and differences of the toxicity profiles, and resistance mechanisms will be determined with the completion of the ongoing studies.

PARP inhibitors are active as a single agent in *BRCA* mutation positive tumors while causing minimal side effects via synthetic lethality. Tumors with other defects in the HR pathway, including PTEN defects, Fanconi anemia protein defects, ATM abnormalities, RAD51 dysfunction, and EMSY defects, may be sensitive to single agent PARP inhibitors through synthetic lethality as well. Search for biomarkers to identify tumors that are more likely to respond to PARP inhibitors by gene profiling, including microarrays that identify BRCAness, is ongoing. In addition, screening for sensitivity and resistance to PARP inhibitors, for example by measuring PAR in PBMCs *ex vivo* or by genome analysis, may allow for improved selection of patients in and out of the *BRCA* population. Trials are underway to evaluate the efficacy of PARP inhibitors in other HR defective mutations aside from *BRCA*. PARP inhibitors are also being tested in combination with chemotherapy that induces SSBs because of the reliance of those damaged cells on the BER for repair. With time, we will discover to what extent PARP inhibitors provide a tool for targeting tumors and minimizing toxic effects of cancer therapy.

Additional areas of research include investigating off-target effects of the PARP inhibitors. This is suggested by the olaparib data that showed a greater response in the 400 mg group over the 100 mg group, despite PARP inhibition plateauing at 90% at doses higher than 60 mg BID. The superiority of the higher dose may be secondary to an off-target effect or trapping of PARP by the inhibitor. Possibly, the NAD+ site, which is a common receptor of other proteins, may be causing an unknown effect. In addition, PARP 1 leads to down regulation of mTOR and to autophagy as part of the cellular metabolic regulation. PARP inhibitors may cause decreased autophagy and overall cellular dysregulation. Research into the mechanisms of resistance to PARP inhibitors will help in the better understanding of this promising class of agents.

Caution should be raised regarding secondary malignancy. PARP repairs DNA mutations that occur daily. By inhibiting PARP long term, inability to repair these mutations may lead to malignancy, similar to what occurs in the presence of *BRCA* mutations. These nonspecific aspects of PARP inhibitors could lead to unpredicted side effects. Most of the current studies are in metastatic patients. Without diligent vigilance, secondary malignancies associated with the use of a PARP inhibitor may not become evident until this agent moves into the adjuvant and prevention settings.

PARP inhibitors are an exciting new class of agents, particularly for *BRCA* mutation tumors and other tumors with HR defects. *BRCA* mutations are seen in cancers other than breast and ovarian. Trials in prostate cancer and pancreatic cancer are ongoing. Over the next few years, we will gain better understanding of the PARP inhibitors and their role in cancer treatment.

References

1. Hughes-Davies L, Huntsman D, Ruas M, Fuks F, Bye J, Chin SF, *et al.* EMSY links the BRCA2 pathway to sporadic breast and ovarian cancer. *Cell* 2003 Nov 26; 115(5): 523–535.
2. Turner N, Tutt A, Ashworth A. Hallmarks of 'BRCAness' in sporadic cancers. *Nat Rev Cancer* 2004 Oct; 4(10): 814–819.
3. Esteller M, Silva JM, Dominguez G, Bonilla F, Matias-Guiu X, Lerma E, *et al.* Promoter hypermethylation and BRCA1 inactivation in sporadic breast and ovarian tumors. *J Natl Cancer Inst* 2000 Apr 5; 92(7): 564–569.
4. Baldwin RL, Nemeth E, Tran H, Shvartsman H, Cass I, Narod S, *et al.* BRCA1 promoter region hypermethylation in ovarian carcinoma: a population-based study. *Cancer Res* 2000 Oct 1; 60(19): 5329–5333.
5. D'Andrea AD. The Fanconi Anemia/BRCA signaling pathway: disruption in cisplatin-sensitive ovarian cancers. *Cell Cycle* 2003 Jul-Aug; 2(4): 290–292.
6. Taniguchi T, Tischkowitz M, Ameziane N, Hodgson SV, Mathew CG, Joenje H, *et al.* Disruption of the Fanconi anemia-BRCA pathway in

cisplatin-sensitive ovarian tumors. *Nat Med* 2003 May; 9(5): 568–574.

7. Venkitaraman AR. A growing network of cancer-susceptibility genes. *N Engl J Med* 2003 May 8; 348(19): 1917–1919.
8. McCabe N, Turner NC, Lord CJ, Kluzek K, Bialkowska A, Swift S, *et al.* Deficiency in the repair of DNA damage by homologous recombination and sensitivity to poly(ADP-ribose) polymerase inhibition. *Cancer Res* 2006 Aug 15; 66(16): 8109–8115.
9. Mendes-Pereira AM, Martin SA, Brough R, McCarthy A, Taylor JR, Kim JS, *et al.* Synthetic lethal targeting of PTEN mutant cells with PARP inhibitors. *EMBO Mol Med* 2009 Sep; 1(6–7): 315–322.
10. Hoeijmakers JH. Genome maintenance mechanisms for preventing cancer. *Nature* 2001 May 17; 411(6835): 366–374.
11. Venkitaraman AR. Linking the cellular functions of BRCA genes to cancer pathogenesis and treatment. *Annu Rev Pathol* 2009; 4: 461–487.
12. Konstantinopoulos PA, Spentzos D, Karlan BY, Taniguchi T, Fountzilas E, Francoeur N, *et al.* Gene expression profile of BRCAness that correlates with responsiveness to chemotherapy and with outcome in patients with epithelial ovarian cancer. *J Clin Oncol* 2010 Aug 1; 28(22): 3555–3561.
13. Cancer Genome Atlas Research Network. Integrated genomic analyses of ovarian carcinoma. *Nature* 2011 Jun 30; 474(7353): 609–615.
14. Farmer H, McCabe N, Lord CJ, Tutt AN, Johnson DA, Richardson TB, *et al.* Targeting the DNA repair defect in BRCA mutant cells as a therapeutic strategy. *Nature* 2005 Apr 14; 434(7035): 917–921.
15. Bryant HE, Schultz N, Thomas HD, Parker KM, Flower D, Lopez E, *et al.* Specific killing of BRCA2-deficient tumours with inhibitors of poly(ADP-ribose) polymerase. *Nature* 2005 Apr 14; 434(7035): 913–917.
16. Chambon P, Weill JD, Mandel P. Nicotinamide mononucleotide activation of new DNA-dependent polyadenylic acid synthesizing nuclear enzyme. *Biochem Biophys Res Commun* 1963 Apr 2; 11: 39–43.
17. Durkacz BW, Omidiji O, Gray DA, Shall S. (ADP-ribose)n participates in DNA excision repair. *Nature* 1980 Feb 7; 283(5747): 593–596.
18. Farzaneh F, Zalin R, Brill D, Shall S. DNA strand breaks and ADP-ribosyl transferase activation during cell differentiation. *Nature* 1982 Nov 25; 300(5890): 362–366.

19. Wang M, Wu W, Rosidi B, Zhang L, Wang H, Iliakis G. PARP-1 and Ku compete for repair of DNA double strand breaks by distinct NHEJ pathways. *Nucleic Acids Res* 2006; 34(21): 6170–6182.
20. Easton D, Ford D, Peto J. Inherited susceptibility to breast cancer. *Cancer Surv* 1993; 18: 95–113.
21. Basham VM, Lipscombe JM, Ward JM, Gayther SA, Ponder BA, Easton DF, *et al.* BRCA1 and BRCA2 mutations in a population-based study of male breast cancer. *Breast Cancer Res* 2002; 4(1): R2.
22. Johns Hopkins Pathology. Ovarian Cancer. Johns Hopkins Pathology; Available from: http://www.ovariancancer.jhmi.edu/hereditary.cfm.
23. http://www.cancer.gov/cancertopics/pdq/genetics/prostate/HealthProfessional/page2. Prostate Cancer Susceptibilty Loci.
24. http://www.cancer.gov/cancertopics/factsheet/Risk/BRCA. BRCA1 and BRCA2: Cancer Risk and Genetic Testing.
25. Gonzalez-Angulo AM, Timms KM, Liu S, Chen H, Litton JK, Potter J, *et al.* Incidence and outcome of BRCA mutations in unselected patients with triple receptor-negative breast cancer. *Clin Cancer Res* 2011 Mar 1; 17(5): 1082–1089.
26. Leegte B, van der Hout AH, Deffenbaugh AM, Bakker MK, Mulder IM, ten Berge A, *et al.* Phenotypic expression of double heterozygosity for BRCA1 and BRCA2 germline mutations. *J Med Genet* 2005 Mar; 42(3): e20.
27. Genetics M. Myriad Genetics *BRCA test kit* 2010.
28. Tommasi S. BRCA1 mutations and polymorphisms in a hospital-based consecutive series of breast cancer patients from Apulia, Italy. *Mutat Res* 2005 2005; 587: 359–405.
29. Society AC. National Guideline Clearinghouse. 2007.
30. James PA, Doherty R, Harris M, Mukesh BN, Milner A, Young MA, *et al.* Optimal selection of individuals for BRCA mutation testing: a comparison of available methods. *J Clin Oncol* 2006 Feb 1; 24(4): 707–715.
31. van Harssel JJ, van Roozendaal CE, Detisch Y, Brandao RD, Paulussen AD, Zeegers M, *et al.* Efficiency of BRCAPRO and Myriad II mutation probability thresholds versus cancer history criteria alone for BRCA1/2 mutation detection. *Fam Cancer* 2010 Jun; 9(2): 193–201.
32. Grann VR, Jacobson JS, Thomason D, Hershman D, Heitjan DF, Neugut AI. Effect of prevention strategies on survival and quality-adjusted survival of women with BRCA1/2 mutations: an updated decision analysis. *J Clin Oncol* 2002 May 15; 20(10): 2520–2529.

33. Lin C, Sasaki T, Strumwasser A, Harken A. The case against BRCA 1 and 2 testing. *Surgery* 2011 Jun; 149(6): 731–734.

34. Futreal PA, Soderkvist P, Marks JR, Iglehart JD, Cochran C, Barrett JC, *et al.* Detection of frequent allelic loss on proximal chromosome 17q in sporadic breast carcinoma using microsatellite length polymorphisms. *Cancer Res* 1992 May 1; 52(9): 2624–2627.

35. Russell SE, Hickey GI, Lowry WS, White P, Atkinson RJ. Allele loss from chromosome 17 in ovarian cancer. *Oncogene* 1990 Oct; 5(10): 1581–1583.

36. Dobrovic A, Simpfendorfer D. Methylation of the BRCA1 gene in sporadic breast cancer. *Cancer Res* 1997 Aug 15; 57(16): 3347–3350.

37. Mancini DN, Rodenhiser DI, Ainsworth PJ, O'Malley FP, Singh SM, Xing W, *et al.* CpG methylation within the 5' regulatory region of the BRCA1 gene is tumor specific and includes a putative CREB binding site. *Oncogene* 1998 Mar 5; 16(9): 1161–1169.

38. Wilson CA, Ramos L, Villasenor MR, Anders KH, Press MF, Clarke K, *et al.* Localization of human BRCA1 and its loss in high-grade, non-inherited breast carcinomas. *Nat Genet* 1999 Feb; 21(2): 236–240.

39. Umetani N, Mori T, Koyanagi K, Shinozaki M, Kim J, Giuliano AE, *et al.* Aberrant hypermethylation of ID4 gene promoter region increases risk of lymph node metastasis in T1 breast cancer. *Oncogene* 2005 Jul 7; 24(29): 4721–4727.

40. Turner NC, Reis-Filho JS, Russell AM, Springall RJ, Ryder K, Steele D, *et al.* BRCA1 dysfunction in sporadic basal-like breast cancer. *Oncogene* 2007 Mar 29; 26(14): 2126–2132.

41. Lamers E. Microarray based expression profi ling of BRCA1 mutated human tumours using a breast-specifi c platform to identify a profile of BRCA1 deficiency. *Breast Cancer Res* 2010; 12(Suppl 1): poster 44.

42. Sorlie T, Tibshirani R, Parker J, Hastie T, Marron JS, Nobel A, *et al.* Repeated observation of breast tumor subtypes in independent gene expression data sets. *Proc Natl Acad Sci U S A* 2003 Jul 8; 100(14): 8418–8423.

43. Jazaeri AA, Yee CJ, Sotiriou C, Brantley KR, Boyd J, Liu ET. Gene expression profiles of BRCA1-linked, BRCA2-linked, and sporadic ovarian cancers. *J Natl Cancer Inst* 2002 Jul 3; 94(13): 990–1000.

44. Hennessy BT. Somatic BRCA status in ovarian tumors. *J Clin Oncol* [abstract]. 2009 May 20 Supplement; 27(155): 5528.

45. Hennessy BT, Timms KM, Carey MS, Gutin A, Meyer LA, Flake DD, 2nd, *et al.* Somatic mutations in BRCA1 and BRCA2 could expand the number

of patients that benefit from poly (ADP ribose) polymerase inhibitors in ovarian cancer. *J Clin Oncol* 2010 Aug 1; 28(22): 3570–3576.

46. Geisler JP, Hatterman-Zogg MA, Rathe JA, Buller RE. Frequency of BRCA1 dysfunction in ovarian cancer. *J Natl Cancer Inst* 2002 Jan 2; 94(1): 61–67.

47. Hilton JL, Geisler JP, Rathe JA, Hattermann-Zogg MA, DeYoung B, Buller RE. Inactivation of BRCA1 and BRCA2 in ovarian cancer. *J Natl Cancer Inst* 2002 Sep 18; 94(18): 1396–1406.

48. Rouleau M, Patel A, Hendzel MJ, Kaufmann SH, Poirier GG. PARP inhibition: PARP1 and beyond. *Nat Rev Cancer* 2010 Apr; 10(4): 293–301.

49. Simbulan-Rosenthal CM, Ly DH, Rosenthal DS, Konopka G, Luo R, Wang ZQ, *et al.* Misregulation of gene expression in primary fibroblasts lacking poly(ADP-ribose) polymerase. *Proc Natl Acad Sci U S A* 2000 Oct 10; 97(21): 11274–11279.

50. Frizzell KM, Gamble MJ, Berrocal JG, Zhang T, Krishnakumar R, Cen Y, *et al.* Global analysis of transcriptional regulation by poly(ADP-ribose) polymerase-1 and poly(ADP-ribose) glycohydrolase in MCF-7 human breast cancer cells. *J Biol Chem* 2009 Dec 4; 284(49): 33926–33938.

51. Chiang YJ, Hsiao SJ, Yver D, Cushman SW, Tessarollo L, Smith S, *et al.* Tankyrase 1 and tankyrase 2 are essential but redundant for mouse embryonic development. *PLoS One* 2008; 3(7): e2639.

52. Haince JF, Kozlov S, Dawson VL, Dawson TM, Hendzel MJ, Lavin MF, *et al.* Ataxia telangiectasia mutated (ATM) signaling network is modulated by a novel poly(ADP-ribose)-dependent pathway in the early response to DNA-damaging agents. *J Biol Chem* 2007 Jun 1; 282(22): 16441–16453.

53. Haince JF, McDonald D, Rodrigue A, Dery U, Masson JY, Hendzel MJ, *et al.* PARP1-dependent kinetics of recruitment of MRE11 and NBS1 proteins to multiple DNA damage sites. *J Biol Chem* 2008 Jan 11; 283(2): 1197–1208.

54. Hoyer-Hansen M, Jaattela M. AMP-activated protein kinase: a universal regulator of autophagy? *Autophagy* 2007 Jul-Aug; 3(4): 381–383.

55. Veuger SJ, Hunter JE, Durkacz BW. Ionizing radiation-induced NF-kappaB activation requires PARP-1 function to confer radioresistance. *Oncogene* 2009 Feb 12; 28(6): 832–842.

56. Martin-Oliva D, Aguilar-Quesada R, O'Valle F, Munoz-Gamez JA, Martinez-Romero R, Garcia Del Moral R, *et al.* Inhibition of poly(ADP-ribose) polymerase modulates tumor-related gene expression,

including hypoxia-inducible factor-1 activation, during skin carcinogenesis. *Cancer Res* 2006 Jun 1; 66(11): 5744–5756.

57a. Iglehart JD, Silver DP. Synthetic lethality--a new direction in cancer-drug development. *N Engl J Med* 2009 Jul 9; 361(2): 189–191.

57b. Murai J, Huang SN, Das BB, Ranaud A, Zhang Y, Doroshow JH, Ji J, Takeda S, Pommier Y. Trapping of PARP1 and PARP2 by Clinical PARP Inhibitors. *Can Research* 2012 Nov 1; 72(21): 5588–5599.

58. Neuhausen SL, Marshall CJ. Loss of heterozygosity in familial tumors from three BRCA1-linked kindreds. *Cancer Res* 1994 Dec 1; 54(23): 6069–6072.

59. Munoz-Gamez JA, Martin-Oliva D, Aguilar-Quesada R, Canuelo A, Nunez MI, Valenzuela MT, *et al.* PARP inhibition sensitizes p53-deficient breast cancer cells to doxorubicin-induced apoptosis. *Biochem J* 2005 Feb 15; 386(Pt 1): 119–125.

60. Tentori L, Graziani G. Chemopotentiation by PARP inhibitors in cancer therapy. *Pharmacol Res* 2005 Jul; 52(1): 25–33.

61. Bennardo N. Alternative-NHEJ is a mechanistically distinct pathway of mammalian chromosome break repair. *PLoS Genet* 2008; doi: 10.1371/journal.pgen.1000110.

62. Wedge SR, Porteus JK, May BL, Newlands ES. Potentiation of temozolomide and BCNU cytotoxicity by O(6)-benzylguanine: a comparative study *in vitro*. *Br J Cancer* 1996 Feb; 73(4): 482–490.

63. Liu X, Shi Y, Guan R, Donawho C, Luo Y, Palma J, *et al.* Potentiation of temozolomide cytotoxicity by poly(ADP)ribose polymerase inhibitor ABT-888 requires a conversion of single-stranded DNA damages to double-stranded DNA breaks. *Mol Cancer Res* 2008 Oct; 6(10): 1621–1629.

64. Jiricny J. The multifaceted mismatch-repair system. *Nat Rev Mol Cell Biol* 2006 May; 7(5): 335--346.

65. Zaremba T, Curtin NJ. PARP inhibitor development for systemic cancer targeting. *Anticancer Agents Med Chem* 2007 Sep; 7(5): 515–523.

66. Maxwell JA, Johnson SP, McLendon RE, Lister DW, Horne KS, Rasheed A, *et al.* Mismatch repair deficiency does not mediate clinical resistance to temozolomide in malignant glioma. *Clin Cancer Res* 2008 Aug 1; 14(15): 4859–4868.

67. Cheng CL, Johnson SP, Keir ST, Quinn JA, Ali-Osman F, Szabo C, *et al.* Poly(ADP-ribose) polymerase-1 inhibition reverses temozolomide resistance in a DNA mismatch repair-deficient malignant glioma xenograft. *Mol Cancer Ther* 2005 Sep; 4(9): 1364–1368.

68. Hegi ME, Diserens AC, Gorlia T, Hamou MF, de Tribolet N, Weller M, *et al.* MGMT gene silencing and benefit from temozolomide in glioblastoma. *N Engl J Med* 2005 Mar 10; 352(10): 997–1003.

69. Zhang YW, Regairaz M, Seiler JA, Agama KK, Doroshow JH, Pommier Y. Poly(ADP-ribose) polymerase and XPF-ERCC1 participate in distinct pathways for the repair of topoisomerase I-induced DNA damage in mammalian cells. *Nucleic Acids Res* 2011 May; 39(9): 3607–3620.

70. Smith LM, Willmore E, Austin CA, Curtin NJ. The novel poly(ADP-Ribose) polymerase inhibitor, AG14361, sensitizes cells to topoisomerase I poisons by increasing the persistence of DNA strand breaks. *Clin Cancer Res* 2005 Dec 1; 11(23): 8449–8457.

71. Donawho CK, Luo Y, Penning TD, Bauch JL, Bouska JJ, Bontcheva-Diaz VD, *et al.* ABT-888, an orally active poly(ADP-ribose) polymerase inhibitor that potentiates DNA-damaging agents in preclinical tumor models. *Clin Cancer Res* 2007 May 1; 13(9): 2728–2737.

72. Burkle A, Chen G, Kupper JH, Grube K, Zeller WJ. Increased poly(ADP-ribosyl)ation in intact cells by cisplatin treatment. *Carcinogenesis* 1993 Apr; 14(4): 559–561.

73. Miknyoczki SJ, Jones-Bolin S, Pritchard S, Hunter K, Zhao H, Wan W, *et al.* Chemopotentiation of temozolomide, irinotecan, and cisplatin activity by CEP-6800, a poly(ADP-ribose) polymerase inhibitor. *Mol Cancer Ther* 2003 Apr; 2(4): 371–382.

74. Calabrese CR, Almassy R, Barton S, Batey MA, Calvert AH, Canan-Koch S, *et al.* Anticancer chemosensitization and radiosensitization by the novel poly(ADP-ribose) polymerase-1 inhibitor AG14361. *J Natl Cancer Inst* 2004 Jan 7; 96(1): 56–67.

75. Rajan A. A Phase I Combination Study of Olaparib (AZD 2208; KU-0059436) and Cisplatin Plus Gemcitabine in Adults With Solid Tumors. TAT [abstract]. 2010.

76. Fong PC, Boss DS, Yap TA, Tutt A, Wu P, Mergui-Roelvink M, *et al.* Inhibition of poly(ADP-ribose) polymerase in tumors from BRCA mutation carriers. *N Engl J Med* 2009 Jul 9; 361(2): 123–134.

77. Fong PC, Yap TA, Boss DS, Carden CP, Mergui-Roelvink M, Gourley C, *et al.* Poly(ADP)-ribose polymerase inhibition: frequent durable responses in BRCA carrier ovarian cancer correlating with platinum-free interval. *J Clin Oncol* 2010 May 20; 28(15): 2512–2519.

78. Audeh MW, Carmichael J, Penson RT, Friedlander M, Powell B, Bell-McGuinn KM, *et al.* Oral poly(ADP-ribose) polymerase inhibitor olaparib in patients with BRCA1 or BRCA2 mutations and recurrent

ovarian cancer: a proof-of-concept trial. *Lancet* 2010 Jul 24; 376(9737): 245–251.

79. Kaye S, *et al.* Phase II study of the oral PARP inhibitor olaparib (AZD2281) versus liposomal doxorubicin in ovarian cancer patients with BRCA1 and/or BRCA2 mutations. ESMO 2010 [abstract]. 2010.
80. Gelman K. Can we define tumors that will respond to PARP inhibitors? A phase II correlative study of olaparib in advanced serous ovarian cancer and triple negative breast cancer Journal of Clinical Oncology [abstract]. 2010;28(15 supplement):abstract 3002.
81. Tutt A, Robson M, Garber JE, Domchek SM, Audeh MW, Weitzel JN, *et al.* Oral poly(ADP-ribose) polymerase inhibitor olaparib in patients with BRCA1 or BRCA2 mutations and advanced breast cancer: a proof-of-concept trial. *Lancet* 2010 Jul 24; 376(9737): 235–244.
82. Kummar S, Kinders R, Gutierrez ME, Rubinstein L, Parchment RE, Phillips LR, *et al.* Phase 0 clinical trial of the poly (ADP-ribose) polymerase inhibitor ABT-888 in patients with advanced malignancies. *J Clin Oncol* 2009 Jun 1; 27(16): 2705–2711.
83. Kummar S. A phase I study of ABT-888 (A) in combination with metronomic cyclophosphamide (C) in adults with refractory solid tumors and lymphomas. *J Clin Oncol* 2010; 28(15s): ASCO abstr 2605.
84. ClinicalTrials.gov. ABT-888 With Cyclophosphamide in Refractory BRCA-Positive Ovarian, Primary Peritoneal or Ovarian High-Grade Serous Carcinoma, Fallopian Tube Cancer, Triple-Negative Breast Cancer, and Low-Grade Non-Hodgkin's Lymphoma 2011.
85. Ihnen M. Therapeutic Advantage of Chemotherapy Drugs in Combination with PARP Inhibitor PF-01367338 (AG-014699) in Human Ovarian Cancer Cells. EORTC [abstract]. 2010.
86. Plummer R, Jones C, Middleton M, Wilson R, Evans J, Olsen A, *et al.* Phase I study of the poly(ADP-ribose) polymerase inhibitor, AG014699, in combination with temozolomide in patients with advanced solid tumors. *Clin Cancer Res* 2008 Dec 1; 14(23): 7917–7923.
87. Plummer R. First and final report of a phase II study of the poly(ADP-ribose) polymerase (PARP) inhibitor, AG014699, in combination with temozolomide (TMZ) in patients with metastatic malignant melanoma (MM) *J Clin Oncol* 2006; 24(18s): ASCO abstr 8013.
88. ClinicalTrials.gov. A Study Of Poly (ADP-Ribose) Polymerase Inhibitor PF-01367338 In Combination With Several Chemotherapeutic Regimens. 2010.
89. hooshieroncologygroup.org. 2010.

90. O'Shaughnessy J, Osborne C, Pippen JE, Yoffe M, Patt D, Rocha C, *et al.* Iniparib plus chemotherapy in metastatic triple-negative breast cancer. *N Engl J Med* 2011 Jan 20; 364(3): 205–214.

91. Sanofi-aventis. Sanofi-aventis Reports Top-line Results from Phase III Study with BSI-201 in Metastatic Triple-Negative Breast Cancer. Sanofi-aventis Press Release [Press Release]. 2011 January 27, 2011.

92. Ji J. Pharmacodynamics of Four Reported Inhibitors of Poly(ADP-ribose)Polymerase: ABT888, AZD 2281, MK 4827, BSI-201. AACR poster. 2011.

93. Boss DS, Schwartz GK, Middleton MR, Amakye DD, Swaisland H, Midgley RS, *et al.* Safety, tolerability, pharmacokinetics and pharmacodynamics of the oral cyclin-dependent kinase inhibitor AZD5438 when administered at intermittent and continuous dosing schedules in patients with advanced solid tumours. *Ann Oncol* 2010 Apr; 21(4): 884–894.

94. Kummar S, Raffeld M, Juwara L, Horneffer YR, Strassberger A, Allen D, *et al.* Multihistology, target-driven pilot trial of oral topotecan as an inhibitor of hypoxia-inducible factor-1{alpha} (HIF-1{alpha}) in Advanced Solid Tumors. *Clin Cancer Res* 2011 Jun 14; doi: 10.1158/1078-0432.CCR-11-0682.

95. ClinicalTrials.gov. MK 4827. clinicaltrialsgov. 2010.

96. Sandhu SK, *et al.* First-in-human trial of a poly(ADP-ribose) polymerase (PARP) inhibitor MK-4827 in advanced cancer patients (pts) with antitumor activity in BRCA-deficient and sporadic ovarian cancers. ASCO 2010 [abstract]. 2010: abstract 3001.

97. http://www.ncbi.nlm.nih.gov/pmc/articles/PMC3009437/. Cisplatin as neoadjuvant chemotherapy in triple negative breast cancer: Exciting early results. 2010.

98. Sakai W, Swisher EM, Karlan BY, Agarwal MK, Higgins J, Friedman C, *et al.* Secondary mutations as a mechanism of cisplatin resistance in BRCA2-mutated cancers. *Nature* 2008 Feb 28; 451(7182): 1116–1120.

99. Norquist B, Wurz KA, Pennil CC, Garcia R, Gross J, Sakai W, *et al.* Secondary somatic mutations restoring BRCA1/2 predict chemotherapy resistance in hereditary ovarian carcinomas. *J Clin Oncol* 2011 Jun 27 ; 29(22): 3008–3015. doi: 10.1200/JCO.2010.34.2980.

100. Jacquemont C, Taniguchi T. Proteasome function is required for DNA damage response and fanconi anemia pathway activation. *Cancer Res* 2007 Aug 1; 67(15): 7395–7405.

101. Rottenberg S, Jaspers JE, Kersbergen A, van der Burg E, Nygren AO, Zander SA, *et al.* High sensitivity of BRCA1-deficient mammary tumors to the PARP inhibitor AZD2281 alone and in combination with platinum drugs. *Proc Natl Acad Sci U S A* 2008 Nov 4; 105(44): 17079–17084.

102. Bunting SF, Callen E, Wong N, Chen HT, Polato F, Gunn A, *et al.* 53BP1 inhibits homologous recombination in Brca1-deficient cells by blocking resection of DNA breaks. *Cell* 2010 Apr 16; 141(2): 243–254.

103. Issaeva N, Thomas HD, Djurenovic T, Jaspers JE, Stoimenov I, Kyle S, *et al.* 6-thioguanine selectively kills BRCA2-defective tumors and overcomes PARP inhibitor resistance. *Cancer Res* 2010 Aug 1; 70(15): 6268–6276.

Chapter 16

EML4-ALK Fusion Gene and Therapy with ALK-Targeted Agents in Non-Small Cell Lung Cancer

Vimal Patel and Biren Saraiya
Department of Medicine, UMDNJ-Robert Wood Johnson Medical School, New Brunswick, New Jersey, USA
saraiybi@umdnj.edu

16.1 Introduction

Lung cancer remains the leading cause of cancer death in the United States as it is estimated that 157,300 Americans will die of the disease in 2010 [1]. The overall 5-year survival rate in the United States now approaches only 15%. Historically, the histologic appearance of lung cancer has been used as a parameter to guide treatment, distinguishing patients with small-cell lung cancer (SCLC) from those with non-small-cell lung cancer (NSCLC). Over the past decade, the standard of care for patients with advanced NSCLC remained chemotherapy with a platinum-containing doublet [2]. Recently, the distinction of specific NSCLC subtypes has become important, with pemetrexed and bevacizumab approved for advanced nonsquamous NSCLC [3, 4].

Handbook of Therapeutic Biomarkers in Cancer
Edited by Sherry X. Yang and Janet E. Dancey
Copyright © 2013 Pan Stanford Publishing Pte. Ltd.
ISBN 978-981-4364-65-2 (Hardcover), 978-981-4364-66-9 (eBook)
www.panstanford.com

Within the last decade, there has been a rapid influx of knowledge with respect to the key molecular biomarkers involved in the oncogenesis of NSCLC. This field was paved initially with the success of the ABL tyrosine kinase inhibitor imatinib for the treatment of chronic myeloid leukemia [5]. By understanding how specific genetic lesions promote proliferation in cancer cells, the original paradigms of nonspecific chemotherapy have shifted toward more targeted therapies leading to dramatic improvements in response [6]. In NSCLC, this was first illustrated in a small subset of patients having activating mutations within the epidermal growth factor receptor (*EGFR*) leading to the initiation and maintenance of lung cancer. Lynch *et al.* demonstrated the correlation between specific mutations in the *EGFR* gene with clinical responsiveness to the tyrosine kinase inhibitor gefitinib [7]. Treatment of these patients with reversible EGFR tyrosine kinase inhibitors (TKIs), has led to a superior response rate and progression-free survival when compared with conventional chemotherapy [6, 8].

Subsequently, Soda *et al.* in 2007 identified another "driver mutation" in a very small subset of patients with NSCLC: the fusion between the 5′ portion of *EML4* (echinoderm microtubule-associated like-protein 4) and the 3′ portion of *ALK* (anaplastic lymphoma kinase) [9]. These fusions are the result of inversions within the short arm of chromosome 2 and currently appear to be novel to NSCLC [10, 11]. The importance of this discovery is highlighted by new agents that target this unique chimeric protein. In this chapter, we will review the biology and treatment of *EML4-ALK* and other *ALK* gene rearrangements in NSCLC.

16.2 The Identification of *EML4-ALK* in NSCLC

Anaplastic lymphoma kinase translocations in NSCLC were first detected in 2007 with the use of a retroviral complementary DNA expression library from a resected NSCLC tumor diagnosed in a 62-year-old Japanese male smoker whose tumor was negative for the *KRAS* and *EGFR* mutations. The *EML4* and *ALK* genes both map to the short arm of chromosome 2p (2p21 and 2p23 respectively) in opposite orientations separated by approximately 12 megabases (Mb) (see Fig. 16.1) [9]. The segment encompassing these two loci must become inverted to generate the *EML4-ALK* fusion gene. In this particular specimen, the *EML4* gene was interrupted at a position 3.6 kb downstream of exon 13, inverted, and ligated approximately 300 bp upstream of *ALK* exon 20 [9]. Soda *et al.* successfully amplified this genome fragment containing the fusion point between the *EML4* and *ALK* genes, proving the presence of inv(2)(p21p23) in the cancer cells.

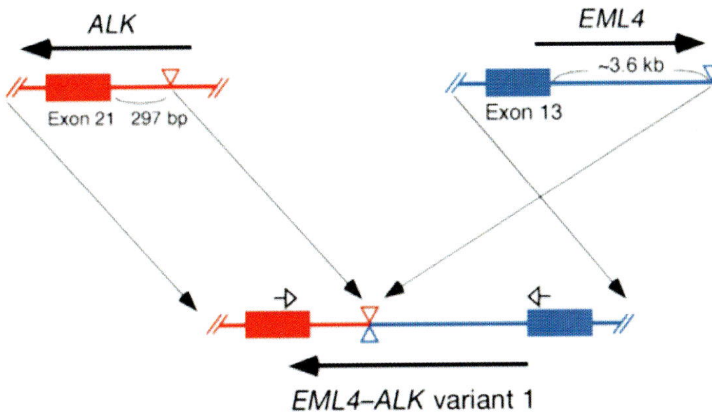

Figure 16.1 Both the ALK gene and the EML4 gene map to chromosome 2p, but have opposite orientations [9]).

16.2.1 The Structure and Function of EML4

Microtubules switch between growing and shrinking phases. These dynamic structures are regulated by microtubule-associated proteins (MAPs) [12]. The most abundant MAP in dividing sea urchin eggs is the echinoderm microtubule-associated protein (EMAP) [13]. There are at least five mammalian homologues of EMAP, or EMLs (EMAP-like proteins), that have been identified. These EMLs share two unique domains that are highly conserved: the hydrophobic EMAP like (HELP) domain near the amino-terminal portion and multiple WD40 domains within the carboxy-terminal portion.

EML4 encodes a 120 kDa EML that has been shown to bind polymerized microtubules *in vitro* [12]. The function of EML4 is likely regulated by differential phosphorylation of serine/threonine residues during the cell cycle. Studies have shown that ELM4 is essential for microtubule formation and stabilization during mitosis [13]. Silencing the gene results in a significant reduction of the proliferation rate of HeLe cells *in vitro* and *EML4*-negative cells show a modified microtubule network, demonstrating that EML4 is necessary for normal cell proliferation [12].

16.2.2 The Structure and Function of ALK

ALK is a receptor tyrosine kinase (RTK) belonging to the insulin receptor (IR) superfamily of RTKs. Based on homology, it forms a subfamily with lymphocyte tyrosine kinase (LTK). ALK was first identified in 1994 as a chimeric protein created by the translocation between chromosomes (2;5)(p23:q35) in patients with anaplastic large cell lymphoma (ALCL). This fusion protein was shown to consist of the amino-terminal portion of nucleophosmin (NPM) ligated to the cytoplasmic domain of a previously unidentified receptor tyrosine kinase [14].

The full-length *ALK* gene was subsequently cloned in 1997 by two groups [15, 16] and encodes a 1620 amino acid protein with features typical of a receptor tyrosine kinase including an extracellular domain, a hydrophobic single pass transmembrane region and an intracellular kinase domain. The extracellular region has a unique combination of domains consisting of an amino-terminal signal peptide, 2 MAM (meprin, A5 protein, and receptor protein tyrosine phosphatase mu) domains, an LDLa (low-density lipoprotein class a) motif and a large glycine-rich putative ligand-binding domain. These domains are highly conserved throughout evolution, but their exact functions have yet to be characterized. The key intracellular kinase domain includes the three-tyrosine motif, YxxxYY, within the activation loop. As in other kinases of the same family, these residues represent major autophosphorylation sites regulating its activity [17]. Lastly, within the intracellular carboxy-terminal portion of ALK are binding sites for Src homology-2 and phospholipase C-γ [18].

The function of the ALK receptor remains poorly understood. Recently, two growth factors have been identified as possible ALK ligands, pleiotrophin (PTN) and midkine [19, 20]. These ligands are thought to activate ALK, resulting in dimerization, autophosphorylation, and signaling through the RAS-ERK, JAK3-STAT3, and PI3-K/Akt pathway (see Fig. 16.2) [21]. Activation of these pathways leads to cell proliferation, survival, and migration. Due to the high expression of ALK mRNA throughout the nervous system during mouse embryogenesis [15, 16], it has been postulated that mammalian ALK plays a key role in the development and function of key cells within the nervous system.

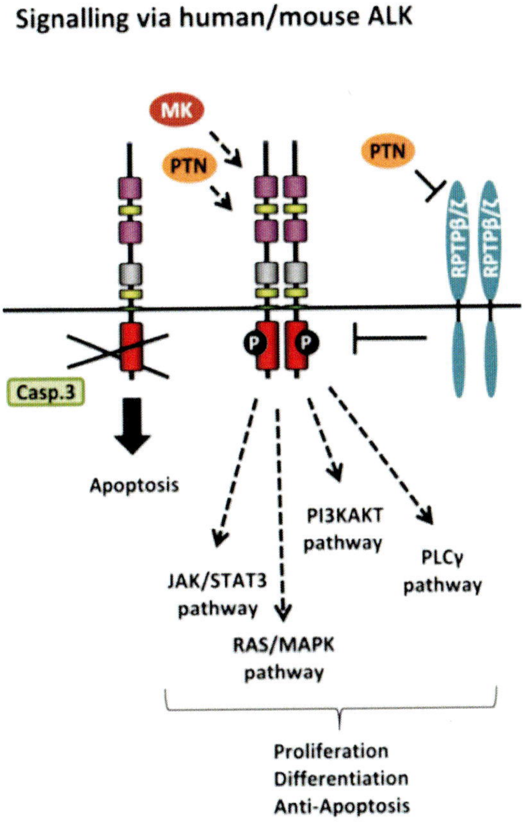

Figure 16.2 Signaling via ALK.

16.2.3 The ALK Gene Rearrangements in Cancer

Many receptor tyrosine kinases have been implicated in oncogenesis due to genetic abnormalities that range from point mutations, gene amplifications, and chromosomal translocations. Whatever the mechanism, the activation of the kinase catalytic domain results in the growth factor-independent proliferation and anti-apoptotic signaling. The most common mechanism of constitutive ALK activation results from chromosomal translocations or inversions involving the 2p23 chromosomal locus where the ALK gene is located. The fusion of various genes at the 5′ end with the cytoplasmic domain of ALK at the 3′ end creates oncogenic ALK fusion genes. Since the initial discovery of the NPM-ALK fusion protein in patients with anaplastic large cell lymphoma, many other ALK chimeras have been described in various tumors in addition to *EML4-ALK* in NSCLC, including, inflammatory myofibroblastic tumors [22], diffuse large B-cell lymphomas [23], and squamous cell carcinoma of the esophagus [24].

All of these ALK fusion proteins share two common features. First, the promoter of the 5′ gene partner controls the expression of the fusion protein. This gene is typically widely expressed in normal tissues and thus leads to ectopic expression of the ALK catalytic kinase domain. Second, the 5′ gene partners have oligomerization domains that mediate self-association of the ALK fusion protein. These domains induce ligand-independent dimerization of the ALK kinase domain, leading to autophosphorylation and constitutive kinase activation. The subsequent aberrant signal transduction cascades potentially result in malignant transformation [25, 26].

16.2.4 The Structure of EML4-ALK and Other Non-EML4 Translocation Partners

To date, the breakpoint within the *ALK* gene always occurs in exon 20 of the gene and thus all *EML4-ALK* fusion proteins contain the entire cytoplasmic tyrosine kinase domain of the receptor (see Fig. 16.3). Soda *et al.* initially described variants 1 and 2, in which exons 13 and 20, respectively, of *EML4* are fused to exon 20 of the *ALK* gene [9]. Truncations of *ELM4* occurring at exons 2, 6, 13, 14, 15, 18, and 20 yield the other isoforms (V3a/b, V4, V5a/b, V6, V7, V8 and V9) of the *EML4-ALK* chimera [10, 27–35]. Despite the various breakpoints within *EML4*, the coiled-coil domain at the amino-terminal end of protein is maintained in each variant and is likely involved in the dimerization of the fusion protein which leads to ligand-independent activation [9]. Of all the variants, the most common ones isolated are variants 1 and 3a/b, representing 33% and 29% of NSCLC, respectively, but the clinical significance of this (if any) has yet to be determined [36].

In *ALK* rearranged NSCLC, *EML4* is not the sole fusion partner with *ALK*. Two other genes, *TFG* and *KIF5B*, have been identified as fusion partners from patients with NSCLC [28, 37]. The function of TFG (TRK-fused gene) has yet to be characterized but interestingly the fusion protein *TFG-ALK* has also been described in anaplastic large cell lymphoma [38]. KIF5B is a component of a motor protein complex that is associated with microtubules and mediates the transport of organelles within eukaryotic cells. Similar to EML4, both TFG and KIF5B have dimerization domains near the amino-terminal end necessary for malignant transformation. The presence of these non-EML4 partners for ALK has implications for the method used to detect *ALK* rearrangements in NSCLC and the clinical significance of these different fusions needs to be further explored.

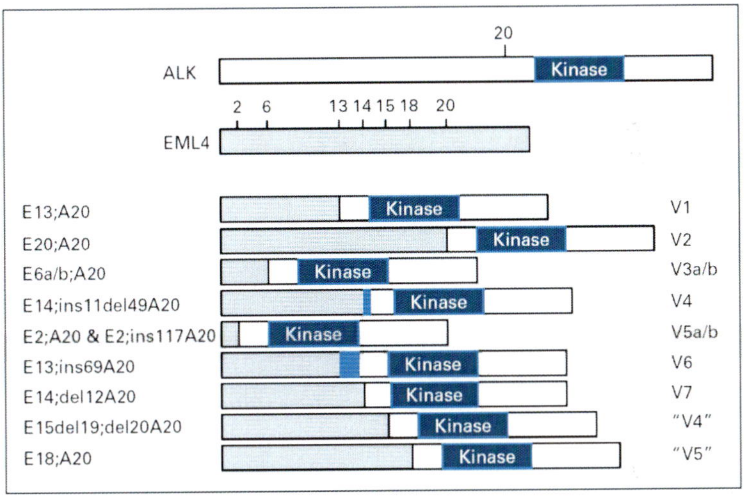

Figure 16.3 Schematic representation of *EML4-ALK* translocations in NSCLC [31].

16.2.5 The Transforming Activity of EML4-ALK

The oncogenic potential of *EML4-ALK* was confirmed by expression in 3T3 fibroblasts that acquired the ability to grow as transformed cell foci in culture [9]. Subcutaneous injections of the transfected 3T3 cells into nude mice also formed tumors [9]. On the contrary, the *EML4-ALK* kinase inactive mutants (K589M) in which Lys 589 in the ATP-binding site of the kinase domain is replaced with Met, showed no ability to form tumors thus confirming the catalytic activity of the kinase domain as being essential.

To further assess the role of *EML4-ALK* in the pathogenesis of NSCLC, transgenic mice specifically expressing the fusion protein in lung epithelial cells driven by the *SPC* (surfactant protein C) promoter were engineered [39]. *EML4-ALK* transgenic mice were found to develop hundreds of adenocarcinoma nodules in both lungs with a very short latency period and with 100% penetrance. The tumor burden and survival of these transgenic mice were improved with a potent oral ALK inhibitor (Novartis 2,4-pyrimidinediamine derivative reported in example 3-39 of PCT WO2005016894), providing further evidence that aberrant ALK tyrosine kinase is required for malignant transformation [39].

16.3 Clinical and Pathologic Features of *EML4-ALK*

With the validation of the *EML4-ALK* oncogene in patients with NSCLC in multiple series published since its initial discovery in 2007, key demographic, pathologic, and epidemiologic features have been associated with this molecular abnormality. Those who are positive for the rearrangement tend to be younger than the general lung cancer population [30, 35, 40–42]. In one of the largest series of *EML4-ALK*-positive patients in NSCLC, the median age of those with the translocation was 53 compared with 66 in those lacking the mutation [40]. Interestingly, many tumors that have *ALK* rearrangements (anaplastic large cell lymphomas, inflammatory myofibroblastic tumors, and neuroblastomas) are also more common in a younger population. There are no dominant gender differences in studies where gender is documented [30, 40, 43].

Another factor that appears to be correlated with *ALK* rearrangements in NSCLC is smoking history. There is a strong trend toward *ALK* positivity being more common among those with little or no smoking history. While the initial smaller studies did not suggest association with smoking status, recent larger studies have consistently shown that the oncogene is associated with a never or light smoking history (<10 pack-years) [9, 29, 30, 32, 33, 35, 40–44]. For instance, Shaw *et al.* published that out of 85 patients categorized as never or light smokers, 19 were positive for *EML4-ALK*, whereas all 56 patients with a significant (>10 pack-years) smoking history were negative for the translocation [38].

With respect to histology, the overwhelming majority of NSCLCs that harbor the fusion oncogene are adenocarcinomas [10, 29, 32, 41, 44]. The other histological subtypes, like large cell carcinoma and squamous cell carcinoma, are rarely seen in *ALK*-rearranged NSCLC patients [28, 40, 41, 43]. A variety of histological features ranging from acinar to mostly signet-ring cell nests have been described. The acinar subtype appears to be more prevalent in Asian patients with adenocarcinomas [28, 43], whereas the signet-ring cell histology is predominantly seen in Western patients [40, 41]. The signet-ring cell histology is frequently seen in gastric cancers and rarely in cancers of other organs such as the lung. The clinical significance of signet-ring cells in *EML4-ALK* transformed NSCLC remains to be determined.

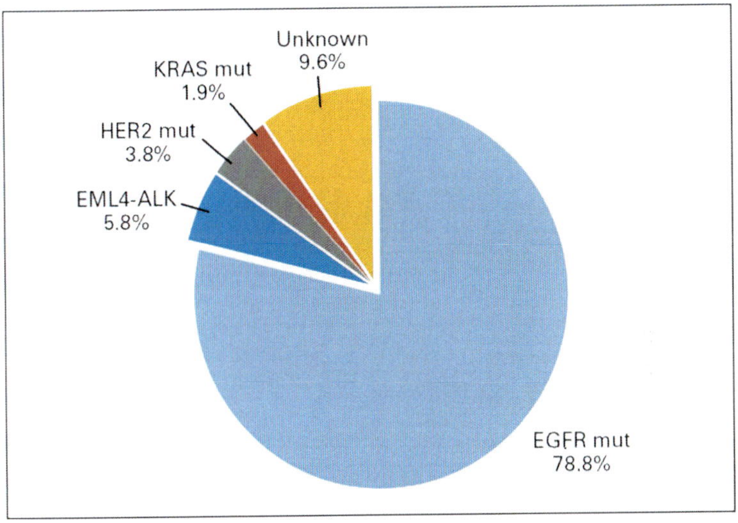

Figure 16.4 Oncogenic driver mutations in East Asian never-smokers with lung [45].

At the molecular level, the presence of *EML4-ALK* is essentially mutually exclusive to that of *EGFR* or *KRAS* mutations in NSCLC [9, 10, 30, 32, 35, 40, 42, 44], however there are very rare exceptions where a NSCLC tumor will harbor both driver mutations [29, 33]. The relative lack of co-occurrence of *EGFR* or *KRAS* mutations with *ALK* rearrangements may reflect the biology of the tumor as there may be a lack of selection pressure for a second oncogenic driver mutation if one already exists.

The *ALK* rearrangement in NSCLC is a relatively rare event with the incidence in largely unselected series ranging anywhere from 2% to 7% [9–11, 29, 32–34, 37, 44]. In the initial description of the *EML4-ALK* oncogene, 5 out of 75 lung tumors (6.7%) had evidence of the *ALK* rearrangement by RT-PCR [9]. Similarly, but in a select population of East Asian never smokers with adenocarcinomas, Sun *et al.* noted that 5.8% (3 out of 52) harbored the *EML4-ALK* fusion, second after the more common oncogenic driver mutation in *EGFR* that was harbored in 78.8% of patients [45]. Knowledge of the clinicopathologic features associated with *EML4-ALK*-positive tumors has enabled researchers to enrich for this defined subset of patients. In the study by Shaw *et al.* in which patients were selected based on features associated with *EGFR* mutations including female gender, Asian ethnicity, never/light smoking history and adenocarcinoma histology, 13% were found to harbor the *ELM4-ALK* fusion [40]. Within the group of patients who were never/light smokers, the frequency of *EML4-ALK* was 22% and when looking at the subgroup of patients who were both never/light smokers and *EGFR* wild-type the frequency increased to 33%. Thus, one in three patients with NSCLC may harbor the *EML4-ALK* fusion if the patient population is enriched with characteristics associated with the *EGFR* mutation but who are actually *EGFR* wild-type.

16.4 Methods of Detection

The accurate identification of a patient with NSCLC positive for the ALK gene rearrangement using any given screening technique is critical for both outcome assessment and therapy. ALK gene rearrangements may be detected in tumor specimens using four techniques: DNA sequencing, reverse transcriptase-PCR (RT-PCR), immunohistochemistry (IHC), and fluorescence *in situ* hybridization (FISH). None of these assays have yet proven to be easily adopted by diagnostic molecular pathology laboratories. DNA sequencing is considered the "gold standard," but the optimal screening method between the other three techniques still needs validation. Although these methods are highly specific and extensively validated in anaplastic large cell lymphoma [46], each has significant challenges when applied to the detection of ALK rearrangements in lung adenocarcinomas.

16.4.1 Reverse Transcriptase-PCR Based Detection

One of the more commonly used screening strategies has been RT-PCR on archival tissue. Given that *EML4* and *ALK* map in opposite orientations within the chromosome 2p, primers designed to amplify the fusion points of *EML4-ALK* would not be expected to yield products from normal cells or cancer cells without the inv(2)(p21p23). In theory, this is a sensitive and rapid method for potentially detecting mutant translocations. Initially, only a few primer sets were tested, but with the recent discovery of at least 9 *EML4-ALK* variants, a number of multiplex assays incorporating different primer sets were designed to detect all in-frame fusions between *EML4* and *ALK* [9, 10, 29]. Despite the fact that different breakpoints in the *EML4* gene may be detected, novel fusions proteins, like TFG and KIF5B, may not be. Depending on the frequency of these non-EML4 fusion partners, the sensitivity of RT-PCR for detecting ALK rearranged NSCLC would be reduced. Another disadvantage of this technique is that RT-PCR requires good-quality mRNA. Most biopsy specimens are stored as formalin-fixed paraffin embedded (FFPE) tissues. The RNA obtained from these samples is highly degraded and makes PCR difficult. Even though PCR from FFPE tissue has been challenging, newer assays are being developed for *EML4-ALK* fusion gene transcripts suitable for use with commonly available FFPE tissues of limited size [47]. Lastly, there are published data indicating that RT-PCR may yield positive results for the EML4-ALK fusion in 15% of non-tumor lung tissue [33] as well as positive results without a corresponding chromosome 2p rearrangement as detected by FISH [33, 34]. Although the data are subject to interpretation, it suggests the possibility of obtaining false positives and hence a reduced specificity of this test. Despite the disadvantages, many still advocate RT-PCR in detecting *EML4-ALK* given the advent of newer high-throughput multiplex assays [10, 47].

16.4.2 Immunohistochemistry Based Detection

Immunohistochemistry (IHC) is a reliable and relatively easy diagnostic tool to detect pathologic proteins in paraffin-embedded tissues without the loss of the cytologic and architectural features. Several antibodies specific for ALK have been developed to diagnose ALK rearranged (NPM-ALK) anaplastic large cell lymphomas with high sensitivity and specificity [46]. Many researchers have had difficulty in detecting ALK fusion proteins by IHC in lung adenocarcinomas likely due to instability of *EML4-ALK* or due to weak transcriptional activity of the promoter-enhancer region of *EML4* compared with that of the *NPM* promoter [33]. As a result, over 30% of *ALK* rearranged lung adenocarcinomas are not detected by standard immunohistochemistry. In one study using standard protocols, only 40% of the FISH-confirmed ALK rearranged lung adenocarcinomas stained positively for ALK fusion proteins. However, by adding an amplification step with tyramide to the protocol, that number improved to 80% [41]. Similarly, another study has shown an even higher sensitivity with the addition of an intercalated antibody-enhanced polymer to the standard antibodies [28, 48]. Despite the addition of these amplification steps though, there are still a number of false negatives. Recently, Mino-Kenudson *et al.* reported a new immunohistochemical assay with novel rabbit monoclonal antibodies (clones D5F3 and D9E4; Cell Signaling Technology, Danvers, MA) with higher sensitivity for ALK protein expression than the conventional mouse monoclonal antibody (clone ALK1; Dako USA) [49]. Their test was both highly sensitive (100%) and highly specific (99%) as ALK protein expression was detected in all of the 22 *ALK*-rearranged lung adenocarcinomas and was not detected in the 131 non-*ALK* rearranged lung adenocarcinomas [49]. In 13 of the 22 cases positive for *ALK* rearrangements (59%), the ALK protein expression could only be detected with the novel antibodies. They were also able to show that the ALK protein expression was much lower in *ALK* rearranged NSCLC then compared with *ALK* rearranged anaplastic large cell lymphoma thus accounting for the reduced sensitivity of conventional immunohistochemistry [49]. These results support the potential role for immunohistochemistry in detecting *ALK* rearrangements in NSCLC.

16.4.3 Fluorescence *in situ* Hybridization Based Detection

A more specific way to detect *ALK* rearrangements is by using fluorescence *in situ* hybridization (FISH) of probes that flank the *ALK* fusion point. FISH has been performed with the commercially available LSI ALK Dual Color break-apart rearrangement probe (Abbott Molecular, Abbot Park, IL) [33, 40, 50]. This assay employs one probe at the 3′ end of ALK (Spectrum Orange (red)-labeled 250 kb probe) and a second probe at the 5′ end of ALK (Spectrum Green (green)-labeled 300 kb probe) that when hybridized against normal nuclei yield a merged (green-orange fluorescent) signal detected through light microscopy. However, any interchromosomal or intrachromosomal rearrangement involving *ALK* and *ELM4* (or any other fusion partner) would be detected by a "split" (green and orange fluorescent) signal. Although FISH is a sensitive and specific diagnostic test to detect these *ALK* rearrangements, it is not entirely flawless. The interpretation of a "split" signal through the introduction of a gap between the two probes can be difficult since the *EML4* and *ALK* loci are closely mapped on 2p and the chromosome inversion is often associated with deletion around the ALK 5′ breakpoint resulting in the 5′ probe failing to hybridize [29, 34]. This pattern is different than the widely split pattern of interchromosomal lesions found in anaplastic large cell lymphoma. In addition to observer error and aberrant probe hybridization (where intended targeted sequences do not hybridize with the probe or the probe hybridizes to nonspecific sequences), nuclear truncation on sectioning leading to the loss of signals and background noise where weak signals mimic nonspecific noise are other drawbacks that contribute to false positive and false negative results. Also, this technique would not be able to differentiate between the various *EML4-ALK* fusion variants however the significance of this remains to be seen. Currently, FISH remains the diagnostic method used as an eligibility criterion in current clinical trials that are accruing patients positive for ALK-rearrangements [51]. These studies deem a specimen to be positive if 15% of the scored tumor cells have split *ALK* 5′ and 3′ signals or have isolated 3′ signals. This cutoff was recently validated in a study by where the positivity via FISH was seen at a minimum of approximately 22% of cells in ALK-positive tumors [52]. This study also noted that both the sensitivity and specificity of this assay improved to 100% when 4 or more tumor areas were examined, including approximately 60 tumor cells [52]. Further studies will need to be performed to validate these results and to standardize the criteria for the detection of *ALK* rearrangements by FISH.

16.4.4 Potential Concerns Independent of the Method of Detection

Each of the three screening platforms used to diagnose *ALK* rearrangements in NSCLC have their inherent advantages and disadvantages. Another challenge, independent of the assay used is tumor heterogeneity with respect to *ALK* rearrangements. This is a well-known phenomenon and will likely impact the capability to detect this molecular biomarker, especially from small biopsies and fine needle aspirates. In one study, 2.7% of NSCLCs cases were noted to be positive in 603 tissue microarray-screen tumors by using separate FISH probes for both *EML4* and *ALK* [34]. However, the *ALK* rearrangements were only detected in 50–100% of cells in positive tumors [34]. Perner *et al.* thus concluded that these rearrangements were a late event in tumorigenesis thus resulting in the co-proliferation of both positive and negative clones. The Camidge *et al.* study also noted that not all cells within a positive tumor display positive signal patterns by FISH [52]. They showed that positivity for *ALK* rearrangements occurred in 22.25% to 86.62% of cells that labeled as *ALK*-positive tumors [52]. The FISH assay had significant background noise as well with up to 11% of the cells in both the non-tumor tissue and the *ALK*-negative tumors having detectable *ALK* rearrangements [52]. This study concluded that this "noise" or false positivity is assay related rather than true biology. To support this, they found no evidence of discrete foci of *ALK* gene rearrangements within the tumor, which one might expect to see if distinct positive and negative subclones truly existed, when a series of four sliding fields were tracked across positive tumors. Although, their data suggest that the variations in the percentage of positive cells in an *ALK* rearranged tumor do not reflect biology, it will need to be confirmed with retrospective clinical data.

In addition to tumor heterogeneity, another concern for detecting *ALK* rearrangements may be that metastases may have a molecular profile that is different from the primary tumor itself. Also, the primary tumor and metastases may have an evolving molecular profile after first-line therapy and subsequent therapies [53]. Further investigations will need to be performed to answer these questions.

16.5 Outcomes with Current Standard NSCLC Therapies

Currently, the prognostic role of *ALK*-rearrangements in NSCLC remains to be determined. There are only a handful of retrospective studies—each with very small sample sizes—that have examined the outcomes of patients harboring the *EML4-ALK* fusion gene. Overall, these limited data show that the clinical response of EGFR tyrosine kinase inhibitors in *EML4-ALK*-positive patients with advanced or metastatic disease more closely follows that of wild-type (*EML4-ALK* negative) patients rather than patients with *EGFR* mutations. In the series published by Camidge *et al.*, the *EML4-ALK*-positive patients who were treated with EGFR tyrosine kinase inhibitor (TKI) therapy (n = 3) had a progression-free survival (PFS) of 5.3 months while the wild-type patients (*EML4-ALK* negative, *EGFR/KRAS* mutation negative) treated with the same TKI therapy (n = 6) had a PFS of 6.5 months [52]. Similarly, Shaw *et al.* reported that none of the 10 patients with NSCLC in their series who harbored the *EML4-ALK* fusion had a clinical response to EGFR TKI therapy, while 3 out of 23 (13%) patients who were wild-type had a partial response [40]. The time to progression (TTP) for patients who were *EML4-ALK* positive versus those who were wild-type were similar, 5 months and 6 months, respectively [40]. Although these data appear to show that *EML4-ALK*-positive patients predict resistance to EGFR tyrosine kinase inhibitors, the sample size is much too small for the two cohorts in these retrospective studies to make these conclusions.

Similarly, it appears that patients with advanced or metastatic NSCLC with *EML4-ALK* do not have better clinical outcomes with platinum-based chemotherapy then those patients who are wild-type. In the Camidge *et al.* series, the PFS of the cohort of *EML4-ALK*-positive patients (n = 7) was shorter than those patients who were *EML4-ALK* negative (n = 22), 3.3 months versus 7 months [52]. Likewise, the response rate in the Shaw *et al.* series was 25% for those who were *EML4-ALK* positive (n = 12) and 35% (n = 34) for those who were wild-type [40]. This did not translate into any significant difference in the time to progression in either group treated with platinum-based chemotherapy (8–10 months) [40]. Despite this, the median overall survival (OS) of the *EML4-ALK*-positive cohort was

20 months compared with 16 months for the wild-type cohort [40]. The results were not statistically significant as it was not powered to detect a difference in overall survival. In addition, the analysis was limited by the retrospective design of the study, the short duration of follow-up, the small number of events, the demographics of the 2 cohorts and the number and type of therapies received. For instance, 41% of the *EML4-ALK*-positive patients went on to participate in a clinical trial with an ALK inhibitor, which likely influenced the outcome of this cohort [40]. In another retrospective study of surgical outcomes in China, a similar improved trend in overall survival was seen in the *EML4-ALK*-positive cohort (n = 12) compared with the *EML4-ALK*-negative cohort (n = 91) [35]. The *EML4-ALK*-negative cohort had a median overall survival of 50.5 months, whereas the *EML4-ALK*-positive cohort did not have enough events to determine the median survival [35]. Given the limited data available, it is still unclear what the prognostic implications are for those with *ALK* rearranged NSCLC. Further retrospective and prospective analyses will need to be performed.

16.5.1 Preclinical ALK Targeted Therapies in NSCLC

Several compounds have been shown to inhibit the kinase activity of ALK thereby suppressing the growth of cells expressing the NPM-ALK fusion protein commonly seen in ALCL. With the initial groundbreaking discovery of the *EML4-ALK* translocation in NSCLC, Soda *et al.* hinted that these same compounds may also be used to inhibit the *EML4-ALK* fusion protein in NSCLC. Initial preclinical studies with the NPM-ALK inhibitors, WHI-P154 and NVP-TAE684, were shown to be effective in the regression of established tumors, both *in vitro* and in mouse xenografts [9, 54, 55]. Another small molecule TKI targeting MET (mesenchymal epithelial transition growth factor), a member of the IR family of TKs, showed antitumor effects on cell and tumor models of NPM-ALK driven ALCL at nanomolar concentrations [54, 55]. Based on these findings, this oral ATP-competitive dual ALK/MET inhibitor, PF-2341066 (better known as crizotinib) became the first ALK inhibitor to date to undergo clinical trial testing for safety and efficacy in patients with ALK-rearranged NSCLC.

16.5.2 Clinical Studies with Crizotinib

In the recently reported phase I study, patients with solid tumors refractory to standard therapy were enrolled in a dose-escalation study of crizotinib [56]. The maximum tolerated dose of crizotinib was 250 mg twice daily. Major toxicities were nausea, vomiting, diarrhea, fatigue, transaminitis and visual disturbances that were particularly noted during changes in ambient lighting [56]. Two patients with ALK rearranged NSCLC who were treated with crizotinib during dose escalation had dramatic improvements which prompted the enrollment of an expanded cohort of patients with ALK rearranged NSCLC patients, prospectively. Among the 82 patients (most of whom were heavily pretreated) Kwak *et al.* observed an overall response rate (ORR) of 57% and a disease control rate (defined as the percentage of patients with a complete or partial response or stable disease) of 90% [51]. These results are impressive when compared with the 10% response rate typically seen with second-line chemotherapy of NSCLC. Of 82 patients, 77% continued to receive crizotinib after data cutoff with a median duration of treatment of 6.4 months and a probability of PFS of 72% at 6 months [51]. After only 3 years from the start of this phase I/II trial, a phase III trial (ClinicalTrials.gov number, NCT00932893) randomizing 318 patients with ALK-rearranged NSCLC who have received one prior line of chemotherapy to either crizotinib or second-line chemotherapy with pemetrexed or docetaxel is currently underway. Patients who are not eligible for the phase III trial or who were randomized to chemotherapy and had progressive disease, could be enrolled in a companion single-arm phase II trial (ClinicalTrials.gov number, NCT00932451) using crizotinib. Given the initial phase I/II data, the results of the phase III trial will likely change the paradigms for this specific subset of patients with NSCLC.

16.5.3 Resistance to Crizotinib and Emergence of New ALK Inhibitors

Unfortunately, a portion of *ALK*-rearranged NSCLC tumors are either refractory to small molecule receptor tyrosine inhibitors or become resistant after an initial response, as seen in chronic myeloid leukemia with imatinib mesylate. Recently, Choi *et al.* presented a case report where a 28 year-old man with chemotherapy refractory,

ALK rearranged metastatic NSCLC progressed after an initial partial regression of his tumor burden over 5 months despite sustained administration of crizotinib [57]. Analysis of sputum samples before treatment with crizotinib and after relapse, identified two de novo mutations within the kinase domain of *EML4-ALK*. BA/F3 cell lines transfected with these secondary mutants of *EML4-ALK* (C1156Y and L1196M) were shown to have a markedly reduced sensitivity to two ALK inhibitors [57]. The amino acid substitutions of the cysteine at position 1156 (N-terminal side to αC helix mutation) and the leucine at position 1196 ("gatekeeper" mutation) occur at positions close to the ATP-binding pocket and likely allosterically hinder the binding of crizotinib to the oncokinase [57]. Recently, an additional 16 mutations within various motifs have been uncovered by Zhang *et al.* using an accelerated *in vitro* mutageneis screen in BA/F3 cells [58]. The elucidation of the crystal structure of the ALK kinase domain with these newly discovered mutations will provide information for the development of second-generation ALK inhibitors.

A candidate second-generation ALK inhibitor is AP26113 and has been studied in preclinical models [58]. Studies demonstrate that AP26113 overcame all the mutations that conferred resistance to crizotinib both *in vitro* and *in vivo* [58]. In a second study, Rivera *et al.* directly compared AP26113 with crizotinib in a series of ALK-dependent cell culture and *in vivo* models [59]. In each model, AP26113 was at least 10 times more potent than crizotinib [59]. This new inhibitor exhibited approximately 100-fold selectivity for ALK-positive cell lines, whereas crizotinib only had a 10-fold selectivity [59]. The promising preclinical data support the clinical evaluation of AP26113 in *ALK* rearranged NSCLC.

16.6 Future Directions

The *EML4-ALK* fusion oncogene defines a new molecular subset of NSCLC with distinct clinicopathologic features. Although current series predict that the genetic translocation occurs in 2–7% of patients with NSCLC, the potential to help tens of thousands of patients in the United States alone, and possibly even more in subsets of the Asian population, exists due to the rather sizable incidence of NSCLC. Further analyses will need to be performed to definitively answer the prognostic role of the *ALK*-rearrangement in NSCLC. However, it is clear that this biomarker is a predictive marker for response with ALK inhibitors, as crizotinib has shown clinical activity in highly refractory patients that harbor this driver mutation. Interest in the development of small molecule ALK inhibitors has recently heightened and will continue to expand. Given the initial success of crizotinib, the ALK pathway status should be assessed at diagnosis (and likely at relapse) to identify patients who will respond to this treatment. The results of the current phase III trial with crizotinib will likely change the standard of care of these patients in the first-line setting. New questions will emerge regarding how to integrate these ALK inhibitors into front-line chemotherapy regimens. The development of resistance is an emerging problem. Second- and subsequent-generation ALK inhibitors as well as monoclonal antibody inhibitors will need to be developed. The promise of such agents brings us closer to personalized treatment of lung cancer.

Acknowledgments

This chapter would not have been possible without the support of many of our colleagues. We would like to express our deepest gratitude to Dr. Joseph Aisner and Dr. Antoinette Tan for their invaluable assistance, guidance, and review of this work.

References

1. Jemal, A., *et al.*, Cancer Statistics, 2010. *CA Cancer J Clin*, 2010. **60**(5): p. 277–300.
2. Schiller, J.H., *et al.*, Comparison of Four Chemotherapy Regimens for Advanced Non–Small-Cell Lung Cancer. *N Engl J Med*, 2002. **346**(2): p. 92–98.
3. Sandler, A., *et al.*, Paclitaxel–Carboplatin Alone or with Bevacizumab for Non–Small-Cell Lung Cancer. *N Engl J Med*, 2006. **355**(24): p. 2542–2550.
4. Scagliotti, G.V., *et al.*, Phase III Study Comparing Cisplatin Plus Gemcitabine With Cisplatin Plus Pemetrexed in Chemotherapy-Naive Patients With Advanced-Stage Non–Small-Cell Lung Cancer. *J Clin Oncol*, 2008. **26**(21): p. 3543–3551.
5. Druker, B.J., *et al.*, Efficacy and Safety of a Specific Inhibitor of the BCR-ABL Tyrosine Kinase in Chronic Myeloid Leukemia. *N Engl J Med*, 2001. **344**(14): p. 1031–1037.

6. Mok, T.S., *et al.*, Gefitinib or Carboplatin–Paclitaxel in Pulmonary Adenocarcinoma. *N Engl J Med*, 2009. **361**(10): p. 947–957.
7. Lynch, T.J., *et al.*, Activating Mutations in the Epidermal Growth Factor Receptor Underlying Responsiveness of Non–Small-Cell Lung Cancer to Gefitinib. *N Engl J Med*, 2004. **350**(21): p. 2129–2139.
8. Maemondo, M., *et al.*, Gefitinib or Chemotherapy for Non–Small-Cell Lung Cancer with Mutated EGFR. *N Engl J Med*, 2010. **362**(25): p. 2380–2388.
9. Soda, M., *et al.*, Identification of the transforming *EML4-ALK* fusion gene in non-small-cell lung cancer. *Nature*, 2007. **448**(7153): p. 561–566.
10. Takeuchi, K., *et al.*, Multiplex reverse transcription-PCR screening for *EML4-ALK* fusion transcripts. *Clin Cancer Res*, 2008. **14**(20): p. 6618–6624.
11. Fukuyoshi, Y., *et al.*, *EML4-ALK* fusion transcript is not found in gastrointestinal and breast cancers. *Br J Cancer*, 2008. **98**(9): p. 1536–1539.
12. Houtman, S.H., *et al.*, Echinoderm microtubule-associated protein like protein 4, a member of the echinoderm microtubule-associated protein family, stabilizes microtubules. *Neuroscience*, 2007. **144**(4): p. 1373–1382.
13. Pollmann, M., *et al.*, Human EML4, a novel member of the EMAP family, is essential for microtubule formation. *Exp Cell Res*, 2006. **312**(17): p. 3241–3251.
14. Morris, S., *et al.*, Fusion of a kinase gene, ALK, to a nucleolar protein gene, NPM, in non-Hodgkin's lymphoma. *Science*, 1994. **263**(5151): p. 1281–1284.
15. Iwahara, T., *et al.*, Molecular characterization of ALK, a receptor tyrosine kinase expressed specifically in the nervous system. *Oncogene*, 1997. **14**(4): p. 439–449.
16. Morris, S.W., *et al.*, ALK, the chromosome 2 gene locus altered by the t(2;5) in non-Hodgkin's lymphoma, encodes a novel neural receptor tyrosine kinase that is highly related to leukocyte tyrosine kinase (LTK). *Oncogene*, 1997. **14**(18): p. 2175–2188.
17. Tartari, C.J., *et al.*, Characterization of Some Molecular Mechanisms Governing Autoactivation of the Catalytic Domain of the Anaplastic Lymphoma Kinase. *J Biol Chem*, 2008. **283**(7): p. 3743–3750.
18. Degoutin, J., M. Vigny, and J.Y. Gouzi, ALK activation induces Shc and FRS2 recruitment: Signaling and phenotypic outcomes in PC12 cells differentiation. *FEBS Lett*, 2007. **581**(4): p. 727–734.

19. Stoica, G.E., *et al.*, Identification of Anaplastic Lymphoma Kinase as a Receptor for the Growth Factor Pleiotrophin. *J Biol Chem*, 2001. **276**(20): p. 16772–16779.

20. Stoica, G.E., *et al.*, Midkine Binds to Anaplastic Lymphoma Kinase (ALK) and Acts as a Growth Factor for Different Cell Types. *J Biol Chem*, 2002. **277**(39): p. 35990–35998.

21. Chiarle, R., *et al.*, The anaplastic lymphoma kinase in the pathogenesis of cancer. *Nat Rev Cancer*, 2008. **8**(1): p. 11–23.

22. Griffin, C.A., *et al.*, Recurrent involvement of 2p23 in inflammatory myofibroblastic tumors. *Cancer Res*, 1999. **59**(12): p. 2776–2780.

23. Arber, D.A., L.-H. Sun, and L.M. Weiss, Detection of the t(2; 5)(p23; q35) chromosomal translocation in large B-cell lymphomas other than anaplastic large cell lymphoma. *Hum pathol*, 1996. **27**(6): p. 590–594.

24. Du, X.-L., *et al.*, Proteomic profiling of proteins dysregulted in Chinese esophageal squamous cell carcinoma. *J Mol Med*, 2007. **85**(8): p. 863–875.

25. Webb, T.R., *et al.*, Anaplastic lymphoma kinase: role in cancer pathogenesis and small-molecule inhibitor development for therapy. *Exp Rev Anticancer Ther*, 2009. **9**(3): p. 331–356.

26. Ardini, E., *et al.*, Anaplastic Lymphoma Kinase: Role in specific tumours, and development of small molecule inhibitors for cancer therapy. *Cancer Lett*, 2010. **299**(2): p. 81–94.

27. Choi, Y.L., *et al.*, Identification of novel isoforms of the *EML4-ALK* transforming gene in non-small cell lung cancer. *Cancer Res*, 2008. **68**(13): p. 4971–4976.

28. Takeuchi, K., *et al.*, *KIF5B-ALK,* a novel fusion oncokinase identified by an immunohistochemistry-based diagnostic system for ALK-positive lung cancer. *Clin Cancer Res*, 2009. **15**(9): p. 3143–3149.

29. Koivunen, J.P., *et al.*, *EML4-ALK* fusion gene and efficacy of an ALK kinase inhibitor in lung cancer. *Clin Cancer Res*, 2008. **14**(13): p. 4275–4283.

30. Wong, D.W., *et al.*, The *EML4-ALK* fusion gene is involved in various histologic types of lung cancers from nonsmokers with wild-type *EGFR* and *KRAS*. *Cancer*, 2009. **115**(8): p. 1723–1733.

31. Horn, L., and W. Pao, *EML4-ALK:* Honing in on a new target in non-small-cell lung cancer. *J Clin Oncol*, 2009. **27**(26): p. 4232–4235.

32. Inamura, K., *et al.*, *EML4-ALK* fusion is linked to histological characteristics in a subset of lung cancers. *J Thorac Oncol*, 2008. **3**(1): p. 13–17 10.1097/JTO.0b013e31815e8b60.

33. Martelli, M.P., *et al.*, *EML4-ALK* rearrangement in non-small cell lung cancer and non-tumor lung tissues. *Am J Pathol*, 2009. **174**(2): p. 661–670.
34. Perner, S., *et al.*, *EML4-ALK* fusion lung cancer: A rare acquired event. *Neoplasia*, 2008. **10**(3): p. 298–302.
35. Zhang, X., *et al.*, Fusion of *EML4* and *ALK* is associated with development of lung adenocarcinomas lacking EGFR and KRAS mutations and is correlated with ALK expression. *Mol Cancer*, 2010. **9**(1): p. 188.
36. Sasaki, T., *et al.*, The biology and treatment of *EML4-ALK* non-small cell lung cancer. *Eur J Cancer*, 2010. **46**(10): p. 1773–1780.
37. Rikova, K., *et al.*, Global survey of phosphotyrosine signaling identifies oncogenic kinases in lung cancer. *Cell*, 2007. **131**(6): p. 1190–1203.
38. Hernandez, L., *et al.*, *TRK*-fused gene *(TFG)* is a new partner of *ALK* in anaplastic large cell lymphoma producing two structurally different *TFG-ALK* translocations. *Blood*, 1999. **94**(9): p. 3265–3268.
39. Soda, M., *et al.*, A mouse model for *EML4-ALK*-positive lung cancer. *Proc Natl Acad Sci U S A*, 2008. **105**(50): p. 19893–19897.
40. Shaw, A.T., *et al.*, Clinical Features and Outcome of Patients With Non-Small-Cell Lung Cancer Who Harbor *EML4-ALK*. *J Clin Oncol*, 2009. **27**(26): p. 4247–4253.
41. Rodig, S.J., *et al.*, Unique clinicopathologic features characterize ALK-rearranged lung adenocarcinoma in the western population. *Clin Cancer Res*, 2009. **15**(16): p. 5216–5223.
42. Sakairi, Y., *et al.*, *EML4-ALK* fusion gene assessment using metastatic lymph node samples obtained by endobronchial ultrasound-guided transbronchial needle aspiration. *Clin Cancer Res*, 2010. **16**(20): p. 4938–4945.
43. Inamura, K., *et al.*, *EML4-ALK* lung cancers are characterized by rare other mutations, a *TTF-1* cell lineage, an acinar histology, and young onset. *Mod Pathol*, 2009. **22**(4): p. 508–515.
44. Kazuya, S., *et al.*, *EML4-ALK* fusion transcripts, but no *NPM-, TPM3-, CLTC-, ATIC-, or TFG-ALK* fusion transcripts, in non-small cell lung carcinomas. *Lung cancer* (Amsterdam, Netherlands), 2008. **61**(2): p. 163–169.
45. Sun, Y., *et al.*, Lung adenocarcinoma from east asian never-smokers is a disease largely defined by targetable oncogenic mutant kinases. *J Clin Oncol*, 2010. **28**(30): p. 4616–4620.

46. Cataldo, K.A., *et al.*, Detection of t(2;5) in anaplastic large cell lymphoma: comparison of immunohistochemical studies, FISH, and *RT-PCR* in paraffin-embedded tissue. *Am J Surg Pathol*, 1999. **23**(11): p. 1386.

47. Danenberg, P.V., *et al.*, A novel *RT-PCR* approach to detecting *EML4-ALK* fusion genes in archival *NSCLC* tissue. ASCO Meeting Abstracts, 2010. **28**(15_suppl): p. 10535.

48. Jokoji, R., *et al.*, Combination of morphological feature analysis and immunohistochemistry is useful for screening of *EML4-ALK*-positive lung adenocarcinoma. *J Clin Pathol*, 2010. **63**(12): p. 1066–1070.

49. Mino-Kenudson, M., *et al.*, A novel, highly sensitive antibody allows for the routine detection of *ALK*-rearranged lung adenocarcinomas by standard immunohistochemistry. *Clin Cancer Res*, 2010. **16**(5): p. 1561–1571.

50. Boland, J.M., *et al.*, Anaplastic lymphoma kinase immunoreactivity correlates with *ALK* gene rearrangement and transcriptional up-regulation in non-small cell lung carcinomas. *Hum pathol*, 2009. **40**(8): p. 1152–1158.

51. Kwak, E.L., *et al.*, Anaplastic lymphoma kinase inhibition in non–small-cell lung cancer. *N Engl J Med*, 2010. **363**(18): p. 1693–1703.

52. Camidge, D.R., *et al.*, Optimizing the detection of lung cancer patients harboring anaplastic lymphoma kinase *(ALK)* gene rearrangements potentially suitable for ALK inhibitor treatment. *Clin Cancer Res*, 2010. **16**(22): p. 5581–5590.

53. Hirsch, F.R., *et al.*, The tissue is the issue: Personalized medicine for non-small cell lung cancer. *Clin Cancer Res*, 2010. **16**(20): p. 4909–4911.

54. Christensen, J.G., *et al.*, Cytoreductive antitumor activity of PF-2341066, a novel inhibitor of anaplastic lymphoma kinase and c-Met, in experimental models of anaplastic large-cell lymphoma. *Mol Cancer Ther*, 2007. **6**(12): p. 3314–3322.

55. Zou, H.Y., *et al.*, An orally available small-molecule inhibitor of c-Met, PF-2341066, exhibits cytoreductive antitumor efficacy through antiproliferative and antiangiogenic mechanisms. *Cancer Res*, 2007. **67**(9): p. 4408–4417.

56. Kwak, E.L., *et al.*, Clinical activity observed in a phase I dose escalation trial of an oral c-met and *ALK* inhibitor, PF-02341066. ASCO Meeting Abstracts, 2009. **27**(15S): p. 3509.

57. Choi, Y.L., *et al.*, *EML4-ALK* Mutations in Lung Cancer That Confer Resistance to *ALK* Inhibitors. *N Engl J Med*, 2010. **363**(18): p. 1734–1739.

58. Zhang, S., *et al.*, *AP26113,* a potent ALK inhibitor, overcomes mutations in *EML4-ALK* that confer resistance to PF-02341066. Proceedings from the 101st Annual Meeting of the American Association for Cancer Research, 2010, 2010: p. April 17–21; Washington, DC. Philadelphia (PA): AACR; 2010. Abstract #LB-298.

59. Rivera, V.M., *et al.*, Efficacy and pharmacodynamic analysis of AP26113, a potent and selective orally active inhibitor of anaplastic lymphoma kinase *(ALK).* Proceedings from the 101st Annual Meeting of the American Association for Cancer Research, 2010: p. April 17–21; Washington, DC. Philadelphia (PA): AACR; 2010. Abstract #3623.

Chapter 17

BRAF-Targeted Therapy in Metastatic Melanoma

Noori Kim and April Deng
Department of Pathology, University of Massachusetts School of Medicine, Worcester, MA 01605, USA
April.deng@umassmemorial.org

17.1 Introduction

Melanoma is a malignant tumor that arises from transformed melanocytes, the cells that make melanin-pigment in the skin, but can involve the eyes, ears, gastrointestinal tract, oral and genital mucous membranes, etc. It is the deadliest of all skin cancers with high potential of metastasizing to any of the body's organs [1]. The melanoma incidence continues to increase with over 68,000 new invasive melanoma cases and over 46,000 cases of melanoma *in situ* estimated in the United States for 2010 [2]. While detected non-metastatic early stage lesions are cured by surgery, metastases often lead to the grave prognosis with a diminished median survival of less than 6 months [3]. Until recently, there were only two therapies approved by the Food and Drug Administration including

Handbook of Therapeutic Biomarkers in Cancer
Edited by Sherry X. Yang and Janet E. Dancey
Copyright © 2013 Pan Stanford Publishing Pte. Ltd.
ISBN 978-981-4364-65-2 (Hardcover), 978-981-4364-66-9 (eBook)
www.panstanford.com

high dose interleukin-2 and dacarbazine [4]. However, randomized trials concluded that the median survival among patients treated with dacarbazine to be less than 8 months [5a, 5b, 6]. Moreover, the response rates for both drugs are disappointing, ranging 10–20%, and neither showed improved survival [7–9]. Combination chemotherapies with multiple agents have only shown moderate increases in the response rates with the caveat of associated increased toxicities [8, 10]. Of note, immunologic agents such as those that target T-cell activation like ipilimumab have shown some promising results in preliminary clinical trials yet are accompanied by severe adverse events [11, 12].

Given the unconvincing results of chemotherapy for patients with malignant melanoma, other avenues for therapies continue to be studied, including targeting the signal-transduction pathways activated by oncogenes as has been investigated in other cancers. Furthermore, the heterogeneity of melanoma allows for different patterns of oncogenic mutations, over expression, and genomic amplification [3]. When a group in 2002 demonstrated a very high mutation rate in the signaling kinase BRAF in cutaneous melanoma, there was immediate interest and anticipation for pharmacologic interventions as was seen in gastrointestinal stromal tumors with KIT mutations treated with imatinib or non-small cell lung cancers with EGFR mutations treated with gefitinib [13]. Recently published multiple clinical trials on BRAF-targeted therapies did reveal the breakthrough results and promised new hope for treating late stage malignant melanoma.

17.2 BRAF and the MAPK Pathway

B-RAF is a member of the Raf Kinase family of serine/threonine-specific protein kinase. This protein plays an important role in regulating the mitogen-activated protein kinase (MAP) signaling pathway, which affects cell division, differentiation and secretion. The MAPK pathway is notably the most recognized and studied pathway out of all the signaling pathways. Virtually all melanomas have high constitutive activity in the MAPK pathway (3). It is important to note that constitutively activating somatic missense mutations in BRAF were discovered in a wide variety of human cancers, including papillary thyroid cancer, cholangiocarcinoma, colorectal cancer, and borderline ovarian cancer [14]. The pathway primarily regulates cell growth and is activated by growth factors released in the local microenvironment [15]. Proceeding along the transduction of extra cellular signals into the cell, the major branch point in the MAPK pathway occurs with RAS, which then activates downstream pathways, including the BRAF-ERK and phosphtidylinositol-3-kinase (PI3K)-AKT pathways [16]. Extracellular signals, including hormones, cytokines and growth factors, interact with receptors to activate the RAS family, which then acts via adaptor proteins to recruit RAF proteins to the cell membrane where they are activated [17].

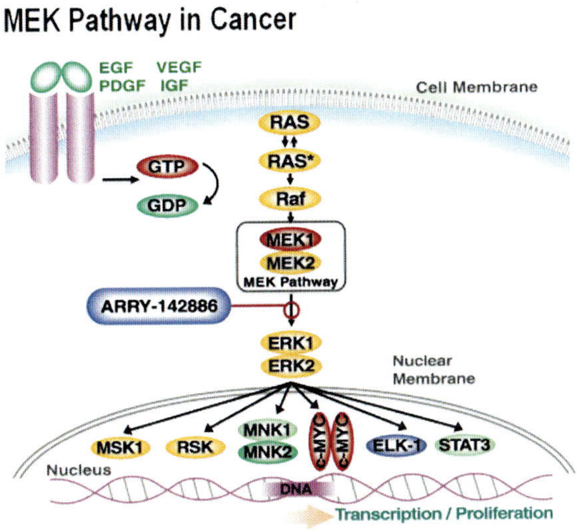

Figure 17.1 The Raf/MEK/ERK signaling pathway.

The RAF family of serine/threonine kinases has three isoforms, ARAF, BRAF, and CRAF. While ARAF and CRAF require both activation by RAS and phosphorylation, BRAF's N-terminus is constitutively active, needing only RAS and thus fewer steps for activation [18]. Once activated, RAF can then sequentially stimulate MEK1 and MEK2, which then turn on ERK1 and ERK2. These proceed to activate the cytoplasm targets or travel to the nucleus to finally phosphorylate transcription factors [19–21].

Specifically, the structure of MEK 1/2 with their proline-rich segments in the carboxyl-terminal is thought to play a role in the activation by RAF [15]. The RAF/MEK interaction is required for the phosphorylation of ERK1/2, which is crucial to cell cycle progression through the G1/S phase [22]. While activation of ERK 1/2 is dependent on the adhesion to the extracellular matrix, the BRAF V600E mutation forgoes this step. Therefore, ERK 1/2 is able to be activated without the ECM and the constitutively activated mutated BRAF allows for up regulated progression through the cell-cycle [23]. Constitutive activity in the Ras/Raf/MEK/ERK MAPK pathway contributes to the oncogenic phenotype of melanoma through its effects on cell proliferation, invasion and survival.

17.3 BRAF V600E Mutation in Melanoma

As already mentioned, the most common mutation to be reported in melanoma is in BRAF with at least 60% of all melanomas possessing activating mutations in the BRAF gene [13]. Among these, the BRAF V600E mutation, consisting of a valine to glutamic acid substitution, make up over 80% of the BRAF mutations [13]. More than 75 somatic mutations in the BRAF gene have been identified in melanoma, and all mutations at V600 in Exon 15 constitutively activate BRAF. The hypothesized mechanism for the uncontrolled activation is the increased accessibility of the activation segment when a small hydrophobic amino acid at 600 (valine) is switched to a hydrophilic residue (glutamic acid) [24]. The mutated BRAF corresponds to a constitutively active MAPK pathway and thus promoting cell cycle entry and cell proliferation through the up-regulation of cyclin D1 expression and down-regulation of the cyclin-dependent kinase inhibitor, p27(Kip1) [25].

The pathophysiology of how the activated BRAF V600E mutation leads to abnormal melanocyte growth and thus melanoma may not be clear as a multitude of factors are involved. However, there is evidence that one of the pro-apoptotic proteins of the BCL-2 family, BIM, plays a role in the BRAF-mediated pathway [26]. The BCL-2 family of proteins is an important regulator of apoptosis including molecules that promote cell growth or cell death [27]. BIM induces apoptosis by antagonizing the anti-apoptotic proteins, BCL-2, BCL-XL and MCL-1 [28]. In relation to melanomas with the BRAF V600E mutation, the BRAF/MEK/ERK signaling pathway has been shown to inhibit the expression of pro-apoptotic BIM in mouse and human melanocytes and, importantly, human melanoma cells [26]. Correspondingly, pharmacological inhibition of BRAF V600E or MEK 1/2 led to upregulated BIM expression [26]. Moreover, among the three BIM isoforms, BIM(S), BIM (EL), and BIM (L), BIM(S) has been demonstrated to be the dominant and potent isoform in apoptosis induction [29, 30]. While BRAF V600E mutation expression inhibits BIM(S), BRAF V600E inhibition conversely induces preferential splicing towards the BIM(S) isoform [29]. These studies reinforce the potential target in BIM and the BCL-2 family as an adjunct to BRAF-targeted therapy.

Interestingly, BRAF mutations are commonly associated with melanomas limited to intermittently sun-exposed skin while those lesions occurring in chronically sun-exposed sites or completely unrelated to sun exposure sites have a low frequency of these mutations [31]. A microdissection study of more than 200 primary melanomas that correlated the genotypic NRAS and BRAF mutation status to the tumor behavior showed additional differentiating characteristics of mutated BRAF tumors. BRAF mutated melanomas showed a higher rate of ulceration compared with melanomas with mutated NRAS and wild type genotypes [32]. In addition, melanomas with either mutation of NRAS, BRAF, or both showed more advanced AJCC stages than the wild type [32].

Another important association to consider is the mc1r gene, a key determinant of pigmentation [33]. For non-chronic sun-induced damage melanomas studied in American and Italian populations, there was an association between MC1R variants with BRAF mutations [34, 35]. MC1R is a G-protein coupled receptor that binds to a-melanocyte-stimulating hormone, initiating a signaling cascade that regulates melanocyte homeostasis [36]. Interestingly, wild-type MC1R favors the production of eumelanin over pheomelanin in response to UV radiation while variant isoforms can lead to the accumulation of pheomelanin [22]. This becomes relevant as pheomelanin increases the production of UV-induced free radicals and can act as a carcinogen via oxidative stress and DNA instability [22].

Although in general, the presence of BRAF mutation in early melanoma shows no association with disease-free interval or overall survival, the mutation's presence in metastatic disease is associated with poorer survival from the time of first metastasis or time from first resected metastasis [37–39]. The BRAF V600E mutation is postulated to occur early in melanoma development and introduction of BRAF V600E expression in primary human melanocytes showed an irreversible growth arrest suggestive of senescence [40, 41]. Thus, it seems that with the BRAF mutation occurring early on in melanocyte proliferation, there must be counteracting mechanisms to constitutively activating the MAPK pathway and protect the cells from malignant change.

Yet, importantly, approximately 80% of benign nevi also have the V600E mutation, indicating that the mutated BRAF pathway alone is not sufficient for malignant transformation [41]. In experimental animal models, it was when the combination of BRAF mutation with a deletion in the phosphatase and tensin homolog (PTEN) did invasive melanoma develop [24, 42]. PTEN is crucial in the signal transduction pathway, functioning as a tumor-suppressor gene in melanoma [16, 43]. It is a negative regulator of AKT, another serine-threonine kinase critical in apoptosis. Mutation or deletion of PTEN is found in up to 60% of all melanomas and loss of PTEN was reported in over 30% of melanomas [44]. Overall, evidence highlights the potential of multi-targeted molecular therapy, with combined inhibition of both BRAF/MEK and PI3K/AKT pathways of particular interest in malignant melanoma.

17.4 Sorafenib and PLX4032

Sorafenib was initially developed to inhibit RAF1 and binds preferentially to the inactive form of RAF [45]. It is found to be considerably less potent against BRAF and even less so against the mutated V600E BRAF. On the other hand, it does have activity against the VEGFR family, PDGFRB, flt-3, and KIT [16, 45]. Yet, sorafenib failed to show significant anti-tumor activity in advanced melanoma patients in a phase II randomized trial [46]. In addition, randomized studies of combination therapy using chemotherapy with sorafenib did not show any survival benefit to previously treated patients, while a follow-up trial with combination therapy in treatment-naïve patients did illustrate promising trends although the results were statistically insignificant [10, 47].

Notably in preclinical studies, the highly selective BRAF inhibitor PLX4032 and its analogue PLX4720 showed that at low micromolar doses the MAP kinase pathway could be inhibited when the V600E

BRAF mutation was present. Then, high nanomolar concentrations of the drug could essentially block the V600E BRAF mutated cells from proliferating *in vitro* [4]. Importantly, PLX4032 only inhibits RAF in cells with the BRAF V600E mutation [48]. Normally, RAF proteins form dimers and binding of a drug to one RAF in the dimer can subsequently activate the other given that there is normal RAS activity [48]. Yet, in these mutant BRAF cells, there is insufficient RAS activation to create RAF dimers, which allows for BRAF inhibiting drugs to abrogate RAF activity and ERK signaling pathway [48]. Thus, identifying the mutation status of the cancer prior to initiating any BRAF-targeted therapy is integral in treatment options.

As an aside, PLX4032 was initially formulated as a crystalline preparation. However, due to poor bioavailability, a new micro precipitated bulk powder form was made available [4]. Recently, Flaherty *et al.* reported the results of the phase I trial of PLX4032. In the phase II dose-escalation trial, 49 of the 55 patients enrolled with solid tumors had melanoma, and it was determined that the recommended phase II drug dose, defined as the highest dose at which no more than 1 of 6 patients had dose-limiting side effects, to be 960 mg twice daily [4]. Of note, out of 16 of those patients with melanoma in the dose-escalation group who specifically carried the V600E BRAF mutation and had received 240 mg or more of PLX4032 twice daily, 10 had partial responses and 1 with a complete response [4]. The tumor response occurred early in the treatment course. The median response duration was 9 months, and among the patients receiving the phase II dose, the median progression-free survival period was more than 7 months, placing encouraging implications on the natural course of the disease [49]. The most common dose-limiting side effects included arthralgia, nausea, skin rash, photosensitivity, fatigue, cutaneous squamous-cell carcinoma, pruritus, and palmar-plantar dysesthesia [4].

Most concerning side effect was the development of the cutaneous neoplasms, which is thought to be an upstream activation of the RAS mutations in pre-existing skin lesions [50]. Alterations in the dimerization of the BRAF kinase by selective BRAF inhibitors can lead to paradoxical activation of the MAPK pathway in the squamous cells [48, 51, 52]. Specifically, in the discussed phase I and II trials, 38% of the 48 patients with melanoma who received the highest doses of drug developed well-differentiated squamous cell carcinoma of the skin within the first few months of the therapy, at which point the lesions were successfully excised and therapy could continue [4]. Histopathologic examination of the lesions revealed usually keratoacanthoma or well-differentiated SCC that had no or very low metastatic potential [49]. These secondary skin cancers usually appeared on sun-exposed areas of the skin, lending additional evidence that there were probably pre-existing oncogenic mutations for these to occur [53]. Importantly, no patient required the discontinuation of the PLX4032 treatment due to this side effect [4].

In the second arm of this study, an additional 32 patients with metastatic melanoma harboring the V600E BRAF mutation were enrolled in the extension phase [4]. They were all treated with 960 mg of PLX4032 twice daily and a total of 26 of the 32 patients (81%) had a response. Alongside evaluating the cutaneous lesions for response, selected patients underwent tumor biopsies with subsequent tumor levels of cell proliferation marks such as phosphorylated ERK, cyclin D1, and Ki-67 to be measured to determine if PLX4032 was, indeed, targeting the MAPK pathway, which was proven to be the case [4]. Recently, a phase II trial analyzing vemurafenib or PLX4032 treatment among 132 previously treated patients with BRAF-mutant melanoma resulted a 52.3% response rate with a median response duration of 6.8 months and a median progression free survival time of 6.2 months [54]. Following this, a randomized phase III clinical trial comparing vemurafenib (PLX4032) versus dacarbazine among 675 previously untreated patients with BRAF V600E-mutant metastatic melanoma further affirmed the efficacy of the BRAF inhibitor. A recently published study comparing dacarbazine and vemurafenib or PLX4032 in treating metastatic melanoma showed an 84% overall survival in the vemurafenib group and 64% in the dacarbazine group. In the interim analysis for overall survival and final analysis for progression-free survival, vemurafenib was associated with a relative reduction of 63% in the risk of death and of 74% in the risk of either death or disease progression, compared with dacarbazine [5b]. Most recently, on August 17, 2011, the US Food and Drug Administration approved vemurafenib tablets (ZELBORAF™, Hoffmann-La Roche Inc.) for the treatment of patients with unresectable or metastatic melanoma with the $BRAF^{V600E}$ mutation as detected by an FDA-approved test.

17.5 Other RAF Inhibitors

Alongside PLX4032, research has led to the development of more second generation selective BRAF inhibitors. XL281 (famotidine) is an oral inhibitor of wild type BRAF, CRAF and mutant BRAF V600E kinases. The phase I study of this particular drug used an unselected patient population and showed overall prolonged disease stabilization in two patients with BRAF V600E mutant papillary thyroid cancer and some minor responses in nine patients with colorectal, ovarian, and non-small-cell lung cancer. Next, a three-arm dose expansion study including melanoma, papillary thyroid, colorectal, and non-small-cell lung cancer patients as well as phase I/II trials have been initiated [50, 55].

GSK 2118436 is an ATP competitive, reversible inhibitor of the mutant BRAF V600E/K/D, wild type BRAF and CRAF kinases. A phase I trial with over 100 patients with primarily melanoma harboring BRAF mutations showed minimal side effects and 9% of them developing cutaneous neoplasms, including low-grade SCC. In the first 61 enrolled patients with metastatic disease and BRAF mutation, 85% showed tumor regression. Clinical responses were seen mainly in lung, liver, bone, and brain metastases at the recommended phase II dose of 150 mg twice daily [50, 56]. Subsequently, a phase II study of GSK 2118436 and phase I study of combination drug therapy with GSK 2118436 and a MEK-inhibitor, GSK 1220212 are in process [50].

Taking into account the complexity of the MAPK pathway and interplay between the signal transduction pathways in general, combination therapies may be an effective approach [50]. Some evidence for this includes that *in vitro* melanoma models shows the combination of BRAF/MEK inhibition increased the levels of apoptosis [57].

At present, MEK is the only known substrate of BRAF and, thus, MEK inhibition becomes a valuable target in affecting the entire BRAF pathway [22]. Preclinical studies using small molecule MEK 1/2 inhibitors reduced proliferation of various cancer cell lines through apoptosis and cell-cycle arrest [58]. The first small-molecule MEK inhibitor CL1040 showed favorable phase I trial results but drug development was later curbed given its poor pharmacodynamics in phase II trials [59, 60]. Another second-generation MEK inhibitor PD0325901 exhibited greater activity than the first-generation MEK inhibitors but was shown to have significant associated drug toxicities in phase II trials [61]. Promisingly, another second-generation MEK inhibitor, AZD6244, is involved in many clinical trials. One patient with malignant melanoma in a phase I trial showed up to 70% tumor shrinkage with this drug [62]. In a phase II two-arm study between AZD6244 and temozolomide in patients with advanced melanoma, anti-tumor activity was confirmed with AZD6244 but no significant differences was observed in the progression free survival times between the two treatments [63]. Once again, this is the further evidence highlighting the considerable role of MEK inhibitors in dual therapy of malignant melanoma.

Interestingly, heat-shock protein 90 (HSP90) is a molecular chaperone that degrades, stabilizes and activates a variety of proteins, including HER2, CRAF, BRAF, AKT, and MET. In melanoma, HSP90 stabilizes CRAF and ARAF and required for BRAF V600E activity. So, this molecule presents another potential target in combination therapy [14, 64].

17.6 BRAF Inhibition Resistance Mechanisms

The major challenge of BRAF-targeted therapy for melanoma is that patients who initially responded to treatment with the specific BRAF inhibitor, PLX4032 experienced relapse, which implicated likely that chronic treatment course was associated with drug resistance. Villanueva *et al.* developed a model of acquired drug resistance by the application of the BRAF inhibitor, 885, to BRAF V600E melanoma cell lines and uncovered interesting findings. Ultimately, chronic treatment of the BRAF inhibitor, 885, to BRAF V600E melanoma cell lines had progressive drug resistance [65].

17.6.1 The Role of RAF Isoforms

The study showed the BRAF V600E melanomas with capability to alternate among the three different RAF isoforms to evade BRAF inhibition and activate the MAPK pathway though an unknown mechanism [65].

Preclinical studies have shown a subset of resistant BRAF V600E mutants demonstrating increased CRAF signaling by BRAF/CRAF heterodimerization so that a shift is made towards CRAF dependency over BRAF. Also, post-transcriptional elevation of CRAF protein levels may decrease intracellular bioavailability of some of the selective RAF inhibitors like AZ628 [66].

In addition, point mutations of the downstream kinase isoform MEK1 result in changes of the allosteric drug binding of the selective RAF inhibitor PLX4720 in mutant melanoma cells in cell culture and *in vivo* [67]. Yet, on the other hand, a study has shown that combination therapy with BRAF and MEK inhibition may allow enhanced apoptosis before BRAF inhibition resistance is acquired [57]. This further suggests that dual targeted treatment may be the strategy of choice to counteract any feedback loops that may allow progression of the signal transduction if one arm of the pathway is inhibited.

Johannessan *et al.* introduced DNA constructs of different kinases into the BRAF V600E cells to screen for particular abnormal kinase activity that may explain drug resistance and found over-expression of mitogen-activated protein kinase 8 or COT [68]. COT over-expression led to constitutive ERK/MEK phosphorylation and MAPK pathway activation, notably under the influence of BRAF inhibition (PLX4720). The results may indicate that BRAF inhibition can actually induce COT over-expression and BRAF inhibitor resistance. Additionally, COT expression was capable of stimulating the MAPK pathway in a RAF-independent but MEK-dependent manner, reiterating the role of combination targeted-therapy between RAF and MEK or COT to account for resistance mechanisms [68].

17.6.2 The Role of IGF-1R and PI3K-AKT Pathway

Furthermore, long-term BRAF targeted treatment was associated with increased IGF-1R and PI3K-AKT activity in BRAF inhibitor resistant melanoma cells, suggesting another potential therapy target in IGF-1R [65]. IGF-1R is important in tumor genesis, resistance to apoptosis as well as to anti-cancer agents and is, in fact, present in all melanocytes and has been shown to participate in the resistance mechanism of therapy in lung and breast cancer [69]. In point, combining MEK and IGF-1R or PI3K inhibitors were more effective inducing cell death of BRAF-inhibitor resistant cells than either used alone [65]. Moreover, the same group did not identify any secondary mutations in the BRAF gene that could explain an inherent resistance to the inhibitor [65].

17.6.3 Amplification of Cyclin-Dependent Kinase

Other proposed mechanisms have been the amplification of the CCND1 gene, which made cyclin D1 over expression, corresponding to independent stimulatory effects of cell-cycle progression via CDK4 or cyclin-dependent kinase 4 [70]. The genomic amplification of CCND1 was documented in 17% of the BRAF V600E-mutated metastatic melanoma samples studied. This association can explain the progression of melanoma in the BRAF-mutants despite targeted therapy through the oncogenic function of cyclin D1 [70].

17.6.4 The Role of Growth Factors

Another target of interest in overcoming drug resistance is platelet-derived growth factor receptor B (PDGFRB) [71]. This particular tyrosine kinase receptor was shown to be over expressed in RAF-resistant cell lines and in biopsy samples of patients with progressive tumors suggesting cellular growth and proliferation through this particular pathway [71]. In addition to this particular activated pathway, another resistance mechanism was found in the same study with N-RAS mutations and the reactivation of the MAPK pathway [71].

17.6.5 The Role of Cytokines

Another hypothetical consideration to evaluate the effectiveness of BRAF inhibitors is the role of inflammation as TNF-alpha is a cytokine that has shown to block apoptosis in melanoma cells when BRAF signaling is inhibited. While initial treatment with BRAF V600E inhibitors had cytotoxic and cytostatic effects on melanoma cells, treatment afterward with TNF-alpha blocked apoptosis and allowed the oncogenic cells to re-enter the cell cycle [72]. This TNF-alpha specific effect is mediated by NF-kB signaling and so when there was NF-kB depletion, TNF-alpha was now unable to allow the survival of BRAF-inhibited cells [22]. This observation is important when considering that activated macrophages secrete TNF-alpha and how their presence may effect BRAF inhibition in certain cellular environments [22].

17.7 Final Thoughts

With more than half of human melanomas having the activating V600E mutation in the BRAF gene, much progress has been made in researching the signal transduction pathways involved in melanoma pathogenesis and especially molecular targeted therapy. Clinical trials continue closely monitor the impact of the BRAF inhibitors on overall survival, disease burden, and clinical quality of life. In addition, trials of combination therapy to improve efficacy in targeting multiple points of the pathogenesis and overcome resistance are underway.

References

1. Habif, T. (2004) *Clinical Dermatology*, 4th Ed. (Mosby, USA).
2. Jemal, A, *et al.* (2011). Global cancer statistics, *CA Cancer J Clin.*, **61**, pp. 69–90.
3. Smalley, K. (2010). Understanding Melanoma Signaling Networks as the Basis for Molecular Targeted Therapy, *Journal of Investigative Dermatology*, **130**, pp. 28–37.
4. Flaherty, K, *et al.* (2010). Inhibition of Mutated, Activated BRAF in Metastatic Melanoma, *N Engl J Med.*, **363**, pp. 809–819.

5a. Chapman, PB, *et al.* (1999). Phase III multicenter randomized trial of the Dartmouth regimen versus dacarbazine in patients with metastatic melanoma, *J Clin Oncol.*, **7**, pp. 2745–2751.

5b. Chapman, PB, *et al.* (2011). Improved survival with vemurafenib in melanoma with BRAF V600E mutation, *N Engl J Med.*, **364**, pp. 2507–2516.

6. Middleton, MR, *et al.* (2000). Randomized phase III study of temozolomide versus dacarbazine in the treatment of patients with advanced metastatic malignant melanoma, *J Clin Oncol.*, **18**, pp. 2351.
7. Atkins, MB, *et al.* (1999). High-dose recombinant interleukin-2 therapy for patients with metastatic melanoma: analysis of 270 patients treated between 1985-1993, *J Clin Oncol.*, **17**, pp. 2105–2116.
8. Middleton, MR, *et al.* (2000). A randomized phase III study comparing dacarbazine, BCNU, cisplatin and tamoxifen with dacarbazine and interferon in advanced melanoma, *Br J Cancer*, pp. 1158–1162.
9. Rosenberg, SA, *et al.* (1994). Treatment of 283 consecutive patients with metastatic melanoma or renal cell cancer using high-dose bolus interleukin 2, *JAMA*, **271**, pp. 907–913.

10. Hauschild, A, *et al.* (2009). Results of a phase III, randomized, placebo-controlled study of sorafenib in combination with carboplatin and paclitaxel as second-line treatment in patients with unresectable stage III or stage IV melanoma, *J Clin Oncol.*, **27**, pp. 2823–2830.
11. Hodi, FS, *et al.* (2010). Improved survival with ipilimumab in patients with metastatic melanoma, *N Engl J Med.*, **363**, pp. 711–723.
12. Robert, C, *et al.* (2011). Ipilimumab plus dacarbazine for previously untreated metastatic melanoma, *N Engl J Med.*, **364**, pp. 2517–2526.
13. Davis, H, *et al.* (2002). Mutations of the BRAF gene in human cancer, *Nature*, **417**, pp. 949–954.
14. Banerji, U, *et al.* (2005). Phase I pharmacokinetic and pharmacodynamic study of 17-allylamino-17-demethoxygeldanamycin in patients with advanced malignancies, *J Clin Oncol.*, **23**, pp. 4152–4161.
15. Robinson, MJ and Cobb, MH. (1997). Mitogen-activated protein kinase pathways, *Curr Opin Cell Biol.*, **9**, pp. 180–186.
16. Haluska, F, *et al.* (2007). The RTK/RAS/BRAF/P13K Pathways in Melanoma: Biology, Small Molecule Inhibitors, and Potential Applications, *Semin Oncol.*, **34**, pp. 546–554.
17. Garnett M and Marais, R. (2004). Guilty as charged: B-RAF is a human oncogene, *Cancer Cell*, **6**, pp. 323–329.
18. Mason, CS, *et al.* (1999). Serine and tyrosine phosphorylations cooperate in Raf-1, but not B-RAF activation, *EMBO J.*, **18**, pp. 2137–2148.
19. Crews, CM, *et al.* (1992). The primary structure of MEK, a protein kinase that phosphorylates the ERK gene product, *Science*, **258**, pp. 478–480.
20. Gardner, AM, *et al.* (1994). MEK-1 phosphorylation by MEK kinase, Raf and mitogen-activated protein kinase: analysis of phospho-peptides and regulation of activity, *Mol Biol Cell.*, **5**, pp. 193–201.
21. Kyriakis, JM, *et al.* (1992). Raf-1 activates MAP kinase-kinase, *Nature*, **358**, pp. 417–421.
22. Cantwell-Dorris, E, *et al.* (2011). BRAF V600E: Implications for Carcinogenesis and Molecular Therapy, *Mol Cancer Ther.*, **10**, pp. 385–394.
23. Conner, SR, *et al.* (2003). Adhesion-dependent activation of the ERK 1/2 cascade is by-passed in melanoma cells, *J Biol Chem.*, **278**, pp. 34548–34554.
24. Dankort, D, *et al.* (2009). Braf(V600E) cooperates with Pten loss to induce metastatic melanoma, *Nat Genet.*, **41**, pp. 544–552.

25. Bhatt, KV, *et al.* (2005). Adhesion control of cyclin D1 and p27Kip1 levels is deregulated in melanoma cells through BRAF-MEK-ERK signaling, *Oncogene*, **24**, pp. 3459–3471.

26. Cartlidge, RA, *et al.* Z (2008). Oncogenic BRAF V600E inhibits BIM expression to promote melanoma cell survival, *Pigment Cell Melanoma Res.*, **21**, pp. 534–544.

27. Youle, RJ and Strasser, A. (2008). The BCL-2 protein family: opposing activities that mediate cell death, *Nat. Rev. Mol. Cell. Biol.*, **9**, pp. 47–59.

28. Willis, SN, *et al.* (2007). Apoptosis initiated when BH3 ligands engage multiple Bcl-2 homologs, not Bax or Bak, *Science*, **315**, pp. 856–859.

29. Jiang, CC, *et al.* (2010). Apoptosis of human melanoma cells induced by inhibition of B-RAF V600E involves preferential splicing of bim(s), *Cell Death and Disease*, **1**, pp. 1–14.

30. O'Connor, L, *et al.* (1998). Bim: a novel member of the Bcl-2 family that promotes apoptosis, *The EMBO Journal*, **17**, pp. 384–395.

31. Curtin, JA, *et al.* (2005). Distinct sets of genetic alterations in melanoma, *N Engl J Med.*, **353**, pp. 2135–2141.

32. Ellenhorst, J, *et al.* (2010). Clinical Correlates of NRAS and BRAF Mutations in Primary Human Melanoma, *Clin Cancer Res.*, **17**, pp. 229–235.

33. Sturm, R, *et al.* (2003). The role of melanocortin-1 receptor polymorphism in skin cancer risk phenotypes, *Pigment Cell Res.*, **16**, pp. 266–272.

34. Fargnoli, M, *et al.* (2008). MC1R variants increase risk of melanomas harboring BRAF mutations, *J Invest Dermatol.*, **128**, pp. 2485–2490.

35. Landi, M, *et al.* (2006). MC1R germline variants confer risk for BRAF-mutant melanoma, *Science*, **313**, pp. 521–522.

36. Abdel-Malek, Z, *et al.* (2008). The melanocortin 1 receptor and the UV response of human melanocytes-a shift in paradigm, *Photochem Photobiol*, **84**, pp. 501–508.

37. Houben, R, *et al.* (2004). Constitutive activation of the Ras-Raf signaling pathway in metastatic melanoma is associated with poor prognosis, *J Carcinog.*, **3**, pp. 6.

38. Long, GV, *et al.* (2010). Clinicopathologic correlates of BRAF mutation status in 207 consecutive patients with metastatic melanoma, *J Clin Oncol.*, **28**, pp. 15s.

39. Shinozaki, M, *et al.* (2004). Incidence of BRAF oncogene mutation and clinical relevance for primary cutaneous melanomas, *Clin Cancer Res.*, **10**, pp. 1753–1755.

40. Michalogiou, C, *et al.* (2005). BRAFE600-associated senescence-like cell cycle arrest of human nevi, *Nature*, **436**, pp. 720–724.

41. Pollock, PM, *et al.* (2003). High frequency of BRAF mutations in nevi, *Nat Genet.*, **33**, pp. 19–20.

42. Stahl, JM, *et al.* (2003). Loss of PTEN promotes tumor development in malignant melanoma, *Cancer Res.*, **63**, pp. 2881–2890.

43. Hwang, PH, *et al.* (2001). Suppression of tumorigenicity and metastasis in B16F10 cells by PTEN/MMAC1/TEP1 gene, *Cancer Lett.*, **172**, pp. 83–91.

44. Tsao, H, *et al.* (2003). PTEN expression in normal skin, acquired melanocytic nevi, and cutaneous melanoma, *J Am Acad Dermatol.*, **49**, pp. 865–872.

45. Flaherty, K and McArthur, G. (2010). BRAF, a Target in Melanoma, *Cancer*, **116**, pp. 4902–4913.

46. Eisen, T, *et al.* (2006). Sorafenib in advanced melanoma: a phase II randomized discontinuation trial analysis, *Br J Cancer*, **95**, pp. 581–586.

47. McDermott, DF, *et al.* (2008). Double-blind randomized phase II study of the combination of sorafenib and dacarbazine in patients with advanced melanoma: a report from the 11715 Study Group, *J Clin Oncol.*, **26**, pp. 2178–2185.

48. Poulikakos, PI, *et al.* (2010). RAF inhibitors transactivate RAF dimmers and ERK signaling in cells with wild-type BRAF, *Nature*, **464**, pp. 427–430.

49. Flaherty, K. (2010). Narrative Review: BRAF Opens the Door for Therapeutic Advances in Melanoma, *Ann Intern Med.*, **153**, pp. 587–591.

50. Arkenau, H, *et al.* (2011). Targeting BRAF for patients with melanoma, *British Journal of Cancer*, **104**, pp. 392–398.

51. Hatzivassiliou, G, *et al.* (2010). RAF inhibitors prime wild-type RAF to activate the MAPK pathway and enhance growth, *Nature*, **464**, pp. 431–435.

52. Heidorn, SJ, *et al.* (2010). Kinase-dead BRAF and oncogenic RAS cooperate to drive tumor progression through CRAF, *Cell*, **140**, pp. 209–221.

53. Bollag, G, *et al.* (2010). Clinical efficacy of a RAF inhibitor needs broad target blockade in BRAF-mutant melanoma, *Nature*, **467**, pp. 596–599.

54. Ribas, A, *et al.* (2011). BRIM-2: an open-label, multicenter phase II study of vemurafenib in previously treated patients with BRAFV600E mutation-positive melanoma, *J Clin Oncol.*, **29**, suppl.8509.abstract.

55. Schwartz, GK, *et al.* (2009). A phase-I study of XL281, a selective oral RAF kinase inhibitor, in patients with advanced solid tumors, *J Clin Oncol.*, **27**, abstract 3513

56. Kefford, R, *et al.* (2010). Phase I/II study of GSK2118436, a selective inhibitor of oncogenic mutant BRAF kinase, in patients with metastatic melanoma and other solid tumors, *J Clin Oncol.*, **28**, abstract 8503.

57. Paraiso, K, *et al.* (2010). Recovery of phospho-ERK activity allow melanoma cells to escape from BRAF inhibitor therapy, *British Journal of Cancer*, **102**, pp. 1724–1730.

58. Solit, D and Rosen, N. (2011). Resistance to BRAF Inhibition in Melanomas, *N Engl J Med.*, **64**, pp. 772–774.

59. Lorusso, PM, *et al.* (2005). Phase I and pharmocodynamic study of the oral MEK inhibitor CI-1040 in patients with advanced malignancies, *J Clin Oncol.*, **23**, pp. 5281–5293.

60. Rinehart, J, *et al.* (2004). Multicenter phase II study of the oral MEK inhibitor, CI-1040 in patients with advanced non-small-cell lung, breast, colon, and pancreatic cancer, *J Clin Oncol.*, **22**, pp. 4456–4462.

61. Lorusso, PM, *et al.* (2010). Phase I pharmacokinetic and pharmacodynamic study of the oral MAPK/ERK kinase inhibitor PD-0325901 in patients with advanced cancers, *Clin Cancer Res.*, **16**, pp. 1924–1937.

62. Adjei, AA, *et al.* (2008). Phase I pharmacokinetic and pharmacodynamic study of the oral, small-molecule mitogen-activated protein kinase kinase ½ inhibitor AZD6244 (ARRY-142886) in patients with advanced cancers, *J Clin Oncol.*, **26**, pp. 2139–2146.

63. Dummer, R, *et al.* (2008). AZD6244 (ARRY-142886) versus temozolomide (TMZ) in patients (pts) with advanced melanoma: an open-label, randomized, multicenter, phase II study, *J Clin Oncol.*, **26**, abstract 9033.

64. da Rocha Dias, S, *et al.* (2005). Activated B-RAF is a Hsp90 client protein that is targeted by the anticancer drug 17-allylamino-17-demethoxygeldanamycin, *Cancer Res.*, **65**, pp. 10686–10691.

65. Villanueva, J, *et al.* (2010). Acquired Resistance to BRAF Inhibitors Mediated by a RAF Kinase Switch in Melanoma Can Be Overcome by Cotargeting MEK and IGF-1R/P13K, *Cancer Cell*, **18**, pp. 683–695.

66. Montagut, C, *et al.* (2008). Elevated CRAF as a potential mechanism of acquired resistance to BRAF inhibition in melanoma, *Cancer Res.*, **68**, pp. 4853–4861.

67. Emery, CM, *et al.* (2009). MEK1 mutations confer resistance to MEK and B-RAF inhibition, *Proc Natl Acad Sci USA*, **106**, pp. 20411–20416.

68. Johannessen, C, *et al.* (2010). COT drives resistance to RAF inhibition through MAP kinase pathway reactivation, *Nature*, **468**, pp. 968–972.

69. Casa, AJ, *et al.* (2008). The type I insulin-like growth factor receptor pathway: a key player in cancer therapeutic resistance, *Front Biosci.*, **13**, pp. 3273–87.

70. Smalley, KS, *et al.* (2008). Increased cyclin D1 expression can mediate BRAF inhibitor resistance in BRAF V600E-mutated melanomas, *Mol Cancer Ther.*, **7**, pp. 2876–2883.

71. Nazarian, R, *et al.* (2010). Melanomas acquire resistance to B-RAF(V600E) inhibition by RTK or N-RAS up regulation, *Nature*, **468**, pp. 973–977.

72. Gray-Schopfer, VC, *et al.* (2007). Tumor necrosis factor-(alpha) blocks apoptosis in melanoma cells when BRAF signaling is inhibited, *Cancer Res.*, **67**, pp. 122–129.

Index